T0336148

Space Sciences Series of ISSI

Volume 52

For further volumes:
www.springer.com/series/6592

Andrew F. Nagy • Michel Blanc •
Charles R. Chappell • Norbert Krupp
Editors

Plasma Sources of Solar System Magnetospheres

Previously published in *Space Science Reviews* Volume 192, Issues 1–4, 2015

Editors
Andrew F. Nagy
University of Michigan
Ann Arbor, MI, USA

Charles R. Chappell
Vanderbilt University
Nashville, TN, USA

Michel Blanc
Institut de Recherche en Astrophysique
et Planétologie
Toulouse, France

Norbert Krupp
Max-Planck Institut für Sonnensystem
Göttingen, Germany

ISSN 1385-7525 Space Sciences Series of ISSI
ISBN 978-1-4939-3543-7 ISBN 978-1-4939-3544-4 (eBook)
DOI 10.1007/978-1-4939-3544-4

Library of Congress Control Number: 2016931193

Springer New York Heidelberg Dordrecht London

Cover Image: Enceladus as a major source for Saturn's magnetosphere. Credit: Margaret Kivelson et al., Does Enceladus govern magnetospheric dynamics at Saturn? Science 311, 1391 (2006)

Printed on acid-free paper

Springer is part of Springer Science+Business Media (www.springer.com)

Contents

DOI 10.1007/978-1-4939-3544-4_1
Reprinted from *Space Science Reviews* Journal, DOI 10.1007/s11214-015-0205-4

Foreword

Michel Blanc[1] · Andrew F. Nagy[2]

Published online: 12 October 2015

In September 1999, ISSI published a volume entitled "Magnetospheric Plasma Sources and Losses", edited by Bengt Hultqvist and collaborators. This volume, which was the result of a two-year preparation and study process within the "ISSI Study Project on Source and Loss Processes", aimed at giving a comprehensive view of what we knew at that time of Earth's magnetospheric plasma sources and losses. To reach this ambitious objective, the team had divided the Earth's plasma environment into specific regions, and in each region the budget of sources and losses of plasma and energetic particles was established, based on the number of spacecraft investigations available. For the sources, essentially two were considered, the ionosphere and the solar wind, and the circulation paths by means of which these two sources feed each region were traced.

Approximately one and a half decade later, it seemed relevant to revisit this issue of plasma sources. Indeed, since the late 1990's comprehensive results from several orbital missions to the intrinsic planetary magnetospheres of the solar system have become available: Galileo at Jupiter, Cassini at Saturn, and more recently Messenger around Mercury. This is the reason why, following a suggestion by Andrew F. Nagy, the directors of ISSI decided to take advantage of this host of space missions to the planets to study the budget of plasma sources not only for the Earth, but this time for all intrinsic magnetospheres in our solar system. To this end, an ISSI workshop gathering over 40 of the best specialists working on these topics was held in Bern from September 23rd to 27th, 2013, with the task of performing a study of magnetospheric plasma sources in the solar system, and of preparing the writing of a comprehensive book on the subject.

In this perspective, the workshop participants had to face and manage the broad diversity of the subject: first, the diversity of the objects to be considered, from Mercury to Neptune, but also, and above all, the diversity in the sources of the plasmas themselves. While for Earth in 1999 we had, and still have a few years later, to consider only two plasma sources,

✉ M. Blanc
michel.blanc@irap.omp.eu

[1] IRAP/OMP, Toulouse, France

[2] University of Michigan, Ann Arbor, MI, USA

ionosphere and solar wind, in our visit to all solar system magnetospheres we had to include new and complex plasma sources: satellites and rings (at the giant planets), or the planetary surface itself (for Mercury), some of which actually happen to be the dominant sources.

This book reports on our findings along this full tour of the solar system performed by the workshop participants. It starts with two introductory chapters which set the stage and provide the basic tools for our visit to solar system magnetospheres.

In the first introductory chapter, Rick Chappell provides a historical perspective on the study of plasma sources at Earth, showing in particular how our understanding of the ionospheric source evolved and how it became understood to be more and more important with time, and has to be considered now at least on equal footings with the solar wind source. His chapter teaches us how the availability of new space data and progress in simulations transformed our view of plasma sources: a lesson certainly to be kept in mind for all other magnetospheres.

In the second chapter, Kanako Seki and her co-authors provide an overview of the main physical processes that are at work in the expression of sources, transport and losses in the different regions of a magnetosphere and at various energies. The chapter also nicely summarizes the main equations used for the description of these processes, and the main types of modeling tools that have been developed to simulate them. In that way it provides us with the "tool box" that we need to start our exploration of the solar system and understand the data and models.

This exploration is performed from closest to the Sun outwards, and therefore starts with the planet Mercury. Jim Raines and co-authors use some of the latest data from the Messenger orbiter to visit the plasma sources and the dynamics of this tiny magnetosphere, where the influence of the solar wind is dominant, but also where the direct interaction of magnetospheric particles and fields with the exosphere and surface of the planet plays a role like nowhere in the solar system, except maybe at Jupiter's satellite Ganymede. The very short time scales within which the magnetospheric configuration and the plasma domains of this magnetosphere are reconfigured are also unique in the solar system.

The next object in our exploration is planet Earth: Dan Welling et al. review the progress made in our understanding of Earth's plasma sources and subsequent transport, acceleration and loss processes since the comprehensive ISSI book of 1999. They consider both the observational and the modelling advances achieved since that time, and establish a new reference for the description of Earth's basic plasma processes.

The book then moves on to the exploration of giant planets, starting with Jupiter. As Bolton et al. write in their introduction to this chapter, the Jupiter system is "a world of superlatives": biggest planet in the solar system, strongest magnetic field, largest magnetosphere, and with the most intense plasma sources. Jupiter is dominated by the Io plasma source, which under the effect of the planet's centrifugal action generates a large plasma disc. Given this, Jupiter is not just the largest magnetosphere, it is also the closest object to a proto-planetary disk we have at hand in our solar system, and to some extent it bridges the gap between planetary sciences and astrophysics. The chapter summarizes the different plasma sources associated with this fascinating object, and the way they are transported and lost from their regions of origin to the outer edges of the magnetosphere.

Saturn, Jupiter's sister planet and our solar system's second gas giant by its size, is described in the next chapter by Blanc et al. Just as the Saturn system is diverse in terms of the objects it includes, its plasma sources display a broad diversity, which has been explored in considerable detail by Cassini since it went into orbit around Saturn in July 2004. The rings and satellites all contribute to its plasma sources, but the space exploration of Saturn revealed, quite unexpectedly, that the dominant source is the tiny satellite Enceladus. Even

Titan, the largest moon in the system and the only one with a thick atmosphere, plays a minor role compared to it. So, like Jupiter, Saturn is dominated by a single source.

Uranus and Neptune, our two ice giants at the outskirts of the solar system, are described by Norbert Krupp in the final chapter of the book. These two planets are by far the least well known, since what we know of them only comes from the fly-bys of Voyager 2 in 1986 and 1989. The Voyager data suggest that plasma at Uranus is produced mainly by its hydrogen corona with a likely complement from its ionosphere, whereas at Neptune the dominant source seems to be associated with its satellite Triton.

In the remainder of his chapter, Norbert Krupp offers a final review of all the planetary plasma sources explored by the book, emphasizing the main similarities and differences between them. Overall, one sees that more or less all the same categories of sources are acting at the different planets, but the dominant ones vary strongly from one planet to another. We hope this book will provide the reader with a good opportunity to visit this variety of sources, and to contemplate the diversity of their expressions.

Before closing this foreword, we would like to thank the Directors and Science Committee of ISSI for their support to this project, and to express our warmest appreciation to the wonderful staff of ISSI, ISSI's science programme manager Maurizio Falanga, Jennifer Fankhauser, Andrea Fischer, Saliba F. Saliba, Irmela Schweizer, Silvia Wenger, and all their colleagues, whose kindness and dedication make ISSI such a convivial and effective place to interact, exchange ideas and work. This book would not have been possible without all of them.

DOI 10.1007/978-1-4939-3544-4_2
Reprinted from *Space Science Reviews* Journal, DOI 10.1007/s11214-015-0168-5

The Role of the Ionosphere in Providing Plasma to the Terrestrial Magnetosphere—An Historical Overview

Charles R. Chappell[1]

Received: 10 March 2015 / Accepted: 25 May 2015 / Published online: 10 July 2015

Abstract Through the more than half century of space exploration, the perception and recognition of the fundamental role of the ionospheric plasma in populating the Earth's magnetosphere has evolved dramatically. A brief history of this evolution in thinking is presented. Both theory and measurements have unveiled a surprising new understanding of this important ionosphere-magnetosphere mass coupling process. The highlights of the mystery surrounding the difficulty in measuring this largely invisible low energy plasma are also discussed. This mystery has been solved through the development of instrumentation capable of measuring these low energy positively-charged outflowing ions in the presence of positive spacecraft potentials. This has led to a significant new understanding of the ionospheric plasma as a significant driver of magnetospheric plasma content and dynamics.

1 Introduction

The early instrumentation used on satellites that probed the Earth's space environment was able to measure the fluxes of the low density, high energy particles found in the magnetosphere or the high density, very low energy particles typical of the ionosphere. Measurements of the high energy radiation belts originally were made with Geiger counters, while the low energy plasma of the ionosphere was measured with retarding potential analyzers and Langmuir probes. As miniaturized channel electron multipliers were developed, the measurement of the full energy range of particles from a few electron volts up to ten's of keV became possible.

As these new instruments were flown on satellites into the magnetosphere and solar wind, it was recognized that there was a similarity in energy between the solar wind particles and particles found in the Earth's plasma sheet and aurora. Early instrumentation did not have the ability to determine ion composition at this medium energy range, and it was assumed that the plasmas in the solar wind and the magnetosphere were both dominated by protons and electrons. Hence the conceptual picture shown in Fig. 1 was developed. In this understanding

✉ C.R. Chappell
rick.chappell@vanderbilt.edu

[1] Vanderbilt University, Nashville, TN, USA

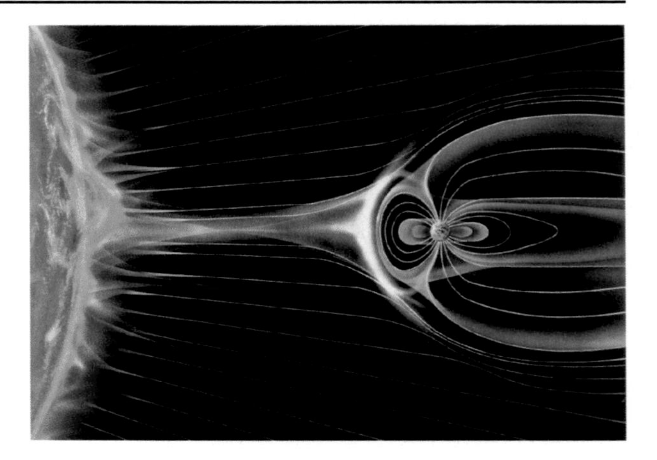

Fig. 1 A schematic image representing the early understanding of the primary role of the solar wind in populating the Earth's magnetosphere with plasma

of our space physics "childhood" the solar wind was seen as the sole source of plasma for the magnetosphere with solar wind particles gaining access into the magnetosphere through the polar cusp on the dayside and through the flanks of the magnetotail into the plasma sheet on the nightside. These keV ions and electrons were thought to be channeled through the magnetic field down into the auroral zone where they collided with the atoms and molecules of the upper atmosphere to create the auroral emissions. The Van Allen radiation belts which were the first major discovery of a magnetospheric plasma population are shown in this figure as a toroidal shaped region surrounding the Earth in the inner magnetosphere.

In this early view of the solar wind/magnetosphere system, the low energy plasma was thought to be confined to the low altitudes of the ionosphere with an extension upward only in the plasmasphere, a roughly donut-shaped region with an outer average boundary of about $L = 4$, extending outward to $L = 6$ in the dusk sector. The plasmasphere was seen to vary with magnetic activity moving inward during disturbed times and growing larger during magnetically quiet times (Gringauz 1963; Carpenter 1963). Early theoretical work by Nishida (1966) and Brice (1967) suggested ways in which the convection electric field in the magnetosphere, combined with the corotation of plasma with the Earth could begin to explain the presence and shape of the plasmasphere. Later measurements from the Orbiting Geophysical Observatory series of spacecraft verified the earlier whistler and satellite measurements and enhanced the understanding of plasmasphere dynamics (Taylor et al. 1965; Brinton et al. 1970; Grebowsky 1970; Chappell 1972).

Although the plasmasphere represented a region of magnetospheric plasma, it was thought to be only an upward extension of the ionosphere and to be of too low an energy to contribute to the dynamic magnetospheric processes that created magnetic storms, the aurora and the radiation belts. Hence, discussions of the plasmasphere in those years were usually placed in ionospheric sessions at the national and international meetings and not in the magnetospheric sessions.

This was the space physics community perception of the solar wind-magnetosphere system through the decade of the 1960s. The locations of the magnetospheric regions of more energetic particles, their energies and unknown composition showed an excellent fit to the idea that the solar wind provided both the energy and the particles for driving the dynamic processes that were observed in the magnetosphere by both space-borne and ground-based measurements. This is what graduate students of that time were taught and these ideas have not gone away easily.

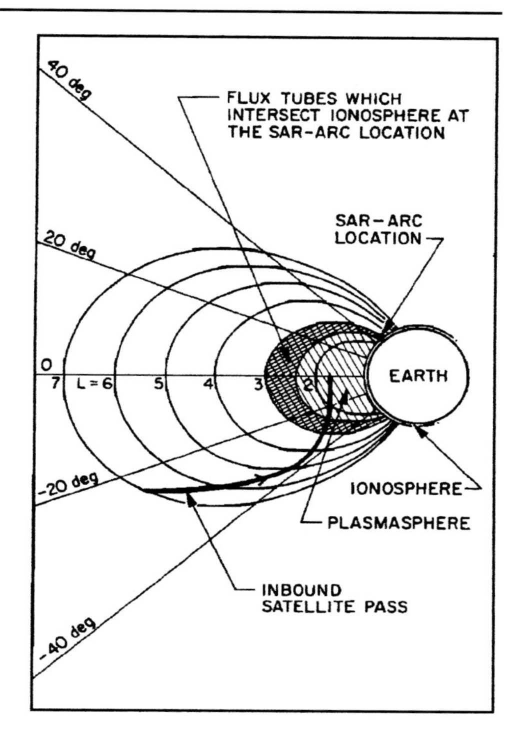

Fig. 2 A sketch of the inner magnetosphere plasma regions showing the overlap of the energetic ions of the ring current with the low energy plasma of the outer plasmasphere. The transfer of energy from the ring current to the plasmasphere results in heating which causes the formation of Stable Auroral Red arcs in the upper atmosphere

2 The Decade of the 1970s

At the end of the 1960s theoretical work at the University of California, San Diego led to the realization that there could be a supersonic escape of light ions from the topside ionosphere. This very low energy ambipolar outflow of H^+ and He^+ ions and electrons was called the polar wind (Axford 1968; Banks and Holzer 1968; Nagy and Banks 1970; Banks et al. 1971, 1974a, 1974b) and predicted significant upward fluxes of the order of 3×10^8 ions/cm^2 sec. This outflow results from the charge separation electric field that is set up between the dominant ionospheric O^+ and the electrons which would then accelerate the minor ions, H^+ and He^+ upward. The polar wind was predicted to be present on all flux tubes in which the plasma content above the ionosphere was still filling and had not yet reached diffusive equilibrium. Given the fact that flux tubes from the pole to the inner plasmapause boundary at $L \sim 2.5$ were predicted to have polar wind outflow, the total magnitude of mass transport into the magnetosphere could be very large, of the order of 10^{25}–10^{26} ions/sec (Moore et al. 1997; Ganguli 1996; Andre and Yau 1997). Measurements by Hoffman et al. (1970) from the ion mass spectrometer on the ISIS satellite confirmed the polar wind outflow showing H^+ and He^+ velocities of 10–20 km/sec and upward fluxes of a few times 10^8 ions/cm^2 sec.

In the early 1970s observations of stable auroral red arcs at the foot of field lines in the vicinity of the plasmapause first suggested an interaction between the energetic protons in the ring current and the low energy H^+ and He^+ ions and electrons near the plasmapause (Chappell et al. 1971; Cole 1965; Cornwall et al. 1971). This was the first identification of a mechanism in which the low energy plasma could potentially affect the dynamics of the energetic plasmas of the magnetosphere. As shown in Fig. 2, energy from the ring current particles could be transformed into heating the cold plasma through coulomb collisions or wave particle interactions and the heat could be transmitted down the flux tube into the atmosphere resulting in heating and causing a resulting emission at 6300 A. Hence, the motion of the plasmapause could influence the dynamics of the inner edge of the ring current.

One of the most significant influences in the magnetospheric community's perception of the role of the ionosphere as a source of plasma for the magnetosphere came from measurements in the early 1970s by the Lockheed group. These measurements showed energetic, keV ions of H^+, He^+ and then O^+ streaming up the magnetic field lines above the auroral zone (Shelley et al. 1972; Sharp et al. 1977). The idea that energetic ions could flow upward into the magnetosphere and that some of them (He^+ and O^+) were definitely of ionospheric origin, was a transforming one. Suddenly, the door was opened to the realization that the low energy plasma of the ionosphere could become energized to the energies characteristic of the magnetosphere and could flow upward into the principal regions of magnetospheric dynamics—the ring current and plasma sheet. This spurred the need for measurements of the composition of energetic particles in the magnetosphere, a need that was met on the ISEE and GEOS set of satellites later in the decade (Shelley et al. 1978; Lennartsson et al. 1979; Young et al. 1982). These new plasma composition instruments verified the presence of ionospheric O^+ in the plasma sheet and ring current. The energy range of this instrument, however, did not effectively include ions with energies below 100 eV, both because of limited geometric factors and because of the typically positive charging of satellites at high altitudes, where surrounding ambient electron densities were not large enough to give a return current to the satellite that could offset the escaping photoelectron current. The resulting positive spacecraft charge prohibited the measurement of the lower energy polar wind ions and hence could not verify their presence out in the magnetosphere.

Thus, at the end of the 1970s interest in the influence of low energy plasma in the magnetosphere had grown. However, its influence was limited to its potential role in destabilizing the hot plasmas of the ring current and radiation belts, which was considered to be a secondary influence on magnetospheric dynamics. It was also accepted that energetic ions of ionospheric origin could contribute to the ring current and plasma sheet populations and could influence magnetospheric dynamics in certain circumstances. However, the ion outflow that was considered was limited to the more energetic outflows that are directly connected to the auroral oval precipitation processes and not to the low energy polar wind fluxes.

3 The Decade of the 1980s

As a result of increasing interest by the atmosphere-ionosphere-magnetosphere community regarding the coupling of these regions in terms of both particles and fields, a new mission, Electrodynamics Explorer, was planned. It grew out of community discussions that were focused by two AGU Chapman Conferences held at Yosemite National Park in 1974 and 1976. As the mission planning progressed, budget decisions led to a limiting of the scope of the mission, becoming Dynamics Explorer (DE). The two spacecraft, one ionospheric and one magnetospheric in coplanar polar orbits, were launched in 1980. They were designed to probe all of the elements of the coupling between the ionosphere and the magnetosphere.

One particular goal of DE was to measure the upward flow of particles from the ionosphere toward the magnetosphere. This goal was realized through measurements such as those from the Lockheed and Marshall Space Flight Center (MSFC) groups, which showed the broad presence of upward flowing ions with energies ranging from the few electron volt polar wind shown in Fig. 3 (Nagai et al. 1984) to the auroral energies of 100's of

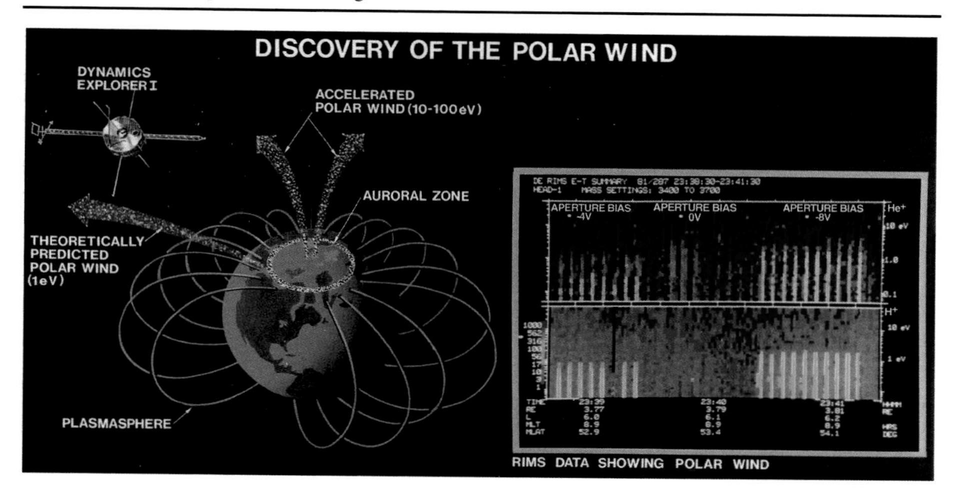

Fig. 3 A segment of data showing the DE Retarding Ion Mass Spectrometer measurements of the low energy polar wind ions flowing upward out of the ionosphere at MLT 8:54, $L = 6.1$, 3.8R_E geocentric altitude. By alternately placing bias potentials of -4, 0 and -8 volts on the entrance aperture of the instrument, the low energy polar wind positive ions with energies of 1 eV can be seen when the spinning instrument looked down the magnetic field line (*yellow bars*, He^+ *top panel* and H^+ *bottom panel*)

eV to 10's of keV (Gurgiolo and Burch 1982; Yau et al. 1985; Lockwood et al. 1985; Chandler et al. 1991). The low energy polar wind was found to be flowing out of the polar cap where it was intermixed with more energetic particles flowing up from the polar cusp. The ions associated with the auroral zone were easier to measure than the polar wind because of their higher energies which could overcome the positive potential of the spacecraft.

In this same time period, Cladis (1986) showed theoretically how very low energy outflowing ions could become energized to >10 eV through a centrifugal acceleration caused by the ions flowing along the curving magnetic field in the polar cap and through the cross-tail convection electric field. This energization allowed the low energy ions to be pushed farther out into the magnetospheric tail where other acceleration processes could energize them even more. The lower and higher energy outflows overlapped, especially in the polar cap and were sent outward into the lobes of the magnetotail and possibly the plasma sheet. In sum, there was a large amount of outflow when the polar wind and the auroral zone outflows were added together. Initial estimates showed total fluxes out of the ionosphere of 10^{25}–10^{26} ions per second.

The magnitude of the total ion outflow, both the low energy polar wind and the higher energy auroral zone ions, led to the first idea that there might be enough ions flowing from the ionosphere to the magnetosphere to fill up the different regions of the magnetosphere to the observed levels. Chappell et al. (1987) looked into this possibility; their concept is shown in Fig. 4. In the inner magnetosphere, the upward flowing polar wind would fill up the flux tubes inside of the plasmapause location as the tubes continued to circulate with the Earth and not intersect the magnetopause where plasma could be lost. Just outside of the plasmapause, upward flowing polar wind particles could be convected to the magnetopause and lost to the magnetosheath on the dayside.

At higher L-shells, the outward flowing polar wind ions would be swept over the polar cap where they could become energized as they drifted through the polar cusp or later by the higher altitude centrifugal acceleration process. These ions would flow out through the lobes of the magnetotail and, depending on the Bz component of the solar wind magnetic

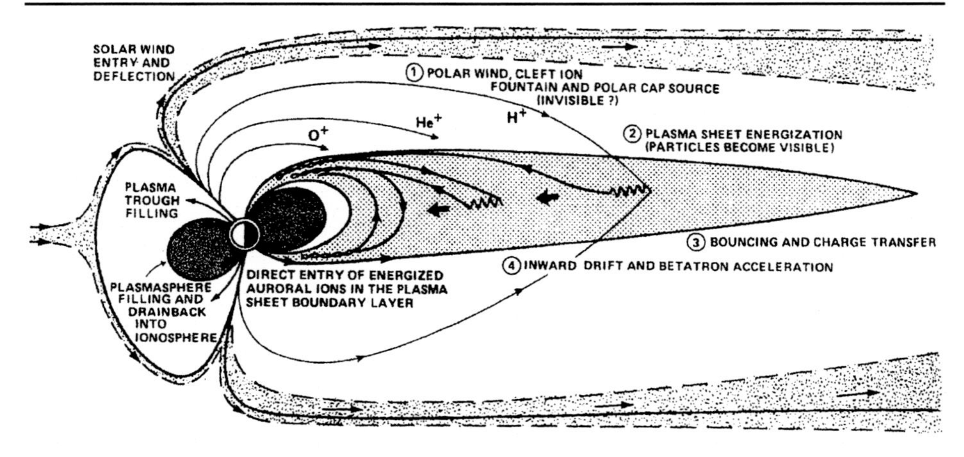

Fig. 4 A sketch of the flow of polar wind and auroral ions upward out of the ionosphere and into the outer magnetosphere. The polar wind ions move through the lobes of the tail into the downstream plasma sheet and the auroral zone ions move directly upward into the inner plasma sheet. In combination, they represent a significant ionospheric source of plasma for the magnetosphere

field at the magnetopause, could drift into the plasma sheet or escape anti-sunward through the lobes of the tail. The low energy polar wind portion of these outflowing ions would not be measurable because of the positive potentials that develop on spacecraft in the tail lobes and outer plasma sheet. These particles would only become visible after they became energized in the plasma sheet because of their movement through the cross-tail potential or because of magnetic reconnection processes. After their energization they would "appear" in the plasma sheet with higher energies. In addition to the polar wind particles, the more energetic upflowing ions from the polar cusp could also be swept across the polar cap into the plasma sheet with the upflowing ions from the nightside auroral zone moving directly upward into the more near-Earth plasma sheet.

Chappell et al. (1987) utilized the DE information on the magnitude of the polar wind and auroral zone outflow and followed the approximate motion of the ions out through the magnetotail lobes, into the plasma sheet, and subsequently into the ring current. Using the approximate volumes of the plasma sheet and ring current and estimating the residence time that an ion would spend drifting through these regions, the densities of the plasma sheet and ring current ions caused by ion outflow could be calculated. It was found that the densities predicted for the lobes of the magnetotail, the plasma sheet and the ring current matched the observed densities very well.

In summary, there appeared to be enough plasma flowing out of the ionosphere to adequately fill up the major regions of the magnetosphere. But does this really happen? Since it was virtually impossible to measure the low energy polar wind ions in the lobes of the tail because of the positive spacecraft potential, this filling mechanism from the ionosphere was not significantly embraced by the magnetospheric community. Consequently, the concept of solar wind access through the polar cusp and the flanks of the magnetotail remained the dominant explanation of the magnetospheric plasma source mechanism at the end of the 1980s.

4 The Decade of the 1990s

One of the uncertainties of the Chappell et al. (1987) paper had been not knowing the more exact trajectories of the ions as they flowed upward through the changing magnetic field and

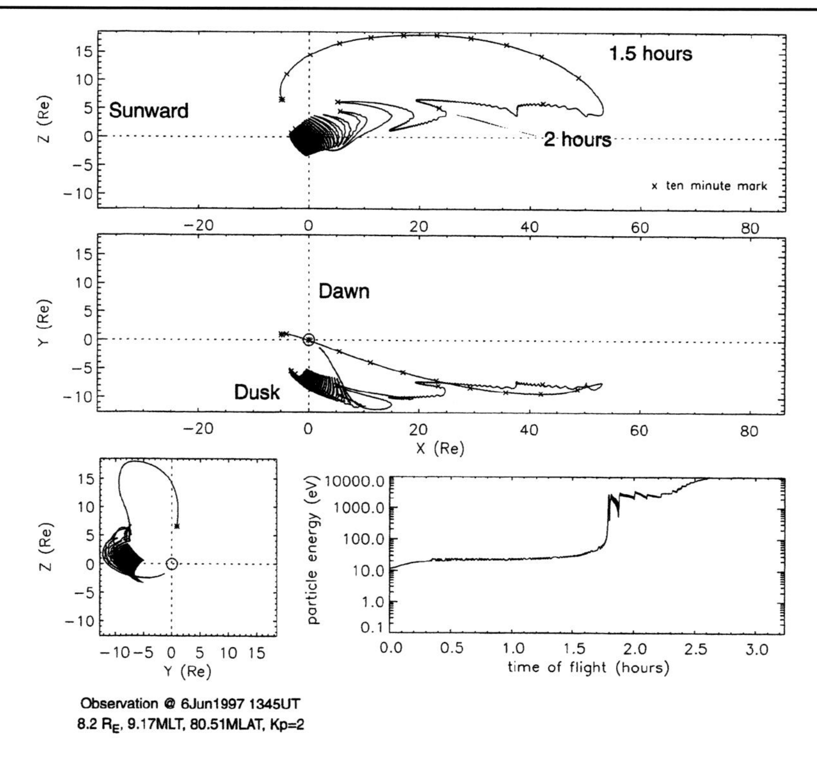

Fig. 5 Results of the tracing of the flow trajectories of a low energy ion which leaves the ionosphere and moves through the magnetosphere, gaining energy as it moves through the cross-tail potential in the magnetosphere. By following the time marks on the energy plot in the *lower right*, and matching them with the two time tags on the trajectory in the *upper plot*, one can see how the initially low energy ion gains energies representative of the plasma sheet and ring current

convection electric field. Delcourt et al. (1993) completed a set of ion trajectory calculations that showed more clearly how ions starting at different locations in the mid and high latitude ionosphere would move into the different regions of the magnetosphere. An example of this trajectory study is shown in Fig. 5. In this figure, an H^+ ion which represents a classical polar wind ion that has been centrifugally accelerated flows out through the lobe of the tail and into the duskside plasma sheet. As it curvature drifts from midnight toward dusk in the cross-tail potential, it is accelerated to energies characteristic of the plasma sheet after it enters that region. As it drifts farther earthward, its energy is increased to 10 keV, characteristic of the ring current region. The first three plots in the figure show the XZ, XY and YZ planes respectively. The times shown on the first plot can be matched with times in the fourth plot of energy versus time to see how the particle gains energy as it moves through the plasma sheet and ring current regions.

The surprising result of this analysis by Delcourt et al. (1993) is not only that the ions drift through the different major regions of the magnetosphere, but that they become energized to the level of energy characteristic of each region. Hence, the same particle can become part of several different major magnetospheric regions as it drifts through the magnetosphere. The Delcourt et al. (1993) ion trajectories were run for the different major upflowing ions, H^+, He^+ and O^+ to demonstrate how they move through the magnetosphere. Different initial pitch angles for the outflowing particles can also be used. The different masses and pitch

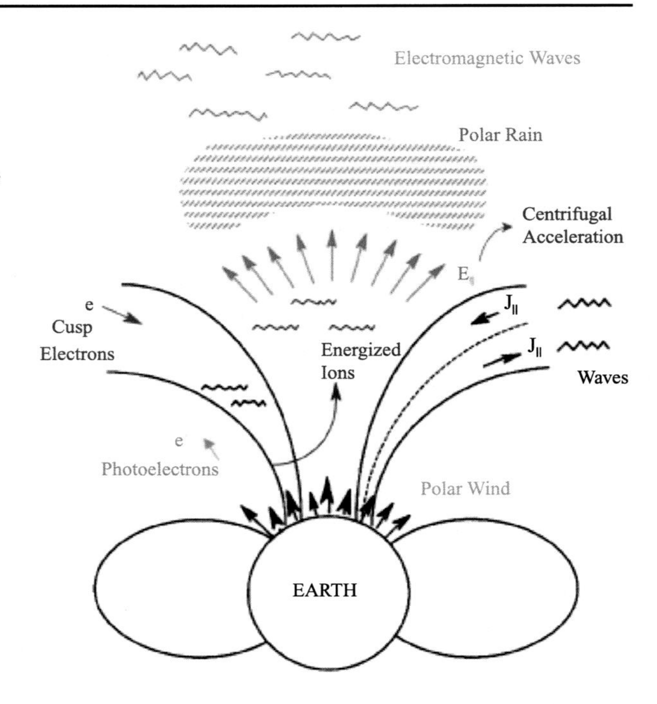

Fig. 6 A sketch illustrating the different acceleration processes that can come into play as a "classical" polar wind ion moves along its flow path through the ionosphere and magnetosphere, becoming the "generalized" polar wind

angles have a significant influence on the energization and fate of each individual particle. This suggests that ionospheric exit location, energy, pitch angle and mass are important in calculating where the particle will go in the magnetosphere and what different energies it will obtain.

Another advancement that took place in this decade was the further development of the polar wind models by scientists at Utah State University (Demars and Schunk 1994; Barakat et al. 1995; Schunk and Sojka 1997). They continued to add important elements to the equations that explained the origin of the outflowing ions and the subsequent energization as the ions moved upward and across the polar cusp, polar cap and nightside auroral oval as shown in Fig. 6. The model was run for different flux tubes that are convecting through the high latitude magnetosphere from sunlight to darkness at the foot of the field line and for different magnetic activity levels, seasons and solar activity conditions. This modeling work began to close the identity gap between the initial classical polar wind results and the polar cusp and auroral zone related outflow. It also filled in the understanding of how a single ion can be influenced by a number of different flow and energization mechanisms as it moves upward into the magnetosphere.

In concert with these modeling advancements, the International Solar Terrestrial Physics (ISTP) mission gave multi-point measurements of the magnetosphere and solar wind from the Polar, Geotail and Wind spacecraft. The Polar satellite in particular measured the major outflow region from mid to high latitudes at altitudes of 5000 km to more than 9 earth radii. This spacecraft was instrumented in a way that covered both the low energy and more energetic ion composition and dynamics. The MSFC Thermal Ion Dynamics Experiment (TIDE) design was optimized to measure the mass composition and dynamics of the low energy polar wind plasma (Moore et al. 1995). The spacecraft also contained a plasma neutralizer that successfully limited the satellite positive potential when it was operated. The TIDE instrument had six opening apertures that were quite large (2 cm by 5 cm) in order to

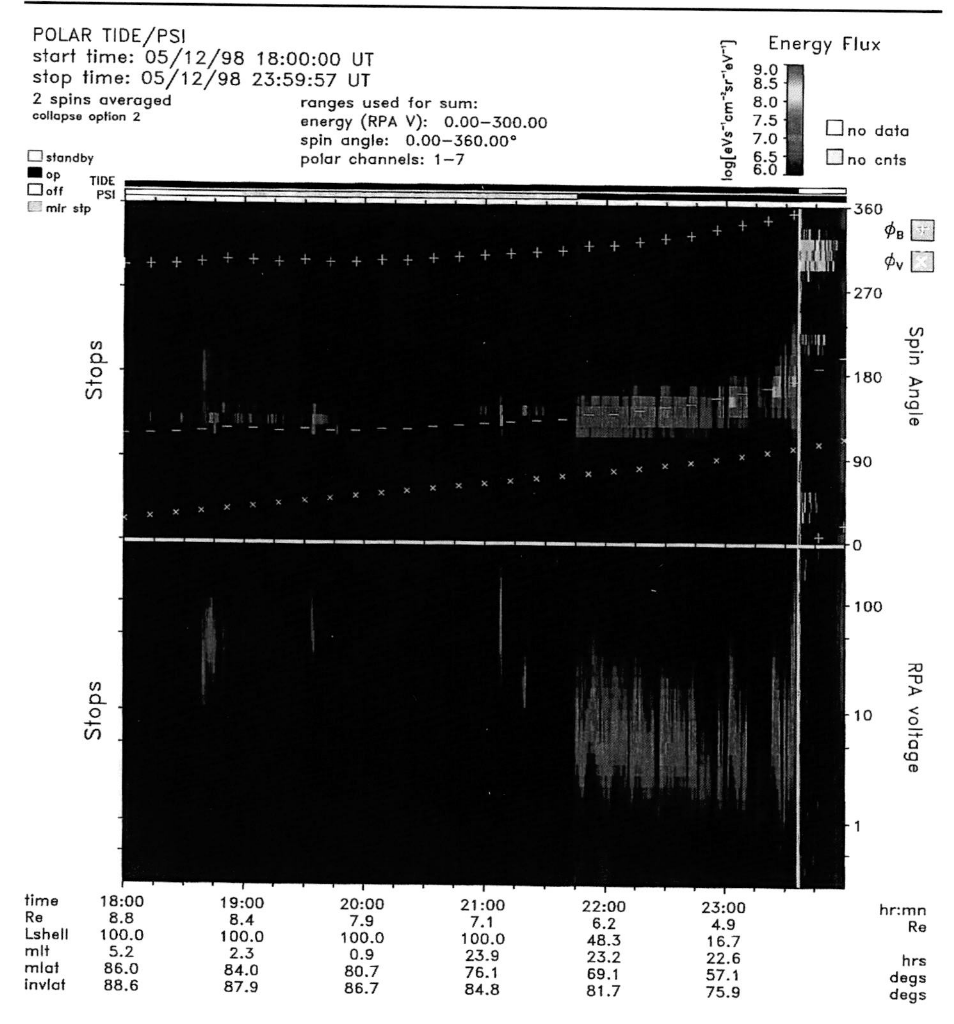

Fig. 7 A TIDE pass across the polar cap and magnetotail lobe with the Plasma Source Instrument turning on during the pass at 2145 UT. Outflows along the field line (*upper panel*) of less than 10 eV (*lower panel*) are hidden view until PSI is activated

get a geometric factor that is large enough to be able to measure the angular distribution of the low velocity polar wind ions. This necessarily large geometric factor makes it difficult if not impossible to use a single instrument to measure the full energy spectrum of ions from an eV to 10's of keV. It has been the case that many succeeding missions have not included an instrument that is specifically designed to measure the characteristics of the low energy plasma; hence this area of knowledge has not been as effectively advanced.

It is important to note that the Polar spacecraft potential was typically positive by 1 or 2 volts even at the low altitudes of 5000 km and this is large enough to prevent the full measurement of the outflowing polar wind. At high altitudes, spacecraft potentials reached 10's of volts positive in the lobe of the tail. With the spacecraft potential control, the TIDE measurements have demonstrated that the polar wind is present in this broad altitude range as shown in Fig. 7 (Moore et al. 1997; Su et al. 1998; Chappell et al. 2000). However, it was

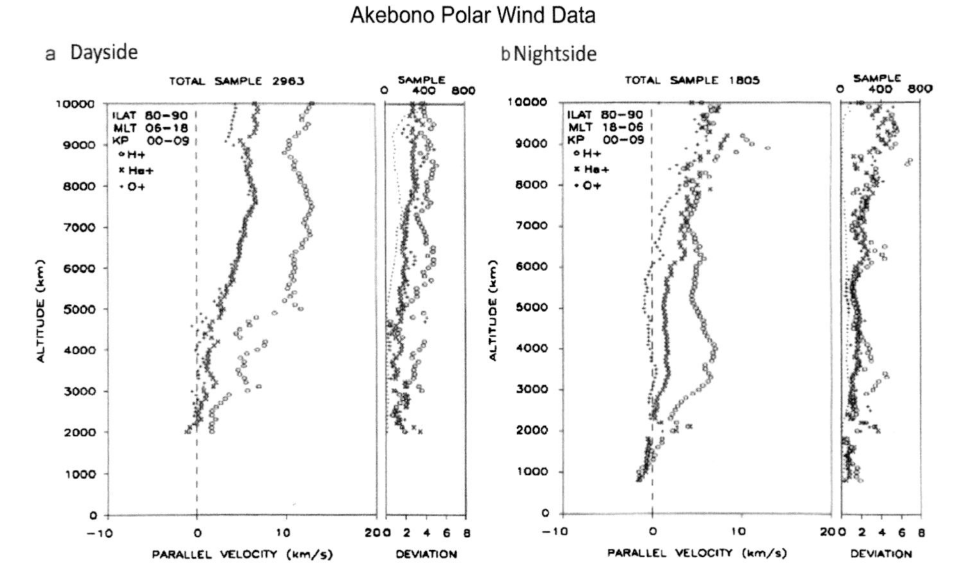

Fig. 8 Akebono SMS data on the outflow velocities of the polar wind versus altitude on the dayside and nightside of the ionosphere

only possible to get glimpses of the polar wind when the plasma source neutralizer was operating. The neutralizer operation was limited during the mission because of the concern by other experimenters about interference from the ionized xenon plasma cloud created by the neutralizer source. Any future spacecraft without potential control will have its direct measurements of the cold flowing particle characteristics compromised.

An extensive set of measurements of outflowing H^+, He^+ and O^+ ions was also made by the suprathermal mass spectrometer (SMS) ion mass spectrometer on the Akebono satellite in this time period. Akebono had a lower altitude orbit than Polar. Figure 8 shows example results from Akebono with upward velocities in the 5–15 km per second range. The Akebono results showed very large fluxes of O^+, which probably reflects the contribution from energization processes in the polar cusp and subsequent drift across the polar cap (Abe et al. 1993, 1996; Yau and Andre 1997; Andre and Yau 1997; Cully et al. 2003a). The measured outflow fluxes were 1–3 $\times$ 10^8 ions/cm^2 sec, as theory had predicted, with a total flux estimated at 10^{25}–10^{26} ions/sec.

Therefore, the experimental verification of the presence of the low energy polar wind at low and high altitudes continued. In addition the increasingly effective modeling of the polar wind generation in the topside ionosphere and its transit upward into the magnetosphere became more accurate. This transit was found to take the low energy ions to regions which strongly influence magnetospheric dynamics and thus set an even stronger foundation for needing to understand the role of the ionosphere as a source of magnetospheric plasma. Yet, somewhat surprisingly, the magnetospheric community acceptance of this idea still remained limited.

5 The Decade of the 2000s

The enhanced measurement techniques for characterizing outflowing ions combined with the better modeling understanding of their origin in the ionosphere and their transport to the

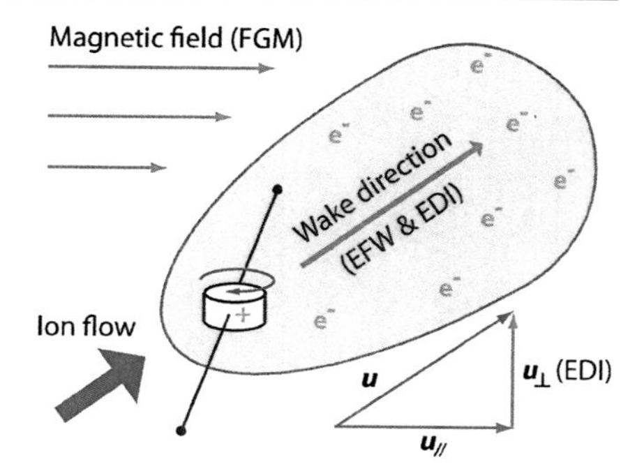

Fig. 9 A schematic drawing of the Cluster spacecraft showing the wake of the spacecraft and the two independent electric field measurements which could be combined with the magnetic field data to determine the flow velocity (*u* parallel) of the very low energy plasma in the magnetotail. Combining the flow velocity with the measured density gives the flux of the low energy ions

magnetosphere were used in two studies to do a more refined assessment of how effectively the ionosphere can act as a source for magnetospheric plasmas. Cully et al. (2003b) used the Akebono measurements of ion outflow together with some basic transport trajectories and calculated the adequacy of the ionospheric source to fill the magnetosphere. He found the ionosphere to be a significant source of magnetospheric plasma, which varied depending on the level of magnetic activity.

Huddleston et al. (2005) refined the Chappell et al. (1987) calculations by using the new TIDE data from Polar with adjustments for spacecraft potential influences coupled with the ion trajectory mapping of Delcourt et al. (1993). Measured fluxes and pitch angles of ions in different invariant latitude/magnetic local time boxes at 5000 km altitude were transported using the Delcourt et al. (1993) trajectory mapping and summed to show how the filling of the magnetosphere could take place. The low energy polar wind outflows combined with the more energetic auroral zone ion outflow showed again that the ionosphere was a significant source of plasma to the magnetosphere, which could be the dominant source under certain magnetic storm conditions. However the question of whether the low energy plasma really flows up through the "empty" lobes of the magnetotail and intersects the plasma sheet still remained unanswered. Limited measurements from Polar showed the presence of this "lobal wind" (Liemohn et al. 2005), so there was reason for optimism about the presence and influence of the polar wind out in the magnetosphere.

Then came the Cluster mission with its 4 spacecraft flying in relatively close formation through the magnetosphere with perigee of $4R_E$ and apogee of $19R_E$. These spacecraft carried the Composition and Distribution Function Analyzer (CODIF), a mass spectrometer capable of measuring the H^+, He^+ and O^+ with energies ranging from 40 eV to 40 keV, (Reme et al. 2001) as well as two independent measurements of the electric field using a set of probes on wire booms (Gustafsson et al. 1997) and a new energetic electron emitting device which could inject and "catch" the electrons after they had spiraled back to the spacecraft (Paschmann et al. 1997). By measuring their spatial displacement during their spiral motion in the magnetic field, the electric field could be derived. In addition there was a spacecraft potential control device which was operated periodically (Tokar et al. 2001). It was this unique, complete set of instruments (see Fig. 9) combined with some impressive scientific detective work examining the location of the spacecraft wake shape that led to the clear demonstration of the extensive presence of the cold flowing polar wind ions moving throughout the lobes of the geotail and then into the plasma sheet at distances that depended on the character of the solar wind flow and magnetic field direction (Engwall et al. 2006, 2009a, 2009b; Nilsson et al. 2010; Haaland et al. 2012; Andre and Cully 2012). This suite of measurements was able to successfully determine the density, flow velocity, and fluxes of the low energy outflowing ions and to pull back the

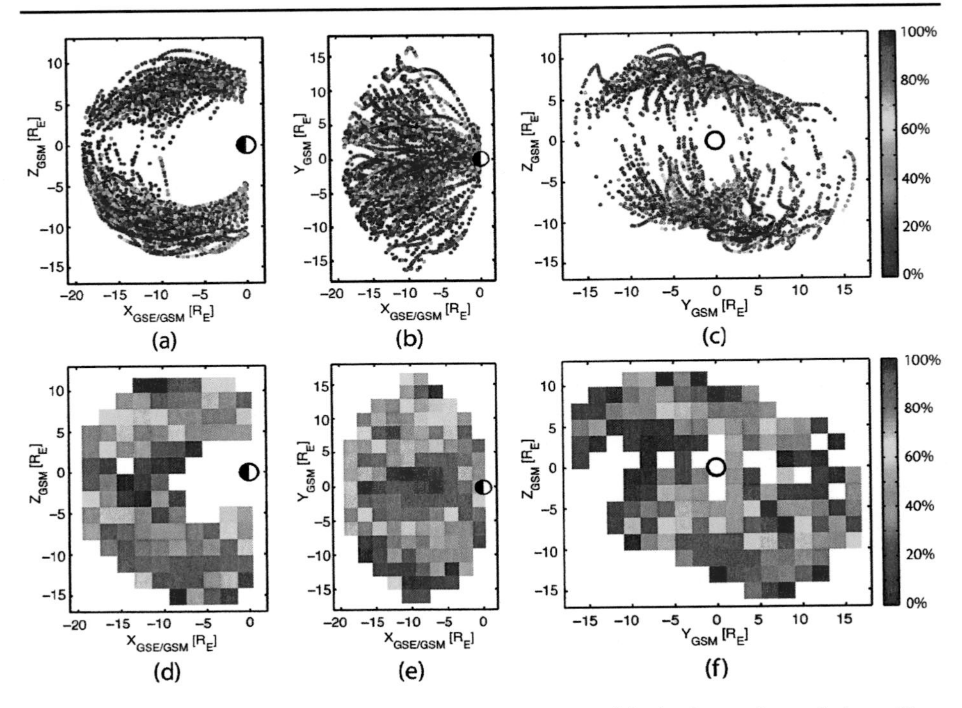

Fig. 10 Shows a collection of Cluster orbits through the magnetotail in the three orthogonal planes. These orbits are *color coded* to show the percentage probability that low energy outflow was seen (see *color bar*). The lower set of orbits have been averaged over a larger spatial area

curtain of spacecraft potential that had obscured the measurement of the presence of these ions in the magnetosphere for four decades.

Figure 10 of Engwall et al. (2009a) shows the portions of the cluster orbits where the low energy polar wind was seen. The orbits are shown in the XZ, XY and YZ planes with a color scale along the orbit that shows the probability that the low energy flowing ions have been measured. The probabilities are significant and the ions are located where the trajectory models would predict. The lower part of the figure shows an average of the individual measurements presented in larger spatial blocks with the probability of observing the outflowing ions shown by the color scale. Figure 11 shows the ion density and flow directions of the low energy polar wind flows in the XZ and XY planes averaged into spatial blocks. Again, the flow directions and densities are as expected from the ionospheric polar wind and ion trajectory calculations. The measured fluxes in the geotail lobes are also as expected from the polar wind models. Mapped back to the topside ionosphere along the flux tubes, the outflow fluxes are in the $1\text{–}3 \times 10^{8}$ ions/cm^{2} sec range, as predicted by the theory, with the total outflowing flux of ions at $10^{25}\text{–}10^{26}$ as shown in Fig. 12.

Do these newly unveiled polar wind outflows in the lobe of the magnetospheric tail enter the plasma sheet? Fig. 13 from Haaland et al. (2012) shows the flow pattern for polar wind ions that are injected from the ionosphere for three different cases. In each case, the north/south component of the solar wind magnetic field is changed, increasing more southward from case a to case c. Note that with the southward component increasing as would be expected during magnetic storms, progressively more of the polar wind outflow is transported into the plasma sheet region where it can become energized, populating the plasma

Fig. 11 A cut through the tail of the magnetosphere in the XZ plane with arrows showing the magnitude and direction of the low energy ion outflow

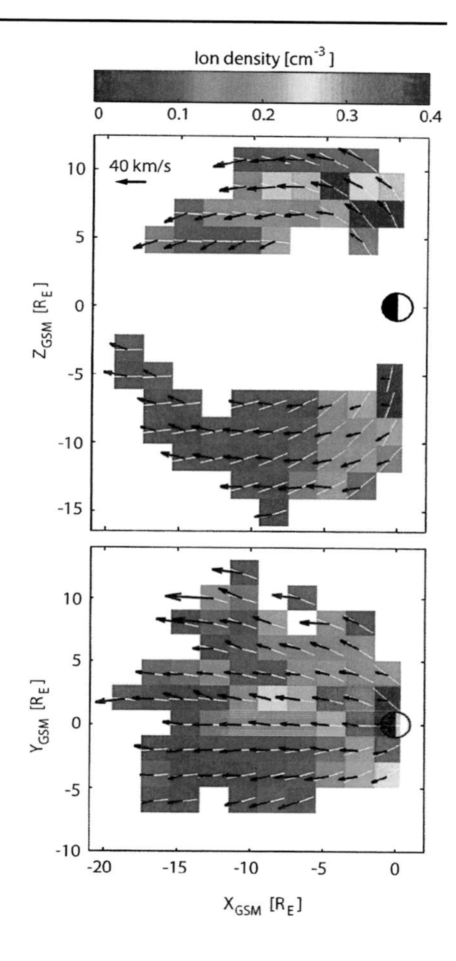

Fig. 12 A plot of the total ion outflow in the magnetotail as a function of K_p as measured by Cluster, Akebono and DE-1. The measured outflow is very similar for the three spacecraft

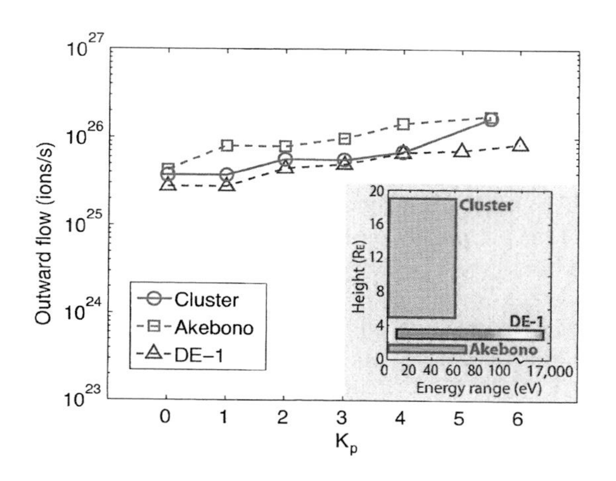

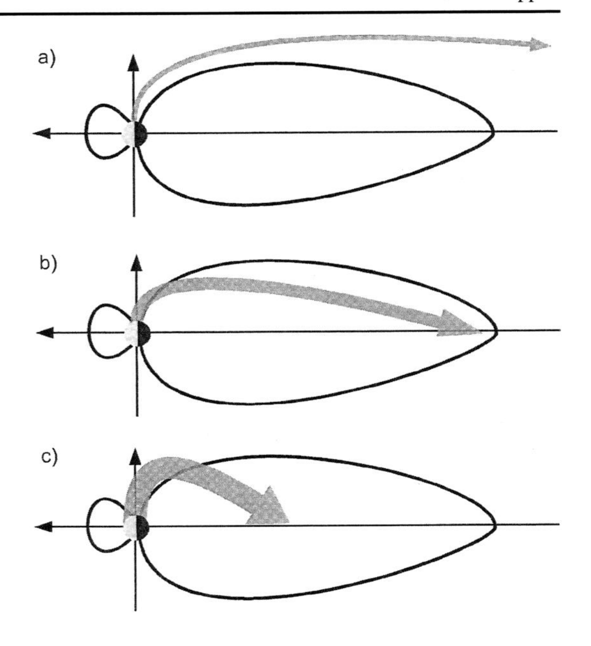

Fig. 13 Shows the calculated flow direction of the low energy ion outflow based on the flow velocity and the cross-tail potential. Based on the increasing strength of the southward component of the solar wind magnetic field from case (**a**) to (**c**), more of the outflow goes into the plasma sheet and less is lost down the tail

sheet. During southward IMF, more than 90 % of the polar wind ions enter the plasma sheet where they can gain energy typical of the plasma sheet particles and become "visible" to the medium energy range instruments given that the spacecraft potential no longer prohibits their observation.

The strength of the ionospheric outflow is known to be affected by the variation in the characteristics of the solar wind as it impinges on the magnetosphere. Changes in solar wind pressure can enhance the magnitude of the outflow (Pollock et al. 1990; Moore et al. 1999, and Cladis 2000) and will play a role in the future determination and modeling of the ion outflow. In addition to the H^+ and He^+ ions expected in the classical polar wind, extensive Cluster measurements using the CODIF instrument by Kistler et al. (2010), show large amounts of O^+ flowing out of the ionosphere caused by a variety of energization processes such as energization from particle precipitation and waves related to the polar cusp (Liao et al. 2010; Mouikis et al. 2010). Although this paper has concentrated on the outflow related more to the classical polar wind, it is clearly the sum of polar wind, polar cusp and nightside auroral zone outflows that combine to make up the total ionospheric mass transport outward into the magnetosphere.

With the flight of new instrumentation that has been able to image the characteristics of the low energy He^+ ions using their emission in 304 angstrom radiation, an enhanced perspective of the more dense low energy plasma the plasmasphere has become possible (Burch 2001). By following the changes in the distribution of these emissions, the evolution of the plasmasphere shape can be determined. During changing magnetic storm times, the plasmasphere is seen to be compressed in the nightside region and distended into tail-like plumes on the dusk side (Goldstein et al. 2003; Goldstein and Sandel 2005). These plumes which range from large and smooth to highly spatially variable extend out toward the magnetopause and have been measured as they cross geosynchronous orbit by the LANL spacecraft (Borovsky and Denton 2008). These "motion picture" images of the changing plasmasphere have verified dynamics measurements that were derived from the earlier spacecraft such as OGO, ISEE, and Polar and give a broader understanding of their evolving nature. Additional measurements showed that these regions of enhanced density were able to drift out to the dayside magnetopause where they are expected to effect the rate of magnetospheric reconnection with the incoming solar wind (Su et al. 2012). As the low energy plasma drifts outward toward the magnetopause, the ionospheric signature of these enhanced density flux tubes can be seen simultaneously (Foster et al. 2014) and their motion can be tracked as they move northward in the ionosphere and through the cusp region.

Hence, the low energy ionospheric-origin plasma can not only act as a source for the more energetic plasma in the magnetosphere; its presence in the magnetopause reconnection region can influence the rate of reconnection and the resultant strength of the magnetospheric convection electric field. Elphic et al. (1997) have shown that the low energy plasma that was convected sunward to the magnetopause and magnetosheath can be carried back toward the tail where it may re-enter the magnetosphere and contribute to the plasma sheet.

In summary, the low energy ionospheric plasma can flow upward from the polar regions directly into the plasma sheet region where it is energized to populate the plasma sheet and ring current or it can flow upward at lower latitudes where it initially fills the low energy plasmasphere and subsequently can be peeled off of the outer dusk region during storms, convected to the magnetopause, can enter the magnetosheath, and be carried back into the tail where it is available to become energized and to contribute to the population of the plasma sheet and ring current in a different way. This latter process adds a new dimension to the ionospheric source through the process of possible recirculation of the low energy plasma through the magnetosphere.

In addition to these new measurements, results from the NSF Geospace Environment Modeling program have brought the modeling of the mass coupling between the ionosphere and the magnetosphere more into focus. The generalized polar wind models, which include all of the energization elements that affect the outflowing ions, are being developed by several groups (Barakat and Schunk 2006; Glocer et al. 2009). The outflow models are now being coupled to the magnetospheric models with the generalized polar wind outflow becoming the source input to the magnetospheric models with promising results (Winglee 2000; Moore et al. 2005; Fok 1999; Welling and Ridley 2010; Welling et al. 2011; Nilsson et al. 2013; Glocer et al. 2009; Brambles et al. 2010).

As an example, the work of Welling and Ridley (2010) adds the ionospheric source to the magnetospheric models and gives an important result that shows the dominant role that the ionosphere can play during southward Bz conditions in the solar wind. In Fig. 14 the relative contribution of the ionosphere (blue) and the solar wind (red) can be seen in two cuts through the magnetosphere—the $Y = 0$, and the $Z = 0$ planes. The two panels on the left show the case in which the IMF Bz is southward and on the right when the IMF Bz is northward. In the two panels on the left, the ionospheric source is dominant in contrast to the two panels on the right where the solar wind contribution is more dominant. In addition to the issue of the plasma population, the inclusion of the ionospheric source in these models brings about a better agreement with a number of measured magnetospheric parameters such as the magnitude of the cross polar cap potential. The prospects for these merged modeling studies are very positive.

6 Summary

The advancement of both measurement and modeling of the ionospheric outflow and its transport throughout the magnetosphere has led to a new understanding of this fundamentally important mass coupling process in the Earth's space environment. Not only does the

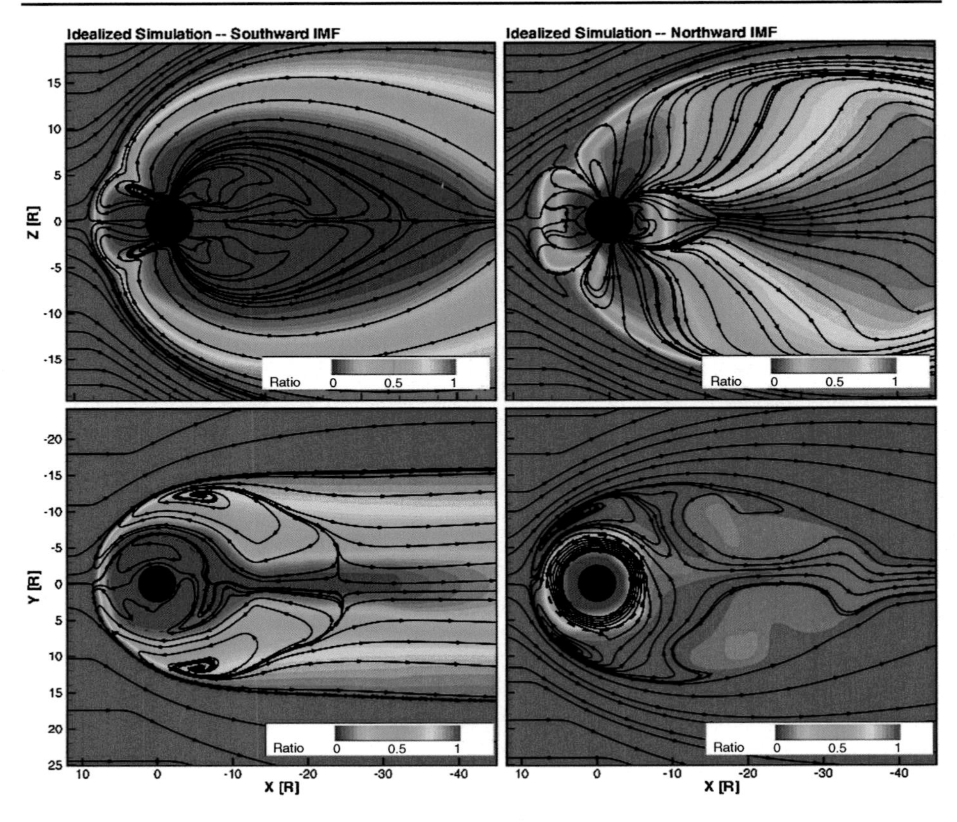

Fig. 14 Shows modeling results which include an ionospheric source. The *top panels* show the Y = 0, XZ plane and the *bottom panels* show the Z = 0, XY(equatorial) plane. The *two panels on the left* show the case for IMF Bz southward and *on the right* for IMF Bz northward. The *color contours* represent the ratio of the ionosphere to solar wind plasma as shown on the *color bar with blue indicating* all ionosphere and *red indicating* all solar wind. *Black lines* show plasma streamlines

low energy polar wind plasma flow upward from the ionosphere to the magnetosphere as predicted by early measurements and models more than 30 years ago; it is found that as it moves through the magnetosphere, it becomes energized to the levels that are characteristic of the regions through which it moves. This transport and energization is in agreement with the plasma characteristics found in the polar cap, magnetotail lobes, plasma sheet, ring current and warm plasma cloak. This filling process is shown in Fig. 15 from Chappell et al. (2008). The general motion from the ionosphere upward through the polar cap and lobe of the geotail and into the plasma sheet region can be seen. The ultimate fate of an upflowing ionospheric ion is determined by the location of its exit from the ionosphere, its energy, pitch angle and mass, and the action of acceleration processes that affect it as it flows into the magnetosphere. The energy that it gains as it moves through the polar cap (blue) and lobes (green) will determine how far back into the tail it flows and where it enters the plasma sheet. This entry point then determines its future energization.

As indicated in Fig. 15, ions which enter the plasma sheet closer to the Earth and on the dawn side of the tail, will be energized less and will drift earthward to form the warm plasma cloak region as indicated in yellow. Ions which are carried farther back into the tail

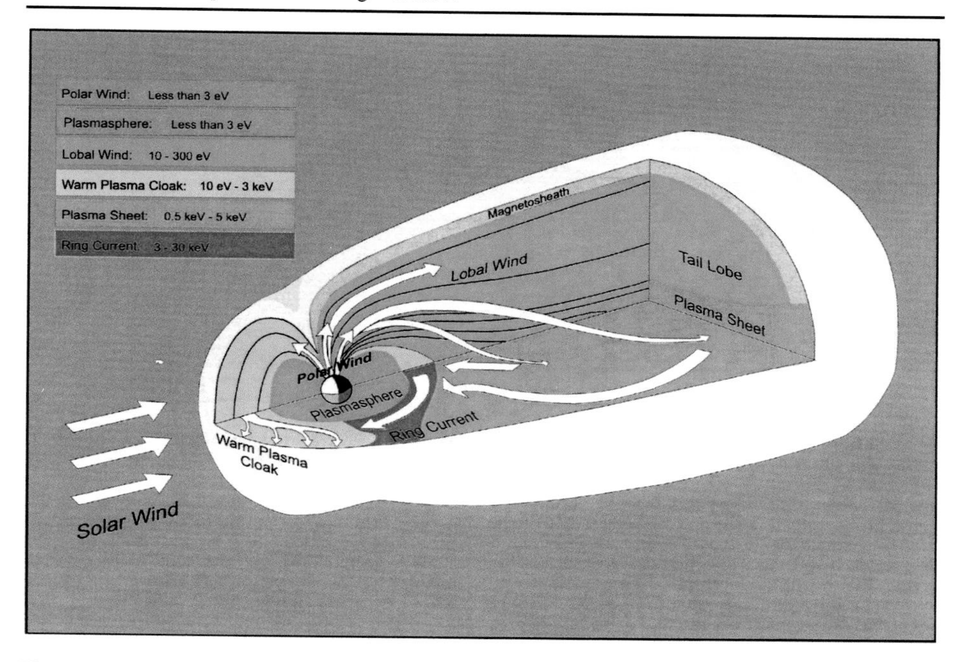

Fig. 15 A sketch of the magnetosphere showing the motion and energization of low energy ionospheric plasma as it moves through the magnetosphere. Beginning at a few eV of energy in the ionosphere, it is transported from the polar cap, through the lobes of the tail into the plasma sheet and then into the warm plasma cloak or the ring current depending on the point of entry into the plasma sheet

enter the plasma sheet region at greater distances. These ions move through the neutral sheet region at points where the magnetospheric magnetic field is more distended (non-dipolar) which causes them to have more curvature drift and hence more movement across the cross-tail potential drop. These ions are energized to plasma sheet energies (pink) and then drift earthward gaining further energy through betatron acceleration to become part of the ring current (red). This view of magnetospheric filling is still not widely understood or appreciated within the space physics community. The recent Engwall et al. (2009a) measurements that have thoroughly solved the mystery of the invisible low energy plasma transport into the magnetosphere regrettably remain unknown to many magnetospheric physicists.

Given the progress with understanding the role of the ionosphere in populating the magnetosphere, there is a different image evolving of the Earth's magnetosphere as shown in Fig. 16. The role of the solar wind in shaping the magnetosphere and delivering energy to the magnetosphere through merging at the nose, convection throughout the magnetosphere and reconnection in the tail is still in place. However, the source of the magnetospheric plasmas now takes on a different perspective. As shown by the green areas, representing ion outflow, the contributors to filling the magnetosphere now include the ionosphere in combination with the solar wind. The dominance of the ionosphere is a function of the characteristics of the solar wind at the nose of the magnetosphere. Both the plasma pressure, which has been statistically related to the amount of polar outflow, and the direction of the IMF, which determines the magnitude of the convection, influence the strength and ultimate energy of the ionospheric source particles.

As a corollary to the more realistic merged modeling, the future measurement of low energy plasma in the magnetosphere must be done more effectively than recent missions

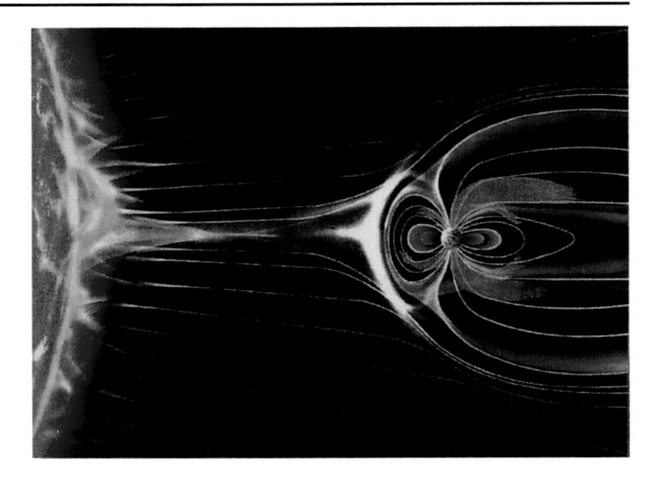

Fig. 16 A revised sketch of the solar wind-magnetosphere region reflecting the addition of the ionospheric source *in green* flowing upward in the polar regions. In contrast to our earlier ideas about the solar wind origin of magnetospheric plasma, the ionosphere is now understood to make a significant and often dominant contribution

have done. It will be imperative that future missions have an instrument with a large enough geometric factor to be able to measure the mass, pitch angle and energy distributions of the very low energy (less than a few eV) ions so that the nature of the processes that transform them can be observed. In addition, there will have to be a potential control device on the spacecraft that will permit the low energy particles to reach the spacecraft without phase space distortion so that their measurement can be made. Combining the very low energy ion measurements into a single instrument with the medium to high energy measurements will most likely not be an acceptable approach. In times of tight budgets for NASA missions, the fundamental role that the ionosphere plays in populating the magnetosphere will have to be taken into consideration when the instruments are selected for the spacecraft.

Finally, it should be noted that the lessons learned from the terrestrial magnetosphere about the role of the ionosphere, should be remembered when considering the plasma sources of other solar system magnetospheres. In the case of Jupiter, the moon Io is certainly a very important source of magnetospheric plasma, but there are some indications from model calculations (Nagy et al. 1986) and measurements (McComas et al. 2007) that Jupiter's ionosphere is also playing a role in populating its magnetosphere. Since this is the case, it becomes very important to facilitate the interaction of those scientists who work on ionosphere-magnetosphere coupling at the Earth with those planetary scientists involved in understanding the coupled plasma and field environments of the planets. This ISSI Workshop has been an important step in bringing these groups together in order to share knowledge on measuring and modeling the fundamentally important coupling within the ionosphere and magnetosphere of different planetary systems.

References

T. Abe et al., EXOS D (Akebono) suprathermal mass spectrometer observations of the polar wind. J. Geophys. Res. **98**, 11,191 (1993)

T. Abe et al., Observations of polar wind and thermal ion outflow by Akebono/SMS. J. Geomagn. Geoelectr. **48**, 319 (1996)

M. Andre, C.M. Cully, Low-energy ions: A previously hidden solar system particle population. Geophys. Res. Lett. **39**, L03101 (2012)

M. Andre, A.W. Yau, Theories and observations of ion energization and outflow in the high latitude magnetosphere. Space Sci. Rev. **80**, 27 (1997)

I. Axford, The polar wind and terrestrial helium budget. J. Geophys. Res. **73**(21), 6855 (1968)
P.M. Banks, T.E. Holzer, The polar wind. J. Geophys. Res. **73**, 6846 (1968)
P.M. Banks, A.F. Nagy, W.I. Axford, Dynamical behavior of thermal protons in the mid-latitude ionosphere and magnetosphere. Planet. Space Sci. **19**(9), 1053 (1971)
P.M. Banks, R.W. Schunk, W.J. Raitt, Temperature and density structure of thermal proton flows. J. Geophys. Res. **79**(31), 4691 (1974a)
P.M. Banks, R.W. Schunk, W.J. Raitt, NO^+ and O^+ in the high latitude F-region. Geophys. Res. Lett. **1**(6), 239 (1974b)
A.R. Barakat, R.W. Schunk, A three-dimensional model of the generalized polar wind. J. Geophys. Res. **111**(A12), 1978 (2006)
A.R. Barakat, I.A. Barghouthi, R.W. Schunk, Double-hump H^+ velocity distribution in the polar wind. Geophys. Res. Lett. **22**, 1857 (1995)
J.E. Borovsky, M.H. Denton, A statistical look at plasmaspheric drainage plumes. J. Geophys. Res. **113**, A09221 (2008)
O.J. Brambles et al., Effects of causally driven cusp O^+ outflow on the storm time magnetosphere-ionosphere system using a multifluid global simulation. J. Geophys. Res. **115**, A00J04 (2010)
N.M. Brice, Bulk motion of the magnetosphere. J. Geophys. Res. **72**, 5193 (1967)
H.C. Brinton et al., Altitude variations of ion composition in the midlatitude trough region: Evidence for upward plasma flow, NASA Preprint, X-621-70-311 (1970)
J.L. Burch, Views of the Earth's magnetosphere with the IMAGE satellite. Science **291**, 619 (2001)
D.L. Carpenter, Whistler evidence of a "knee" in the magnetospheric ionization density profile. J. Geophys. Res. **68**, 1675 (1963)
M.O. Chandler, T.E. Moore, J.H. Waite, Observations of polar ion outflows. J. Geophys. Res. **96**, 1412 (1991)
C.R. Chappell, Recent satellite measurements of the morphology and dynamics of the plasmasphere. Rev. Geophys. Space Phys. **10**(4), 951 (1972)
C.R. Chappell, K.K. Harris, G.W. Sharp, Ogo 5 measurements of the plasmasphere during observations of stable auroral red arcs. J. Geophys. Res. **76**, 2357 (1971)
C.R. Chappell, T.E. Moore, J.H. Waite Jr., The ionosphere as a fully adequate source of plasma for the Earth's magnetosphere. J. Geophys. Res. **92**, 5896 (1987)
C.R. Chappell et al., The adequacy of the ionospheric source in supplying magnetospheric plasma. J. Atmos. Sol.-Terr. Phys. **62**, 421 (2000)
C.R. Chappell et al., Observations of the warm plasma cloak and an explanation of its formation in the magnetosphere. J. Geophys. Res. **113**, A09206 (2008)
J.B. Cladis, Parallel acceleration and transport of ions from polar ionosphere to plasma sheet. Geophys. Res. Lett. **13**, 893 (1986)
J.B. Cladis, Observations of centrifugal acceleration during compression of magnetosphere. Geophys. Res. Lett. **27**, 915 (2000)
K.D. Cole, Stable auroral red arcs, sinks for energy of Dst main phase. J. Geophys. Res. **70**(7), 1689 (1965)
J.M. Cornwall, F.V. Coroniti, R.M. Thorne, Unified theory of SAR arc formation at the plasmapause. J. Geophys. Res. **76**(19), 4428 (1971)
C.M. Cully et al., Akebono/Suprathermal Mass Spectrometer observations of low-energy ion outflow: Dependence on magnetic activity and solar wind conditions. J. Geophys. Res. **108**, 1093 (2003a)
C.M. Cully et al., Supply of thermal ionospheric ions to the central plasma sheet. J. Geophys. Res. **108**, 1092 (2003b)
D.C. Delcourt, J.A. Sauvaud, T.E. Moore, Polar wind ion dynamics in the magnetotail. J. Geophys. Res. **98**, 9155 (1993)
H.G. Demars, R.W. Schunk, A multi-ion generalized transport model of the polar wind. J. Geophys. Res. **99**, 221 (1994)
R.C. Elphic et al., The fate of the outer plasmasphere. Geophys. Res. Lett. **24**, 365 (1997)
E. Engwall et al., Low-energy (order 10 eV) ion flow in the magnetotail lobes inferred from spacecraft wake observations. Geophys. Res. Lett. **33**, 6110 (2006)
E. Engwall et al., Survey of cold ionospheric outflows I the magnetotail. Ann. Geophys. **27**, 3185 (2009a)
E. Engwall et al., Earth's ionospheric outflow dominated by hidden cold plasma. Nat. Geosci. **2**, 24 (2009b)
M.-C. Fok, Storm time modeling of the inner plasma sheet/outer ring current. J. Geophys. Res. **104**, 14557 (1999)
J.C. Foster et al., Storm time observations of plasmasphere erosion flux in the magnetosphere and ionosphere. Geophys. Res. Lett. **41**, 762 (2014)
S.B. Ganguli, The polar wind. Rev. Geophys. **34**, 311 (1996)
A. Glocer et al., Modeling ionospheric outflows and their impact on the magnetosphere: Initial results. J. Geophys. Res. **114**, A05216 (2009)

J. Goldstein, B.R. Sandel, The global pattern of evolution of plasmaspheric drainage plumes, in *Inner Magnetosphere Interactions: New Perspectives from Imaging*. AGU Geophysical Monograph Series, vol. 159 (2005)
J. Goldstein et al., IMF-driven plasmasphere erosion of 10 July 2000. Geophys. Res. Lett. **30**, 1146 (2003)
J.M. Grebowsky, Model study of plasmapause motion. J. Geophys. Res. **75**, 4329 (1970)
K.I. Gringauz, The structure of the ionized gas envelope of Earth from direct measurements in the USSR of local charged particle concentrations. Planet. Space Sci. **11**, 281 (1963)
C. Gurgiolo, J.L. Burch, DE-1 observations of the polar wind—A heated and an unheated component. Geophys. Res. Lett. **9**, 945 (1982)
G. Gustafsson et al., The electric field and wave experiment for the cluster mission. Space Sci. Rev. **79**, 137 (1997)
S. Haaland et al., Estimating the capture and loss of cold plasma from ionospheric outflow. J. Geophys. Res. **117**, A07311 (2012)
J.H. Hoffman et al., Studies of the composition of the ionosphere with a magnetic deflection mass spectrometer. Int. J. Mass Spectrom. Ion Phys. **4**, 315 (1970)
M. Huddleston et al., An examination of the process and magnitude of ionospheric plasma supply to the magnetosphere. J. Geophys. Res. **110**, 12,202 (2005)
L.M. Kistler et al., Cusp as a source for oxygen in the plasma sheet during geomagnetic storms. J. Geophys. Res. **115**, A03209 (2010)
W. Lennartsson et al., Some initial ISEE-1 results on the ring current composition and dynamics during the magnetic storm of December 11, 1977. Geophys. Res. Lett. **6**, 483 (1979)
J. Liao et al., Statistical study of O+ transport from the cusp to the lobes with Cluster CODIF data. J. Geophys. Res. **115**, A00J15 (2010)
M.W. Liemohn et al., Occurrence statistics of cold, streaming ions in the near-earth magnetotail: Survey of Polar-TIDE observations. J. Geophys. Res. **110**, A07211 (2005)
M. Lockwood et al., A new source of suprathermal O^+ ions near the dayside polar cap boundary. J. Geophys. Res. **90**, 4099 (1985)
D.J. McComas, F. Allegrini, F. Bagenal, Diverse plasma populations and structures in Jupiter's magnetotail. Science **318**, 5848, 217 (2007)
T.E. Moore et al., The thermal ion dynamics experiment and plasma source instrument. Space Sci. Rev. **71**, 409 (1995)
T.E. Moore et al., High altitude observations of the polar wind. Science **277**, 349 (1997)
T.E. Moore et al., Ionospheric mass ejection in response to a CME. Geophys. Res. Lett. **26**, 2339 (1999)
T.E. Moore et al., Plasma sheet and (nonstorm) ring current formation from solar and polar wind sources. J. Geophys. Res. **110**, A02210 (2005)
C.G. Mouikis et al., H+ and O+ content of the plasma sheet at 15–19 Re as a function of geomagnetic and solar activity. J. Geophys. Res. **115**, A00J16 (2010)
T. Nagai et al., First measurements of supersonic polar wind in the polar magnetosphere. Geophys. Res. Lett. **11**, 669 (1984)
A.F. Nagy, P.M. Banks, Photoelectron fluxes in the ionosphere. J. Geophys. Res. **75**(31), 6260 (1970)
A.F. Nagy, A.R. Barakat, R.W. Schunk, Is Jupiter's ionosphere a significant plasma source for its magnetosphere? J. Geophys. Res. **91**, 351 (1986)
H. Nilsson et al., Centrifugal acceleration in the magnetotail lobes. Ann. Geophys. **28**, 569 (2010)
H. Nilsson et al., Hot and cold ion outflow: Observations and implications for numerical models. J. Geophys. Res. **118**, 1 (2013)
A. Nishida, Formation of plasmapause, or magnetospheric plasma knee, by the combined action of magnetospheric convection and plasma escape from the tail. J. Geophys. Res. **71**, 5669 (1966)
G. Paschmann et al., The electron drift instrument for Cluster. Space Sci. Rev. **79**, 233 (1997)
C.J. Pollock et al., A survey of upwelling ion event characteristics. J. Geophys. Res. **95**, 18,969 (1990)
H. Reme et al., First multispacecraft ion measurements in and near the Earth's magnetosphere with the identical Cluster ion spectrometry (CIS) experiment. Ann. Geophys. **19**, 1303 (2001)
R.W. Schunk, J.J. Sojka, Global ionosphere-polar wind system during changing magnetic activity. J. Geophys. Res. **11**, 625 (1997)
R.D. Sharp et al., Observation of an ionospheric acceleration mechanism producing energetic (keV) ions primarily normal to the geomagnetic field direction. J. Geophys. Res. **82**, 3324 (1977)
E.G. Shelley, R.G. Johnson, R.D. Sharp, Satellite observations of energetic heavy ions during a geomagnetic storm. J. Geophys. Res. **77**, 6104 (1972)
E.G. Shelley et al., Plasma composition experiment on ISEE-A. Trans. Geosci. Electron. **16**, 266 (1978)
Y.J. Su et al., Polar wind survey with the thermal ion dynamics Experiment/Plasma source instrument suite aboard POLAR. J. Geophys. Res. **103**(29), 305 (1998)
Y.J. Su et al., Plasmaspheric material on high-latitude open field lines. J. Geophys. Res. **106**, 6085 (2012)

H.A. Taylor et al., Positive ion composition in the magnetosphere obtained from the OGO-A satellite. J. Geophys. Res. **70**, 5769 (1965)

K. Tokar et al., Active spacecraft potential control for Cluster-implementation and first results. Ann. Geophys. **19**, 1289 (2001)

D.T. Welling, A.J. Ridley, Exploring sources of magnetospheric plasma using multispecies MHD. J. Geophys. Res. **115**, A04201 (2010)

D.T. Welling et al., The effects of dynamic ionospheric outflow on the ring current. J. Geophys. Res. **116**, A00J19 (2011)

R.M. Winglee, Mapping of ionospheric outflows into the magnetosphere for varying IMF conditions. J. Atmos. Terr. Phys. **62**, 527 (2000)

A.W. Yau, M. Andre, Sources of ion outflow in the high latitude ionosphere. Space Sci. Rev. **80**, 1 (1997)

A.W. Yau et al., Energetic auroral and polar ion outflow at DE 1 altitudes: Magnitude, composition, magnetic activity dependence, and long-term variations. J. Geophys. Res. **90**, 8417 (1985)

D.T. Young, H. Balsiger, J. Geiss, Correlations of magnetospheric ion composition with geomagnetic and solar activity. J. Geophys. Res. **87**(A11), 9077 (1982)

DOI 10.1007/978-1-4939-3544-4_3
Reprinted from *Space Science Reviews* Journal, DOI 10.1007/s11214-015-0170-y

A Review of General Physical and Chemical Processes Related to Plasma Sources and Losses for Solar System Magnetospheres

K. Seki[1] · A. Nagy[2] · C.M. Jackman[3] · F. Crary[4] · D. Fontaine[5] · P. Zarka[6] · P. Wurz[7] · A. Milillo[8] · J.A. Slavin[2] · D.C. Delcourt[5] · M. Wiltberger[9] · R. Ilie[2] · X. Jia[2] · S.A. Ledvina[10] · M.W. Liemohn[2] · R.W. Schunk[11]

Received: 5 March 2015 / Accepted: 28 May 2015 / Published online: 1 August 2015

Abstract The aim of this paper is to provide a review of general processes related to plasma sources, their transport, energization, and losses in the planetary magnetospheres. We provide background information as well as the most up-to-date knowledge of the comparative studies of planetary magnetospheres, with a focus on the plasma supply to each region of the magnetospheres. This review also includes the basic equations and modeling methods commonly used to simulate the plasma sources of the planetary magnetospheres. In this paper, we will describe basic and common processes related to plasma supply to each region of the planetary magnetospheres in our solar system. First, we will describe source processes in Sect. 1. Then the transport and energization processes to supply those source plasmas to various regions of the magnetosphere are described in Sect. 2. Loss processes are also important to understand the plasma population in the magnetosphere and Sect. 3 is dedicated to the explanation of the loss processes. In Sect. 4, we also briefly summarize the basic

✉ K. Seki
seki@stelab.nagoya-u.ac.jp

1 Solar-Terrestrial Environment Laboratory, Nagoya University, Nagoya, Aichi 464-8601, Japan

2 Department of Atmospheric, Oceanic and Space Sciences, University of Michigan, Ann Arbor, MI 48109, USA

3 Department of Physics and Astronomy, University of Southampton, Southampton, SO17 1BJ, UK

4 Laboratory for Atmospheric and Space Physics, University of Colorado, Boulder, CO 80303, USA

5 Laboratoire de Physique des Plasmas (CNRS-EP-UPMC-UPSud), Ecole Polytechnique, 91128 Palaiseau Cedex, France

6 LESIA, Observatoire de Paris, CNRS, UPMC, Université Paris-Diderot, 5 place Jules Janssen, 92195 Meudon, France

7 Physics Institute, University of Bern, Sidlerstrasse 5, 3012 Bern, Switzerland

8 Institute of Space Astrophysics and Planetology, INAF, Rome, Italy

9 High Altitude Observatory, National Center for Atmospheric Research, Boulder, CO 80212, USA

10 Space Sciences Lab., University of California, Berkeley, CA 94720, USA

11 Center for Atmospheric and Space Sciences, Utah State University, 4405 Old Main Hill, Logan, UT 84322, USA

equations and modeling methods with a focus on plasma supply processes for planetary magnetospheres.

Keywords Planetary magnetosphere · Plasma sources · General processes

1 Sources

There are three possible sources of plasma for a planetary magnetosphere. The first one is the surface of solid bodies (planet and/or its satellites), the second one is the planetary (or satellite in the unique case of Titan) atmosphere/ionosphere, if any, and the third source is the plasma from the solar atmosphere, i.e., the solar wind. In this section, we will review processes related to each the sources, i.e., surface in Sect. 1.1, ionosphere in Sect. 1.2, and solar wind in Sect. 1.3.

1.1 Surface

In this subsection, we will review source and loss processes related to the planetary or satellite surface as illustrated in Fig. 1. Topics to be explained include ion-induced sputtering, chemical sputtering, photon stimulated desorption, micro-meteoroid impact vaporization, adsorption by surface, ion-sputtering and radiolysis in icy surfaces, sputter yields from water ice, binding energies, sticking and bouncing, and energy distributions.

1.1.1 Ion-Induced Sputtering

The impact of energetic ions or neutrals (typically of keV/nucleon energies) onto a solid surface causes the removal of atoms, ions and molecules from the top surface. This process is referred to in the literature as sputtering, in particular nuclear sputtering when nuclear interaction between the impacting ion and the surface atoms cause the particle release, or electronic sputtering when the electronic interaction results in particle release, as discussed below for icy surfaces. Sputtering is a well-studied phenomenon in material science (e.g. Behrisch and Eckstein 2007).

The energy distribution for particles sputtered from a solid, $f(E_e)$, with the energy E_e of the sputtered particle, has originally been given by Sigmund (1969) and adapted for planetary science (Wurz and Lammer 2003; Wurz et al. 2007)

$$f(E_e) = \frac{6E_b}{3 - 8\sqrt{E_b/E_c}} \frac{E_e}{(E_e + E_b)^3} \left\{ 1 - \sqrt{\frac{E_e + E_b}{E_c}} \right\} \tag{1}$$

where E_b is the surface binding energy of the sputtered particle, typically in the eV range, and E_c is the cut-off energy. The cut-off energy E_c, which is the maximum energy that can be imparted to a sputtered particle by a projectile particle with energy E_i, is given by the limit imposed by a binary collision between a projectile atom or ion, with mass m_1 and the target atom, with mass m_2 (to be sputtered) as

$$E_c = E_i \frac{4m_1 m_2}{(m_1 + m_2)^2} \tag{2}$$

An example of energy distributions based on Eq. (1) is shown in Fig. 2.

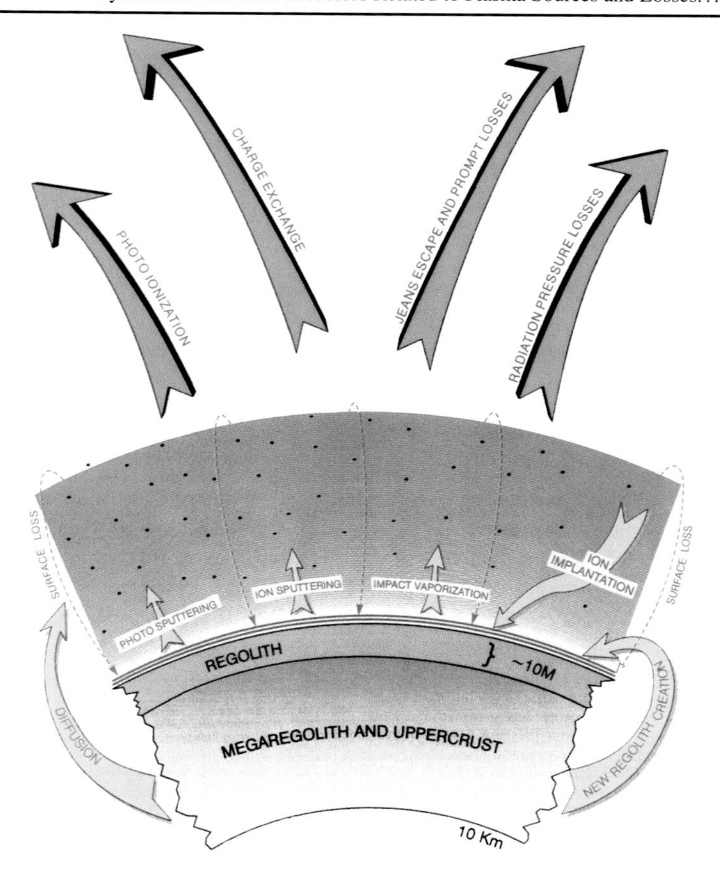

Fig. 1 A schematic illustration of the surface sources and sinks for the exosphere (from Killen and Ip 1999)

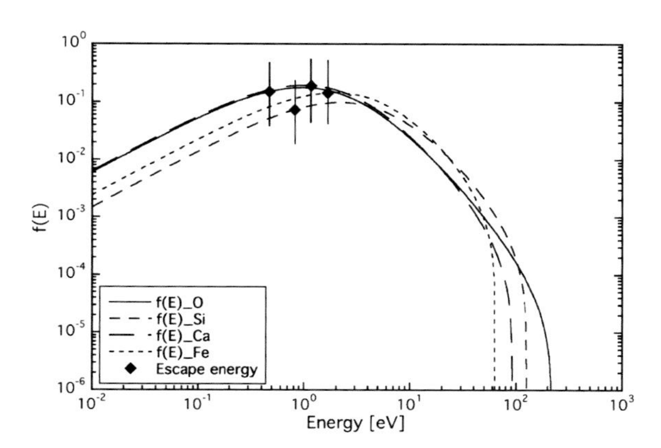

Fig. 2 Energy distribution for sputtered O, Si, Ca, and Fe atoms according to Eq. (1) using incident protons of 1 keV energy (from Wurz et al. 2007)

The energy imparted to the sputtered atoms and molecules is significant with respect to typical escape energies from planetary objects and a considerable fraction of the sputtered particles do not return to the planetary surface (Wurz et al. 2007, 2010).

The polar angle distribution of sputtered atoms is generally described by a $\cos^k(\varphi_e)$ law (Hofer 1991), where the exponent k, is usually between 1 and 2, and depends on the structure of the surface and φ_e is the ejection angle relative to the normal. For the rough surfaces typically encountered in planetary application $k = 1$ is usually chosen (Cassidy and Johnson 2005; Wurz et al. 2007).

The sputter yield is the average number of atoms or molecules removed from the solid surface per incident particle. Ion sputtering releases all species from the surface into space reproducing more or less the local surface composition on an atomic level. Preferential sputtering of the different elements of a compound will lead to a surface enrichment of those elements with low sputtering yields in the top-most atomic layers. However, the steady-state composition of the flux of sputtered atoms will reflect the average bulk composition. Thus, particle sputtering, when operative, will give us compositional information about the refractory elements of the bulk surface.

Sputter yields for the different species can be obtained using the TRIM.SP simulation software (Biersack and Eckstein 1984; Ziegler et al. 1984; Ziegler 2004); see also the recent review on computer simulation of sputtering by Eckstein and Urbassek (2007). TRIM, like many other simulation programmes for sputtering, assumes that the collisions between atoms can be approximated by elastic binary collisions described by an interaction potential. The energy loss to electrons is handled separately as an inelastic energy loss. For typical rock (regolith) surface compositions, the total sputter yield, i.e., all species sputtered from the surface taken together, is about 0.12 atoms per incoming solar wind ion at 400 $\mathrm{km\,s^{-1}}$, considering protons and alpha particles only (Wurz et al. 2007). This sputter yield is the integral over all emission angles and all energies of sputtered particles. The 5 % alpha particles in the solar wind contribute about 30 % to the sputter yield. Heavier ions in the solar wind do not contribute to the sputtering because of their low abundance in the solar wind (Wurz et al. 2007). CMEs can cause increased sputtering of surface material, because their ion density can be much larger than that of the regular solar wind. In addition, alpha particles are often more abundant in the CME plasma, which increases the sputter yield even more.

1.1.2 Chemical Sputtering

When a surface is bombarded with chemically reactive species, chemical alterations in the surface material have to be considered. Chemical reactions between the rock (or regolith grain) and the surface and impacting ions may form species, which are more loosely bound to the surface and thus more easily sputtered. This causes an increase of the sputtering yield or allows for some other release from the surface. This process is usually referred to as chemical sputtering in the literature. In the context of planetary science Potter (1995) considered chemical sputtering for the first time to explain the Na exosphere on Mercury. When a solar wind proton hits the mineral surface processes like the following

$$2\mathrm{H} + \mathrm{Na_2SiO_3} \rightarrow 2\mathrm{Na} + \mathrm{SiO_2} + \mathrm{H_2O} \tag{3}$$

may occur that liberate the Na from the mineral compound. If this happens on the surface, or the liberated Na migrates to the surface, the liberated Na can be released from the surface also by thermal desorption or photon stimulated desorption. This process was successfully implemented in a 3D model to explain the three dimensional structure of the Na exosphere of Mercury with very good agreement with observations for the spatial distribution and the density (Mura et al. 2009).

1.1.3 Photon Stimulated Desorption

Photon-stimulated desorption (PSD), sometimes also referred to as photon sputtering, is the removal of an atom or molecule by an ultraviolet photon absorbed at the surface, via an electronic excitation process at the surface. PSD is highly species selective, and works efficiently for the release of Na and K from mineral surfaces. Also water molecules are removed from water ice very efficiently via PSD. PSD is considered the major contributor for the Na and K exospheres of Mercury and the Moon (Killen et al. 2007; Wurz et al. 2010). Since PSD releases only Na and K from the mineral matrix it is not very important for the overall erosion of the surface since it will cease once the surface is void of Na and K. The situation is different for PSD of water for an icy object, where the PSD process can remove the major surface species.

The flux of material removed by PSD, Φ_i^{PSD}, of a species i from the surface can be calculated by the convolution of the solar UV photon flux spectrum, $\Phi_{ph}(\lambda)$, with the wavelength-dependent PSD-cross section, $Q_i(\lambda)$,

$$\Phi_i^{\mathrm{PSD}} = f_i N_s \int \Phi_{ph}(\lambda) Q_i(\lambda) d\lambda \tag{4}$$

where N_s is the surface atom density, and f_i is the species fraction on the grain surface. Equation (4) can be approximated as

$$\Phi_i^{\mathrm{PSD}} = \frac{1}{4} f_i N_s \Phi_{ph} Q_i \tag{5}$$

where the factor $1/4$ gives the surface-averaged value. The experimentally determined PSD-cross section for Na is $Q_{\mathrm{Na}} = (1\text{–}3) \cdot 10^{-20}\ \mathrm{cm}^2$ in the wavelength range of 400–250 nm (Yakshinskiy and Madey 1999) and for K the PSD-cross section is $Q_K = (0.19\text{–}1.4) \cdot 10^{-20}\ \mathrm{cm}^2$ in the same wavelength range (Yakshinskiy and Madey 2001). Equation (4) can also be written in terms of the PSD yield, Y_i^{PSD}, per incoming photon

$$\Phi_i^{\mathrm{PSD}} = \frac{1}{4} f_i \Phi_{ph} Y_i^{\mathrm{PSD}} \tag{6}$$

which has been determined for water by Westley et al. (1995) in the laboratory. The PSD-yield of water is found to be temperature dependent

$$Y_{\mathrm{H_2O}}^{\mathrm{PSD}} = Y_0 + Y_1 \exp\left(-\frac{E_{\mathrm{PSD}}}{k_B T}\right) \tag{7}$$

with $Y_0 = 0.0035 \pm 0.002$, $Y_1 = 0.13 \pm 0.10$, $E_{\mathrm{PSD}} = (29 \pm 6) \cdot 10^{-3}$ eV, and k_B is the Boltzmann constant (Westley et al. 1995). The temperature dependence is very similar to the one for sputtering of ice (see below), which was found later.

1.1.4 Micro-Meteoroid Impact Vaporisation

The impact of micro-meteorites on a planetary surface will volatilise a certain volume of the solid surface, which contributes to the exospheric gas at the impact site. At Mercury, for example, about one to two orders of magnitude more material than the impactor is released because of the high impact speed for meteorites (Cintala 1992).

The ratio of the maximum ejecta velocity to the primary impact velocity decreases with increasing impact speed. The measured temperature in the micro-meteorite produced vapour cloud is in the range of 2500–5000 K. Eichhorn (1978a, 1978b) studied the velocities of impact ejecta during hypervelocity primary impacts and found that the velocity of the ejecta increases with increasing impact velocity and with decreasing ejection angle, with the ejection angle measured with respect to the plane of the target surface. Such ejecta temperatures are significantly higher than typical dayside surface temperatures, but the corresponding characteristic energies are still lower than for particles that result from surface sputtering. In general, the simulated gaseous material from micro-meteorite vaporisation assumes a thermal distribution (e.g. Wurz and Lammer 2003), i.e., a Maxwellian-like energy distribution with an average gas temperature of about 4000 K. For a rocky planetary object in the solar wind the contributions to the exosphere from ion sputtering and from micro-meteorite impact are about the same (Wurz et al. 2007, 2010).

Most of the meteorites falling onto a planetary object are very small, see for example Bruno et al. (2007) for the Moon and Müller et al. (2002) for Mercury. Micro-meteorite bombardment can be regarded as a continuous flux of small bodies onto the surface, and thus as a steady contribution to the exosphere. However, occasionally larger meteorites may fall onto a surface causing a much larger release of particles into the exosphere. Such a scenario was studied for Mercury by Mangano et al. (2007). They found that for a meteorite of 0.1 m radius an increase in the exospheric density by a factor 10–100, depending on species, for about an hour over the density from sputtering should be observed.

1.1.5 Adsorption by Surface

Most of the material released from the surface falls back onto it. Depending on the species, the surface, and the surface temperature the particle may stick or may bounce back into the exosphere. Metal atoms, chemical radicals and similar species will stick to the surface because they become chemically bound, i.e., their sticking coefficient is $S = 1$. For example a sputtered oxygen atom will stick, i.e., will form a chemical bond with the atom it lands on. Similarly, metal atoms will bind chemically to the surface site they land on. Exception are the alkali metals, where Na and K are observed often in exospheres, and which are transiently adsorbed on mineral surfaces. The probability adsorption (sticking) on silicate surface was measured Yakshinskiy and Madey (2005) as function of the surface temperature. For sodium they found $S_{\mathrm{Na}}(100\ \mathrm{K}) = 1.0$, $S_{\mathrm{Na}}(250\ \mathrm{K}) = 0.5$ and $S_{\mathrm{Na}}(500\ \mathrm{K}) = 0.2$, and for potassium they found $S_{\mathrm{K}}(100\ \mathrm{K}) = 1.0$ and $S_{\mathrm{K}}(500\ \mathrm{K}) = 0.9$. Non-reactive chemical compounds will only stick to the surface when they freeze onto it, which is important mostly for icy moons and planetary objects further out in the solar system. For example O_2 will not freeze onto the surfaces of the icy moons of Jupiter, thus remain in the atmosphere after they have been released from the surface. The same is true for noble gases.

1.1.6 Ion-Sputtering and Radiolysis in the Icy Surface

In the outer solar system it is quite common to encounter icy moons embedded in the planetary magnetosphere, hence, subjected to ion bombardment. The ion impacts onto a water-icy surface can cause sputtering, ionization and excitation of water-ice molecules. Following electronic excitations and ionization water-ice molecules can get dissociated; chemical reactions among the water-dissociation products result in the formation of new molecules (e.g. O_2, H_2, OH and minor species) that are finally ejected from the surface to the moon's exosphere in a two-phase process (e.g., Johnson 1990).

1.1.7 Sputter Yields from Water Ice

These processes have been extensively studied and simulated in laboratory (e.g., Shi et al. 1995; Johnson 1990, 2001; Baragiola et al. 2003). The energy deposited to a solid by the impacting ion, called stopping power, has two components: electronic excitation of molecules predominant at higher energies and momentum transfer collisions (elastic sputtering) predominant at lower energies (Sigmund 1969; Johnson et al. 2009). Famà et al. (2008) obtained through laboratory data fitting the total sputter yields (i.e., number of neutrals released after the surface impact of one ion) for different incident ions at different energies. They discriminated the contributions due to the two components that produce the release of H_2O (direct ion sputtering) and of O_2 and H_2 (electronic sputtering and radiolysis). The total sputter yield Y depends on the type, j, and energy, E_j, of the impacting ion and the surface temperature, T, and it can be written in the following form:

$$Y^j_{total}(E_j, T) = Y^j_{\mathrm{H_2O}}(E_j) + Y^j_{diss}(E_j, T) \tag{8}$$

where $Y^j_{\mathrm{H_2O}}(E_j)$ is the sputtering yield of the H_2O molecules, given by:

$$Y^j_{\mathrm{H_2O}}(E) = 1/U_o \cdot \left(\frac{3}{4\pi^2 C_0} a S^j_n(E) + \eta\left(S^j_e(E)\right)^2\right) \cos^{-f}(\vartheta) \tag{9}$$

where $U_o = 0.45$ eV is the surface sublimation energy, C_0 is the constant of the differential cross section $d\sigma$ for elastic scattering in the binary collision approximation (Sigmund 1969), a is an energy-independent function of the ratio between the mass of the target molecules and of the projectile (Andersen and Bay 1981), S^j_n is the nuclear stopping cross section, S^j_e is the electronic stopping cross section, η is a factor that gives the proportionality between electronic sputtering and $(S^j_e(E))^2/U_o$, θ is the incidence angle, and f is an exponent of the angular dependence of the yield (Famà et al. 2008).

$Y^j_{diss}(E_j, T)$ in Eq. (8) is the yield associated to the loss of O_2 and H_2, produced on ice after its irradiation by energetic ions, given by:

$$Y^j_{diss}(E, T) = 1/U_o \cdot \left(\frac{3}{4\pi^2 C_0} a S^j_n(E) + \eta\left(S^j_e(E)\right)^2\right) \frac{Y_1}{Y_0} e^{-\frac{E_a}{k_B T(lat,\varphi)}} \cos^{-f}(\vartheta) \tag{10}$$

where Y_1 and Y_0 are fitting parameters obtained by laboratory data elaboration (see Famà et al. 2008). Only this second term is temperature dependent. Laboratory measurements have shown that H_2O molecules dominate the total release yield at lower temperatures (<120 K) and O_2 and H_2 at higher (>120 K) temperatures (Johnson 2001).

Since H_2 is eventually lost from ice stoichiometrically, and since the measurements used by Famà et al. (2008) referred to water-equivalent molecules, the total yield for the O_2 ejection can be expressed as follows:

$$Y^j_{\mathrm{O_2}} = \left[m_{\mathrm{H_2O}}/(m_{\mathrm{O_2}} + 2m_{\mathrm{H_2}})\right] \cdot Y^j_{diss}(E_j, T) = 0.5 \cdot Y^j_{diss}(E_j, T) \tag{11}$$

where $m_{\mathrm{H_2O}}, m_{\mathrm{O_2}}$ and $m_{\mathrm{H_2}}$ are the molecular masses of a water, oxygen and hydrogen, respectively (Plainaki et al. 2015). The total number, N_i, of the released molecules of type i depends on the product of the energy spectrum of the ion fluxes impacting the surface with the energy dependent yield:

$$N_i = \int_E \sum_j dF^j/dE_j \cdot Y^j_i dE_j \tag{12}$$

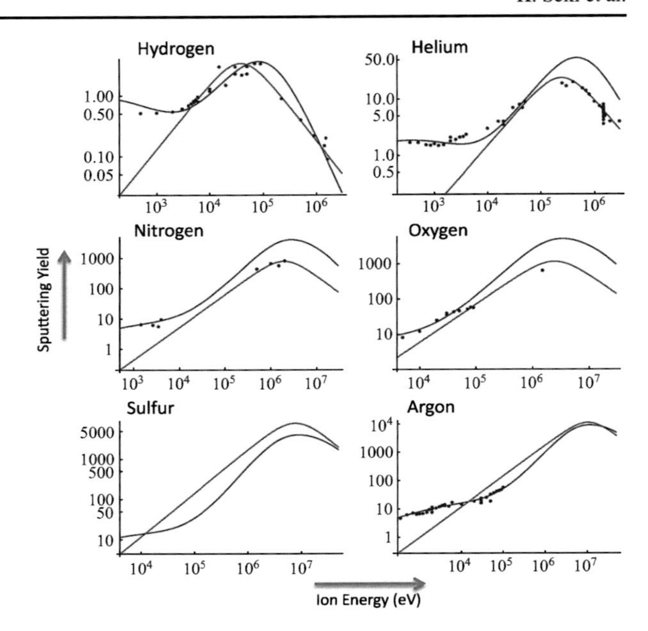

Fig. 3 Yields for the released molecules as a function of energy and impact ion species. Empirical derived functions by Famà et al. (2008) (*blue*) / by Johnson et al. (2009) (*red*) reproduce low/high energies (Cassidy et al. 2013)

Some example of laboratory measured sputtering yields of water ice (see website http://people.virginia.edu/~rej/h2o.html) as a function of energy and impact ion species are shown in Fig. 3 together with the theoretical functions by Famà et al. (2008) function (blue) and the Johnson et al. (2009) function (red).

The sputter yields obtained by laboratory simulations could be different (lower or higher) in the planetary environments since the aggregation status and the purity of the surface material could be different from the sample prepared in the laboratory. Also important is the radiative history of the ice, in fact, the irradiation enhances the sputter yield (Teolis et al. 2005).

1.1.8 Binding Energies, Sticking and Bouncing

The kinetic energy of a water molecule ejected from the surface is affected mainly by the surface binding energy and secondarily by the energy or mass of the impacting ion (Johnson 1990, 1998). Although the sublimation energy of H_2O is 0.45 eV/molecule, the sputtered particle energy distributions for molecular ices tend to have maxima at lower energies than a collision cascade prediction with surface binding energy equal to the normal sublimation energy (Brown and Johnson 1986; Boring et al. 1984; Brown et al. 1984; Haring et al. 1984). Several explanations for this phenomenon have been proposed; the surface may be strongly disrupted with many atoms or molecules leaving at once without experiencing the same binding energy as a single atom leaving a planar surface (Roosendaal et al. 1982; Reimann et al. 1984). In addition, the surface region may be electronically and collisionally excited and the interatomic or intermolecular forces are lower as a result of that excitation (Reimann et al. 1984). The assumption of an 'effective' binding energy for the H_2O molecules equal to $E_b = 0.054$ eV, which was experimentally obtained in the past (Boring et al. 1984; Haring et al. 1984) seems appropriate.

The H_2O and the O_2 molecules released from the surface are set up to ballistic trajectories until they either return to the surface of the body or they escape. Upon return to the surface, the H_2O molecules stick, while the O_2 molecules get thermalized and bounce back to continue their ballistic travel until electron-impact (see next section) ionizes them (Plainaki et al. 2012, 2013). The average kinetic energy that the O_2 molecules have after impacting the surface is about $k_B T$, where k_B is the Boltzmann constant and T is the surface temperature.

1.1.9 Energy Distributions

The emitted O_2 molecules have a complex energy distribution consisting of two components. The distribution of the O_2 molecules that escape the gravity of an icy moon (e.g., Ganymede) is assumed to be described by an empirical function (Johnson et al. 1983; Brown et al. 1984) used also in earlier modelling (Plainaki et al. 2012, 2013; Cassidy et al. 2007; Shematovich et al. 2005):

$$\frac{dF}{dE_e} = a_n E_{O_2}/(E_e + E_{O_2})^2 \quad (13)$$

where $E_{O_2} = 0.015$ eV (Shematovich et al. 2005), a_n is the normalization factor and E_e is the energy of the ejected O_2 molecules.

The O_2 molecules that have had at least one contact with the surface form a Maxwellian velocity distribution function with a temperature equal to the surface temperature. On the basis of the above, the overall energy distribution of the exospheric O_2 can be considered to be mainly thermal exhibiting however the high energy tail in Eq. (13) (De Vries et al. 1984).

The energy distribution of the sputtered water molecules is similar to the regolith ion sputtering distribution, given by Sigmund (1969) as discussed in Eq. (1). The major difference to sputtering of rock is that the 'effective' binding energy, E_b, is equal to 0.054 eV (Johnson et al. 2002). The binding energy E_b influences significantly the energy spectrum at low energies, while the high energy tail of the distribution is affected mainly by $E_c(E_i)$ (see Eq. (2)).

Finally, since the energetic and heavy ions of giant planets' magnetospheres can produce a release of up to 1000 water molecules per impacting ion after the interaction with the icy moon surfaces (see Fig. 3), the ion sputtering process of water is often a major contributor to the exosphere population for the outer solar system moons, where surface temperatures are generally around 80–150 K and the solar illumination is low. The spatial distributions of the exospheres are expected to depend mainly on the illumination of the moon's surface, which determines the moon's surface temperature responsible for the efficiency of radiolysis (Famà et al. 2008). At these low temperatures, in fact, the averaged expected contribution of sublimated water-ice to the moon's exospheric density is important only locally, i.e., at small altitudes above the subsolar point (Smyth and Marconi 2006; Marconi 2007; Plainaki et al. 2010). The high rate of release of particles at relatively high energy, produce a net escape from the moon and high surface erosion rates; for example, the erosion rate of the icy moons embedded in the Jupiter magnetospheric plasma radiation is estimated in the range of 0.01–0.1 μm per year (Cooper et al. 2001; Paranicas et al. 2002). Usually, H_2 formed in ice diffuses and escapes much more efficiently than O_2 at the relevant temperatures in the outer solar system, and, in turn, escapes from the icy moons because of their relatively weak gravitational fields (Cassidy et al. 2010). Therefore, the irradiation of icy surfaces can preferentially populate the magnetosphere with hydrogen (Lagg et al. 2003; Mauk et al. 2003), leaving behind an oxygen-rich satellite surface (e.g., Johnson et al. 2009).

1.2 Ionosphere

1.2.1 Ionization Processes

Solar extreme ultraviolet (EUV) radiation and particle, mostly electron, precipitation are the two major sources of energy input and ionization in solar system ionospheres (for details see Schunk and Nagy 2009). Relatively long wavelength photons (>90 nm) generally cause dissociation, while shorter wavelengths cause ionization; the exact distribution of these different outcomes depends on the relevant cross sections and the atmospheric species.

Radiative transfer calculations of the solar EUV energy deposition into the thermospheres are relatively simple, because absorption is the only dominant process. Taking into account the fact that the incoming photon flux and absorption cross sections depend on the wavelength and the different absorbing neutral species have different altitude variations, the decrease in the intensity of the incoming flux after it travels an incremental distance ds_λ is:

$$d\mathcal{I}(z,\lambda,\chi) = -\sum_s n_s(z)\sigma_s^a(\lambda)\mathcal{I}(z,\lambda)ds_\lambda \tag{14}$$

where $\mathcal{I}(z,\lambda,\chi)$ is the intensity of the solar photon flux at wavelength λ and altitude z, $n_s(z)$ is the number density of the absorbing neutral gas, s, $\sigma_s^a(\lambda)$ is the wavelength dependent absorption cross section of species s and ds_λ is the incremental path length in the direction of the flux. Integration of Eq. (14) leads to the following expression for the solar flux as a function of altitude and wavelength:

$$\mathcal{I}(z,\lambda,\chi) = \mathcal{I}_\infty(\lambda)\exp\left[-\int_\infty^z \sum_s n_s(z)\sigma_s^a(\lambda)ds_\lambda\right] \tag{15}$$

where, $\mathcal{I}_\infty(\lambda)$ is the flux at the top of the atmosphere and the integration is to be carried out along the optical path. The argument of the exponential in Eq. (15) is defined as the optical depth sometimes also called optical thickness, τ, thus:

$$\tau(z,\lambda,\chi) = \int_\infty^z \sum_s n_s(z)\sigma_s^a(\lambda)ds_\lambda \tag{16}$$

and thus Eq. (15) can be written as:

$$\mathcal{I}(z,\lambda,\chi) = \mathcal{I}_\infty(\lambda)\exp\left[-\tau(z,\lambda,\chi)\right] \tag{17}$$

Once the ionizing solar photon flux is known, the photoionization rate for a given ion species $P_s(z,\chi)$ can be written as:

$$P_s(z,\chi) = n_s(z)\int_0^{\lambda_z} \mathcal{I}_\infty(\lambda)\exp\left[-\tau(z,\lambda,\chi)\right]\sigma_s^i(\lambda)d\lambda \tag{18}$$

where λ_{si} is the ionization wavelength threshold and $\sigma_s^i(\lambda)$ is the wavelength dependent ionization cross section for species s. Figure 4 shows an example of the production rates calculated for Saturn.

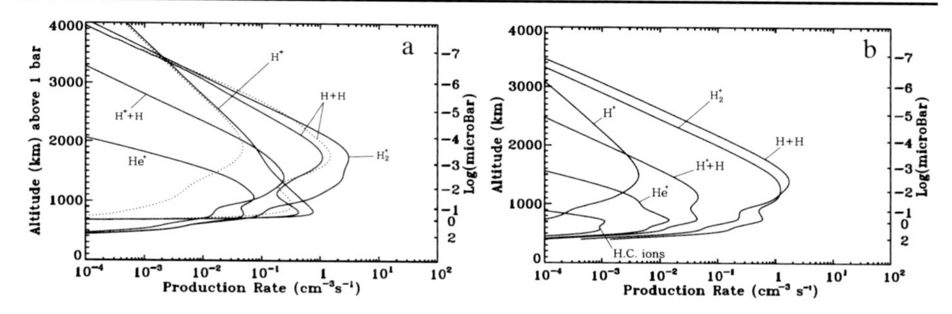

Fig. 4 Calculated production rates for Saturn for a solar zenith angle of 27°. Panel **a** shows the direct photo-production rates and **b** shows the secondary production rates by the resulting photoelectrons. Note that the electron impact ionization rates are very significant at the lower altitudes (from Kim et al. 2014)

1.2.2 Electron Transport

The transport calculations for electrons in an atmosphere are more difficult than those for EUV radiation because scattering and local sources play an important role. In a collisionless plasma the motion of charged particles in a magnetic field can be considered to consist of a combination of a gyrating motion around the field line and the motion of the instantaneous center of this gyration called the guiding center. When the radius of gyration is small compared to the characteristic dimensions of the field line (the case in many ionospheres), one can just concentrate on the motion of the guiding center. Furthermore in most ionospheric applications steady state condition can be assumed; if one further neglects the presence of external electric fields and the divergence of the magnetic field, the equation for the electron flux Φ simplifies down to:

$$\alpha \frac{\partial \Phi}{\partial x} = \sqrt{\frac{m}{2\varepsilon}} \frac{\delta \Phi}{\delta t} \tag{19}$$

where α is the pitch angle with respect to the magnetic field, r the distance along the field line, m_e is the mass of the electron, ε is the energy of the electron and $\frac{\delta \Phi}{\delta t}$ denotes collision induced changes in the flux. This equation is usually solved by dividing the flux into a number of equal angular components or streams. The so called two stream approach is the most commonly used approach and it has been shown, using Monte Carlo calculations (Solomon 1993), that given all the uncertainties associated with differential scattering cross sections, it is generally sufficient to consider only two streams.

Once the electron flux is established, as a function of altitude, the electron impact ionization rate P_s of ion species, s, is given by the following relation:

$$P_s(z) = n_s(Z) \int_{\varepsilon_{sl}}^{\infty} \Phi(z, \varepsilon) \sigma_s^i(\varepsilon) d\varepsilon \tag{20}$$

where ε_{sl} is the ionization energy threshold for species s.

The transport of superthermal ions and neutral gas particles is even more complicated than that of electrons because additional processes, such as charge exchange and ionization are involved. Recent approaches to obtain 3D values of these ion and/or neutral fluxes have used the so-called direct simulation Monte Carlo (DSMC) method (Bird 1994). This approach is well suited to address this problem and as increasing computing resources become available good, comprehensive and accurate solutions are becoming available. Here again once the ion/neutral fluxes are obtained, the impact ionization rate can be directly calculated using an equation analogous to Eq. (20) above.

1.2.3 Loss Processes and Ion Chemistry

The area of science concerned with the study of chemical reactions is known as chemical kinetics. A chemical reaction in which the phase of the reactants does not change is called a homogeneous reaction and in the solar system upper atmospheres and ionospheres these reactions dominate. Dissociative recombination of O_2^+ with an electron is a typical, so called stoichiometric, reaction:

$$O_2^+ + e \rightarrow O + O \tag{21}$$

Reactions that can proceed in both directions are called reversible. Charge exchange between an ion and parent atom and accidentally resonant charge exchange between H and O are such reactions:

$$H^+ + H \rightleftarrows H + H^+ \tag{22}$$

$$O^+ + H \rightleftarrows H^+ + O \tag{23}$$

The reactions indicated by Eqs. (21), (22), and (23) are called elementary reactions, because the products are formed directly from the reactants. O^+, for example can recombine with an electron directly, via radiative recombination, but this process is very slow. In most cases atomic ions recombine via a multi-step process. Two examples of such recombination, via multiple-step processes, are:

$$\begin{aligned} &O^+ + N_2 \rightarrow NO^+ + O \\ &NO^+ + e \rightarrow N + O \end{aligned} \tag{24}$$

and

$$\begin{aligned} &H^+ + H_2O \rightarrow H_2O^+ + H_2O \\ &H_2O^+ + H_2O \rightarrow H_3O^+ + OH \\ &H_3O^+ + e \rightarrow H_2O + OH \end{aligned} \tag{25}$$

The two-step process indicated in Eq. (24) is important in the terrestrial E-region, and the multi-step one indicated by Eq. (25) is very important in the ionospheres of Jupiter and Saturn.

Given the typical thermospheric and ionospheric temperatures the only chemical reactions likely to occur are the so-called exothermic ones. These are reactions that result in zero or positive energy release. Thus, for example, the reaction of H^+ in the ionospheres of Saturn or Jupiter does not take place with ground state H_2, because the ionization potential of H_2 is larger than that of H. However, if H_2 is in a vibrational state of 4 or higher, the reaction becomes exothermic and can proceed. This is potentially very important in the ionospheres of Jupiter and Saturn (McElroy 1973; Majeed and McConnell 1996). Similarly, in the terrestrial thermosphere the reaction between ground state N and O_2 is very slow, because of the high activation energy that is needed, but the reaction with the excited atomic nitrogen, in the ^2D state is rapid and important. For concrete values for chemical reaction reference data can be found in literature (e.g., Schunk and Nagy 1980, 2009; Nagy et al. 1980; Anicich 1993; Fox and Sung 2001; Terada et al. 2009).

1.2.4 Ionospheric Outflows

When a planet has a global intrinsic field, the ions originating in the ionosphere can escape to space from high-latitude regions such as the cusp/cleft, auroral zone, and polar cap. It is observationally known that ions of ionospheric origin can be one of the most important sources of the plasma in the terrestrial magnetosphere especially in the near-Earth regions (see Chappell 2015 for more details). The outflowing ions along the magnetic field can be categorized into several types of ion outflows, i.e., the polar wind, bulk ion upflow, ion conics, and beams. Detailed reviews of observational aspects and theories of ionospheric outflows can be found in the literature (e.g., Yau and Andre 1997; Andre and Yau 1997; Moore and Horwitz 2007; Chappell 2015). Here we briefly summarize important types of ionospheric ion outflows from a magnetized planet or satellite with atmosphere. A good schematic illustration of these outflows can be found in Fig. 1 of Moore and Horwitz (2007).

Polar Wind The polar wind refers to low-energy ion outflows along the open magnetic field lines in the polar ionosphere, mainly caused by an ambipolar electric field formed by the separation of ions and electrons. To achieve charge neutrality with the lighter and faster upflowing electrons, ambient ions are accelerated by the ambipolar electric field. The polar wind has larger flux in the dayside region, where the outflowing photoelectrons can contribute to the ambipolar electric field. However, the controlling factor of the polar wind outflow rate is still under debate (e.g., Kitamura et al. 2012). A variety of modeling efforts have been made for the polar wind (e.g., Banks and Holzer 1969; Ganguli 1996; Schunk and Sojka 1997; Tam et al. 2007). Observations showed a large flux of O^+ polar wind, which was not expected by classical theories (e.g., Abe et al. 1996; Yau et al. 2007). Possible additional acceleration mechanisms include the mirror force, pressure gradient, and centrifugal acceleration by plasma convection in the curved magnetic field. The acceleration mechanisms of the polar wind ions can be ubiquitous in the ionospheres of magnetized planets or satellites.

Bulk Ion Upflow The bulk ion upflows refer to the upward ion flow in the low-altitude ionosphere around the F region, which is observed in the auroral zone and cusp (e.g., Ogawa et al. 2008). The bulk ion upflows do not significantly contribute to the outflow flux from the ionosphere, since their energy is usually less than 1 eV and well below the escape energy of heavy ions such as O^+, O_2^+, and NO^+. On the other hand, they are considered important to transport these heavy ions to the high-altitude ionosphere to enable them to undergo additional acceleration in the auroral region and cusp. The mechanisms that cause the bulk ion upflow include the electron heating driven by soft electron precipitation, Joule heating of ions, and frictional ion heating.

Ion Conics Ion conics are named after the typical shape of the velocity distribution function of ion outflows caused by transverse acceleration in terms of the local magnetic field. The transverse ion heating with typical energies from thermal to a few keV are often seen in the cusp region and the auroral zone. The resultant heated ions are called TAIs (transversely accelerated ions). They are often accompanied by electron precipitation, electron density depletions, and a variety of different resonant waves, such as lower hybrid (LH) waves or broadband extremely low frequency (BBELF) waves (Norqvist et al. 1996; Frederick-Frost et al. 2007). Once the ions are heated transversely to the magnetic field, the mirror force can accelerate them further upward by conserving kinetic energy. Thus the resultant ion velocity distribution functions at high altitudes show conical shapes. Various types of ion conics have been observed in the terrestrial ionosphere (e.g., Øieroset et al.

1999). This same process can occur and create ion conics, when there is an energy input, such as electron precipitation, into a planetary ionosphere under an open magnetic field line geometry.

Ion Beam It has been observationally shown that there exist parallel electric fields in the auroral region in both the upward and downward current regions. Their significance for auroral acceleration had been widely discussed (e.g., Mozer et al. 1977; McFadden et al. 1999). The formation of parallel electric field has been also studied theoretically (e.g., Brown et al. 1995; Wu et al. 2002). The static electric potential drop typically up to several kV accelerates electrons downward and cause discrete auroras. The same parallel electric field can accelerate ions upward. The resultant ion outflows become mostly field-aligned energetic beams. It is suggested that a distributed field-aligned potential drop produced self-consistently from a balance between magnetospheric hot ion and electron populations, soft electron precipitations, and transverse heating of ionospheric ions. When the magnetospheric population has significant differential anisotropy between the ion distribution and the electron distribution, significant parallel potential drops can develop (Wu et al. 2002).

1.3 Solar Wind

In addition to the sources detailed above, the solar wind can act as a plasma source for magnetospheres. The character of the solar wind changes significantly with increasing radial distance from the Sun, and this, combined with the contrasting obstacles presented by various planetary magnetospheres, leads to a large variation in solar wind-magnetosphere dynamics and in the degree to which the solar wind can act as a plasma source for a given magnetosphere. The electron density, flow speed, and magnetic field strength in the solar wind near the orbit of the Earth are known to be about 7 cm^{-3}, 450 km/s, and 7 nT, respectively. The solar wind mostly consists of protons, but contains about 3–4 % of He^{2+} (Cravens 1997; Gombosi 1998; Schunk and Nagy 2009).

It is well known that interplanetary magnetic field (IMF) lines become increasingly tightly wound with distance from the Sun, as modelled by Parker (1958). The average angle that the interplanetary field lines make with respect to the radial direction increases from $\sim$20° at Mercury's orbital distance of $\sim$0.4 AU (Kabin et al. 2000) through $\sim$45° at Earth (Thomas and Smith 1980), $\sim$80° at Jupiter (Forsyth et al. 1996) to $\sim$83° at Saturn's orbital distance of $\sim$9.5 AU (Jackman et al. 2008). The IMF strength also changes with radial distance, with the strength of the B_R component decreasing approximately as r^{-2}. For example, the IMF at Mercury is much stronger than at Saturn (Burlaga 2001), and this has implications for solar wind-magnetosphere coupling.

The form of interaction between the solar wind and magnetosphere changes with the IMF orientation depending on whether the IMF has a parallel or anti-parallel component to the planetary magnetic field at the subsolar magnetopause. The parallel (anti-parallel) case corresponds to the northward (southward) IMF condition at Earth and vice versa at Jupiter and Saturn where the planetary dipoles are oppositely directed to Earth. In addition to magnetic reconnection between the planetary field and IMF, other important physical processes in terms of the solar wind entry into the magnetosphere include the magnetic reconnection, anomalous diffusion across the magnetopause caused by the Kelvin-Helmholtz instability (KHI), and kinetic Alfvén waves.

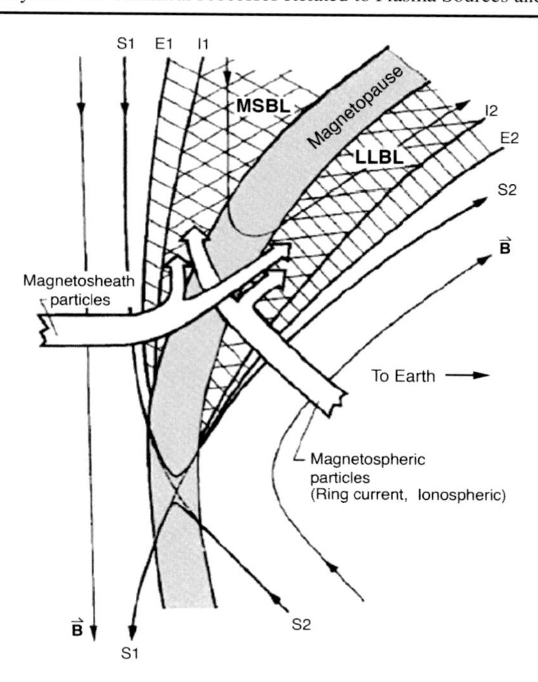

Fig. 5 Reconnection geometry for Earth from Gosling et al. (1990). The *left-hand side* shows a southward magnetosheath field and the *right-hand side* the northward magnetospheric field. The current layer is shown as the *shaded boundary* in the center of these two regions. As the fields reconnect (where the two separatrices, S1 and S2 cross) and thread the magnetopause, a region of "open" field allows the entry of magnetosheath particles into the magnetosphere. Additionally, a portion of the population is reflected. Both populations are energized by the process of interacting with the current sheet. On the *right* the magnetospheric counterpart is transmitted through the boundary and a population is again reflected, again both are energized. As the reconnection continues the fields convect away from the site (up and down in this figure), carrying the plasma with them. Owing to the velocity filter effect a layer of electrons is seen further away from the magnetopause on both sides (between E1 and I1 and I2 and E2). Once the ions "catch up," a layer of both electrons and ions is then seen (within I1 and I2). (Caption from McAndrews et al. 2008)

1.3.1 Magnetic Reconnection

The solar wind is thought to enter planetary magnetospheres primarily through magnetic reconnection at the magnetopause (Dungey 1961). Reconnection at the magnetopause accelerates and directs a mixture of magnetosheath and magnetospheric plasma along newly opened magnetic flux tubes down into the cusp (see review by Paschmann et al. 2013). The anti-sunward flow in the magnetosheath carries these open flux tubes downstream where they are assimilated into the lobes of the magnetotail (Caan et al. 1973). Much of the plasma injected down into the cusp mirrors and flows upward into the high latitude magnetotail to form the plasma mantle (Rosenbauer et al. 1975; Pilipp and Morfill 1978). The plasma in this region then $\boldsymbol{E} \times \boldsymbol{B}$ drifts down into the equatorial plasma sheet. In this manner magnetic reconnection between the IMF and planetary magnetic field transfers mass, energy and momentum from the solar wind into the magnetosphere. This dayside reconnection at the Earth (Gosling et al. 1990; McAndrews et al. 2008) is illustrated schematically in Fig. 5.

The rate of magnetopause reconnection is modulated strongly by the magnitude and orientation of the IMF relative to the planetary field and plasma conditions in the magnetosheath adjacent to the magnetopause. More specifically, low-latitude reconnection at

Earth's magnetopause is strongly controlled by the magnetic shear angle across the magnetopause with the highest rates being observed for the largest shear angles when the interplanetary magnetic field (IMF) has a strong southward component (Sonnerup 1974; Fuselier and Lewis 2011). This is called the "half-wave rectifier effect" (Burton et al. 1975a). The ultimate reason that reconnection at Earth requires large shear angles, ~90 to 270°, is the high average Alfvenic Mach number at 1 AU, i.e., ~6–8 (Slavin et al. 1984). These high Mach numbers result in a high-β magnetosheath and, generally, thin, weak plasma depletion layers (PDLs) adjacent to the magnetopause (Zwan and Wolf 1976). The typically high-β magnetosheaths at the Earth and the outer planets cause the magnetic fields on either side of the magnetopause to differ largely in magnitude. Under these circumstances, reconnection is only possible for large shear angles, typically larger than 90° (Sonnerup 1974). In contrast, the presence of a strong PDL in the inner magnetosheath naturally leads to magnetic fields of similar magnitude on either side of the magnetopause. For low-β magnetosheaths and well developed PDLs observed at Mercury (Gershman et al. 2013), the near equality of the magnetic field on either side of the magnetopause will allow reconnection to occur for arbitrarily low shear angles (DiBraccio et al. 2013; Slavin et al. 2014) such as observed, for example, across heliospheric current sheets (Gosling et al. 2005) or terrestrial low latitude boundary layer under northward IMF (Phan et al. 2005), where the magnetic fields are also nearly equal on both sides.

At Earth an extensive literature exists describing the empirical relationships between the upstream solar wind and IMF (e.g. Perreault and Akasofu 1978; Bargatze et al. 1985; Burton et al. 1975b). These relationships are all based upon the general formula to calculate the magnetopause reconnection voltage that is:

$$\Phi = v_{sw} B_{perp} L \tag{26}$$

where v_{sw} is the solar wind velocity, B_{perp} is the magnitude of the perpendicular component of the IMF (such that $V_{sw} B_{perp}$ is the motional solar wind electric field), and L is the width of the solar wind channel perpendicular to B_{perp}, in which the IMF can reconnect with closed planetary field lines.

The length, L, depends in some way on the properties of the interplanetary medium, and is most frequently taken as some function of the "clock angle" of the IMF. Studies have shown that while dayside reconnection (at Earth) is certainly much weaker for northward than for southward IMF, it does not switch off entirely until the clock angle falls below ~30° (Sandholt et al. 1998; Grocott et al. 2003). Such empirical functions to quantify the rate of dayside reconnection have in turn been applied at Saturn (Jackman et al. 2004) and Jupiter (Nichols et al. 2006) and integrated over time to estimate the amount of flux opened through reconnection at the dayside.

In recent years, the debate about what determines the reconnection rate at the dayside has intensified, in part due to the wealth of spacecraft data at planets such as Mercury, Jupiter and Saturn, which all represent vastly different parameter spaces and thus are likely to differ from the terrestrial magnetosphere in terms of their level of solar wind-magnetosphere coupling (Slavin et al. 2014). A comprehensive study by Borovsky et al. (2008) for Earth found that the reconnection rate is controlled by four local plasma parameters: B_s (the magnetic field strength in the magnetosheath), B_m (the magnetic field strength in the magnetosphere), ρ_s (the plasma mass density in the magnetosheath), and ρ_m (the plasma mass density in the magnetosphere).

Scurry and Russell (1991) argued that dayside reconnection at the outer planets should have a negligible influence as it would be impeded by the high Mach number regimes there.

This argument was countered by the observations of McAndrews et al. (2008) for Saturn and Grocott et al. (2009) for Earth. Subsequently Lai et al. (2012) interpreted a lack of observation of FTEs at Saturn as lack of reconnection. Most recently, Masters et al. (2012) proposed that the plasma beta conditions adjacent to Saturn's magnetopause can restrict the regions over which reconnection can operate. By way of contrast, reconnection at Mercury's dayside has been found to be much more intense than Earth, is independent of the magnetic field shear angle, and varies inversely with magnetosheath plasma β (DiBraccio et al. 2013). Furthermore, large flux transfer events, relative to Mercury's small magnetosphere, occur at Mercury's magnetopause with typical frequencies of 1 every 8 to 10 s (Slavin et al. 2012b; Imber et al. 2014).

MESSENGER observations at Mercury have found that the rate of magnetic reconnection at the dayside magnetopause is on average three times larger than at Earth (Slavin et al. 2009; DiBraccio et al. 2013). A schematic illustration of Mercury's magnetosphere based on MESSENGER observations can be found in Fig. 1 of Slavin et al. (2009). Further, the rate of reconnection at the magnetopause appears independent of IMF direction with high reconnection rates being measured even for small shear angles (DiBraccio et al. 2013; Slavin et al. 2014). These results at Mercury regarding the relationship between low upstream M_A, plasma-β, magnetic shear angle, and reconnection rate parallel the recent developments regarding PDL formation under low M_A (Farrugia et al. 1995) and reconnection as a function of plasma-β (Phan et al. 2013) at Earth. At Earth the typically high-β magnetosheath limits fast reconnection to IMF orientations that have a southward component, i.e. magnetic shear angles across the magnetopause larger than 90° (i.e. the half-wave rectifier effect). However, during encounters with coronal mass ejections at Earth, the upstream M_A approaches values typical of what is seen at Mercury and similar effects are seen; i.e., low-beta magnetosheaths and high reconnection rates even for small magnetic shears across the magnetopause (Lavraud et al. 2013).

1.3.2 Kelvin-Helmholtz Instability (KHI)

Another important mechanism of plasma entry from the solar wind to the magnetosphere is anomalous diffusion across the magnetopause at low latitudes, i.e., around the equatorial plane. The solar wind plasma needs to be transported in the direction perpendicular to the local magnetic field to realize the diffusion. It is observationally known that the flank plasma sheet of Earth's magnetosphere becomes colder and denser than usual during prolonged periods of northward IMF (e.g., Terasawa et al. 1997; Borovsky et al. 1998). One mechanism to cause the anomalous diffusion can be represented by the Kelvin-Helmholtz instability (KHI), which is driven by a flow shear between the magnetosheath (shocked solar wind) and the magnetosphere. KHI itself is basically an MHD instability, while the non-linear evolution of KHI vortex can facilitate the cross field diffusion and the mixing of the solar wind and magnetospheric plasmas inside the rolled-up vortex.

A number of mechanisms have been proposed that would cause the plasma mixing inside the vortex. One of the candidate mechanisms is magnetic reconnection inside the vortex triggered by vortex roll-up in the presence of finite in-plane component of the magnetic field (e.g., Nykyri and Otto 2001; Nakamura et al. 2008). Once the magnetosheath and magnetospheric field lines are reconnected, the detached plasma from the solar wind can be transported inside the magnetosphere. Another idea to realize the mixing is turbulent transport of solar wind plasma across the field line for the inhomogeneous density case of KHI (e.g., Matsumoto and Hoshino 2006). When the density gradient between the magnetosheath and magnetosphere sides is large, the secondary instability is excited at the density interface inside the vortex and the laminar flow is changed to turbulence. The secondary instability is

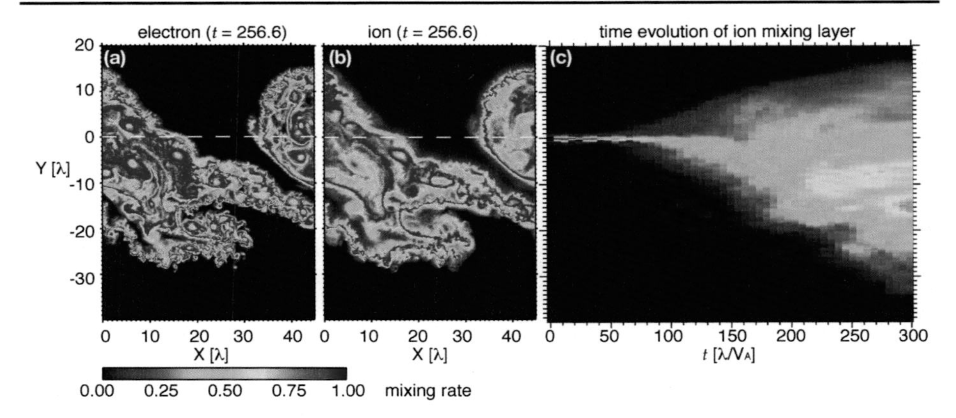

Fig. 6 An example of PIC (particle in cell) simulations of KHI for inhomogeneous density case with the density ratio of 0.1. The initial velocity shear layer was located at $Y = 0$, whose width was set to λ. Color codes show the mixing rate of magnetosheath particles. The mixing rate is defined so that it is maximized ($= 1$) when the magnetosheath-origin particles from $Y > 0$ at $t = 0$ occupy the simulation cell equally with the magnetospheric population from $Y \leq 0$. Snapshots of spatial distribution of the mixing rate at $t = 256.6$ for electrons and ions are shown in panels (**a**) and (**b**), respectively. Panel (**c**) presents the time evolution of the mixing layer. (Adopted and modified from Matsumoto and Seki 2010)

a kind of Rayleigh-Taylor instability (RTI) where the centrifugal force by the rotation motion inside the vortex acts as a gravitational force in the regular RTI. Development of the secondary instability creates a thin, winding, and elongated interface of the solar wind and magnetospheric plasmas. PIC simulation results show that the turbulent electrostatic fields excited by the secondary RTI facilitate an efficient mixing of collisionless plasmas across the field lines. Figures 6a and 6b show an example of such an elongated mixing interface for electrons and ions, respectively (adopted from Matsumoto and Seki 2010).

These proposed nonlinear theories of KHI provide plausible mechanisms for solar wind transport across the magnetopause. On one hand, a remaining problem has been to explain the cold dense plasma sheet formation with KHI. Another question has been how to form a broad mixing layer of several Earth radii observed at Earth (Wing and Newell 2002), since the proposed mixing is basically limited insider the vortex whose size is expected to be much smaller if one consider a simple KHI vortex without nonlinear vortex paring. Based on large-scale MHD and PIC simulations, Matsumoto and Seki (2010) showed that rapid formation of a broad plasma turbulent layer can be achieved by forward and inverse energy cascades of the KHI. Figure 6 shows an example of the full particle simulations. The forward cascade is triggered by growth of the secondary Rayleigh-Taylor instability excited during the nonlinear evolution of the KHI, while the inverse cascade is accomplished by nonlinear mode couplings between the fastest growing mode of the KHI and other KH unstable modes. As a result of the energy transport by the inverse cascade, the growth rate of the largest vortex allowed in the system reaches a value of 3.7 times greater than that of the linear growth rate and it can create the boundary layer extended over several Earth radii (Fig. 6c).

The KHI is also considered important in Saturn's magnetosphere (e.g., Masters et al. 2009, 2010; Delamere et al. 2013). Given that the corotating flows in the magnetosphere have the opposite (same) directions compared to the shocked solar wind flow in the dawn (dusk) side dayside magnetopause, the occurrence of KHI is expected to be highly asymmetrical, i.e., the dawn side magnetopause has a favorable condition to KHI excitation. Observations of kilometric radiation suggested that the KHI at Saturn's magnetopause tends to

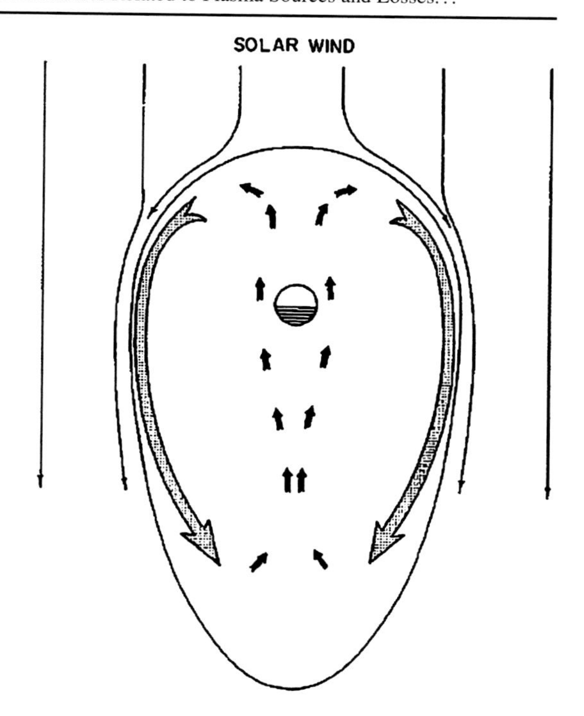

Fig. 7 Schematic of the viscous cycle (from Axford and Hines 1961). This is a view down on to the equatorial plane with the solar wind blowing from *top* to *bottom* of the diagram

occur in the morning sector (Galopeau et al. 1995). Based on the 3-D MHD simulations, Fukazawa et al. (2007a) show that the KHI vortex is more pronounced for the northward IMF case than the southward case. However, the effects of KHI on the plasma mixing and transport in Saturn's magnetosphere are still far from understood.

2 Transport and Energization of Plasma

There are a number of methods by which plasma can be transported and energized within magnetospheres. We refer the reader to Jackman et al. (2014a) for a comprehensive review of transport and loss processes in the magnetospheres of Mercury, Earth, Jupiter and Saturn. In this section we describe major transport and energization processes which are important to understand how to populate various parts of planetary magnetospheres.

2.1 Axford/Hines Cycle

A key transport mechanism, thought to be at work in slowly-rotating magnetospheres, is the so-called viscous interaction driven model (Axford and Hines 1961; Axford 1964). This involves momentum transfer from the solar wind to the magnetotail via quasi-viscous interaction, particularly at the low-latitude magnetopause. It is illustrated schematically in Fig. 7. This cycle can drive circulation within a closed magnetosphere, provided an appropriate tangential-drag mechanism exists. A major mechanism to enable this interaction is the Kelvin-Helmholtz instability described in Sect. 1.3, driven by flow shear at the magnetopause, which may also be coupled with magnetic reconnection (e.g. Hasegawa et al. 2004; Nykyri et al. 2006).

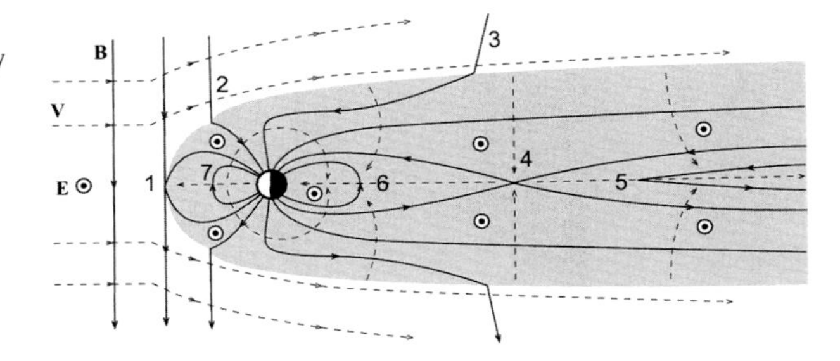

Fig. 8 Schematic diagram showing the stages of the Dungey cycle for the case of Earth's magnetosphere (courtesy Steve Milan)

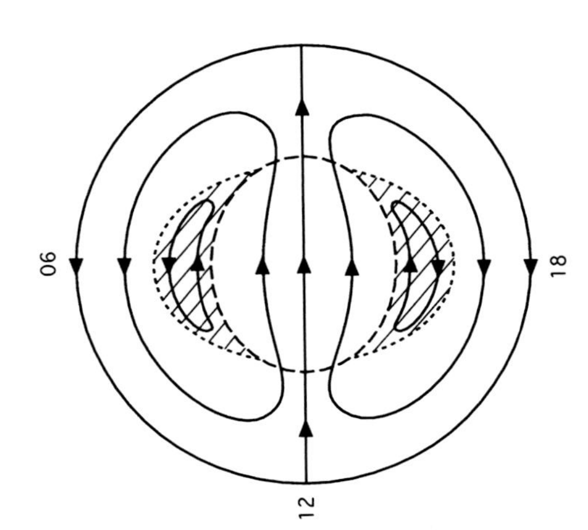

Fig. 9 Northern high-latitude ionospheric flow associated with a combination of Dungey and viscous cycle (after Cowley 1982). The hatched region indicates convection driven by the boundary layers in which magnetic flux tubes remain closed during the cycle, while the remainder of the flow is associated with the reconnection process

2.2 Dungey Cycle

A second transport mechanism driven by solar wind interaction is the Dungey cycle. In this cycle, dayside reconnection opens magnetic flux, and the solar wind interaction carries these open magnetic field lines from dayside to nightside, where they are stretched out to form the tail lobes (defined here as the open field line region, while noting that centrifugal confinement of plasma to the equator in rapidly rotating systems can alter this picture somewhat (e.g. Hill and Michel 1976; Ray et al. 2009)). As they are stretched out down-tail, open field lines sink in towards the center plane of the tail, where they reconnect again, closing the flux that was opened on the dayside. The "Dungey cycle timescale" refers to the length of time from the opening of the field lines at the dayside to the closing of the field lines on the nightside. Figure 8 shows the stages involved in the Dungey cycle for the case of Earth, where the timescale is ~1 hour (Cowley 1982). The Dungey cycle is also known to operate strongly in the slowly rotating magnetosphere of Mercury, with a timescale of just ~1–2 minutes (Siscoe and Christopher 1975; Slavin et al. 2012a). The relative importance of the Dungey cycle at the rapidly rotating magnetospheres of Jupiter and Saturn is a topic of some debate. Badman and Cowley (2007) estimated that when active, the Dungey cycle timescale at Jupiter is of order several weeks, whereas at Saturn the timescale is ~1 week or more (Jackman et al. 2004). Figure 9 illustrates the combination of the Dungey and viscous-cycle flows in the Earth's ionosphere.

2.3 Rotational Driven Transport and Vasyliunas Cycle

The role of rotation in a planetary magnetosphere may be estimated by considering the superposition of dawn-dusk electric field resulting from the solar wind flow and the radial electric field imposed by the planetary ionosphere (Brice and Ioannidis 1970). The resulting potential is

$$\Phi = -\eta v_{sw} B_{sw} r \sin\varphi - \frac{\Omega B_0 R^3}{r} \tag{27}$$

where v_{sw} and B_{sw} are the solar wind speed and magnetic field, η the efficiency with which the solar wind field penetrates into the magnetosphere, and B_0, R and Ω the planetary equatorial magnetic field, radius and rotation rate. This implies that the plasma will $\boldsymbol{E} \times \boldsymbol{B}$ drift along closed paths and in the sense of planetary rotation within a distance

$$r_0 = \sqrt{\frac{\Omega B_0 R^3}{\eta v_{sw} B_{sw}}} \tag{28}$$

For the Earth, this approximation suggests a corotating region inside of 4 R_E, reasonably consistent with the observed size of the Earth's plasmasphere. For Jupiter and Saturn, however, the same calculation suggests a size of over 150 and 50 planetary radii, respectively. This would be larger than the actual size of these planetary magnetospheres. In practice, the observed corotating region occupies most, but not all, of these planetary magnetospheres. Nor are the flows at a rigid corotation speed. At Jupiter they begin to depart from corotation somewhere near the orbit of Europa (10 R_J) (McNutt et al. 1979; Krupp et al. 2001) and at Saturn the flows are 10–20 % of full corotation as close to the planet as 4 R_S (Wilson et al. 2009). An example of application of Eq. (27) to Jupiter's case can be found in Fig. 5 of Delamere and Bagenal (2010).

This corotational flow results in a dramatically different distribution of plasma along magnetic field lines and allows internal plasma sources to drive magnetospheric dynamics. The distribution of plasma along a magnetic field line is determined by the gravitational, centrifugal and ambipolar electric potentials (Siscoe 1977; Bagenal and Sullivan 1981)

$$\begin{aligned} n_\alpha &= n_{\alpha,0} \exp\left[-\frac{U(\lambda) + q_\alpha \Phi(\lambda)}{kT_\alpha}\right] \\ U(\lambda) &= -\frac{GMm_\alpha}{LR\cos^2\lambda} + \frac{m_\alpha}{2}\Omega^2 L^2 R^2 \cos^6\lambda \end{aligned} \tag{29}$$

and the requirement of charge neutrality $\sum q_\alpha n_\alpha = 0$. The above equations assume a dipole magnetic field and isotropic Maxwellian velocity distributions, but can be appropriately modified to treat any magnetic field geometry, as well as non-Maxwellian distributions (e.g. anisotropic Maxwellians (Huang and Birmingham 1992), kappa distributions (Meyer-Vernet et al. 1995), etc.).

When we consider the electric potential inside the synchronous orbit:

$$\left(\frac{2GM}{3\Omega^2}\right)^{1/3} = \left(\frac{2}{3}\right)^{1/3} R_{sync} \tag{30}$$

where R_{sync} is the radius of synchronous orbit, the potential has a maximum at the equator. Outside this distance, there is a potential minimum at the equator and a local maximum at a latitude:

$$\cos^8 \lambda = \frac{2}{3}\left(\frac{R_{sync}}{LR}\right)^3 \tag{31}$$

As a result, ions produced in the equatorial magnetosphere and inside this "critical distance" will freely precipitate into the planetary atmosphere, while those produced farther from the planet are equatorially trapped. In the case of the Earth, the critical distance would be 5.75 R_E. Since this is outside the corotating plasmasphere, no such equatorial trapping occurs in the Earth's magnetosphere. In contrast, trapping may occur outside 1.96 R_J at Jupiter and 1.62 R_S at Saturn. Thus, the plasma in virtually all of these magnetospheres is equatorially trapped. This "critical distance" has also been identified as a limit for stable orbits of charged dust particles, in the limit $m/q \to 0$ (Northrop and Hill 1983) and in simulations of ions produced over Saturn's ring plane (Luhmann et al. 2006).

In addition to allowing equatorial trapping, the mid-latitude potential minimum also results in a minimum in electron density. While the exact location of this minimum depends on the ambipolar field, and therefore on the abundance and temperature of the various species, calculations using typical, observed values place it close to the latitude given in Eq. (31). At these latitudes, due to their lower mass, protons are expected to be the most abundant species even though they are not at the equator. An increase in proton abundance with latitude has been observed by the Cassini spacecraft at Saturn (Thomsen et al. 2010), but no clear minimum has been reported, probably due to the very low densities present at these latitudes. At Jupiter, protons represent only a few percent of the equatorial ions and mass-resolved observations are unavailable.

This mid-latitude density minimum and the predominance of protons have strong implications for magnetosphere-ionosphere coupling. The dynamical processes of the low-latitude magnetosphere are connected to the planetary ionosphere through field-aligned currents. These currents are limited by the availability of charge carriers and are therefore sensitive to the electron density profile along a field line. By finding solutions to a one-dimensional Vlasov equation, Ray et al. (2009) showed that the current-voltage relation along a Jovian field line differs significantly from the traditional Knight relation (Knight 1973) (see Eqs. (39) and (40)). The saturation current may be one to two orders of magnitude lower and depends on the conditions at the electron density minimum rather than the equator. Other aspects of magnetosphere-ionosphere coupling are mediated by MHD waves. Wave velocities and propagation times are sensitive to the plasma properties along the field lines. As a result, many aspects of magnetosphere-ionosphere coupling at Jupiter and Saturn depend on the poorly measured mid-latitude plasma.

In the presence of equatorial trapping, any plasma sources in the magnetosphere must be balanced by some loss process. In the case of Jupiter and Saturn, plasma is produced by the ionization of neutrals from satellites (primarily Io and Enceladus), rings and the planetary exospheres. Recombination is not an efficient loss process, and charge exchange does not result in a net removal of ions. The main loss process balancing these plasma sources is centrifugally-driven, radial transport. The corotating plasma experiences an outward, centrifugal force. To first order, this is balanced by magnetic tension. Field lines are stretched under the condition:

$$\frac{1}{\mu_0}(\vec{\nabla} \times \vec{B}) \times \vec{B} = \rho \Omega^2 r. \tag{32}$$

This result in a current sheet which resembles that of the Earth's magnetotail in some ways, but which is present at all local times. The stretching of the field lines can be roughly approximated by

$$\frac{B_r}{B_z} \sim \frac{H}{r} \frac{\Omega^2 r^2}{2V_A^2} \tag{33}$$

where V_A is the Alfvén speed, H the thickness of the current sheet and B_r the radial field immediately above or below the sheet.

This balance of centrifugal force and magnetic tension is unstable. The situation is analogous to the magnetized Rayleigh-Taylor instability, where a denser fluid is above a less dense one. In this case, the centrifugal force replaces gravity, and radial transport is driven by a denser plasma inside a less dense plasma (Krupp et al. 2004 and references therein). Time scales for this instability are of order the rotation period of the planet, but may be partially stabilized by considerations such as the Coriolis force and coupling to the ionosphere (Pontius 1997).

In the inner and middle magnetosphere, interchange appears to be the key method by which mass can be transported within magnetospheres. It is a process whereby cool, dense plasma can move outward, to be replaced by hotter, more tenuous plasma moving inward, resulting in a net outward transport of mass. This has been observed both at Jupiter (Thorne et al. 1997; Kivelson et al. 1997; Krupp et al. 2004 and references therein) and Saturn (Hill et al. 2005; Burch et al. 2005). The phenomena are less well-measured at Jupiter, since their typical duration there is shorter and below the 80-s time resolution of the Galileo plasma instrument in almost all cases. Typically, the inward-moving flux tubes are characterized by an abrupt increase in magnetic pressure, the disappearance of thermal plasma, and the presence of a hot, energetic particle population. In the case of older (or more inward transported flux tube) events, flux tubes may contain a mixture of low energy plasma diffusing in and energetic particles curvature-gradient drifting out. Much older events are surrounded by a time-dispersed signature in keV and higher energy particles. This is a result of the superposition of the corotating flow and the particles' curvature-gradient drift (in the direction of corotation for ions and opposite it for electrons). The corresponding outward motion of cold, dense plasma has not been reported.

For the rapidly rotating magnetospheres of the outer planets with their large moon- derived plasma sources, the "planetary wind" or "Vasyliunas cycle" is of critical importance (Hill et al. 1974; Michel and Sturrock 1974; Vasyliunas 1983). This Vasyliunas cycle is driven not by the solar wind, but by the energy transferred to internally generated plasma by the fast rotation of these planets. The plasma created deep inside the magnetosphere is accelerated by magnetic stresses from the ionosphere, gains energy, and moves outward from the planet. Centrifugal forces cause the field lines to stretch. These stretched field lines can form a thin current sheet, across which the closed field lines reconnect. This reconnection simultaneously shortens the field line and (like the Dungey cycle), releases plasma down the tail in the form of a "plasmoid". The stages of this cycle, as viewed in an inertial frame of reference, are illustrated in Fig. 10, the picture originally put forward by Vasyliunas (1983).

2.4 Field-Aligned Potential Drop

Many efforts in theories, simulations and observations showed the role played by magnetic-field-aligned electric fields at different locations in the Heliosphere. Significant insights of field-aligned processes, such as particle acceleration, parallel electric fields and currents and their relationships come from numerous observations in the terrestrial magnetosphere at different altitudes along magnetic field lines during the last 50 years. To give examples among others, a few missions that contributed to this field after some of the pioneering spacecraft

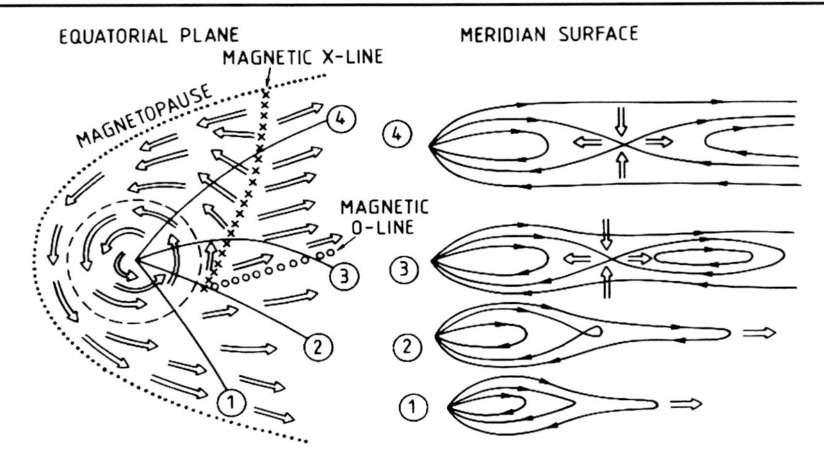

Fig. 10 Flow pattern (*left*) and field configuration (*right*) expected for a steady-state planetary wind, first proposed for Jupiter by Vasyliunas (1983)

have flown (see the review by Mozer et al. 1980) are listed hereafter. The long-term US program "Defense Meteorological Satellite Program" (DMSP) maintains satellites orbiting at low altitude (830 km) since 1971. In the decades 1980–2000, the Swedish missions VIKING, FREJA and the NASA mission "Fast Auroral Snapshot Explorer" (FAST) were designed to achieve measurements with excellent time and space resolutions at mid-altitudes (from about 400 to 4000 km altitude). The ESA multi-spacecraft pioneering mission CLUSTER has been exploring all latitudes and longitudes between typically 4 and 20 Earth radii since 2000 over a time period of more than 15 years. The signatures identified in the terrestrial case provide guidelines to interpret observations in other magnetospheres.

In planetary magnetospheres, where plasmas are collisionless in most regions, the mobility of electrons along magnetic field lines is very high as compared to perpendicular motions mostly driven by large-scale electric fields, magnetic or pressure gradients. Therefore, this high field-aligned mobility contributes to cancel out any potential drop that would appear along magnetic field lines. However, from the mid-70s, observations revealed a secondary peak in the energy spectrum of precipitating electrons in the terrestrial auroral region. Evans (1974) interpreted it as the acceleration by a field-aligned potential difference. Numerous observations have then provided evidence of particle acceleration by parallel electric fields and different processes have been invoked. We first recall that field-aligned particle acceleration does not necessarily imply parallel electric fields, an example being the Fermi acceleration. We then present some of the main classes of processes involving quasi-static and transient parallel electric fields.

2.4.1 Fermi Acceleration

The Lagrangian formulation of mechanics describes the particle motion through "generalized coordinates" and associated "generalized momentum". It allows in particular an easy derivation of the conservation laws for cyclic motions. In magnetized environments, particles are rotating around the magnetic field. The first adiabatic invariant associated to this cyclotron motion is μ:

$$\mu = \frac{1}{2} \frac{m v_\perp^2}{B} \tag{34}$$

where m is the particle mass and $v_\perp$ its velocity in the direction perpendicular to the magnetic field B. μ shows that the perpendicular velocity increases with the magnetic field. It is conserved if the magnetic field does not vary in time or evolves slowly relative to the gyration period. At time scales much larger than the cyclotron motion, the particle motion is represented by the guiding center of this cyclotron motion. In an approximately dipolar planetary magnetic field, the magnetic field magnitude increases along magnetic field lines from the apex towards the planet. The conservation of the first adiabatic invariant μ shows that the mirror points are located at the points where the magnetic field is equal to B_m, such that:

$$\frac{1}{B_m} = \frac{(\sin\alpha_0)^2}{B_0} \tag{35}$$

where α_0 and B_0 are the particle pitch-angle and magnetic field magnitude at a given point along the magnetic field line, for example at the apex. The pitch-angle, α, is the angle between the particle velocity and the magnetic field. The location of the mirror points does not depend on the particle energy but only on its pitch-angle. If the particles do not cross another medium with different properties before reaching their mirror points, they remain trapped in the magnetosphere describing this bouncing motion along magnetic field lines.

The Fermi acceleration along magnetic field lines is related to the second adiabatic invariant. The second adiabatic invariant, also called longitudinal invariant, associated with this bounce motion is I:

$$I = \int_{M_S}^{M_N} p_\parallel dl \tag{36}$$

where $p_\parallel$ is the particle momentum $(mv_\parallel)$ in the direction parallel to the magnetic field, dl an elementary distance along the curved magnetic field line, M_N and M_S, the magnetic mirror points in each hemisphere, and the integral is taken along the bounce motion. If the magnetic field does not vary in time or evolves slowly relative to the particle bounce motion, the second adiabatic invariant is conserved. An order of magnitude is given by

$$I \approx mv_\parallel L_{SN} \tag{37}$$

where $\langle v_\parallel \rangle$ is the average velocity in the direction parallel to the magnetic field and L_{SN}, the total length along the magnetic field line between the two mirror points. If for an external cause, the distance between the two mirror points decreases, the conservation of I implies that $v_\parallel$ increases: this is the so-called Fermi acceleration along magnetic field lines and it does not involve any parallel electric fields. In planetary magnetosphere, this occurs for example during compression events or substorms. More generally, the Fermi acceleration is considered as an efficient process to explain particle acceleration at shocks or the acceleration of cosmic rays.

2.4.2 Parallel Electric Fields, Currents and Particle Acceleration

While most magnetospheric particles remain bouncing back and forth along magnetic field lines between their mirror points, only particles with mirror points located at ionospheric altitudes or below will reach the ionosphere. Their pitch-angle at the field line apex (see Eq. (33)) will be smaller than a maximum pitch-angle α_c, half-angle of the so-called loss cone:

$$(\sin\alpha_c)^2 = \frac{B_0}{B_I} \tag{38}$$

where B_I and B_0 are the magnetic field magnitude at the ionospheric end and at the apex of the magnetic field line. The loss cone is small: for a dipolar magnetic field decreasing with the cube of the distance, the loss-cone angle is of the order of a few degrees at a distance of 10 planetary radii. In planetary magnetospheres, particles within the loss cone are lost from the magnetosphere due to collisions with the upper atmosphere. These precipitating particles also have the fundamental property to be the only magnetospheric particles capable of carrying field-aligned currents between the magnetosphere and the ionosphere. Conversely, the mirror force is favorable for ionospheric particles. All ionospheric particles that could be extracted from the ionosphere reach the magnetosphere and contribute to carry currents.

Highly conductive magnetic field lines provide an electrodynamic coupling between magnetosphere and ionosphere by connecting both plasmas, by transmitting perpendicular electric fields and by circulating field-aligned currents. Both media, ionosphere and magnetosphere, permanently undergo independent large-scale or local processes that modify their electric field and current distribution at a given time. These modifications are transmitted in the conjugate medium through field-aligned currents where they cause a modification of the electrodynamic parameter distribution, which is transmitted to the conjugate medium trough field-aligned currents in a self-consistent feed-back process. If the required current density is larger than the density available from magnetospheric current carriers, then the coupling is imperfectly achieved and both media are partially disconnected. In this case, the generation of parallel electric fields represents a way to achieve the required current circulation given that the particle acceleration contributes to the increase in the field-aligned current density to the required value. Such parallel electric fields can be associated with quasi-static structures or with transient processes such as waves.

2.4.3 Quasi-Static Parallel Electric Fields

All developed magnetospheres show evidence of accelerated particles, as for example accelerated electrons precipitating into ionosphere and responsible for auroral light emissions. In the terrestrial magnetosphere, observations show auroral electrons accelerated to keV energies; they move faster than the local Alfvén speed, so that they cannot stay in phase with Alfvén waves. This result led Knight (1973) to consider a simple quasi-static model for field-aligned currents carried by ionospheric and magnetospheric electrons accelerated by a quasi-steady parallel electric potential. From the conservation of the energy and of the first adiabatic invariant, he derived a general current-voltage relationship. For applications to auroral magnetic field lines, where:

$$\frac{e\Delta V}{kT_I} \gg 1 \quad \text{and} \quad \frac{e\Delta V}{kT_0} \ll \frac{B_I}{B_0}$$

it simplifies to:

$$j_{\|} \sim -en_0\sqrt{\frac{kT_0}{2\pi m_0}}\left(1+\frac{e\Delta V}{kT_0}\right) \tag{39}$$

and, if $\frac{e\Delta V}{kT_0} \gg 1$, it becomes:

$$j_{\|} \sim -en_0\sqrt{\frac{kT_0}{2\pi m_0}}\left(\frac{e\Delta V}{kT_0}\right) \tag{40}$$

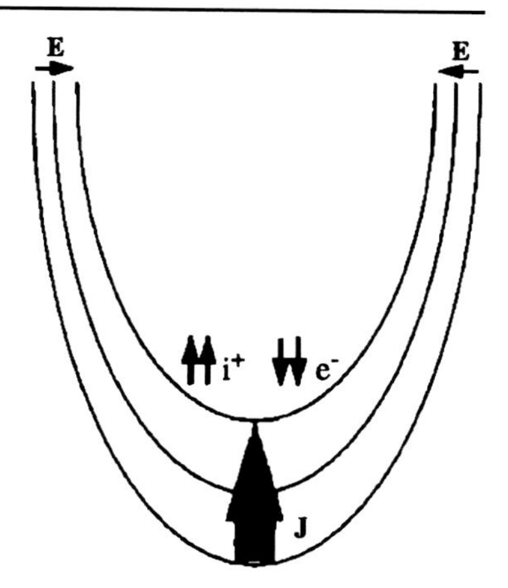

Fig. 11 A schematic illustration of the upward current region adapted from Carlson et al. (1998)

where k is the Boltzmann's constant, e and m_e the electron mass and charge, n_0 and T_0 respectively the magnetospheric electron density and temperature, T_I is the ionospheric temperature, ΔV is the total potential drop between the ionosphere and the magnetosphere: $\Delta V = E_I - E_0$, B_I and B_0 respectively the ionospheric and magnetospheric magnetic fields.

This relation provides an estimate of the field-aligned current density that the plasma can carry between the ionosphere and the magnetosphere without any parallel electric fields ($\Delta V = 0$). It also shows that the presence of a potential drop allows increasing this threshold value to much larger current densities if required for other reasons (e.g., current continuity, mismatch between the ionosphere and the magnetosphere). Field-aligned currents associated with a positive potential drop are directed upward, which corresponds to auroral observations. Improvements were presented by Chiu and Schulz (1978), who took into account the motion of the ions in such a potential structure and their contribution to field-aligned currents.

Following similar steps, Lyons (1980) demonstrated that discontinuities with div $E \neq 0$ in the large-scale electric field pattern could generate large-scale regions of field-aligned currents, associated with parallel electric fields and electron acceleration. Such discontinuities are known to exist near magnetospheric boundaries (boundaries of the plasma sheets, boundary layers, etc. in large-scale plasma flow inhomogeneities. A discontinuity with: div $E < 0$ (>0) would account for upward (downward) field-aligned currents. A typical width of such structures would be of the order of 100 km in the terrestrial ionosphere, i.e. about 0.01 Earth radius.

Acceleration structures are observed at smaller scales in the auroral zone. For instance, accelerated electron precipitations are observed with a typical shape of inverted V and with widths about ten times smaller ($\sim$0.001 Earth radius in the terrestrial ionosphere) than the preceding effect. Such acceleration structures are interpreted as the acceleration due to a U-shaped field-aligned upward potential structure, as illustrated in Fig. 11 adapted from Carlson et al. (1998). The magnetic field near the planet is highly incompressible, resulting in nearly electrostatic structures. Downgoing field-aligned electrons crossing the middle of the structure will gain an energy corresponding to the total upward potential drop, but only a fraction of it if they cross the sides. This effect produces the well-known inverted-V shape for the acceleration structure observed by spacecraft flying below it. Spacecraft crossing at higher altitudes (near the top of Fig. 11) will detect outflowing ions accelerated at energies corresponding to the potential drop below the spacecraft and thus with the typical inverted V shape for the same reasons. They will also observe large convergent electric fields near the edges of the structure, as shown in Fig. 11. These electric structures are not detected below the spacecraft, implying the presence of an electrostatic shock associated with parallel electric fields at intermediate altitudes as shown in Fig. 11 (see a review by Mozer and Hull 2001). Precipitating electrons and outflowing ions carry upward currents.

Diverging electrostatic shocks are also observed and produce the opposite effects with up-going electrons accelerated to somewhat lower energies than the preceding case, and carrying downward currents. More details can be found in a review by Marklund (2009).

De Keyser et al. (2010) proposed a different mechanism to explain the existence of small-scale quasi-static bipolar (convergent or divergent) electric fields. They considered the case of the field-aligned boundary between a dense region of hotter particles and a diluted region of colder particles, as for example the boundary between the plasmasheet and the lobes. This boundary is approximated as a tangential discontinuity which has a finite thickness of the order of the largest Larmor radius, i.e. that of the hotter ions. The transition width differs for each species and is related to their Larmor radius. The difference between the Larmor radii of the hot ions and the hot electrons will produce a charge separation and thus a polarization electric field perpendicular to the interface. The same occurs for the cold ions and electrons, but their Larmor radii are much shorter. In the absence of any potential structure across the interface, this polarization electric field displays a wider region (related to the hot ion Larmor radius) of smaller magnitude and a smaller region (related to the hot electron Larmor radius) of larger magnitude directed in the opposite direction, so that the integrated electric field over the interface cancels out. This produces the bipolar electric field structure. The presence of a potential across the interface attracts or repels ions and electrons depending on its sign, which in both cases results in a monopolar electric field structure, also observed. The mapping in the ionosphere of this magnetospheric electric field distribution and the closure of the currents in the ionosphere lead to the generation of parallel electric fields and currents.

These quasi-static models are very useful in explaining the observed particle acceleration, field and current signatures related to quasi-static structures. However, they cannot explain observations of transient or highly time-dependent features in the distribution of electric fields and currents.

2.4.4 Transient Acceleration

Accelerated particles and large currents are factors capable of triggering instabilities and of generating waves through wave-particle interactions. These waves contribute to modify in turn the initial particle distribution by energy and pitch-angle scattering of the resonant particles, or by energy and momentum propagation to other regions. As a result, the initial electric currents and fields are modified.

Wave-Particle Interactions and Radiation In ideal MHD, shear Alfvén waves propagate with perpendicular electric fields. They have the property to carry field-aligned currents. When perpendicular scales become too small, the ideal MHD approximation is no longer fulfilled, the waves become dispersive and a parallel electric field appears in so-called kinetic Alfvén waves. In the topside terrestrial ionosphere, parallel electric fields can become very important at altitude below a few Earth's radii (Alfvén resonator). The same is true above Jupiter's ionosphere (Ergun et al. 2006). Numerical simulations suggest that Alfvén waves should evolve towards small scales, with the appearance of a filamentary structure resulting in electrostatic structures such as strong Double Layers (DLs) (Mottez and Génot 2011). High resolution remote sensing of the Io-Jupiter magnetic flux tube based on radio waves observations have demonstrated the existence of strong DLs (up to $\sim$1.5 keV amplitude), which were found to move upwards along the magnetic flux tube at the plasma sound velocity (Hess et al. 2007, 2009).

Paschmann et al. (2003) reviewed typical effects at different frequencies occurring in regions of upward and downward currents of the terrestrial auroral zone. Briefly, electron solitary waves or ELF electric field turbulence are found in downward field-aligned region, associated with divergent electric fields and up-going field-aligned electrons. This is the source region of VLF saucers (whistler emissions) and among the first radio emissions observed in the auroral zone. Large-amplitude ion cyclotron waves and electric field turbulence are found in upward current regions, associated with convergent electric fields and precipitating "inverted-V" events. This is also the source region of auroral radiation, powerful emissions observed in the auroral zones of magnetized planets.

One of the most powerful emissions is the auroral radiation observed above the auroral zone of the magnetized planets. These emissions are primarily driven by precipitating electrons accelerated to keV energies. The generation mechanism is well identified as the Cyclotron Maser Instability (Wu and Lee 1979) and has been extensively studied (see review by Treumann 2006). In situ observations, especially by Viking and FAST, have shown that the source regions are the acceleration regions described in Fig. 11, which are strongly depleted in cold plasma ($f_{pe}/f_{ce} < 0.1$ to 0.3) due to the parallel electric field structure (Roux et al. 1993). The instability appears to be most efficiently driven by quasi-trapped energetic electrons, i.e. keV electrons with velocity mostly perpendicular to the magnetic field. However, this quasi-trapped electron population lies in a region of velocity space which should be empty in a simple adiabatic theory, thus its presence in the auroral zone was suggested to be due to time-varying (or space-varying) parallel electric fields (Louarn et al. 1990). The above filamented Alfvén waves are good candidates, consistent with the filamentary structure of the depleted sources of auroral radio radiation.

Reconnection Acceleration Magnetic reconnection is a well-known example of transient situations. The simplest concept involves a configuration with a "X-point" in a 2D geometry, where the magnetic field vanishes. More complicated configurations are considered with 3D geometries, with guide field. In the "frozen-in" conditions where $\boldsymbol{E} + \boldsymbol{V} \times \boldsymbol{B} = 0$, all points of a given magnetic field line will remain magnetically connected during their motion at the velocity V. The magnetic reconnection implies that the magnetic field line has been modified or broken and the existing connection region reconnected with another one. This leads to a global reconfiguration of the magnetic structure. Reconnection is generally considered as the result of a local departure from the "frozen-in" conditions and involves parallel electric fields. The triggering factors differ on the plasma types, near the Sun or in planetary magnetospheres; it is generally difficult to predict the time and location where they occur. One of the distant signatures, well-identified onboard spacecraft, is again the particle acceleration. It is observed in the perpendicular direction mainly near the central part of the plasmasheet or in the parallel direction along the separatrices (Paschmann 2008).

On the magnetopause, reconnection can be accompanied by the development of vortices due to the Kelvin-Helmholtz instability (KHI). This process is known to occur at Earth (see e.g. Hasegawa et al. 2009), Mercury (Sundberg et al. 2012), and Saturn (Delamere et al. 2013). Parallel acceleration of electrons is caused by K-H waves, as strongly suggested at Saturn by the observation of Cyclotron Maser radio emission from the morningside sector of the magnetosphere (Galopeau et al. 1995).

2.5 Non-Adiabatic Acceleration

It is sometimes said that the motion of charged particles is nonadiabatic when the second adiabatic invariant (viz., the action integral $I \equiv m \int V_{\parallel} ds$ associated with the particle bounce motion; see Eq. (33)) is not conserved. This may be the case for instance during substorm dipolarization of the magnetic field lines that can lead to different particle energization depending upon their bounce phase; hence, the formation of bouncing ion clusters (e.g., Mauk 1986). However, in the most general case, the motion of charged particles is defined as being nonadiabatic when the first adiabatic invariant (i.e., the magnetic moment associated with the particle gyromotion Eq. (34)) is not conserved. This may occur either when the length scale of the field variation is comparable to or smaller than the ion Larmor radius (spatial nonadiabaticity) or when the time scale of the field variation is comparable to or smaller than the ion cyclotron period, i.e., temporal nonadiabaticity (e.g., Northrop 1963). Under such conditions, the guiding center approximation is not appropriate to investigate the motion of charged particles and a description based on the full equation of motion is necessary. In the steady state terrestrial magnetosphere, the guiding center approximation may be used to characterize the transport of charged particles in the lobes where substantial centrifugal acceleration (up to a few tens of eV) due to $\boldsymbol{E} \times \boldsymbol{B}$ convection of the magnetic field lines may be obtained (e.g., Cladis 1986). The guiding center approximation also is appropriate in the nearly dipolar region of the inner magnetosphere. As a matter of fact, in this region of space, the second adiabatic invariant often is conserved as well so that an adiabatic bounce-averaged description may be adopted to explore the dynamics of, e.g., ring current and radiation belt particles (e.g., Fok et al. 2006). As for the third adiabatic invariant associated with the particle azimuthal drift about the planet, it is often violated; hence, prominent radial diffusion of the particles takes place.

2.5.1 Spatial Nonadiabaticity

At large distances in the equatorial magnetotail, the magnetic field significantly varies on the length scale of the particle Larmor radius and a gyro-averaged description such as that of the guiding center cannot be applied. To characterize the particle behavior, Sergeev et al. (1983) introduced a scaling parameter K defined as the minimum field line curvature radius-to-maximum particle Larmor radius ratio. Sergeev et al. (1983) demonstrated that, as K becomes smaller than $\sim$8, deviations from an adiabatic behavior gradually develop as identified by, e.g., the injection of trapped particles into the loss cone. In a subsequent study, Sergeev et al. (1993) identified the latitude in the auroral zone where the parallel flux becomes comparable to the perpendicular one, as the projection at low altitudes of the nonadiabaticity threshold in the magnetotail (for given particle species and energy). This latitudinal boundary that is referred to as "Isotropy Boundary" forms a convenient proxy to remotely probe the distant tail topology from low-altitude measurements, as shown for instance by Newell et al. (1998).

The fact that particles may not perform a regular helical motion and actually behave in a nonadiabatic manner in the distended Earth's magnetotail was already uncovered in the pioneering work of Speiser (1965). In the case of a pure neutral sheet such as the self-consistent one of Harris (1962) with opposite magnetic field orientations on either side of the midplane, Speiser (1965) showed that particles execute rapid oscillations about the midplane and are subsequently lost into the flanks. In the case of a quasi-neutral sheet with a small magnetic field component normal to the midplane, such as that due to the Earth's dipole field, Speiser (1965) showed that the above oscillations are coupled with a slow rotation of the oscillation plane so that particles may be turned back toward the planet instead of traveling into the flanks. Sonnerup (1971) considered the action integral $I_Z \equiv m \int V_Z dZ$ (see Eq. (36)) to characterize the behavior put forward by Speiser (1965) since particle orbits do have some regularity (although not in an adiabatic sense).

Using Poincaré surfaces of section or, equivalently, phase space mapping upon crossing of the midplane, Chen and Palmadesso (1986) examined the dynamics of charged particles in the magnetotail in a more systematic manner. In this latter study, it was shown that the above Speiser orbits actually form one of three distinct classes of nonadiabatic orbits. That is, in the Speiser regime, particles do not experience significant pitch angle scattering upon crossing the neutral sheet and those originating from regions of strong magnetic field may return to such regions after neutral sheet crossing; hence, their denomination as "transient" particles. In the second class of orbits, particles experience prominent pitch angle scattering upon crossing of the neutral sheet. Accordingly, particles originating from regions of strong magnetic fields may remain temporarily trapped near the midplane, while those trapped near the midplane may escape after crossing of the neutral sheet; hence, their denomination as "quasi-trapped" particles. Finally, a third class of orbits consists of particles that remain trapped near the midplane, an example of them being the ideal case of particles with 90° pitch angle at equator (see Fig. 4 of Chen and Palmadesso 1986). Chen and Palmadesso (1986) showed that the phase space is systematically partitioned according to these three distinct orbit classes and that the Speiser regime becomes predominant for specific values of the (normalized) Hamiltonian.

Following the approach of Sonnerup (1971), Büchner and Zelenyi (1989) developed a comprehensive interpretation framework of the particle dynamical behaviors. The formalism put forward in this latter study relies on a piecewise description of the particle motion, considering that it can be viewed as a succession of $I_Z \equiv m \int V_Z dZ$ conserving sequences (see Eq. (36)). In this interpretation framework, at some point during transport toward the neutral sheet, particles cross a phase space separatrix that delineates two different dynamical regimes (viz., crossing and non-crossing of the midplane). In the course of these separatrix crossings, small quasi-random jumps of the invariant I_Z occur as put forward by Neishtadt (1987) (see Fig. 14 of Büchner and Zelenyi 1989). In this approach, the Speiser regime (also referred to as "transient") corresponds to a negligible net change of I_Z; hence, its denomination as "quasi-adiabatic". In contrast, in the above quasi-trapped regime (also referred to as "cucumber-like" in Büchner and Zelenyi 1989), particles are subjected to significant net changes of I_Z. To describe these I_Z changes, Büchner and Zelenyi (1989) introduced a parameter κ defined as the square root of the minimum field line curvature radius-to-maximum Larmor radius ratio (see, e.g., Eq. (41) of Büchner and Zelenyi 1989). This latter κ parameter, that is now commonly used to characterize the adiabatic character of the particle motion, is the square root of the K parameter of Sergeev et al. (1983). It is also comparable with the dimensionless Hamiltonian used by Chen and Palmadesso (1986) since one has $2H \equiv \kappa^{-4}$. According to the analysis of Büchner and Zelenyi (1989), the particle motion turns nonadiabatic for $\kappa < 3$ (equivalently, $K < 8$ in Sergeev et al. 1983), and the above regimes with transient (Speiser) and quasi-trapped behaviors are obtained for $\kappa < 1$ (a κ regime that is also referred to as the current sheet limit). Between $\kappa > 3$ and $\kappa < 1$, there exists an intermediate regime where particles do not oscillate about the midplane (because of Larmor radii smaller than the field reversal length scale) but their motion is chaotic.

Delcourt et al. (1994) further explored this intermediate $1 < \kappa < 3$ regime, considering a centrifugal perturbation of the particle motion near the magnetotail midplane. The interpretation framework developed in this latter study is that the adiabatic (magnetic moment conserving) sequences upon approach and exit of the neutral sheet are separated by a critical cyclotron turn during which an impulsive centrifugal force (due to the enhanced field line elongation) perturbs the cyclotron motion of the particles. This so-called Centrifugal Impulse Model that describes a single (prototypical) crossing of the field reversal leads to a characteristic three-branch pattern of magnetic moment variations, viz., (i) at small pitch angles, large magnetic moment enhancements regardless of the particle gyration phase, (ii) at large pitch angles, negligible magnetic moment changes and (iii) at intermediate pitch angles, either magnetic moment enhancement or damping depending upon gyration phase. As κ decreases from 3 toward 1, this three-branch pattern gradually expands in velocity space, consistently with the results of Sergeev et al. (1983) (see Fig. 1 of Delcourt et al. 1996). Repeated crossings of the field reversal (equivalently, repeated applications of the three-branch pattern of magnetic moment variations) lead to a chaotic behavior with prominent dependence upon initial phase of gyration since magnetic moment enhancement and damping are obtained at small and intermediate pitch angles, respectively.

In this respect, using single-particle simulations in a model magnetic field of the magnetotail, Ashour-Abdalla et al. (1992) suggested that the $\kappa \approx 1$ regime leads to enhanced particle trapping and duskward drift, a feature referred to as the "wall" region. This $\kappa \approx 1$ regime lies in the mid-tail at the transition between the nearly dipolar region where the particle motion is adiabatic ($\kappa > 3$) and the distant tail where one has $\kappa < 1$. It corresponds to the onset ($K = 8$) of nonadiabaticity examined by Sergeev et al. (1983) and the "wall" feature is thus at odds with the "Isotropy Boundary" interpretation framework discussed above with particle injection into the loss cone and subsequent precipitation. However, the three-branch pattern obtained with the Centrifugal Impulse Model suggests that the two behaviors coexist, the "wall" feature corresponding to large magnetic moment enhancements at (relatively) small pitch angles while the "Isotropy Boundary" follows from damping of the magnetic moment at intermediate pitch angles.

The nonadiabatic features discussed above are of paramount importance for the development of thin current sheets that are essential magnetotail elements at Earth and at other planets. In the terrestrial magnetosphere, in situ observations from GEOTAIL, CLUSTER and THEMIS have revealed a number of magnetic field features in the tail current sheet such as flapping, flattening, tilting, waving, twisting and bifurcation. This current sheet can become very thin (with a thickness comparable to the ion inertial length), yielding a metastable state that can lead to current sheet disruption as observed during the expansion phase of substorms (e.g., Mitchell et al. 1990). The formation of nongyrotropic distribution functions in these nonadiabatic regimes also leads to nonzero off-diagonal terms in the pressure tensor and allows for a current sheet equilibrium that does not require a prominent pressure gradient along the tail axis (e.g., Ashour-Abdalla et al. 1994). As for the predominant Speiser regimes obtained within specific $\kappa < 1$ intervals, they follow from resonance between the fast particle oscillation about the midplane (imposed by the opposite orientations of the magnetic field above and below the midplane) and the slow gyromotion (imposed by the small magnetic field component normal to the midplane). In this Speiser regime, particles are subjected to prominent energization owing to large displacement along the dawn-to-dusk convection electric field. This efficient Speiser acceleration can thus lead to large particle flux within limited intervals at high energies (small κ); hence, the formation of "beamlets" traveling down to low altitudes as reported in CLUSTER observations (see, Keiling et al. 2004).

2.5.2 Temporal Nonadiabaticity

It was mentioned above that during the expansion phase of substorms, the second adiabatic invariant may not be conserved (Mauk 1986). Indeed, the short-lived electric field induced by dipolarization of the magnetic field lines can lead to significant energization of particles that are located in the equatorial vicinity while those located at low altitudes may remain unaffected. Here, violation of the second adiabatic invariant is due to temporal variations of the magnetic field on the time scale of the particle bounce period. Note that this second adiabatic invariant may be violated because of spatial variations of the magnetic field as well, as is the case for instance near the frontside magnetopause where particles evolve from bouncing about the equatorial plane to bouncing about the field minimum in the outer cusp region (Shabansky 1971; Delcourt and Sauvaud 1999).

Still, temporal variations of the magnetic field can also lead to violation of the first adiabatic invariant, a behavior that is obtained whenever the magnetic field varies significantly on a time scale comparable to the particle gyro-period. In this regard, it was shown by Delcourt et al. (1990) that, during dipolarization of the magnetic field lines, violation of the first adiabatic invariant may be obtained for heavy ions (O^+) that have cyclotron periods of a minute or so in the terrestrial mid-tail. As a result, while protons with small gyro-periods behave in an adiabatic manner (with respect to the first invariant), O^+ may experience prominent nonadiabatic energization, in a like manner to spatial nonadiabaticity, where protons and O^+ ions may exhibit $\kappa > 3$ and $\kappa \leq 1$, respectively.

Unlike the energization by the large-scale convection electric field that is constrained by the magnitude of the cross-polar cap potential drop (typically, in the 50 kV–150 kV range) so that ions drifting over a few R_E across the steady state magnetotail can gain at most a few tens of keV, there is no well defined limit for the energization that can be achieved from the induced electric field (Heikkila and Pellinen 1977; Pellinen and Heikkila 1978). Delcourt et al. (1990) actually showed that O^+ energization up to the 100 keV range is readily obtained during substorm reconfiguration of the magnetic field lines. Since this energization occurs in a nonadiabatic manner and goes together with prominent enhancement of the particle magnetic moment, it radically changes the long-term behavior of the particles that may evolve from an open drift path (i.e., connected to the dayside magnetopause) to injection into the ring current and rapid gradient drift around the planet owing to the large energy gain realized (see Fig. 5 of Delcourt 2002).

At Earth, a variety of in situ measurements suggest that such a mass-to-charge dependent energization is at work during substorm dipolarization. Post-dipolarization spectra obtained for O^+ can be significantly harder than those of protons (Ipavich et al. 1984; Nosé et al. 2000). Observations of energetic neutral atoms by Mitchell et al. (2003) also reveal repeated injections of energetic (above 100 keV) O^+ in conjunction with auroral break-ups, while no similar injections are obtained for protons. The (temporally) nonadiabatic heating at work here increases when the inductive electric field increases or if the ions are located further away from the inner dipolar region in the equatorial magnetotail, and it may actually occur in regions where spatial adiabaticity is achieved (viz., $\kappa > 3$). Note also that prominent fluctuations of the magnetic field on short time scales may somewhat alter this description and lead to significant nonadiabatic heating of protons as well, as displayed in the GEOTAIL data analysis of Ono et al. (2009). From a general viewpoint, temporal nonadiabaticity critically depends upon the characteristics of the magnetic field transition and one may expect that Mercury's environment with small temporal scales as compared to those at Earth is characterized by specific nonadiabatic responses.

2.6 Pick-up Acceleration and Mass Loading

Ions produced within a flowing plasma are a significant source of energy and a sink of momentum, as well as being a source of plasma. Although sometimes used more generally, the classical pick-up process occurs when the parent neutrals have a velocity different from the $(\vec{E} \times \vec{B})/B^2$ drift of the local plasma. The new ions are then accelerated by the convection electric field and form a ring-beam distribution in velocity space. This distribution is unstable to the cyclotron maser instability and may result in the generation of electromagnetic ion cyclotron waves.

Neglecting the energy lost to these waves, the ions have an energy, in the plasma frame, of $2mv_{rel}^2$ (where v_{rel} is relative velocity between the source neutrals and local plasma) or four times the ram energy of a background ions of the same mass. In many cases, this can be a significant source of plasma heating. In addition, acceleration by the convection electric field initially causes the newly created ion and electron to move in opposite directions, and their guiding centers become separated by a gyroradius. The resulting "pick-up current" is $\vec{J} = \frac{v_{rel}m}{B}\frac{dn}{dt} = \frac{m}{B^2}\frac{dn}{dt}\vec{E}$. This is often treated as a "pick-up" conductivity (Thomas et al. 2004 and references therein). The pick-up current, flowing across the background magnetic field, also acts to slow, or mass-load the plasma.

In one common case, pick-up acceleration, heating and mass-loading may occur without producing a net source of mass or plasma. If the ions are produced through symmetric charge exchange, $X^+ + X \rightarrow X + X^+$, then the newly ionized particle will be accelerated as any other pick-up ion, producing a ring-beam distribution, heating and mass-loading. However, the reaction will also generate a fast neutral which escapes the system. As a result, there is no net change in the ion density.

3 Losses

In previous sections, we have considered the various sources of plasma and their transport and energization processes to supply magnetospheric plasmas. We next consider the ways in which this material can be lost from the system, to "balance" the mass budget. There are a number of methods by which plasma can be lost from magnetospheres.

3.1 Tail Reconnection and Plasmoids

Magnetic reconnection in a planetary magnetotail is a key mechanism by which magnetic field lines stretch to instability and break, which then allows the release of parcels of mass and plasma called plasmoids, of varying sizes and shapes (Hones 1976, 1977). Observations in the Earth's magnetosphere have shown that plasmoids are typically about 1 to 10 Re in diameter (Ieda et al. 1998; Slavin et al. 2003). Figure 12 shows a schematic of the formation of earthward and tailward-moving plasmoids following reconnection. Figure 13 shows the effect that a large plasmoid has on the surrounding magnetotail at Earth including the magnetic field signatures associated with the traveling compression region in the surrounding lobe regions and the tailward retreat of the near-Earth neutral line (NENL) following ejection (Slavin 1998). In situ observations of tail reconnection include observations of changes in magnetic field topology and plasma flows. In recent years the study of tail reconnection has been extended beyond Earth. Plasmoids have been observed in the magnetotails of Saturn (e.g. Jackman et al. 2007, 2011, 2014b), Jupiter (Russell et al. 1998; Vogt et al. 2010, 2014), and Mercury (Slavin et al. 2009, 2012b; DiBraccio et al. 2013).

In recent years several authors have sought to consider the role of tail reconnection as a loss mechanism for magnetospheric plasma (e.g. Bagenal and Delamere 2011). At Jupiter, Bagenal (2007) highlighted the mismatch between the inferred mass input rate from Io of ~500–100 kg/s and the mass loss rate from plasmoids, estimated at ~30 kg/s. Kronberg

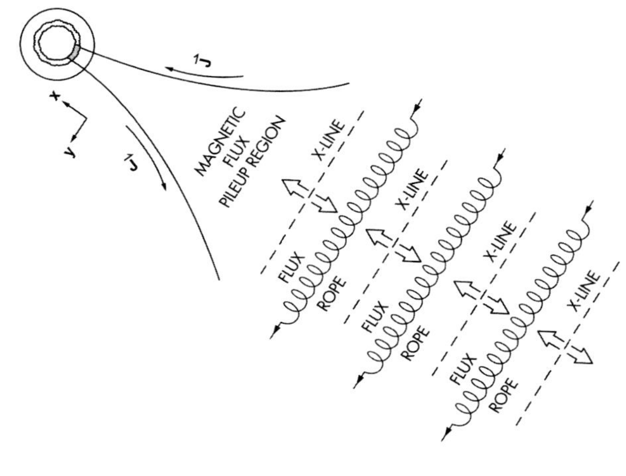

Fig. 12 A schematic diagram of the formation of earthward and tailward-moving plasmoids following reconnection; after Slavin et al. (2003)

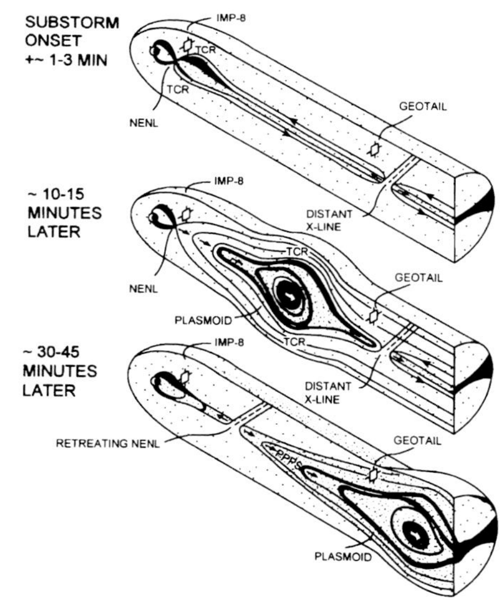

Fig. 13 The formation of single large plasmoid during a substorm is illustrated. Reconnection at a single near-Earth neutral line disconnects the magnetic flux in the downstream plasma sheet to form either magnetic loops or helical magnetic fields depending upon the existence of an east-west magnetic field component for makeup the plasmoid. The plasmoid compresses the surrounding tail lobes as it moves tailward to produce the traveling compression (TCR) signature that is detected when the observing spacecraft are in the magnetotail, but to far from the equatorial region to encounter the plasmoid proper (adapted from Slavin 1998)

et al. (2008) attempted a similar calculation (based on Galileo energetic particle measurements) and found that their inferred mass of $\sim 8 \times 10^5$ kg per plasmoid would require far more plasmoids than had been observed to account for the input. Vogt et al. (2014) completed the most comprehensive study to date at Jupiter, whereby they found that mass loss ranged from $\sim$0.7–120 kg/s. They concluded that while tail reconnection is indeed an active process at Jupiter, it likely cannot account for the mass input from Io, suggesting that additional mass loss mechanisms may be significant. Jackman et al. (2014b) investigated the analogous picture at Saturn. They found an average mass loss rate of $\sim$2.59 kg/s, much less than the $\sim$100 kg/s expected to be loaded into the magnetosphere by the volcanic moon Enceladus.

These studies raise the question: If large-scale reconnection is not sufficient to account for the required loss of material from the tails of Jupiter and Saturn, what other processes/new physics are required to balance the mass budgets? Other loss mechanisms are investigated in the sections below.

3.2 Charge Exchange

In the Earth's magnetosphere, there exist a region called the ring current, where high energetic ions and electrons with energy between hundreds of eV and hundreds of keV are trapped by Earth's dipole-dominated magnetic field (Frank 1967; Williams 1981). In the ring current, the ions (electrons) drift westward (eastward) due to the magnetic drift, and the ring current development causes the decrease in the horizontal magnetic field component at Earth's surface. Thus, the strength of the ring current is often measured by the Dst or SYM-H indices derived from ground-based magnetometer observations (Sugiura 1964; Wanliss and Showalter 2006). If the magnetic field is strong enough and dominated by the dipole component and there are transportation and energization processes to populate high-energy ions in the inner magnetosphere of a planet, a similar ring current is expected to exist at such planets.

One efficient loss mechanism for the terrestrial ring current particles is the charge exchange (see Eqs. (22) and (23)) of the ring current ions with the neutral hydrogen that makes up the geocorona. When the convection weakens, this becomes the dominant process by which ring current ions are removed from the system, depleting the inner magnetosphere of its energetic population. The geocorona is a halo-like extension of the exosphere out to several Earth radii, consisting of relatively cold ($\sim$1000 K), very tenuous neutral hydrogen atoms with densities ranging from thousands of atoms per cubic centimeter at the inner edge of the ring current to less than a hundred at geosynchronous orbit. This cold gas plays a critical role in the energy budget of the Earth's inner magnetosphere since the charge exchange reactions make the exosphere act as an energy sink for ring current particles, replacing a hot ion with a cold one. Singly charged ring current ions can be neutralized after collisions with thermal exospheric hydrogen atoms as described below:

$$\mathrm{H}^+ + \mathrm{H}_{cold} \rightarrow \mathrm{H} + \mathrm{H}^+_{cold} \tag{41}$$

$$\mathrm{O}^+ + \mathrm{H}_{cold} \rightarrow \mathrm{O} + \mathrm{H}^+_{cold} \tag{42}$$

$$\mathrm{He}^+ + \mathrm{H}_{cold} \rightarrow \mathrm{He} + \mathrm{H}^+_{cold} \tag{43}$$

The incident ring current ion picks up the orbital electron of the cold geocoronal hydrogen atom resulting in the formation of an Energetic Neutral Atom (ENA). These particles are not affected by magnetic or electric field forces therefore they are no longer trapped in the geomagnetic field and leave the interaction region in ballistic orbits in the direction of the incident ion velocity at the time of the impact. If the resulting ENA's velocity exceeds the Earth's gravitational escape field, then it is lost into space or precipitates down into the ionosphere. On the other hand, the low energy ENAs populate the plasmasphere. Meinel (1951) first reported the existence of energetic neutral atoms, based on observations of precipitating energetic neutral hydrogen precipitating into the upper atmosphere during auroral substorms. A few years later, Dessler and Parker (1959) were the first to suggest that charge exchange between protons and neutral atmospheric hydrogen atoms would effectively contribute to the decay of the ring current, although the effectiveness of ion removal from the ring current through charge exchange processes was previously investigated by Stuart (1959) and Fite et al. (1958).

Multiply charged ions allow for multiple charge exchange reactions,

$$\mathrm{He}^{++} + \mathrm{H}_{cold} \rightarrow \mathrm{He}^+ + \mathrm{H}^+_{cold} \tag{44}$$

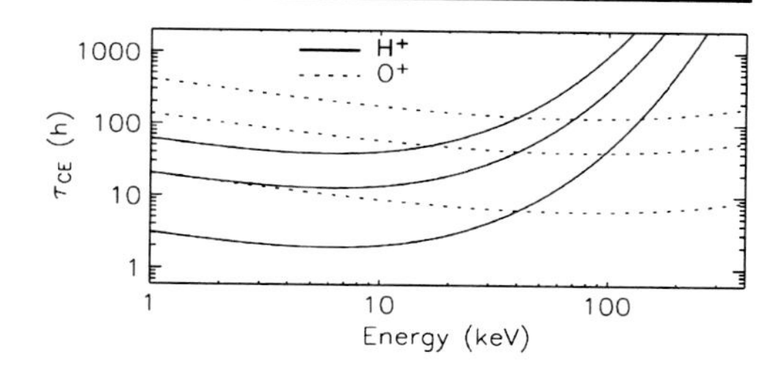

Fig. 14 The mean lifetime for charge exchange decay as a function of energy for O^+ and H^+ species. Figure adopted from Liemohn and Kozyra (2005)

and Spjeldvik and Fritz (1978) showed that the higher charge states of helium and oxygen ions are increasingly important for energies above 100 keV, while at energies below this cut-off the lower charge states are dominant. Energetic neutral atoms generated in the main ring current traversing the inner magnetosphere can be re-ionized. This happens by converting ENAs back into ring current ions albeit on new L shells, undergoing subsequent charge-exchange collisions with geocoronal atoms and generating secondary ENA fluxes that can participate in further ionizing collisions (Bishop 1996). This yields the formation of a secondary ring current close to the Earth, at L shell values of approximately 3, although this is not a large ring current population. Moreover, low pitch angle ions are subject to additional charge exchange collisions with the oxygen atoms in the upper atmosphere.

Solar far-ultraviolet light is reflected off this hydrogen gas (Chamberlain 1963) and so its abundance has been quantified. It has been reported (Fahr 1974; Rairden et al. 1986; Hodges 1994; Østgaard et al. 2003; Fuselier et al. 2010; Zoennchen et al. 2010, 2011; Bailey and Gruntman 2011) that the geocoronal hydrogen density decreases exponentially with radial distance. This means that at large altitudes down the magnetotail, the collisions with the neutral hydrogen become negligible. However, in the ring current region, these collisions become increasingly important and magnetospheric H^+ can be easily removed by charge exchange with the neutral exospheric hydrogen.

The probability of collisions with neutral atoms from the exosphere depends strongly on the energy of the incident particles and is determined by the charge exchange cross sections. Charge exchange cross sections are both energy and species dependent and thus different ring current ion species have different charge exchange lifetimes. A compilation of charge exchange cross sections for various ring current ions can be found in Spjeldvik (1977), Smith and Bewtra (1978), and Orsini and Milillo (1999).

Numerous studies, both based on both observations and numerical modeling show that due to the strong species and energy dependence of the charge-exchange cross sections along with the temporal and spatial dependence of ring current composition, the charge exchange process strongly affects the ring current plasma. Figure 14 shows the profile of charge exchange lifetime as a function of energy and species (Liemohn and Kozyra 2005). Moreover, it is inferred that the charge exchange loss processes are predominantly important after the initial phase of the ring current decay.

The efficiency of ion removal from the ring current through charge exchange depends on several factors: the energy and the species of the ion population as well as the density of the neutral cloud. The latter depends on the changes in the atmospheric temperature and density, the radiation pressure exerted by the solar far ultra violet photons and the strengths of all these interactions determine the structure of the exosphere. Therefore reliable measurements of the geocoronal density are essential in determining the relative importance of charge exchange losses of ring current ions. The majority of geocoronal models report on vastly different densities in the inner magnetosphere (Ilie et al. 2013) and therefore the decay rates and lifetimes for ring current ions are significantly different depending on the neutral density distribution, affecting the amount of ENAs emitted in a given region in space (see Fig. 15).

Keika et al. (2003, 2006), based on measurements of energetic neutral atoms (ENAs) made by the High Energy Neutral Atom (HENA) imager on board the Imager for Magnetopause-to-Aurora Global Exploration (IMAGE) satellite, show that the rate of the charge exchange energy losses is comparable to the ring current decay rate for the intervals of the slow decay, while the loss rate is much smaller than the decay rate in the rapid decay phase, in particular for the early stage of a storm recovery. Similarly, Jorgensen et al. (2001) show that during the fast recovery the measured ENAs can only account for a small portion of the total energy loss and the lifetime of the trapped ions is significantly shorter during the fast recovery phase than during the late recovery phase, suggesting that different processes are operating during the two phases. Furthermore Kozyra et al. (2002) suggested that charge-exchange losses can be solely responsible for the decay of the ring current during the recovery phase only if IMF abruptly turns northward at the end of the main phase.

The neutral gases in the upper atmospheres of Jupiter and Saturn are molecular and atomic hydrogen and thus either as a result of direct ionization or dissociative ionization a significant number of H^+ ions are created. H^+ can only recombine directly via radiative recombination, which is an extremely slow process and thus there must be other ways to remove them otherwise very large ion densities would result. As indicated in Sect. 1.2.3 (Loss Processes and Ion Chemistry), McElroy (1973) suggested some time ago that the following charge exchange would be important in removing H^+:

$$H^+ + H_2\ (v \geq 4) \longrightarrow H_2^+ + H \tag{45}$$

H_2^+ is rapidly transformed to H_3^+ via the following reaction:

$$H_2^+ + H_2 \longrightarrow H_3^+ + H \tag{46}$$

H_3^+ will most likely undergo dissociative recombination and thus this series of reactions removes ions relatively rapidly. There is another way that H^+ can be lost at Jupiter and Saturn (see Eq. (25)), namely by reacting with water molecules, originating in the rings (Connerney and Waite 1984).

3.3 Precipitations into Planets

3.3.1 High Latitudes

As seen in Sect. 2.4, the atmospheric loss cone can be defined at any location by its half-angle $\sin\alpha_{lc} = (B/B_m)^{1/2}$ (see Eq. (34)) where B is the magnetic field amplitude at the position considered and B_m its value at the ionospheric end of the magnetic field line. Charged particles with pitch angle $< \alpha_{lc}$ will precipitate into the planet and be lost for the magnetosphere. The loss cone is permanently fed by new particles resulting from processes such as pitch-angle scattering by electric and magnetic fluctuations (e.g. whistler waves; see Bolton et al. 2004 and references therein). Due to the converging field line geometry, most precipitations occur at relatively high magnetic latitude (~55°–75°).

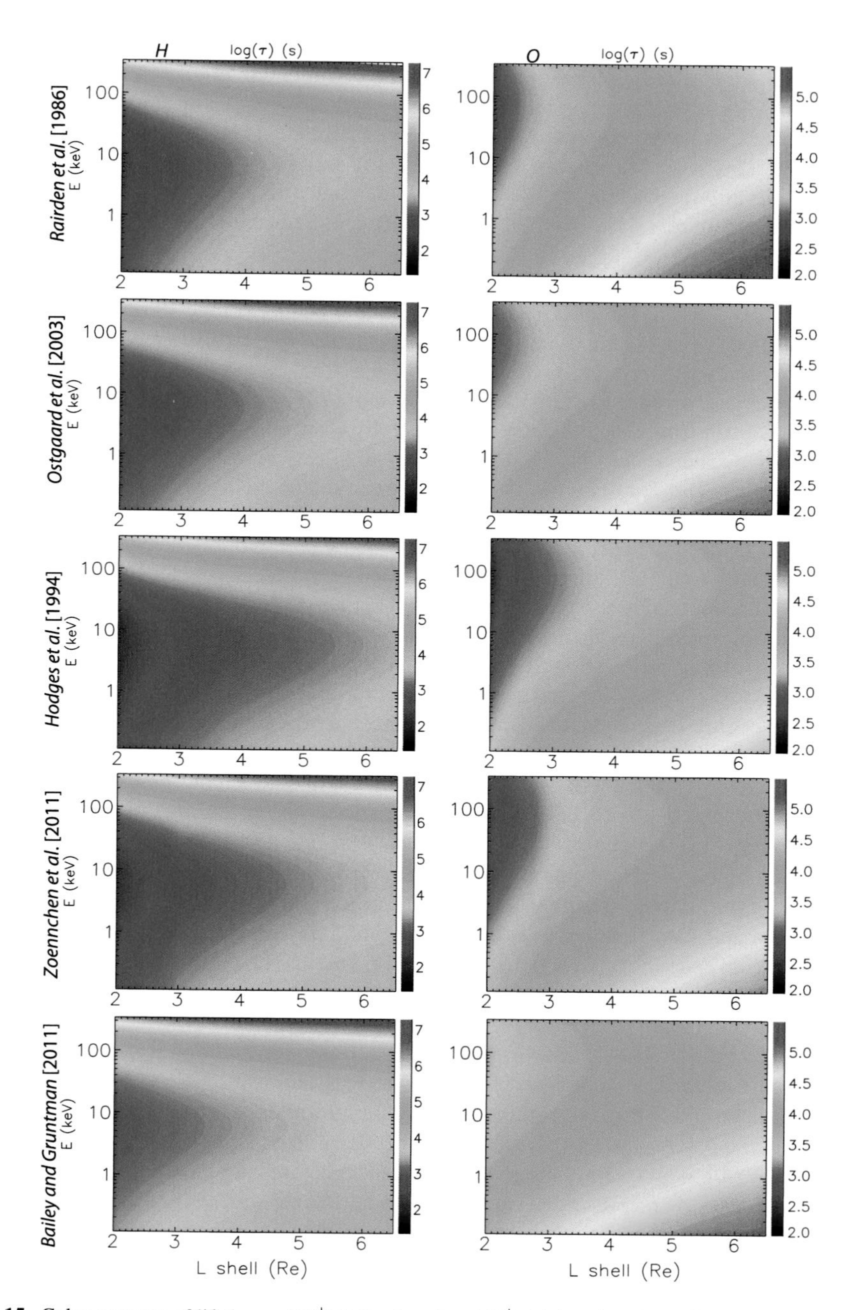

Fig. 15 Color contours of lifetimes of H^+ (*left column*) and O^+ (*right column*) as a function of energy and radial distance from the Earth. From *top* to *bottom* are shown the lifetime predictions from Rairden et al. (1986), Østgaard et al. (2003), Hodges (1994), Zoennchen et al. (2011) and Bailey and Gruntman (2011). The color scale is logarithmic and lifetimes are in seconds. Figure from Ilie et al. (2013)

Auroral Ovals Precipitations of electrons with energy $\geq$0.1 keV and of protons or ions with energy $\geq$ a few keV produce auroras (Birkeland 1910), seen from the ground as curtains of light, and from space as bright variable narrow circumpolar ring, arcs and spots. The precipitating electrons have energies in the range $\sim$100 eV–10 keV for the Earth (Feldstein et al. 2001) and Saturn (Cowley et al. 2004), reaching more than 100 keV for Jupiter (Prangé et al. 1998). This is well above their thermal energy in the magnetosphere or solar wind ($\leq$1 eV), thus strong acceleration is required, as discussed in Sect. 2.4. Total precipitated auroral power is up to $\sim 10^{11}$–10^{12} W for the Earth and Saturn, $\sim 10^{13}$–10^{14} W for Jupiter (Clarke 2012).

X-ray to radio emissions are produced in the high altitude atmosphere (80–300 km on Earth, 10^{-5}–10^{-9} bar at Jupiter) or in the precipitating beam (Prangé 1992; Bhardwaj and Gladstone 2000). The visible aurora is most spectacular on Earth, related to the excitation/deexcitation of O (red and green lines), N (blue line) and N_2 (purple), whereas H-α and H-β lines are very faint at Jupiter. The UV aurora, 10× to 100× more intense than visible ones, result from the collisional excitation (by electrons from a few to 100 keV) and then radiative deexcitation of N_2^+, N, H at Earth, and H (Ly-α) and H_2 (Lyman and Werner bands) at Jupiter. The X-ray aurora on Earth is mainly generated via bremsstrahlung from precipitating electrons, and at Jupiter from the collisional excitation (followed by radiative deexcitation) of deep internal levels of O and S ions by precipitating heavy ions of energy >100 MeV. The X and UV aurorae are often pulsed on timescales of tens of minutes. The IR auroral emission is due to atmospheric Joule heating (followed by radiative cooling). It is emitted as nitrogen lines at Earth and H_3^+ and hydrocarbons lines at Jupiter. As UV absorption by hydrocarbons is strongly frequency-dependent, the comparison between auroral and laboratory H and H_2 UV spectra provides information to deduce the depth at which precipitated energy is deposited and, with an atmospheric model, to derive the nature and energy of precipitating particles. Coherent circularly polarized cyclotron radio emissions are generated below $\sim$1 MHz ($\leq$40 MHz at Jupiter) by the interaction of unstable precipitating (or mirrored) energetic (1–10 keV) electron populations with electromagnetic fluctuations, in a rarefied and magnetized plasma ($f_{pe}/f_{ce} \ll 1$) (Zarka 1998). Their generation causes the diffusion of the electrons in velocity space (Pritchett 1986) in particular into the loss cone, causing further precipitations. Imaging the auroral activity in UV (HST—Prangé et al. 1998), IR (ground-based telescopes—Connerney et al. 1993) and radio (via DE-1 (Huff et al. 1988) or Cassini spacecraft (Cecconi et al. 2009)) permits to map the precipitations and, by projection along the magnetic field, the magnetospheric activity.

Polar Cusps and Satellite-Magnetosphere Interactions In addition to the auroral ovals, at the limit between open and closed field lines at Earth or near the corotation breakdown region at Jupiter and Saturn, signatures of precipitations are also observed at the magnetic footprints of the polar cusps and of satellites embedded in the giant planets' magnetospheres (Waite et al. 2001; Pallier and Prangé 2004). Cusp signatures are around 12:00 LT and reveal sporadic dayside reconnections at timescales between 5 min (at Earth) and 20 min (at Jupiter), causing direct precipitation of accelerated particles in the polar cusps. They are more intense for a southern solar wind B_z at Earth (northern at Jupiter). The auroral input power into the cusp is only $\approx$1 % of the total auroral input power. The magnetic footprints of Io, Ganymede and Europa were detected in UV at Jupiter (Bonfond 2012), as well as that of Enceladus at Saturn (Pryor et al. 2011). Precipitation in the satellites magnetic flux tubes result from the imposed current across the satellite due to the electric field $\boldsymbol{E} = \boldsymbol{v} \times \boldsymbol{B}$ arising from the motion of the satellite (at velocity $\boldsymbol{v} = \boldsymbol{v}_{Keplerian} - \boldsymbol{v}_{corotation}$) across the planetary magnetic field lines. This current is carried by Alfvén waves accelerating electrons. In the Io-Jupiter case, the precipitated power reaches 10^{12} W, i.e., ~10–15 W/m^2 at the satellite ionospheric footprints. This power, within a factor 2 of the solar input, strongly heats the local ionosphere and modifies its properties, such as conductivity (Prangé et al. 1996). Satellite footprints have downstream tails related to currents reaccelerating the magnetospheric plasma downstream of the obstacle.

Magnetospheric particles also precipitate onto the surface of embedded satellites. If the latter possesses a magnetic field, precipitating particles are guided toward the magnetic poles of the satellite, generating satellite auroras as well as significant surface alterations, as for example, in the case of Ganymede's polar caps (Khurana et al. 2007).

3.3.2 Low Latitudes

Radiation Belts and Synchrotron Losses Radiation belts consist of electrons and ions accelerated to very high energies (0.1 to >10 MeV) and brought by radial inward diffusion close to the planet (typically between the surface and ~6 radii), where they bounce between their mirror points. Satellites and rings embedded in the belts cause strong collisional absorption of these energetic particles. Unabsorbed electrons can emit synchrotron radiation, a linearly polarized incoherent nonthermal radiation from high energy electrons in cyclotron motion in a magnetic field. This emission extends over a spectral range from <100 MHz to several GHz in the case of Jupiter, and can thus be imaged by ground-based radiotelescopes (Bolton 2004). Intensity is maximal near the equator (trapped population) and near the poles (mirror points, where the residence time is maximum due to low parallel velocity). The lifetime of an emitting electron is relatively short (10^8 to 10^9 s), during which the perpendicular energy of the particle is radiated away and finally causes precipitation onto the planet at low latitudes (≤50°). At Earth and Saturn, synchrotron emission (yet undetected) may exist at much lower frequency and intensity. Saturn radiation belts are largely absent due to ring absorption, but a small belt was discovered by Cassini between the inner edge of the rings and the planet (<1.4 Rs) (Krimigis et al. 2005).

Precipitations from the Rings Other precipitation into Saturn's ionosphere come from the rings' ionized atmosphere (Luhmann et al. 2006). It is composed of O_2^+ and O^+ ions between ~1.4 and ~2.4 Rs near the equator, resulting from the ionization by sunlight and magnetospheric impacts of the neutral atmosphere due to sputtering, photo-desorption and meteoroid impacts. The ion motions in the planetary quasi-dipolar magnetic field, subject to the corotation electric field, gravitation and collisional scattering, lead to precipitation into the planet at mid-latitudes (30°–40°) of ions created at radial distances within the corotation orbit at ~1.8 Rs. Due to the slight North-South asymmetry of the magnetic field (stronger in the northern hemisphere), precipitation (of energy ≤100 eV) occurs mostly in the southern hemisphere.

4 Basic Equations and Modeling Methods

4.1 MHD (Magnetohydrodynamic) Simulation

The basic equations of magnetohydrodynamics (MHD) are derived in numerous textbooks including those by Chen (1984) and Krall and Trivelpiece (1986) and are traditionally presented in terms of the primitive or state variables; density (ρ), velocity (u), thermal pressure

(P), and magnetic field (B) as

$$\frac{\partial \rho}{\partial t} + \nabla \cdot (\rho \vec{u}) = 0, \tag{47}$$

$$\rho \frac{\partial \vec{u}}{\partial t} + \vec{u} \cdot \nabla \vec{u} + \nabla P - \frac{1}{\mu_0} \nabla \times \vec{B} \times \vec{B} = 0, \tag{48}$$

$$\frac{\partial P}{\partial t} + \gamma \nabla (P \vec{u}) - (\gamma - 1) \vec{u} \nabla \cdot P = 0, \tag{49}$$

$$\frac{\partial \vec{B}}{\partial t} - \nabla \times (\vec{u} \times \vec{B}) = 0. \tag{50}$$

The assumption of ideal gas law has been used to define the pressure equation (49) and the fact that the current density (J) is the curl of the magnetic field has been used to simplify the equations. More importantly in the generalized Ohm's law,

$$\vec{E} = -(\vec{u} \times \vec{B}) + \eta \vec{J} + \frac{1}{en_e} \vec{J} \times \vec{B} - \frac{1}{en_e} \nabla P_e, \tag{51}$$

terms related to the finite resistivity (η), Hall effect ($J \times B$), and electron pressure (P_e) have been neglected to get to Eq. (50). This formulation is commonly referred to as the equations of ideal MHD and it is important to point out that unless some term in the generalized Ohm's law is restored either analytically or numerically it is not possible for magnetic reconnection to occur in a system that obeys the equations of ideal MHD.

Numerical simulation of these equations usually involves discretization in space and time so it is common to formulate the ideal MHD equations in conservative form in order to allow for the direct application of advanced numerical techniques. The algorithm paper by Toth et al. (2012) not only provides a description of the motivation for utilizing conservative formulation but it also provides a more detailed discussion of the Hall and multifluid formulations than can be covered here. The conservative formulation involves equations of the form,

$$\frac{\partial U}{\partial t} + \nabla \cdot \vec{F}(U) = 0, \tag{52}$$

so that on a discrete grid the change of a conserved quantity is simply the sum of fluxes entering and leaving that cell. Recasting the ideal MHD equations in conservative form results in,

$$\frac{\partial \rho}{\partial t} + \nabla \cdot (\rho \vec{u}) = 0, \tag{53}$$

$$\frac{\partial \rho \vec{u}}{\partial t} + \nabla \cdot \left(\rho \vec{u} \vec{u} + \left(P + \frac{B^2}{2\mu_o} \right) - \frac{\vec{B} \vec{B}}{\mu_o} \right) = 0, \tag{54}$$

$$\frac{\partial \mathcal{E}}{\partial t} + \nabla \cdot \left(\vec{u} \left(\mathcal{E} + P + \frac{B^2}{2\mu_o} \right) - \vec{u} \cdot \frac{\vec{B} \vec{B}}{\mu_o} \right) = 0, \tag{55}$$

$$\frac{\partial \vec{B}}{\partial t} + \nabla \cdot (\vec{u} \vec{B} - \vec{B} \vec{u}) = 0, \tag{56}$$

where

$$\mathcal{E} = \frac{P}{\gamma - 1} + \frac{\rho U^2}{2} + \frac{B^2}{2\mu_o} \tag{57}$$

is the total energy density of the plasma element. In this formulation it is clear that the change in momentum density in a given region or computational cell is the result of the momentum entering or leaving the cell combined with the effects of thermal and magnetic pressure forces as well as with magnetic tension. Along with these equations comes an important constraint from Maxwell's equations, namely, the fact that the magnetic field must be divergence free ($\nabla \cdot B = 0$) throughout the entire computation domain for all times. In computational solvers this means using a simple projection scheme, a staggered type mesh (Yee 1966) with the magnetic fluxes defined on the faces and the electric fields on the edges, or the constrained transport 8-wave scheme (Powell et al. 1999). The staggered mesh approach is used by the OpenGGCM (Raeder et al. 2008) and LFM (Lyon et al. 2004) global simulations of the Earth's magnetosphere. The 8-wave solver is one of several methods available in the Space Weather Modeling Framework (SWMF), which has been used for a variety of problems throughout the heliosphere (Toth et al. 2005).

Huba (2005) presents an excellent discussion of the effects of including the Hall term in the MHD equations and the numerical techniques needed to solve them. In the notation of this chapter the inclusion of the Hall term in the generalized Ohm's law results in changes to the energy and induction equations,

$$\begin{aligned} \frac{\partial \mathcal{E}}{\partial t} &+ \nabla \cdot \left(\vec{u} \left(\mathcal{E} + P + \frac{B^2}{2\mu_o} \right) - \vec{u} \cdot \frac{\vec{B}\vec{B}}{\mu_o} \right) \\ &+ \nabla \cdot \left(\vec{u}_H \frac{B^2}{2\mu_o} - 2\frac{1}{\mu_o} \vec{B}(\vec{u}_H \cdot \vec{B}) \right) = 0, \end{aligned} \tag{58}$$

$$\frac{\partial \vec{B}}{\partial t} + \nabla \cdot \big((\vec{u} + \vec{u}_H)\vec{B} - \vec{B}(\vec{u} + \vec{u}_H) \big) = 0, \tag{59}$$

where the "Hall velocity",

$$\vec{u}_H = -\frac{\vec{J}}{ne}, \tag{60}$$

has been introduced to clearly illustrate how the Hall terms enter the system of equations. Since these terms are only present in the energy and induction equations it should be clear that the Hall term only transports the magnetic field and energy. To be clear, this means that the Hall effects are not a transport mechanism for mass or momentum. The inclusion of the Hall term introduces a new wave mode, the whistler mode, into the dynamics of the system. The whistler wave speed is significantly larger than the Alfvén speed. This introduces challenges into numerical computation. Since it is the largest wave speed that governs the time step that can be taken within a numerical solution this limitation can result in significant increases in the computational time to the solution. This can be addressed by sub-cycling the Hall physics on the shorter timescale and calculating the ideal MHD physics on the longer timescale.

Of course, the plasma in the Earth's magnetotail and other plasmas throughout the heliosphere can contain more than one ion species so it is often necessary to utilize the multi fluid formulations of the MHD equations to simulate these plasmas. In the notation of this paper these equations are:

$$\frac{\partial \rho_\alpha}{\partial t} + \nabla \cdot \rho_\alpha \vec{u}_\alpha = 0, \tag{61}$$

$$\begin{aligned} \frac{\partial \rho_\alpha \vec{u}_\alpha}{\partial t} + \nabla(\rho_\alpha \vec{u}_\alpha \vec{u}_\alpha + I P_\alpha) = {} & n_\alpha q_\alpha (\vec{u}_\alpha - \vec{u}_M) \times \vec{B} \\ & + \frac{n_\alpha q_\alpha}{n_e e}(\vec{J} \times \vec{B} - \nabla P_e), \end{aligned} \tag{62}$$

$$\frac{\partial \mathcal{E}_\alpha}{\partial t} + \nabla \cdot \left[(\mathcal{E}_\alpha + P_\alpha)\vec{u}_\alpha\right] = \left[n_\alpha q_\alpha (\vec{u}_\alpha - \vec{u}_M) \times \vec{B} + \frac{\rho_\alpha q_\alpha}{n_e e}(\vec{J} \times \vec{B} - \nabla P_e)\right], \tag{63}$$

$$\frac{\partial \vec{B}}{\theta t} = \nabla \times (\vec{u}_M \times \vec{B}) \tag{64}$$

where the α subscript has been used for the ion species and the term q_α allows for the inclusion of higher charge state ions. Furthermore,

$$\vec{u}_M = \frac{1}{e n_e} \sum_\beta n_\beta q_\beta \vec{u}_\beta \tag{65}$$

is the charge averaged ion velocity and

$$\vec{J} = e n_e (\vec{u}_M - \vec{u}_e) \tag{66}$$

is the current density.

For the electrons, the quasi-neutrality assumption gives,

$$ne = \sum_\beta n\rho, \tag{67}$$

as the electron density. Using the definition of current density presented in Eq. (66) we can obtain the electron velocity. The standard fluid equation,

$$\frac{\partial P_e}{\partial t} = -\gamma \nabla (P_e \vec{u}_e) + (\gamma - 1)\vec{u}_e \nabla \cdot P_e, \tag{68}$$

is used to solve for the electron pressure. As this formulation illustrates it is not mathematically possible to cast the multifluid equations in a purely conservative formulation. Numerical techniques used for single fluid have to be adjusted to deal with this situation (Toth et al. 2012 discuss these issues in more detail). It is also worth noting that the energy equation is only true for the hydrodynamic energy density and not the total energy density. In this system to lowest order all the species move in the perpendicular directions with the $E \times B$ velocity. As the magnetic field changes momentum can be transferred between the species in the plasma.

4.2 Incorporation of Internal Plasma Sources in Global MHD Models

In addition to the solar wind plasma, there are various other sources of plasma present in planetary magnetospheres. Plasma sources internal to a planetary magnetosphere may come from the atmosphere/ionosphere, such as the ionospheric outflows at Earth (Chappell 2015;

Welling et al. 2015, this issue) and the planetary ions produced from the exosphere at Mercury (Raines et al. 2015, this issue). In addition, plasma sources may originate from planetary moons and this is especially the case for the gas giants, Jupiter (Bolton et al. 2015, this issue) and Saturn (Blanc et al. 2015, this issue). Through processes like surface warming, active plumes or surface sputtering by magnetospheric particles, moons of the giant planets may possess significant sources of neutrals. The neutrals originating from the moons can become charged particles through various mass-loading processes, thereby supplying plasma to their parent magnetospheres. It is now well known that Io and Enceladus are the major plasma sources of the magnetospheres of Jupiter and Saturn, respectively. The presence of the internal plasma sources to some degree modifies the plasma distribution and composition within the magnetosphere, and in some cases can significantly affect the configuration and dynamics of the magnetosphere. It is, therefore, important to include the internal plasma sources in modeling the structure and dynamics of planetary magnetospheres. Here we provide an overview of the various approaches adopted to incorporate internal plasma sources in global MHD models.

4.2.1 Impact of Ionospheric Outflows

The Alfvén speed in the high-latitude, low-altitude region above the ionosphere is usually very high. Therefore, including this part of the magnetosphere in global magnetosphere simulations imposes severe constraints on the allowable time step that can be used in numerically solving the MHD equations. As a result, presently most global magnetosphere models exclude this region ("gap region") by placing their simulation inner boundaries at altitudes between a couple of and several planetary radii. The ionosphere is conventionally modeled in a separate module as a two-dimensional spherical surface where the electric potential (thus the electric field) is solved for a given distribution of height-integrated conductivity and field-aligned currents (FACs). The FACs are obtained directly from the MHD model of the magnetosphere by first calculating the currents at or near the simulation inner boundary and then mapping them along the dipole field line down to the ionosphere. The electric field obtained from the ionosphere solver is mapped back along the field lines to the magnetosphere boundary, where the $E \times B$ drift velocity is calculated and used to set the boundary condition for plasma velocity. Given the way in which the coupling between the magnetosphere and the ionosphere is treated in present global magnetosphere models, physical processes responsible for producing the ionospheric outflows usually are not directly included in those models. In such cases, the introduction of ionospheric plasma into magnetosphere simulations typically is enabled through prescription of boundary conditions at the low-altitude boundary of the magnetosphere model, similar to the way in which the solar wind plasma is injected into the simulation domain at the sunward boundary. It is worth noting that this type of treatment does not require significant modifications to the MHD equations and is, therefore, relatively convenient in terms of numerical implementation.

Several different approaches have been adopted for adding ionospheric outflows in global MHD models. A relatively simple method is to set the plasma density to relatively high values at the inner boundary and fix it throughout a simulation run. For example, the multi-fluid MHD model by Winglee et al. (2002) specified constant densities for the light (H^+) and heavy ionospheric species (O^+) at their simulation inner boundary. Pressure gradients and/or other effects (e.g., centrifugal acceleration and numerical diffusion) may drive the ionospheric plasma to flow from the low-altitude boundary into the magnetosphere domain. As such, the ionospheric plasma is added in the simulation in a passive manner in that the outflow parameters are not explicitly set and controlled.

In contrast to the passive method described above, some global models used methods in which the outflow parameters, such as the source location, outflow density and velocity, are explicitly specified at the low-altitude boundary of the magnetosphere model. Several global modeling studies (e.g., Wiltberger et al. 2010; Garcia et al. 2010; Yu and Ridley 2013) performed controlled global simulations to examine the effects of the outflow source location and intensity on the global magnetospheric configuration and dynamics. In these studies, ion outflows were introduced in localized regions, such as the dayside cusp or the nightside auroral zone, and the outflow rates were specified by setting the plasma density and parallel velocity in the boundary conditions.

The choice of outflow parameters may also be made based on empirical outflow models. For example, Brambles et al. (2010) incorporated in the LFM global simulation a driven outflow model based on the empirical model by Strangeway et al. (2005), which was built upon the FAST satellite observations. The empirical model provides a scaling relation between the average outflow flux and the average earthward-flowing Poynting flux, which is calculated directly from the MHD model near the inner boundary. This approach in effect enables a two-way coupling between the magnetosphere and the ionosphere, because the outflow source location and intensity may vary in time depending on the magnetospheric conditions.

More self-consistent implementation of ionospheric outflows may be achieved by coupling a global MHD model with a physics-based ionospheric outflow model. Glocer et al. (2009) coupled the Polar Wind Outflow Model (PWOM) into the SWMF to study the effects of polar wind type outflows on the coupled magnetosphere-ionosphere system. PWOM includes important physical processes responsible for the transport and acceleration of the ionospheric gap region between the magnetosphere and ionosphere. It takes inputs from both the magnetosphere model (FACs and plasma convection pattern) and the upper atmosphere model (neutral densities and neutral winds) to calculate the upwelling and outflowing of ionospheric plasma. In return, the outflow fluxes obtained at the top boundary of the PWOM model are used to set the inner boundary conditions of the magnetosphere model.

4.2.2 Plasma Sources Associated with Planetary Satellites

Different from the Earth's magnetosphere where the magnetospheric plasma comes either from the solar wind or the ionosphere, the bulk of the magnetospheric plasma in the giant planet magnetospheres originate predominantly from planetary satellites. At Jupiter, the major plasma source is the volcanic moon, Io, which supplies plasmas to the Jovian magnetosphere at a rate of 260–1400 kg/s (Bagenal and Delamere 2011). At Saturn, the dominant source of magnetospheric plasma is the icy moon, Enceladus, which produces predominantly water-group ions to the magnetosphere at a rate of 12–250 kg/s (Bagenal and Delamere 2011). At both planets, the presence of internal plasma sources plays a crucial role in shaping the magnetosphere. It is, therefore, essential to include the internal plasma sources associated with the moons in global models of the giant planet magnetospheres.

There are, in general, two types of approaches used for incorporating plasma sources associated with moons. One relies on prescription of boundary conditions, similar to the approach outlined above for incorporating ionospheric outflows into Earth's magnetosphere models. For example, the global MHD model by Ogino et al. (1998) which was first applied to Jupiter and later adapted to Saturn (Fukazawa et al. 2007a, 2007b), does not explicitly include in the simulation domain plasma sources associated with moons. Rather, the model included the internal plasma sources by fixing plasma density and pressure in time at the inner boundary, which was placed outside of the main regions in which moon-associated plasmas are added to the systems. Similarly, in the multi-fluid MHD model applied to Saturn's magnetosphere, Kidder et al. (2009) held the densities of various plasma fluids fixed near their simulation inner boundary to mimic the addition of new plasma from Enceladus.

The other approach used in the modeling of the giant planets' magnetospheres incorporates internal plasma sources associated with moons in an explicit manner. The neutral gases emanating from the moons in the Jovian and Saturnian magnetospheres are distributed in a broad region forming plasma and neutral tori, which mass-load newly created charged particles which then modify the plasma flow in the system via electromagnetic forces (see a review by Szegö et al. 2000). This occurs not only near the vicinities of the moons, but also over extended regions of space. It is desirable to self-consistently take into account this effect in a global magnetosphere model. This can be done by incorporating appropriate source and loss terms into the MHD equations described above. One can derive the mass-loading source terms for MHD using first-principles from the Boltzman equation (Cravens 1997; Gombosi 1998). Terms describing the change of the plasma phase-space distribution due to collisional processes, including ionization, charge-exchange, recombination, and elastic collisions, can be included in the Boltzman equation. Appropriate velocity moments can then be taken to obtain the source terms associated with various mass-loading processes for the continuity, momentum and energy equations of MHD. One advantage of this method over the boundary condition method is that it describes in a self-consistent way the change of mass, momentum and energy of magnetospheric plasma due to mass-loading. This approach has been used in global models of the giant planets' magnetospheres, such as the SWMF applications to Saturn's magnetosphere by Hansen et al. (2005), Jia et al. (2012a, 2012b), and Jia and Kivelson (2012) and the global MHD model of Jupiter's magnetosphere by Chané et al. (2013).

4.3 Hybrid Models

The most common hybrid approach used in simulating space plasmas treats the ions kinetically and the electrons as a massless charge neutralizing fluid. In the hybrid regime, the density, temperatures and magnetic field is such that the ions are essentially collisionless. On the other hand the electrons have relatively small gyroradii and may undergo an order of magnitude or more collisions. Thus the electrons are described as a massless collision-dominated thermal fluid. There are finite electron mass hybrid schemes in existence, which will not be discussed here. Hybrid schemes have been around for many years thus the interested reader should see the reviews by Brecht and Thomas (1988), Lipatov (2002), Winske et al. (2003), and the references therein for historical perspectives. The most recent review is that of Ledvina et al. (2008), where the following brief description is taken from.

The hybrid approach starts with the following assumptions.

(i) Quasi-neutrality is assumed,

$$n_e = \sum_i n_i \tag{69}$$

Thus the displacement current is ignored in Ampere's law (Eq. (74)). This assumption is valid on scales larger than the Debye length. The assumption breaks down when the grid resolution is finer than the Debye length. This also implies that $\nabla \cdot \boldsymbol{J} = 0$, and removes most electrostatic instabilities.

(ii) The Darwin approximation is assumed.

This approximation splits the electric field into a longitudinal part $\boldsymbol{E}_L$ and a solenoidal part $\boldsymbol{E}_T$. Then $\nabla \times \boldsymbol{E}_L = 0$ and $\nabla \cdot \boldsymbol{E}_T = 0$ and $\partial \boldsymbol{E}_T / \partial t$ is neglected in Ampere's law (Eq. (74)). This allows the light waves to be ignored. It also removes relativistic phenomena.

(iii) The mass of the electrons is taken to be zero.
(iv) The electrons collectively act as a fluid.

Thus the electron plasma and gyrofrequencies are removed from the calculations. This means that high frequency modes are not present, such as the electron whistler. By using these last two assumptions there is no longer a physical mechanism to describe the system behavior at small scales. The Debye length and the magnetic skin depth are not meaningful in this scheme. This sets the limit on the cell size that should be used to at least an order of magnitude larger than the electron skin depth c/ω_{pe}. It is possible to use cell sizes less than the ion skin depth but the results are meaningless. The chosen cell size should resolve the ion kinetic effects (e.g. gyroradius and ion skin depth). If the cell size is much larger than the kinetic scales all that is accomplished is the creation of the world's most expensive MHD simulation.

With these assumptions the hybrid scheme solves the following ion momentum and position equations for each particle:

$$\frac{dv}{dt} = \frac{q}{m_i}[\boldsymbol{E} + v \times \boldsymbol{B}] - \eta \boldsymbol{J}_{total} \tag{70}$$

$$\frac{d\boldsymbol{x}}{dt} = \boldsymbol{v} \tag{71}$$

where $\boldsymbol{J}$ is the total current density and η is the plasma resistivity. The electron momentum equation can be written as:

$$\boldsymbol{E} = \frac{1}{n_i e}\left[(\nabla \times \boldsymbol{B}) \times \boldsymbol{B} - \boldsymbol{J}_i \times \boldsymbol{B} - \nabla(n_e T_e) + \eta \boldsymbol{J}_{total}\right] \tag{72}$$

With the electron temperature given by:

$$\frac{\partial T_e}{\partial t} + \boldsymbol{u}_e \cdot \nabla T_e + \frac{3}{2} T_e \cdot \boldsymbol{u}_e = \frac{2}{3n_e} \eta \boldsymbol{J}_{total}^2 \tag{73}$$

Here T_e is the electron temperature and $\boldsymbol{u}_e$ is the electron velocity. Note that (73) does not include the effects of thermal conduction, but that can be added if appropriate. Ampere's law becomes:

$$\nabla \times \boldsymbol{H} = \boldsymbol{J}_i + \boldsymbol{J}_e \tag{74}$$

where $\boldsymbol{J}_i$ and $\boldsymbol{J}_e$ are the ion and electron current densities. The magnetic field is obtained from Faraday's law, given below:

$$\nabla \times \boldsymbol{E} + \frac{\partial \boldsymbol{B}}{\partial t} = 0 \tag{75}$$

The electric field contains contributions from the electron pressure gradient, resistive effects and Hall currents. The scheme correctly simulates electromagnetic plasma modes up to and including the lower portion of the whistler wave spectrum (well below the electron cyclotron frequency, $\omega \ll \omega_{ce}$). Shock formation physics is included, therefore no assumptions or shock capturing techniques are needed to capture a shock. The time step is determined by the ion cyclotron frequency. This comes at the price of the loss of electron particle effects and charge separation. Some small-scale electrostatic effects can be included through the resistivity terms. The resistivity terms can also be used to stabilize the numerical scheme used to solve the equations by adding it in as a small amount of artificial resistivity.

4.4 Magnetosphere-Ionosphere Coupling

The ionosphere-magnetosphere coupling is not a process in itself. It is rather a chain of processes that act as a control loop between the dynamics of the ionospheric and of the magnetospheric plasmas connected by conductive magnetic field lines as shown in Fig. 8. A modification of the transport in one region has consequences on the transport in the conjugate region and that affects in turn the initial transport in the first region. For example, the convection in the magnetosphere results in convection in the ionosphere (see Fig. 9). The plasma dynamics in one region is constrained by the dynamics in the other. For each region, the ionosphere-magnetosphere coupling could be assimilated to some kind of interactive boundary conditions (representing the interaction with the conjugate region) that need to be solved self-consistently with the dynamics of the region considered.

In a first approach, the ionospheric plasma exhibits local-time, latitudinal, seasonal variations but forms a continuous conductive shell embedded in the high-altitude planetary atmosphere. It lies at the footprints of conductive planetary magnetic field lines that connect it to different magnetospheric regions. The polar cap magnetic field lines are open with one footprint in the polar ionosphere and the other end extended to large distances downtail, in the so-called lobes. The lobe plasma is believed to be diluted and therefore does not develop significant couplings with the ionosphere. Near the equator, the magnetic field lines remain fully embedded in the topside ionosphere and do not reach the magnetosphere. Between the polar cap and the equatorial strip, the magnetic field lines are closed with both footprints in the ionosphere and their apex reach the magnetosphere. Near the planet, a region called "plasmasphere" filled with cold plasma of ionospheric origin in corotation with the planet may exist, as well as radiation belts with very energetic particles trapped on closed orbits around the planet. The so-called "plasmasheet" represents the main plasma reservoir in the magnetospheres of Earth, Jupiter and Saturn. The transport mechanisms differ for each planet: they involve corotation, outward diffusion from inner plasma sources or earthward convection of plasma ultimately extracted from external sources (solar wind), but all result in the formation of a dense and hot plasma sheet, confined near the equatorial plane and extending up to large distances down tail. The conductive magnetic field lines allow electric field transmission, current circulation and particle exchanges. The effects of these magnetic-field-aligned processes are enhanced when they involve dense and dynamical regions such as the ionosphere and the plasmasheet, resulting in significant consequences on the dynamics of both regions at large scales as well as at local or transient scales.

The coupled ionosphere-magnetosphere system can be described by a feedback loop derived from various investigations in the terrestrial environment (Vasyliunas 1970; Wolf 1979; Harel et al. 1981; Fontaine et al. 1985; Peymirat and Fontaine 1994) as illustrated in Fig. 13, where the magnetospheric plasma is indicated in the top row. External sources such as the planetary rotation or the solar wind-magnetosphere dynamo contribute to produce large-scale electric fields in the magnetosphere, which combine with the magnetospheric magnetic field to drag this magnetospheric plasma into a large-scale motion. Smaller-scale processes, instabilities, phase space diffusion processes, etc. add smaller-scale motions and contribute to the global and local plasma distribution and current circulation in the magnetosphere.

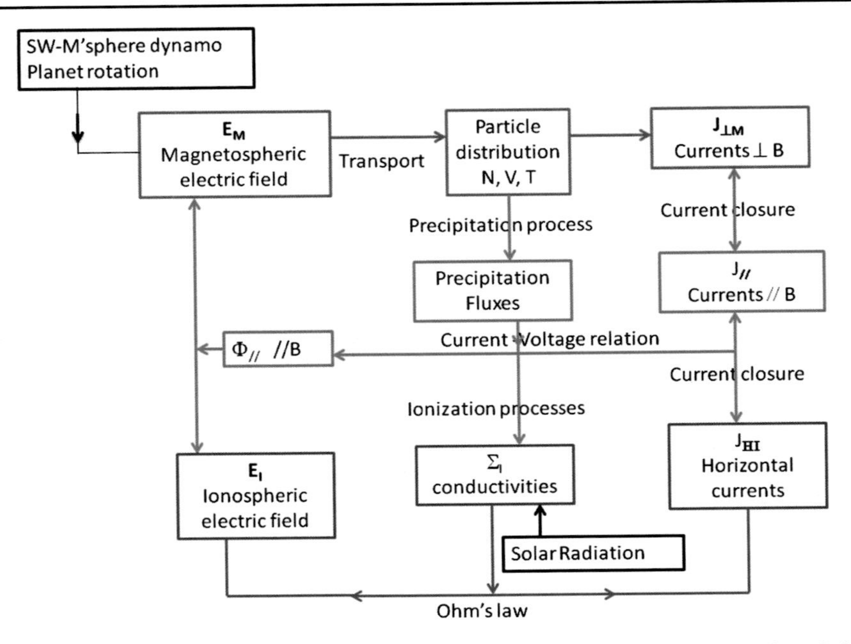

Fig. 16 Feedback loop illustrating the ionosphere-magnetosphere coupling with magnetospheric (*red*), ionospheric (*blue*) and field-aligned (*green*) electrodynamic parameters and processes. *Black rectangles* represent external sources

Field-aligned processes are shown in the second row of Fig. 16. On one hand, the current closure $\nabla \cdot \boldsymbol{j}_M = 0$ in the magnetosphere, where $\boldsymbol{j}_M$ is the magnetospheric current density, implies a current circulation along magnetic field lines $\boldsymbol{j}_\parallel$ down to the ionosphere. On the other hand, particles with pitch-angles smaller than the atmospheric loss cone reach the ionosphere at the footprint of magnetic field lines: they contribute to the field-aligned currents. The mirror effect due to the magnetic field line convergence limits the particle fluxes that reach the ionosphere and thus the field-aligned current density transmitted to the ionosphere. Current-voltage relations, such as those proposed by Knight (1973) (see Eqs. (39) and (40)), predict that parallel potentials can develop and increase the field-aligned current density when the available precipitating fluxes cannot match the current density required by the current closure in the magnetosphere. Ionospheric particles can also escape from the ionosphere, in particular electrons which are very mobile along magnetic field lines. They carry return currents due to a favorable effect of the mirror force from the ionosphere toward the magnetosphere. It is generally difficult to measure particle outflows of ionospheric origin due to their low energy, except if they are accelerated (see Chappell 2015).

The UV and EUV solar radiation contribute to create an ionospheric layer in the high-altitude atmosphere. The dynamics of the ionosphere is governed by the ionospheric Ohm's law:

$$\boldsymbol{j}_I = \boldsymbol{\sigma}(\boldsymbol{E} + \boldsymbol{V}_n \times \boldsymbol{B}) \tag{76}$$

and the ionospheric current closure equation:

$$\nabla \cdot \boldsymbol{j}_I = 0 \tag{77}$$

where $\boldsymbol{j}_I$ is the ionospheric current density, $\boldsymbol{\sigma}$ the ionospheric conductivity tensor, $\boldsymbol{E}_I$ the electric field at ionospheric altitudes, $\boldsymbol{V}_n$ the velocity of the neutral wind, $\boldsymbol{B}$ the magnetic field. In addition of this solar source, the fluxes of energetic magnetospheric precipitating particles into the ionosphere contribute to produce the well-known auroral light emissions and also ionization. The resulting conductivity enhancements and the presence of field-aligned currents modify the distribution of perpendicular electric currents and electric fields at the ionospheric level (bottom row of Fig. 16). This modification is finally transmitted to the magnetosphere via magnetic field lines by taking into account the eventual presence of parallel electric fields. This new electric field distribution modifies in turn the plasma transport in the magnetosphere, which closes the feedback loop.

Finally, any modification/event at large or smaller scales that occurs in one region is transmitted to the other one where it modifies its own dynamics. However, the possibilities of exchanges of particles, momentum, and energy are limited by the plasma configuration in each region. A mismatch between both regions can be overcome by the set up of field-aligned electric fields and currents, in the limit of energy density available in each region. These effects result in parallel particle acceleration and thus in light emissions when accelerated particles precipitate into the ionosphere/upper atmosphere.

4.4.1 *Time-Varying Coupling*

The above description does not only apply to quasi-steady ionosphere-magnetosphere coupling, but works similarly at smaller-scales (see Lysak 1990 for a review). For example, time-varying fluctuations in the magnetosphere or wave-particle interactions occurring during plasma transport may generate Alfvén waves that carry field-aligned currents. These currents close similarly through the ionosphere. They result in fluctuating effects in the ionosphere that will affect auroras, conductivities, electric fields and currents. Fluctuating conditions in the ionosphere are in turn transmitted to the magnetosphere via magnetic field lines and produce fluctuating feedback effects. The superposition of initial and feedback fluctuations can stabilize or destabilize the plasma; it can also give rise to periodic effects as pulsations, formation of multiple arcs, etc.

Small-scale processes such as magnetic reconnection imply a connectivity interruption and reconfiguration for a subset of magnetic field lines in a localized region. On the reconnection time scale, field-aligned processes cannot exist because of connectivity changes and the ionosphere and magnetosphere dynamics are disconnected. This is not the case for the time periods just before and after reconnection: important effects occur in both regions, resulting in enhanced field-aligned couplings, i.e. large field-aligned particle fluxes, electric fields and currents.

4.4.2 *Planet-Moon Interactions*

The interaction of magnetized planets with moons is another example of local feedback processes. It depends on the electrical properties of the moons, or rather of the obstacle, and on the flow characteristics (for a review, see for example Kivelson et al. 2004). The obstacle can be the magnetic field, the atmosphere and ionosphere or the body itself depending of the radial variation of the energy density. The magnetospheric flow is coupled to the planetary ionosphere via magnetic field lines and this coupling drags the magnetospheric plasma at a speed which may differ from the moons' orbital velocity. If the flow velocity in the rest frame of the moon were super-Alfvénic, it would produce a shock wave ahead of the obstacle as in the solar wind/magnetosphere interaction. Inside magnetospheres, the interaction velocity is usually sub-Alfvénic.

In the case of an insulating body, the sub-Alfvénic magnetospheric flow is absorbed by the surface, an initially empty wake appears downstream and the magnetic field exhibits only weak perturbations. Ions can be created from various interaction processes between the magnetospheric particles and the moon, and this so-called ion pickup source contributes to the mass-loading of the magnetospheric flow.

In the case of a conducting body, the sub-Alfvénic magnetospheric flow slows upstream of the body, the planetary magnetic field lines get bent and shear Alfvén waves are launched. These waves carry field-aligned currents and they generate perturbations in the field which are known as Alfvén wings. Alfvén wings form an angle Θ_A with the initial magnetospheric magnetic field:

$$\theta_A = \tan^{-1}\left(\frac{V_M}{V_A}\right) \tag{78}$$

where V_M is the velocity of the magnetospheric flow, and V_A the Alfvén velocity.

They extend down to the planetary ionosphere which allows the current closure. This localized ionosphere-magnetosphere coupling contributes to divert the magnetospheric plasma flow around the conducting body and all along the Alfvén wings. It modifies locally the properties in the magnetically conjugated ionosphere. For example, light emissions in the ionosphere at the magnetic footprints of the Galilean moons in the Jovian magnetosphere represent the signature of this localized ionosphere-magnetosphere coupling.

In the case of a magnetized body, the moon's magnetic field creates a small magnetosphere inside the planetary magnetosphere. Up to now, Ganymede is the only known magnetized moon in the solar system. Although very small, Ganymede's magnetosphere contains features similar to terrestrial and planetary magnetospheres (e.g., Kivelson et al. 1998), as the presence of a magnetopause, innermost regions protected by the internal magnetic field, and auroras (Jia et al. 2009). One of the differences is that polar magnetic field lines from Ganymede's polar region connect the Jovian ionosphere at their other end. They carry field-aligned currents and contribute to a local coupling between the planetary ionosphere and the moon's magnetosphere embedded in the magnetospheric flow.

5 Summary

In this paper, the basic and common processes, related to plasma supply to each region of the planetary magnetospheres in our solar system, were reviewed. In addition to major processes related to the source, transport, energization, loss of the magnetospheric plasmas, basic equations and modeling methods, with a focus on plasma supply processes for planetary magnetospheres, are also reviewed. The topics reviewed in this paper can be summarized as follows: Source Processes related to the surface (Sect. 1.1), ionosphere (Sect. 1.2), and solar wind (Sect. 1.3). Section 2 is dedicated to processes related to the transport and energization of plasma such as Axford/Hines cycle (Sect. 2.1), Dungey cycle (Sect. 2.2), rotational driven transport and Vasyliunas cycle (Sect. 2.3), field-aligned potential drop (Sect. 2.4), non-adiabatic acceleration (Sect. 2.5), and pick-up acceleration and mass loading (Sect. 2.6). In Sect. 3, loss processes related to the tail reconnection and plasmoids (Sect. 3.1), charge exchange (Sect. 3.2), and precipitations into planets (Sect. 3.3) are reviewed. Section 4 contains an overview of basic equations and modeling methods, which includes MHD simulation

(Sect. 4.1), incorporation of internal plasma sources in global MHD models (Sect. 4.2), hybrid models (Sect. 4.3), and magnetosphere-ionosphere coupling (Sect. 4.4). The review provides the basic knowledge to understand various phenomena in planetary magnetospheres described in the following chapters.

Acknowledgements This work was partially supported by Grant-In-Aid for Scientific Research (B) 24340118 from JSPS and MEXT of Japan. KS also thanks for supports from the GEMSIS project at STEL, Nagoya University. CMJ's work at Southampton was supported by a Royal Astronomical Society Fellowship and a Science and Technology Facilities Council Ernest Rutherford Fellowship. FC is funded by JPL contract 1467206. MW is supported by NASA grants NNH12AU10I and NNH11AR82I. The National Center for Atmospheric Research is sponsored by the National Science Foundation. XJ is supported by NASA grants NNX12AK34G and NNX12AM74G, and by NSF grant AGS 1203232.

Ethical Statement This manuscript is prepared to submit to SSR as a review article after discussion at the ISSI workshop in 2013 and never submitted to elsewhere. Contents of this manuscript have nothing to do with the following issues:

- Disclosure of potential conflicts of interest,
- Research involving Human Participants and/or Animals,
- Informed consent.

References

T. Abe et al., Observations of polar wind and thermal ion outflow by Akebono/SMS. J. Geomagn. Geoelectr. **48**, 319 (1996)

H. Andersen, H.L. Bay, in *Sputtering by Particle Bombardment I*, ed. by R. Behrisch (Springer, Berlin, 1981), Chap. 4

M. Andre, A.L. Yau, Theories and observations of ion energization and outflow in the high latitude magnetosphere. Space Sci. Rev. **80**, 27 (1997)

V.G. Anicich, Evaluated bimolecular ion-molecule gas phase kinetics of positive ions for use in modeling planetary atmospheres, cometary comae, and interstellar clouds. J. Phys. Chem. Ref. Data **22**, 1469–1569 (1993)

M. Ashour-Abdalla et al., The formation of the wall region—Consequences in the near-Earth magnetotail. Geophys. Res. Lett. **19**, 1739 (1992)

M. Ashour-Abdalla et al., Consequences of magnetotail ion dynamics. J. Geophys. Res. **99**, 14891 (1994)

W.I. Axford, Viscous interaction between the solar wind and the Earth's magnetosphere. Planet. Space Sci. **12**, 45 (1964)

W.I. Axford, C.O. Hines, A unifying theory of high-latitude geophysical phenomena and geomagnetic storms. Can. J. Phys. **39**, 1433 (1961)

S.V. Badman, S.W.H. Cowley, Significance of Dungey-cycle flows in Jupiter's and Saturn's magnetospheres, and their identification on closed equatorial field lines. Ann. Geophys. **25**, 94 (2007)

F. Bagenal, The magnetosphere of Jupiter: Coupling the equator to the poles. J. Atmos. Sol.-Terr. Phys. **69**, 387 (2007). doi:10.1016/j.jastp.2006.08.012

F. Bagenal, P.A. Delamere, Flow of mass and energy in the magnetospheres of Jupiter and Saturn. J. Geophys. Res. **116**, A05209 (2011). doi:10.1029/2010JA016294

F. Bagenal, J.D. Sullivan, Direct plasma measurements in the Io torus and inner magnetosphere of Jupiter. J. Geophys. Res. **86**, 8447 (1981)

J. Bailey, M. Gruntman, Experimental study of exospheric hydrogen atom distributions by Lyman-alpha detectors on the TWINS mission. J. Geophys. Res. (Space Phys.) **116**(A15), 302 (2011). doi:10.1029/2011JA016531

P.M. Banks, T.E. Holzer, High-latitude plasma transport: The polar wind. J. Geophys. Res. **74**, 6317 (1969). doi:10.1029/JA074i026p06317

R.A. Baragiola et al., Nucl. Instrum. Methods Phys. Res. B **209**, 294 (2003)

L.F. Bargatze, D.N. Baker, R.L. McPherron, E.W. Hones, Magnetospheric impulse response for many levels of geomagnetic activity. J. Geophys. Res. **90**, 6387 (1985). doi:10.1029/JA090iA07p06387

R. Behrisch, W. Eckstein, *Sputtering by Particle Bombardment: Experiments and Computer Calculations from Threshold to MeV Energies* (Springer, Berlin, 2007)

A. Bhardwaj, G.R. Gladstone, Auroral emissions of the giant planets. Rev. Geophys. **38**, 295 (2000)

J.P. Biersack, W. Eckstein, Sputtering of solids with the Monte Carlo program TRIM.SP. Appl. Phys. A **34**, 73 (1984)
G.A. Bird, *Molecular Gas Dynamics and the Direct Simulation of Gas Flows* (Clarendon Press, Oxford, 1994)
K. Birkeland, Sur la déviabilité magnétique des rayons corpusculaires provenant du Soleil. C. R. Acad. Sci. **150**, 246 (1910)
J. Bishop, Multiple charge exchange and ionization collisions within the ring current-geocorona-plasmasphere system: generation of a secondary ring current on inner L shells. J. Geophys. Res. **101**, 17,325 (1996). doi:10.1029/95JA03468
M. Blanc et al., Space Sci. Rev. (2015, this issue)
S.J. Bolton et al., Space Sci. Rev. (2015, this issue)
S.J. Bolton, Jupiter's inner radiation belts, in *Jupiter: The Planet, Satellites, and Magnetosphere*, ed. by F. Bagenal et al. (Cambridge Univ. Press, Cambridge, 2004), p. 671, Chap. 27
S.J. Bolton, R.M. Thorne, S. Bourdarie, I. de Pater, B. Mauk, Jupiter's inner radiation belts, in *Jupiter: The Planet, Satellites, and Magnetosphere*, ed. by F. Bagenal et al.(Cambridge Univ. Press, Cambridge, 2004), p. 671. Chapter 27
B. Bonfond, When Moons create Aurora: the satellite footprints on giant planets, in *Auroral Phenomenology and Magnetospheric Processes: Earth and Other Planets*. AGU Geophysical Monograph Series, vol. 197 (2012), p. 133
J.W. Boring et al., Sputtering of solid SO_2. Nucl. Instrum. Methods B **1**, 321 (1984)
J.E. Borovsky, M.F. Thomnsen, R.C. Elphic, The driving of the plasma sheet by the solar wind. J. Geophys. Res. **103**(A8), 17,617 (1998). doi:10.1029/97JA02986
J.E. Borovsky et al., What determines the reconnection rate at the dayside magnetosphere? J. Geophys. Res. **113**, A07210 (2008). doi:10.1029/2007JA012645
O.J. Brambles et al., Effects of causally driven cusp O^+ outflow on the storm time magnetosphere-ionosphere system using a multifluid global simulation. J. Geophys. Res. **115**, A00J04 (2010). doi:10.1029/2010JA015469
S.H. Brecht, V.A. Thomas, Multidimensional simulations using hybrid particle codes. Comput. Phys. Commun. **48**, 135–143 (1988)
N.M. Brice, G.A. Ioannidis, The magnetospheres of Jupiter and Earth. Icarus **13**, 173 (1970)
W.L. Brown, R.E. Johnson, Sputtering of ices: a review. Nucl. Instrum. Methods B **13**, 295 (1986)
W.L. Brown, W.M. Augustyniak, K.J. Marcantonio, E.N. Simmons, J.W. Boring, R.E. Johnson, C.T. Reimann, Electronic sputtering of low temperature molecular solids. Nucl. Instrum. Methods Phys. Res. B **1**, 307 (1984)
D.G. Brown, J.L. Horwitz, G.R. Wilson, Synergistic effects of hot plasma-driven potentials and wave-driven ion heating on auroral ionospheric plasma transport. J. Geophys. Res. **100**, 17,499 (1995)
M. Bruno, G. Cremonese, S. Marchi, Neutral sodium atoms release from the surfaces of the Moon and Mercury induced by meteoroid impacts. Planet. Space Sci. **55**, 1494 (2007)
J. Büchner, L.M. Zelenyi, Regular and chaotic charged particle motion in magnetotaillike field reversals: 1. Basic theory of trapped motion. J. Geophys. Res. **94**, 11,821 (1989)
J.L. Burch et al., Properties of local plasma injections in Saturn's magnetosphere. Geophys. Res. Lett. **32**, L14S02 (2005). doi:10.1029/2005GRL022611
L.F. Burlaga, Magnetic fields and plasmas in the inner heliosphere: Helios results. Planet. Space Sci. **49**, 1619 (2001)
R.K. Burton, R.L. McPherron, C.T. Russell, Terrestrial magnetosphere—half-wave rectifier of interplanetary electric-field. Science **189**(4204), 717 (1975a)
R.K. Burton, R.L. McPherron, C.T. Russell, An empirical relationship between interplanetary conditions and Dst. J. Geophys. Res. **80**, 4204 (1975b)
M.N. Caan, R.L. McPherron, C.T. Russell, Solar wind and substorm-related changes in the lobes of the geomagnetic tail. J. Geophys. Res. **78**, 8087 (1973)
C.W. Carlson, R.F. Pfaff, J.G. Watzin, Fast Auroral Snapshot (FAST) mission. Geophys. Res. Lett. **25**, 2013 (1998)
T.A. Cassidy, R.E. Johnson, Monte Carlo model of sputtering and other ejection processes within a regolith. Icarus **176**, 499 (2005)
T.A. Cassidy et al., The spatial morphology of Europa's near-surface O_2 atmosphere. Icarus **191**, 755 (2007)
T.A. Cassidy et al., Radiolysis and photolysis of icy satellite surfaces: experiments and theory. Space Sci. Rev. **153**(1–4), 299 (2010)
T.A. Cassidy et al., Magnetospheric ion sputtering and water ice grain size at Europa. Planet. Space Sci. **77**, 64 (2013)
B.L. Cecconi et al., Goniopolarimetric study of the Rev 29 perikrone using the Cassini/RPWS/HFR radio receiver. J. Geophys. Res. **114**, A03215 (2009)

J.W. Chamberlain, Planetary coronae and atmospheric evaporation. Planet. Space Sci. **11**, 901 (1963). doi:10.1016/0032-0633(63)90122-3

E. Chané, J. Saur, S. Poedts, Modeling Jupiter's magnetosphere: Influence of the internal sources. J. Geophys. Res. **118**, 2157 (2013). doi:10.1002/jgra.50258

C.R. Chappell, The role of the ionosphere in providing plasma to the terrestrial magnetosphere—an historical overview. Space Sci. Rev. (2015, in press)

F.F. Chen, Introduction to Plasma Physics and Controlled Fusion. Boom Koninklijke Uitgevers (1984)

J. Chen, P.J. Palmadesso, Chaos and nonlinear dynamics of single-particle orbits in magnetotaillike magnetic field. J. Geophys. Res. **91**, 1499 (1986)

Y.T. Chiu, M. Schulz, Slf-consistent particle and parallel electrostatic electric field distributions in the magnetospheric-ionospheric auroral region. J. Geophys. Res. **83**, 629 (1978)

M.J. Cintala, Impact induced thermal effects in the lunar and Mercurian regoliths. J. Geophys. Res. **97**, 947 (1992)

J.B. Cladis, Parallel acceleration and transport of ions from polar ionosphere to plasma sheet. Geophys. Res. Lett. **13**, 893 (1986)

J.T. Clarke, Auroral processes on Jupiter and Saturn, in *Auroral Phenomenology and Magnetospheric Processes: Earth and Other Planets*. AGU Geophysical Monograph Series, vol. 197 (2012), p. 113

J.E.P. Connerney, J.H. Waite, New model of Saturn's ionosphere with an influx of water. Nature **312**, 136 (1984)

J.E.P. Connerney et al., Images of excited H_3^+ at the foot of the Io flux tube in Jupiter's atmosphere. Science **262**, 1035–1038 (1993)

J.F. Cooper et al., Energetic ion and electron irradiation of the icy Galilean satellites. Icarus **149**, 133 (2001)

S.W.H. Cowley, The causes of convection in the Earth's magnetosphere: A review of developments during the IMS. Rev. Geophys. **20**(3), 531–565 (1982). doi:10.1029/RG020i003p00531

S.W.H. Cowley, E.J. Bunce, R. Prangé, Saturn's polar ionospheric flows and their relation to the main auroral oval. Ann. Geophys. **22**, 1379 (2004)

T.E. Cravens, *Physics of Solar System Plasmas* (Camb. Univ. Press, Cambridge, 1997). doi:10.1017/CBO9780511529467

J. De Keyser, R. Maggiolo, M. Echim, Monopolar and bipolar auroral electric fields and their effects. Ann. Geophys. **28**, 2027 (2010)

A.E. De Vries et al., Synthesis and sputtering of newly formed molecules by kiloelectronvolt ions. J. Phys. Chem. **88**, 4510 (1984)

P.A. Delamere, F. Bagenal, Solar wind interaction with Jupiter 's magnetosphere. J. Geophys. Res. **115**, A10201 (2010). doi:10.1029/2010JA015347

P.A. Delamere et al., Magnetic signatures of Kelvin-Helmholtz vortices on Saturn's magnetopause: Global survey. J. Geophys. Res. **118**, 393 (2013)

D.C. Delcourt, Particle acceleration by inductive electric fields in the inner magnetosphere. J. Atmos. Sol.-Terr. Phys. **64**, 551 (2002)

D.C. Delcourt, J.-A. Sauvaud, Populating of the cusp and boundary layers by energetic (hundreds of keV) equatorial particles. J. Geophys. Res. **104**, 22,635 (1999)

D.C. Delcourt, J.-A. Sauvaud, A. Pedersen, Dynamics of single-particle orbits during substorm expansion phase. J. Geophys. Res. **95**, 20,853 (1990)

D.C. Delcourt, R.F. Martin Jr., F. Alem, A simple model of magnetic moment scattering in a field reversal. Geophys. Res. Lett. **21**, 1543 (1994)

D.C. Delcourt et al., On the nonadiabatic precipitation of ions from the near-Earth plasma sheet. J. Geophys. Res. **101**, 17,409 (1996)

J. Dessler, E.N. Parker, Hydromagnetic theory of geomagnetic storms. J. Geophys. Res. **64**, 2239 (1959). doi:10.1029/JZ064i012p02239

G.A. DiBraccio, J.A. Slavin, S.A. Boardsen, B.J. Anderson, H. Korth, T.H. Zurbuchen, J.M. Raines, D.N. Baker, R.L. McNutt Jr., S.C. Solomon, MESSENGER observations of magnetopause structure and dynamics at Mercury. J. Geophys. Res. **118**, 997 (2013). doi:10.1002/jgra.50123

J.W. Dungey, Interplanetaly magnetic field and the auroral zones. Phys. Rev. Lett. **6**, 47 (1961)

W. Eckstein, H.M. Urbassek, Computer simulation of the sputtering process, in *Sputtering by Particle Bombardment: Experiments and Computer Calculations from Threshold to MeV Energies*, ed. by R. Behrisch, W. Eckstein (Springer, Berlin, 2007), p. 21

G. Eichhorn, Heating and vaporization during hypervelocity particle impact. Planet. Space Sci. **26**, 463 (1978a)

G. Eichhorn, Primary Velocity Dependence of impact ejecta parameters. Planet. Space Sci. **26**, 469 (1978b)

R.E. Ergun et al., S bursts and the Jupiter ionospheric Alfvén resonator. J. Geophys. Res. **111**, A06212 (2006)

D.S. Evans, Precipitation electron fluxes formed by magnetic-field-aligned potential differences. J. Geophys. Res. **79**, 2853 (1974)
H.J. Fahr, The extraterrestrial UV-background and the nearby interstellar medium. Space Sci. Rev. **15**, 483 (1974). doi:10.1007/BF00178217
M. Famà, J. Shi, R.A. Baragiola, Sputtering of ice by low-energy ions. Surf. Sci. **602**, 156 (2008)
C.J. Farrugia, N.V. Erkaev, H.K. Biernat et al., Anomalous magnetosheath properties during Earth passage of an interplanetary magnetic cloud. J. Geophys. Res. **100**, 19245 (1995)
Y.I. Feldstein et al., Structure of the auroral precipitation region in the dawn sector: relationship to convection reversal boundaries and field-aligned currents. Ann. Geophys. **19**, 495 (2001)
W.L. Fite, T.R. Brackman, W.R. Snow, Charge transfer in proton-hydrogen atom collisions. Phys. Rev. **112**, 1161 (1958)
M.-C. Fok et al., J. Geophys. Res. **111** (2006). doi:10.1029/2006JA011839
D. Fontaine et al., Numerical simulation of the magnetospheric convection including the effect of electron precipitation. J. Geophys. Res. **90**, 8343 (1985)
R.J. Forsyth et al., The underlying Parker spiral structure in the Ulysses magnetic field observations, 1990–1994. J. Geophys. Res. **101**, 395 (1996)
J.L. Fox, K.Y. Sung, Solar activity variations of the Venus thermosphere/ionosphere. J. Geophys. Res. **106**, 21305 (2001)
L.A. Frank, On the extraterrestrial ring current during geomagnetic storm. J. Geophys. Res. **72**, 3753 (1967)
K.M. Frederick-Frost et al., SERSIO: Svalbard EISCAT rocket study of ion outflows. J. Geophys. Res. **112**, A08307 (2007). doi:10.1029/2006JA011942
K. Fukazawa, T. Ogino, R.J. Walker, Magnetospheric convection at Saturn as a function of IMF Bz. Geophys. Res. Lett. **34**(1) (2007a). doi:10.1029/2006GL028373
K. Fukazawa, T. Ogino, R.J. Walker, Vortex-associated reconnection for northward IMF in the Kronian magnetosphere. Geophys. Res. Lett. **34** (2007b). doi:10.1029/2007GL031784
S.A. Fuselier, W.S. Lewis, Properties of near-earth magnetic reconnection from in-situ observations. Space Sci. Rev. **160**(1–4), 95 (2011)
S.A. Fuselier et al., Energetic neutral atoms from the Earth's subsolar magnetopause. Geophys. Res. Lett. **371**, 101 (2010). doi:10.1029/2010GL044140
P. Galopeau, P. Zarka, D. Le Quéau, Source location of SKR: the Kelvin-Helmholtz instability hypothesis. J. Geophys. Res. **100**, 26397 (1995)
S.B. Ganguli, The polar wind. Rev. Geophys. **34**, 311 (1996)
K.S. Garcia, V.G. Merkin, W.J. Hughes, Effects of nightside O^+ outflow on magnetospheric dynamics: Results of multifluid MHD modeling. J. Geophys. Res. **115**, A00J09 (2010). doi:10.1029/2010JA015730
D.J. Gershman, J.A. Slavin, J.M. Raines et al., Magnetic flux pileup and plasma depletion in Mercury's subsolar magnetosheath. J. Geophys. Res. **118**, 7181 (2013)
A. Glocer, G. Tóth, T. Gombosi, D. Welling, Modeling ionospheric outflows and their impact on the magnetosphere, initial results. J. Geophys. Res. **114**(A), 05,216 (2009). doi:10.1029/2009JA014053
T.I. Gombosi, *Physics of the Space Environment* (Cambridge University Press, Cambridge, 1998)
J.T. Gosling et al., The electron edge of the low latitude boundary layer during accelerated flow events. Geophys. Res. Lett. **17**(11), 1833 (1990a)
J.T. Gosling, R.M. Skoug, D.J. McComas et al., Magnetic disconnection from the Sun: observations of a reconnection exhaust in the solar wind at the heliospheric current sheet. Geophys. Res. Lett. **32**, L05105 (2005). doi:10.1029/2005GL022406
A. Grocott, S.W.H. Cowley, J.B. Sigwarth, Ionospheric flow during extended intervals of northward but BY-dominated IMF. Ann. Geophys. **21**, 509 (2003)
A. Grocott et al., Magnetosonic Mach number dependence of the efficiency of reconnection between planetary and interplanetary magnetic fields. J. Geophys. Res. **114**, A07219 (2009). doi:10.1029/2009JA014330
K.C. Hansen et al., Global MHD simulations of Saturn's magnetosphere at the time of Cassini approach. Geophys. Res. Lett. **32**, L20S06 (2005). doi:10.1029/2005GL022835
M. Harel et al., Quantitative simulations of a magnetospheric substorm, 1. Model logic and overview. J. Geophys. Res. **86**, 2217–2241 (1981)
R.A. Haring et al., Reactive sputtering of simple condensed gases by keV ions. III. Kinetic energy distributions. Nucl. Instrum. Methods B **5**, 483 (1984)
E.G. Harris, On a plasma sheath separating regions of oppositely directed magnetic fields. Nuovo Cimento **23**, 115 (1962)
H. Hasegawa et al., Transport of solar wind into Earth's magnetosphere through rolled-up Kelvin-Helmholtz vortices. Nature **430**(7001), 755 (2004). doi:10.1038/nature02799
H. Hasegawa et al., Kelvin-Helmholtz waves at the Earth's magnetopause: Multiscale development and associated reconnection. J. Geophys. Res. **114**, A12207 (2009)

W.J. Heikkila, R.J. Pellinen, Localized induced electric field within the magnetotail. J. Geophys. Res. **82**, 1610 (1977)
S. Hess, P. Zarka, F. Mottez, Io-Jupiter interaction, millisecond bursts and field aligned potentials. Planet. Space Sci. **55**, 89 (2007)
S. Hess et al., Electric potential jumps in the Io-Jupiter Flux tube. Planet. Space Sci. **57**, 23 (2009)
T.W. Hill, F.C. Michel, Heavy ions from the Galilean satellites and the centrifugal distortion of the Jovian magnetosphere. J. Geophys. Res. **81**, 4561 (1976)
T.W. Hill, A.J. Dessler, F.C. Michel, Configuration of the Jovian magnetosphere. Geophys. Res. Lett. **1** (1974). doi:10.1029/GL001i001p00003
T.W. Hill et al., Evidence for rotationally-driven plasma transport in Saturn's magnetosphere. Geophys. Res. Lett. **32**, L41S10 (2005)
R.R. Hodges Jr., Monte Carlo simulation of the terrestrial hydrogen exosphere. J. Geophys. Res. **99**, 23,229 (1994). doi:10.1029/94JA02183
W.O. Hofer, Angular, energy, and mass distribution of sputtered particles, in *Sputtering by Particle Bombardment*, ed. by R. Behrisch, K. Wittmaack (Springer, Berlin, 1991), p. 15
E.W. Hones Jr., The magnetotail: its generation and dissipation, in *Physics of Solar Planetary Environments*, ed. by D.J. Williams (AGU, Washington, 1976), pp. 559–571
E.W. Hones Jr., Substorm processes in the magnetotail: comments on "On hot tenuous plasma, fireballs, and boundary layers in the Earth's magnetotail" by L.A. Frank et al. J. Geophys. Res. **82**, 5633 (1977)
T.S. Huang, T.J. Birmingham, The polarization electric field and its effects in an anisotropic, rotating magnetospheric plasma. J. Geophys. Res. **97**, 1511 (1992)
J.D. Huba, Numerical Methods: Ideal and Hall MHD **7**, 26 (2005)
R.L. Huff et al., Mapping of auroral kilometric radiation sources to the Aurora. J. Geophys. Res. **93**, 11445 (1988)
A. Ieda et al., Statistical analysis of the plasmoid evolution with Geotail observations. J. Geophys. Res. **103**(A3), 4453 (1998). doi:10.1029/97JA03240
R. Ilie, R.M. Skoug, H.O. Funsten, M.W. Liemohn, J.J. Bailey, M. Gruntman, The impact of geocoronal density on ring current development. J. Atmos. Sol.-Terr. Phys. **99**, 92 (2013). doi:10.1016/j.jastp.2012.03.010
S.M. Imber et al., MESSENGER observations of large dayside flux transfer events: do they drive Mercury's substorm cycle? J. Geophys. Res. (2014, submitted)
F.M. Ipavich et al., Energetic (greater than 100 keV) $O^{(+)}$ ions in the plasma sheet. Geophys. Res. Lett. **11**, 504 (1984)
C.M. Jackman et al., Interplanetary magnetic field at ~9 AU during the declining phase of the solar cycle and its implications for Saturn's magnetospheric dynamics. J. Geophys. Res. **109**, A11203 (2004). doi:10.1029/2004JA010614
C.M. Jackman et al., Strong rapid dipolarizations in Saturn's magnetotail: In situ evidence of reconnection. Geophys. Res. Lett. **34**(11), L11203 (2007)
C.M. Jackman, R.J. Forsyth, M.K. Dougherty, The overall configuration of the interplanetary magnetic field upstream of Saturn as revealed by Cassini observations. J. Geophys. Res. **113**, A08114 (2008). doi:10.1029/2008JA013083
C.M. Jackman, J.A. Slavin, S.W.H. Cowley, Cassini observations of plasmoid structure and dynamics: Implications for the role of magnetic reconnection in magnetospheric circulation at Saturn. J. Geophys. Res. **116**, A10212 (2011). doi:10.1029/2011JA016682
C.M. Jackman et al., Large-scale structure and dynamics of the magnetotails of Mercury, Earth, Jupiter and Saturn. Space Sci. Rev. **182**(1), 85–154. (2014a). doi:10.1007/s11214-014-0060-8
C.M. Jackman et al., Saturn's dynamic magnetotail: A comprehensive magnetic field and plasma survey of plasmoids and travelling compression regions, and their role in global magnetospheric dynamics. J. Geophys. Res. **119**, 5465–5494 (2014b). doi:10.1002/2013JA019388
X. Jia, M.G. Kivelson, Driving Saturn's magnetospheric periodicities from the atmosphere/ionosphere: Magnetotail response to dual sources. J. Geophys. Res. **117**, A11219 (2012). doi:10.1029/2012JA018183
X.Z. Jia et al., Properties of Ganymede's magnetosphere inferred from improved three-dimensional MHD simulations. J. Geophys. Res. **114**, A09209 (2009). doi:10.1029/2009JA014375
X. Jia et al., Magnetospheric configuration and dynamics of Saturn's magnetosphere: A global MHD simulation. J. Geophys. Res. **117**, A05225 (2012b). doi:10.1029/2012JA017575
X. Jia, M.G. Kivelson, T.I. Gombosi, Driving Saturn's magnetospheric periodicities from the upper atmosphere/ionosphere. J. Geophys. Res. **117**, A04215 (2012a). doi:10.1029/2011JA017367
R.E. Johnson, *Energetic Charged-Particle Interactions with Atmospheres and Surfaces, vol. X*. Phys. Chem. Space (Springer, Berlin, Heidelberg, New York, 1990), p. 19
R.E. Johnson, Sputtering and desorption from icy surfaces, in *Solar System Ices*, ed. by B. Schmitt, C. de Bergh (Kluwer Acad., Dordrecht, 1998), p. 303

R.E. Johnson, Surface chemistry in the Jovian magnetosphere radiation environment, in *Chemical Dynamics in Extreme Environments*, ed. by R. Dessler. Adv. Ser. Phys. Chem., vol. 11 (World Scientific, Singapore, 2001), p. 390, Chap. 8

R.E. Johnson et al., Charged particle erosion of frozen volatiles in ice grains and comets. Astron. Astrophys. **123**, 343 (1983)

R.E. Johnson, F. Leblanc, B.V. Yakshinskiy, T.E. Madey, Energy distributions for desorption of sodium and potassium from ice: the Na/K ratio at Europa. Icarus **156**, 136 (2002)

R.E. Johnson, M.H. Burger, T.A. Cassidy, F. Leblanc, M. Marconi, W.H. Smyth, Composition and detection of Europa's sputter-induced atmosphere, in *Europa*, ed. by R.T. Pappalardo, W.B. McKinnon, K. Khurana (The University of Arizona Press, Tucson, 2009), p. 507

M. Jorgensen, M.G. Henderson, E.C. Roelof, G.D. Reeves, H.E. Spence, Charge exchange contribution to the decay of the ring current, measured by energetic neutral atoms (ENAs). J. Geophys. Res. **106**, 1931 (2001). doi:10.1029/2000JA000124

K. Kabin et al., Interaction of Mercury with the solar wind. Icarus **143**(2), 397 (2000)

K. Keika, M. Nose, K. Takahashi, S. Ohtani, P.C. Brandt, D.G. Mitchell, S.P. Christon, R.W. McEntire, Contribution of ion flowout and charge exchange processes to the decay of the storm-time ring current. AGU Fall Meeting Abstracts, p. A565 (2003)

K. Keika, M. Nose, P.C. Brandt, S. Ohtani, D.G. Mitchell, E.C. Roelof, Contribution of charge exchange loss to the storm time ring current decay: IMAGE/HENA observations. J. Geophys. Res. **111**, 11 (2006). doi:10.1029/2006JA011789

A. Keiling et al., Transient ion beamlet injections into spatially separated PSBL flux tubes observed by Cluster-CIS. Geophys. Res. Lett. **31** (2004)

K.K. Khurana et al., The origin of Ganymede's polar caps. Icarus **191**, 193 (2007)

A. Kidder, R.M. Winglee, E.M. Harnett, Regulation of the centrifugal interchange cycle in Saturn's inner magnetosphere. J. Geophys. Res. **114**, A02205 (2009). doi:10.1029/2008JA013100

R.M. Killen, W.H. Ip, The surface-bounded atmospheres of mercury and the moon. Rev. Geophys. **37**(3), 361 (1999)

R. Killen et al., Processes that promote and deplete the exosphere of Mercury. Space Sci. Rev. **132**, 433 (2007)

Y.H. Kim et al., Hydrocarbon ions in the lower ionosphere of Saturn. J. Geophys. Res. **119**, 384 (2014)

N. Kitamura et al., Photoelectron flows in the polar wind during geomagnetically quiet periods. J. Geophys. Res. **117**, A07214 (2012). doi:10.1029/2011JA017459

M.G. Kivelson et al., Intermittent short-duration magnetic field anomalies in the Io torus: Evidence for plasma interchange? Geophys. Res. Lett. **24**, 2127 (1997)

M.G. Kivelson et al., Ganymede's magnetosphere: Magnetometer overview. J. Geophys. Res. **103**, 19963 (1998)

M.G. Kivelson et al., Moon-magnetosphere interaction: a tutorial. Adv. Space Res. **33**, 2061 (2004)

S. Knight, Parallel electric fields. Planet. Space Sci. **21**, 741 (1973)

J.U. Kozyra, M.W. Liemohn, C.R. Clauer, A.J. Ridley, M.F. Thomsen, J.E. Borovsky, J.L. Roeder, V.K. Jordanova, W.D. Gonzalez, Multistep Dst development and ring current composition changes during the 4–6 June 1991 magnetic storm. J. Geophys. Res. **107**, 1224 (2002). doi:10.1029/2001JA000023

N.A. Krall, A.W. Trivelpiece, *Principles of Plasma Physics* (San Francisco Press, Incorporated, San Francisco, 1986)

S.M. Krimigis et al., Dynamics of Saturn's magnetosphere from MIMI during Cassini's orbital insertion. Science **307**, 1270 (2005)

E.A. Kronberg et al., Comparison of periodic substorms at Jupiter and Earth. J. Geophys. Res. **113**, A04212 (2008). doi:10.1029/2007JA012880

N.A. Krupp et al., Global flows of energetic ions in Jupiter's equatorial plane: First-order approximation. J. Geophys. Res. **106**, 26,017 (2001). doi:10.1029/2000JA900138

N.A. Krupp et al., Dynamics of the Jovian magnetosphere, in *Jupiter: The Planet, Satellites, and Magnetosphere*, ed. by F. Bagenal et al. (Cambridge Univ. Press, Cambridge, 2004), p. 617, Chap. 25

A. Lagg et al., In situ observations of a neutral gas torus at Europa. Geophys. Res. Lett. **30**, 110000 (2003)

H.R. Lai et al., Reconnection at the magnetopause of Saturn: Perspective from FTE occurrence and magnetosphere size. J. Geophys. Res. **117**, A05222 (2012). doi:10.1029/2011JA017263

B. Lavraud, E. Larroque, E. Budnik et al., Asymmetry of magnetosheath flows and magnetopause shape during low Alfven Mach number solar wind. J. Geophys. Res. **118**, 1089 (2013). doi:10.1002/jgra.50145

S.A. Ledvina, Y.-J. Ma, E. Kallio, Modeling and simulating flowing plasmas and related phenomena. Space Sci. Rev. **139**, 143189 (2008). doi:10.1007/s11214-008-9384-6

M.W. Liemohn, J.U. Kozyra, Testing the hypothesis that charge exchange can cause a two-phase decay, in *The Inner Magnetosphere: Physics and Modeling*, ed. by T.I. Pulkkinen, N.A. Tsyganenko, R.H.W. Friedel, vol. 155 (American Geophysical Union Geophysical Monograph Series, Washington, 2005), p. 211

A. Lipatov, *The Hybrid Multiscale Simulation Technology* (Springer, Berlin, 2002)
P. Louarn et al., Trapped electrons as a free energy source for auroral kilometric radiation. J. Geophys. Res. **95**, 5983 (1990)
J.G. Luhmann et al., A model of the ionosphere of Saturn's rings and its implications. Icarus **181**, 465 (2006)
J.G. Lyon, J.A. Fedder, C.M. Mobarry, The Lyon-Fedder-Mobarry (LFM) global MHD magnetospheric simulation code. J. Atmos. Sol.-Terr. Phys. **66**, 1333 (2004). doi:10.1016/j.jastp.2004.03.020
L.R. Lyons, Generation of large-scale regions of auroral currents, electric potentials, and precipitation by the divergence of the convection electric field. J. Geophys. Res. **85**, 17 (1980)
R. Lysak, Electrodynamic couplig of the ionosphere and magnetosphere. Space Sci. Rev. **52**, 33 (1990)
T. Majeed, J.C. McConnell, Voyager electron density measurements on Saturn: Analysis with a time dependent ionospheric model. J. Geophys. Res. **101**, 7589 (1996). doi:10.1029/96JE00115
V. Mangano et al., The contribution of impact-generated vapour to the hermean atmosphere. Planet. Space Sci. **55**(11), 1541 (2007)
M.L. Marconi, A kinetic model of Ganymede's atmosphere. Icarus **190**, 155 (2007)
G.T. Marklund, Electric fields and plasma processes in the auroral downward current region, below, within, and above the acceleration region. Space Sci. Rev. **142**, 1 (2009). doi:10.1007/s11214-008-9373-9
A. Masters et al., Surface waves on Saturn's dawn flank magnetopause driven by the Kelvin-Helmholtz instability. Planet. Space Sci. **57**, 1769 (2009). doi:10.1016/j.pss.2009.02.010
A. Masters et al., Cassini observations of a Kelvin-Helmholtz vortex in Saturn's outer magnetosphere. J. Geophys. Res. **115**(A7), A07225 (2010). doi:10.1029/2010JA015351
A. Masters et al., The importance of plasma b conditions for magnetic reconnection at Saturn's magnetopause. Geophys. Res. Lett. **39**, L08103 (2012). doi:10.1029/2012GL051372
Y. Matsumoto, M. Hoshino, Turbulent mixing and transport of collisionless plasmas across a stratified velocity shear layer. J. Geophys. Res. **111**, A05213 (2006). doi:10.1029/2004JA010988
Y. Matsumoto, K. Seki, Formation of a broad plasma turbulent layer by forward and inverse energy cascades of the Kelvin–Helmholtz instability. J. Geophys. Res. **115**, A10231 (2010). doi:10.1029/2009JA014637
B.H. Mauk, Quantitative modeling of the "convection surge" mechanism of ion acceleration. J. Geophys. Res. **91**, 13,423 (1986)
B.H. Mauk et al., Energetic neutral atoms from a trans-Europa gas torus at Jupiter. Nature **421**, 920 (2003)
H.J. McAndrews et al., Evidence for reconnection at Saturn's magnetopause. J. Geophys. Res. **113**, A04210 (2008). doi:10.1029/2007JA012581
M.B. McElroy, The ionospheres of the major planets. Space Sci. Rev. **14**, 460 (1973)
J. McFadden, C. Carlson, R. Ergun, Microstructure of the auroral acceleration region as observed by FAST. J. Geophys. Res. **104**(A7), 14453 (1999). doi:10.1029/1998JA900167
R.L. McNutt Jr. et al., Departure from rigid co-rotation of plasma in Jupiter's dayside magnetosphere. Nature **280**, 803 (1979)
B. Meinel, Doppler-shifted auroral hydrogen emission. Astrophys. J. **113**, 50 (1951). doi:10.1086/145375
N. Meyer-Vernet, M. Moncuquet, S. Hoang, Temperature inversion in the Io plasma torus. Icarus **116**, 202 (1995)
F.C. Michel, P.A. Sturrock, Centrifugal instability of the Jovian magnetosphere and its interaction with the solar wind. Planet. Space Sci. **22**, 1501 (1974)
D.G. Mitchell et al., Current carriers in the near-earth cross-tail current sheet during substorm growth phase. Geophys. Res. Lett. **17**, 583–586 (1990)
D.G. Mitchell et al., Global imaging of O^+ from IMAGE HENA. Space Sci. Rev. **109**, 63 (2003)
T.E. Moore, J.L. Horwitz, Stellar ablation of planetary atmospheres. Rev. Geophys. **45**, RG3002 (2007). doi:10.1029/2005RG000194
F. Mottez, V. Génot, Electron acceleration by an Alfvénic pulse propagating in an auroral plasma cavity. J. Geophys. Res. **116**, A00K15 (2011)
F.S. Mozer, A. Hull, Origin and geometry of upward parallel electric fields in the auroral acceleration region. J. Geophys. Res. **106**, 5763 (2001)
F.S. Mozer et al., Observations of paired electrostatic shocks in the polar magnetosphere. Phys. Rev. Lett. **38**, 292 (1977)
F.S. Mozer et al., Satellite measurements and theories of low altitude auroral particle acceleration. Space Sci. Rev. **27**, 155 (1980)
M. Müller et al., Estimation of the dust flux near Mercury. Planet. Space Sci. **50**, 1101 (2002)
A. Mura et al., The sodium exosphere of Mercury: Comparison between observations during Mercury's transit and model results. Icarus **200**, 1 (2009)
A.F. Nagy, T.E. Cravens, S.G. Smith, H.A. Taylor Jr., H.C. Brinton, Model calculations of the dayside ionosphere of Venus: Ionic composition. J. Geophys. Res. **85**, 7795–7801 (1980)
T.K.M. Nakamura, M. Fujimoto, A. Otto, Structure of an MHD-scale Kelvin–Helmholtz vortex: Two-dimensional two-fluid simulations including finite electron inertial effects. J. Geophys. Res. **113**, A09204 (2008). doi:10.1029/2007JA012803

A.I. Neishtadt, On the change in the adiabatic invariant on crossing a separatrix in systems with two degrees of freedom. J. Appl. Math. **51**, 586 (1987)
P.T. Newell et al., Characterizing the state of the magnetosphere: Testing the ion precipitation maxima latitude (b2i) and the ion isotropy boundary. J. Geophys. Res. **103**, 4739 (1998)
J.D. Nichols, S.W.H. Cowley, D.J. McComas, Magnetopause reconnection rate estimates for Jupiter's magnetosphere based on interplanetary measurements at ~5AU. Ann. Geophys. **24**, 393 (2006)
P. Norqvist et al., Ion cyclotron heating in the dayside magnetosphere. J. Geophys. Res. **101**, 13,179 (1996)
T.G. Northrop, *The Adiabatic Motion of Charged Particles* (Wiley Interscience, New York, 1963)
T.G. Northrop, J.R. Hill, Stability of negatively charged dust grains in Saturn's ring plane. J. Geophys. Res. **87**, 6045 (1983)
M. Nosé et al., Acceleration of oxygen ions of ionospheric origin in the near-Earth magnetotail during substorms. J. Geophys. Res. **105**, 7669 (2000)
K. Nykyri, A. Otto, Plasma transport at the magnetospheric boundary due to reconnection in Kelvin–Helmholtz vortices. Geophys. Res. Lett. **28**(18), 3565 (2001). doi:10.1029/2001GL013239
K. Nykyri et al., Cluster observations of reconnection due to the Kelvin-Helmholtz instability at the dawnside magnetospheric flank. Ann. Geophys. **24**, 2619 (2006)
Y. Ogawa, K. Seki, M. Hirahara et al., Coordinated EISCAT Svalbard radar and Reimei satellite observations of ion upflows and suprathermal ions. J. Geophys. Res. **113**, A05306 (2008). doi:10.1029/2007JA012791
T.R. Ogino, R.J. Walker, M.G. Kivelson, A global magnetohydrodynamic simulation of the Jovian magnetosphere. J. Geophys. Res. **84**, 47 (1998)
M. Øieroset et al., A statistical study of ion beams and conics from the dayside ionosphere during different phases of a substorm. J. Geophys. Res. **104**, 6987 (1999)
Y.M. Ono et al., The role of magnetic field fluctuations in nonadiabatic acceleration of ions during dipolarization. J. Geophys. Res. **114** (2009). doi:10.1029/2008JA013918
S. Orsini, A. Milillo, Magnetospheric plasma loss processes and energetic neutral atoms. Il Nuovo Cimento **22**(5), 633 (1999)
N. Østgaard, S.B. Mende, H.U. Frey, G.R. Gladstone, H. Lauche, Neutral hydrogen density profiles derived from geocoronal imaging. J. Geophys. Res. **108**, 1300 (2003). doi:10.1029/2002JA009749
L. Pallier, R. Prangé, Detection of the southern counterpart of the north FUV polar cusp. Shared properties. Geophys. Res. Lett. **31**, L06701 (2004)
C. Paranicas et al., The ion environment near Europa and its role in surface energetics. Geophys. Res. Lett. **29**(5), 1074 (2002). doi:10.1029/2001GL014127
E.N. Parker, Dynamics of the interplanetary magnetic field. Astrophys. J. **128**, 664 (1958)
G.S. Paschmann, Recent in-situ observations of magnetic reconnection in near-Earth space. Geophys. Res. Lett. **35**, L19109 (2008). doi:10.1029/2008GL035297
G. Paschmann, S. Haaland, R. Treumann (eds.), *Auroral Plasma Physics* (Kluwer Academic, Dordrecht, 2003)
G. Paschmann, M. Øieroset, T. Phan, In-situ observations of reconnection in space. Space Sci. Rev. **178**, 385–417 (2013). doi:10.1007/s11214-012-9957-2
R.J. Pellinen, W.J. Heikkila, Energization of charged particles to high energies by an induced substorm electric field within the magnetotail. J. Geophys. Res. **83**, 1544 (1978)
P. Perreault, S.I. Akasofu, Study of geomagnetic storms. Geophys. J. R. Astron. Soc. **54**(3), 547 (1978)
C. Peymirat, D. Fontaine, Numerical simulation of the magnetospheric convection including the effect of field-aligned currents and electron precipitation. J. Geophys. Res. **99**, 11155 (1994)
T.D. Phan, M. Oieroset, M. Fujimoto, Reconnection at the dayside low-latitude magnetopause and its nonrole in low-latitude boundary layer formation during northward interplanetary magnetic field. Geophys. Res. Lett. **32**, L17101 (2005). doi:10.1029/2005GL023355
T.D. Phan, G. Paschmann, J.T. Gosling, M. Oieroset, M. Fujimoto, J.F. Drake, V. Angelopoulos, The dependence of magnetic reconnection on plasma b and magnetic shear: evidence from magnetopause observations. Geophys. Res. Lett. **40**, 11 (2013). doi:10.1029/2012GL054528
W.G. Pilipp, G. Morfill, The formation of the plasma sheet resulting from plasma mantle dynamics. J. Geophys. Res. **83**, 5670 (1978)
C. Plainaki et al., Neutral particle release from Europa's surface. Icarus **210**, 385 (2010)
C. Plainaki et al., The role of sputtering and radiolysis in the generation of Europa exosphere. Icarus **218**(2), 956 (2012). doi:10.1016/j.icarus.2012.01.023
C. Plainaki et al., Exospheric O_2 densities at Europa during different orbital phases. Planet. Space Sci. **88**, 42 (2013)
C. Plainaki et al., The H_2O and O_2 exospheres of Ganymede: the result of a complex interaction between the jovian magnetospheric ions and the icy moon. Icarus **245**, 306 (2015)
D.H. Pontius, Coriolis influences on the interchange instability. Geophys. Res. Lett. **24**, 2961 (1997)

A.E. Potter, Chemical sputtering could produce sodium vapour and ice on Mercury. Geophys. Res. Lett. **22**(23), 3289 (1995)

K.G. Powell et al., A solution-adaptive upwind scheme for ideal magnetohydrodynamics. J. Comput. Phys. **154**(2), 284 (1999). doi:10.1006/jcph.1999.6299

R. Prangé, The UV and IR Jovian aurorae. Adv. Space Res. **12**(8), 379 (1992)

R. Prangé et al., Rapid energy dissipation and variability of the Io-Jupiter electrodynamic circuit. Nature **379**, 323 (1996)

R. Prangé et al., Detailed study of FUV Jovian auroral features with the post COSTAR Hubble Faint Object Camera. J. Geophys. Res. **103**, 20195 (1998)

P.L. Pritchett, Electron-cyclotron maser instability in relativistic plasmas. Phys. Fluids **29**, 2919 (1986)

W.R. Pryor et al., The auroral footprint of Enceladus on Saturn. Nature **472**, 331 (2011)

J.D. Raeder et al., Open GGCM simulations for the THEMIS mission. Space Sci. Rev. **141**, 535 (2008). doi:10.1007/s11214-008-9421-5

J. Raines et al., Space Sci. Rev. (2015, this issue)

R.L. Rairden, L.A. Frank, J.D. Craven, Geocoronal imaging with Dynamics Explorer. J. Geophys. Res. **91**, 13,613 (1986). doi:10.1029/JA091iA12p13613

L.C. Ray et al., Current-voltage relation of a centrifugally confined plasma. J. Geophys. Res. **114**, A04214 (2009). doi:10.1029/2008JA013969

C.T. Reimann et al., Ion-induced molecular ejection from D_2O ice. Surf. Sci. **147**, 227 (1984)

H.E. Roosendaal, R.A. Hating, J.B. Sanders, Surface disruption as an observable factor in the energy distribution of sputtered particles. Nucl. Instrum. Methods **194**, 579 (1982)

H. Rosenbauer, H. Grunwaldt, M.D. Montgomery, G. Paschmann, N. Sckopke, Heos 2 plasma observations in the distant polar magnetosphere: the plasma mantle. J. Geophys. Res. **80**, 2723 (1975)

A.A. Roux et al., Auroral kilometric radiation sources: in situ and remote sensing observations from Viking. J. Geophys. Res. **98**, 11657 (1993)

C.T. Russell et al., Localized reconnection in the near Jovian magnetotail. Science **280**, 1061 (1998). doi: 10.1126/science.280.5366.1061

P.E. Sandholt et al., Dayside auroral configurations: Responses to southward and northward rotations of the interplanetary magnetic field. J. Geophys. Res. **103**(20), 279 (1998)

R.W. Schunk, A.F. Nagy, Ionospheres of the terrestrial planets. Rev. Geophys. **18**, 813–852 (1980)

R.W. Schunk, A.F. Nagy, *Ionospheres*, 2nd edn. (Cambridge University Press, Cambridge, 2009)

R.W. Schunk, J.J. Sojka, Global ionosphere–polar wind system during changing magnetic activity. J. Geophys. Res. **102**, 11625 (1997)

L. Scurry, C.T. Russell, Proxy studies of energy transfer to the magnetosphere. J. Geophys. Res. **96**, 9541 (1991)

V.A. Sergeev et al., Pitch-angle scattering of energetic protons in the magnetotail current sheet as the dominant source of their isotropic precipitation into the nightside ionosphere. Planet. Space Sci. **31**, 1147 (1983)

V.A. Sergeev, M. Malkov, K. Mursula, Testing the isotropic boundary algorithm to evaluate the magnetic field configuration of the tail. J. Geophys. Res. **98**, 7609 (1993)

V.P. Shabansky, Some processes in the magnetosphere. Space Sci. Rev. **12**, 299 (1971)

V.I. Shematovich et al., Surface-bounded atmosphere of Europa. Icarus **173**, 480 (2005)

M. Shi, R.A. Baragiola, D.E. Grosjean, R.E. Johnson, S. Jurac, J. Schou, Sputtering of water ice surfaces and the production of extended neutral atmospheres. J. Geophys. Res. **100**, 26387 (1995)

P. Sigmund, Theory of Sputtering. I. Sputtering Yield of Amorphous and Polycrystalline (1969)

G.L. Siscoe, On the equatorial confinement and velocity space distribution of satellite ions in Jupiter's magnetosphere. J. Geophys. Res. **82**, 1641 (1977)

G. Siscoe, L. Christopher, Variations in the solar wind stand-off distance at Mercury. Geophys. Res. Lett. **2**, 158 (1975)

J.A. Slavin, Traveling compression regions, in *New Perspectives in Magnetotail Physics*, ed. by A. Nishida, S.W.H. Cowley, D.N. Baker. AGU Monograph, vol. 105 (AGU, Washington, 1998), pp. 225–240

J.A. Slavin, R.E. Holzer, J.R. Spreiter, S.S. Stahara, Planetary mach cones: theory and observation. J. Geophys. Res. **89**, 2708 (1984)

J.A. Slavin et al., Geotail observations of magnetic flux ropes in the plasma sheet. J. Geophys. Res. **108**(A1), 1015 (2003). doi:10.1029/2002JA009557

J.A. Slavin, M.H. Acuna, B.J. Anderson et al., MESSENGER observations of magnetic reconnection in Mercury's magnetosphere. Science **324**(5927), 606 (2009). doi:10.1126/science.1172011

J.A. Slavin et al., MESSENGER and Mariner 10 flyby observations of magnetotail structure and dynamics at Mercury. J. Geophys. Res. **117**, A01215 (2012a). doi:10.1029/2011JA016900

J.A. Slavin et al., MESSENGER observations of flux transfer events at Mercury. J. Geophys. Res. **117**, A00M06 (2012b). doi:10.1029/2012JA017926

J.A. Slavin et al., MESSENGER observations of Mercury's dayside magnetosphere under extreme solar wind conditions. J. Geophys. Res. **119**, 8087–8116 (2014)
P.H. Smith, N.K. Bewtra, Charge exchange lifetimes for ring current ions. Space Sci. Rev. **22**, 301 (1978). doi:10.1007/BF00239804
W.H. Smyth, M.L. Marconi, Europa's atmosphere, gas tori, and magnetospheric implications. Icarus **181**, 510 (2006)
S.C. Solomon, Auroral electron transport using the Monte Carlo method. Geophys. Res. Lett. **20**, 185 (1993)
B.U.O. Sonnerup, Adiabatic particle orbits in a magnetic null sheet. J. Geophys. Res. **76**, 8211 (1971)
B.U.O. Sonnerup, The magnetopause reconnection rate. J. Geophys. Res. **79**, 1546 (1974). doi:10.1029/JA079i010p01546
T.W. Speiser, Particle trajectory in model current sheets, 1, Analytical solutions. J. Geophys. Res. **70**, 4219 (1965)
W.N. Spjeldvik, Equilibrium structure of equatorially mirroring radiation belt protons. J. Geophys. Res. **82**, 2801 (1977). doi:10.1029/JA082i019p02801
W.N. Spjeldvik, T.A. Fritz, Theory for charge states of energetic oxygen ions in the earth's radiation belts. J. Geophys. Res. **83**, 1583 (1978). doi:10.1029/JA083iA04p01583
R.J. Strangeway et al., Factors controlling ionospheric outflows as observed at intermediate altitudes. J. Geophys. Res. **110**, A03221 (2005). doi:10.1029/2004JA010829
G.W. Stuart, Satellite-measured radiation. Phys. Rev. Lett. **2**, 417 (1959)
M. Sugiura, *Hourly Values of Equatorial Dst for the IGY*. Ann. Int. Geophys. Year, vol. 35 (Pergamon Press, Oxford, 1964), p. 9
T.S. Sundberg et al., MESSENGER orbital observations of large-amplitude Kelvin-Helmholtz waves at Mercury's magnetopause. J. Geophys. Res. **117**, A04216 (2012)
K. Szegö et al., Physics of mass loaded plasmas. Space Sci. Rev. **94**, 429 (2000)
S.W.Y. Tam, T. Chang, V. Pierrard, Kinetic modeling of the polar wind. J. Atmos. Sol.-Terr. Phys. **69**, 1984 (2007)
B.D. Teolis, R.A. Vidal, J. Shi, R.A. Baragiola, Mechanisms of O_2 sputtering from water ice by keV ions. Phys. Rev. B **72**, 245422 (2005). 2005. doi:10.1103/PhysRevB.72.245422
N. Terada, H. Shinagawa, T. Tanaka, K. Murawski, K. Terada, A three-dimensional, multispecies, comprehensive MHD model of the solar wind interaction with the planet Venus. J. Geophys. Res. **114**, A09208 (2009). doi:10.1029/2008JA013937
T. Terasawa et al., Solar wind control of density and temperature in the near-Earth plasma sheet: WIND/GEOTAIL collaboration. Geophys. Res. Lett. **24**(8), 935 (1997). doi:10.1029/96GL04018
B.T. Thomas, E.J. Smith, The Parker spiral configuration of the interplanetary magnetic field between 1 and 8.5 AU. J. Geophys. Res. **85**, 6861 (1980)
N. Thomas, F. Bagenal, T.W. Hill, J.K. Wilson, The Io neutral cloud and plasma torus, in *Jupiter: The Planet, Satellites, and Magnetosphere*, ed. by F. Bagenal et al. (Cambridge Univ. Press, Cambridge, 2004), p. 560, Chap. 23
M.F. Thomsen et al., Survey of ion plasma parameters in Saturn's magnetosphere. J. Geophys. Res. **115**, A10220 (2010). doi:10.1029/2010JA015267
R.M. Thorne et al., Galileo evidence for rapid interchange transport in the Io torus. Geophys. Res. Lett. **24**, 2131 (1997)
G. Toth et al., Space weather modeling framework: a new tool for the space science community. J. Geophys. Res. **110**, 12,226 (2005). doi:10.1029/2005JA011126
G. Toth et al., Adaptive numerical algorithms in space weather modeling. J. Comput. Phys. **231**(3), 870–903 (2012). doi:10.1016/j.jcp.2011.02.006
R.A. Treumann, The electron–cyclotron maser for astrophysical application. Astron. Astrophys. Rev. **13**, 229 (2006)
V. Vasyliunas, Mathematical models of magnetospheric convection and its coupling to the ionosphere, in *Particles and Fields in the Magnetosphere*, ed. by B. McCormac (Reidel, Hingham, 1970)
V.M. Vasyliunas, Plasma distribution and flow, in *Physics of the Jovian Magnetosphere*, ed. by A.J. Dessler (Cambridge Univ. Press, Cambridge, 1983), p. 395
M.F. Vogt et al., Reconnection and flows in the Jovian magnetotail as inferred from magnetometer observations. J. Geophys. Res. **115**, A06219 (2010). doi:10.1029/2009JA015098
M.F. Vogt et al., Structure and statistical properties of plasmoids in Jupiter's magnetotail. J. Geophys. Res. **119**, 821 (2014). doi:10.1002/2013JA019393
J.H. Waite et al., An auroral flare at Jupiter. Nature **410**, 787 (2001)
J.A. Wanliss, K.M. Showalter, High-resolution global storm index: Dst versus SYM-H. J. Geophys. Res. **111**(A2), A02202 (2006). doi:10.1029/2005JA011034
D. Welling et al., Space Sci. Rev. (2015, this issue)

M.S. Westley et al., Photodesorption from low-temperature water ice in interstellar and circumstellar grains. Nature **373**, 405 (1995)
D.J. Williams, Ring current composition and sources: An update. Planet. Space Sci. **29**, 1195 (1981)
R.J. Wilson, R.L. Tokar, M.G. Henderson, Thermal ion flow in Saturn's inner magnetosphere measured by the Cassini plasma spectrometer: A signature of the Enceladus torus? Geophys. Res. Lett. **36**, L23104 (2009). doi:10.1029/2009GRL040225
M. Wiltberger et al., Influence of cusp O^+ outflow on magnetotail dynamics in a multifluid MHD model of the magnetosphere. J. Geophys. Res. **115**, A00J05 (2010). doi:10.1029/2010JA015579
S. Wing, P.T. Newell, 2D plasma sheet ion density and temperature profiles for northward and southward IMF. Geophys. Res. Lett. **29**(9), 1307 (2002). doi:10.1029/2001GL013950
R.M. Winglee et al., Global impact of ionospheric outflows on the dynamics of the magnetosphere and cross-polar cap potential. J. Geophys. Res. **107**, 1237 (2002). doi:10.1029/2001JA000214
D. Winske, L. Yin, N. Omidi, H. Karimabadi, K. Quest, Hybrid simulation codes: Past, present and future—A tutorial, in *Space Plasma Simulation*, ed. by J. Büchner, C. Dum, M. Scholer. Lect. Notes Phys., vol. 615 (2003), pp. 136–165
R.A. Wolf, Ionosphere-magnetosphere coupling. Space Sci. Rev. **17**, 535 (1979)
C.S. Wu, L.C. Lee, A theory of the terrestrial kilometric radiation. Astrophys. J. **230**, 621 (1979)
X.-Y. Wu, J.L. Horwitz, J.-N. Tu, Dynamic fluid kinetic (DyFK) simulation of auroral ion transport: Synergistic effects of parallel potentials, transverse ion heating, and soft electron precipitation. J. Geophys. Res. **107**(A10), 1283 (2002). doi:10.1029/2000JA000190
P. Wurz, H. Lammer, Monte-Carlo simulation of Mercury's exosphere. Icarus **164**, 1 (2003)
P. Wurz et al., The lunar exosphere: the sputtering contribution. Icarus **191**, 486–496 (2007). doi:10.1016/j.icarus.2007.04.034
P. Wurz et al., Self-consistent modelling of Mercury's exosphere by sputtering, micro-meteorite impact and photon-stimulated desorption. Planet. Space Sci. **58**, 1599 (2010)
B.V. Yakshinskiy, T.E. Madey, Photon-stimulated desorption as a substantial source of sodium in the lunar atmosphere. Nature **400**, 642 (1999)
B.V. Yakshinskiy, T.E. Madey, Electron- and photon-stimulated desorption of K from ice surfaces. J. Geophys. Res. **106**, 33303 (2001)
B.V. Yakshinskiy, T.E. Madey, Temperature-dependent DIET of alkalis from SiO_2 films: Comparison with a lunar sample. Surf. Sci. **593**, 202 (2005)
A.W. Yau, M. Andre, Sources of ion outflow in the high latitude ionosphere. Space Sci. Rev. **80**, 1 (1997). doi:10.1023/A:1004947203046
A.W. Yau, T. Abe, W.K. Peterson, The polar wind: Recent observations. J. Atmos. Sol.-Terr. Phys. **69**, 1936 (2007)
K.S. Yee, Numerical solution of initial boundary value problems involving Maxwell's equations in isotropic media. IEEE Trans. Antennas Propag. **14**, 302 (1966)
Y. Yu, A.J. Ridley, Exploring the influence of ionospheric O^+ outflow on magnetospheric dynamics: dependence on the source location. J. Geophys. Res. **118**, 1711 (2013). doi:10.1029/2012JA018411
P. Zarka, Auroral radio emissions at the outer planets: observations and theories. J. Geophys. Res. **103**, 20159 (1998)
J.F. Ziegler, SRIM-2003. Nucl. Instrum. Methods B **219**, 1027 (2004)
J.F. Ziegler, J.P. Biersack, U. Littmark, *The Stopping and Range of Ions in Solids*. Stopping and Ranges of Ions in Matter, vol. 1 (Pergamon Press, New York, 1984)
J.H. Zoennchen, U. Nass, G. Lay, H.J. Fahr, 3-D-geocoronal hydrogen density derived from TWINS Ly-alpha-data. Ann. Geophys. **28**, 1221 (2010). doi:10.5194/angeo-28-1221-2010
J.H. Zoennchen, J.J. Bailey, U. Nass, M. Gruntman, H.J. Fahr, J. Goldstein, The TWINS exospheric neutral H-density distribution under solar minimum conditions. Ann. Geophys. **29**, 2211 (2011)
B.J. Zwan, R.A. Wolf, Depletion of solar-wind plasma near a planetary boundary. J. Geophys. Res. **81**, 1636 (1976)

DOI 10.1007/978-1-4939-3544-4_4
Reprinted from *Space Science Reviews* Journal, DOI 10.1007/s11214-015-0193-4

Plasma Sources in Planetary Magnetospheres: Mercury

J.M. Raines[1] · **G.A. DiBraccio**[1,2] · **T.A. Cassidy**[3] · **D.C. Delcourt**[4] · **M. Fujimoto**[5] · **X. Jia**[1] · **V. Mangano**[6] · **A. Milillo**[6] · **M. Sarantos**[7] · **J.A. Slavin**[1] · **P. Wurz**[8]

Received: 9 March 2015 / Accepted: 25 July 2015 / Published online: 14 October 2015

1 Introduction

The proximity of Mercury to the Sun makes this planet a particularly interesting subject because of the extreme environmental conditions that led to its unique evolutionary history. Mercury's present plasma environment has its foundation in a weak intrinsic global magnetic field that supports a small, but dynamic, magnetosphere. The plasma in Mercury's space environment coexists with the planet's exosphere and strongly interacts with the surface. In fact, Mercury's environment is a complex and tightly coupled system where the magnetosphere, exosphere, and surface are linked by interaction processes that facilitate material production and energy exchange (Killen and Ip 1999; Killen et al. 2007; Milillo et al. 2010). Investigations regarding the coupling of Mercury's magnetosphere with the interplanetary magnetic field (IMF) of the Sun, as well as the planet's interaction with solar radiation (both electromagnetic and particle) and with interplanetary dust, can pro-

✉ J.M. Raines
jraines@umich.edu

[1] Department of Atmospheric, Oceanic and Space Sciences, University of Michigan, Ann Arbor, MI, 48109, USA

[2] Solar System Exploration Divison, NASA Goddard Space Flight Center, Greenbelt, MD, 20771, USA

[3] Laboratory for Atmospheric and Space Physics, University of Colorado, Boulder, CO, 80303, USA

[4] LPP, Ecole Polytechnique-CNRS, Université Pierre et Marie Curie, 4 Place Jussieu, 75252 Paris, France

[5] Institute of Space and Astronautical Science, JAXA, Sagamihara, Kanagawa 252-5210, Japan

[6] Institute of Space Astrophysics and Planetology, INAF, Rome, Italy

[7] Heliophysics Science Division, NASA Goddard Space Flight Center, Greenbelt, MD, USA

[8] Physics Institute, University of Bern, 3012 Bern, Switzerland

vide important clues to the process of planetary evolution (Orsini et al. 2014). The study of Mercury may reveal processes fundamental to the interpretation of exoplanet observations: In fact, many discovered exoplanets are located only a few stellar radii away from their parent star, even closer than Mercury is to the Sun. The resulting effects and type of interactions in these particular situations are among the key questions to be answered in the future.

The first in situ measurements provided by three flybys of Mariner 10 (reviewed in Vilas et al. 1988) in 1974–1975 revealed the weak, intrinsic magnetic field of Mercury that gives rise to its small magnetosphere (for a review, see Russell et al. 1988; Wurz and Blomberg 2001; Slavin et al. 2007). Plasma sheet electrons were measured during the first flyby, though no ion measurements were made due to a hardware failure (Ogilvie et al. 1974). After those measurements, the scientific community had to wait almost 40 years until MErcury Surface, Space ENvironment, GEochemistry, and Ranging (MESSENGER) was launched in August 2004 (Solomon et al. 2007) and became the first spacecraft to obtain systematic measurements by orbiting Mercury. The MESSENGER magnetic field measurements indicate that Mercury's magnetic dipole moment is offset northward from the planet's center by 0.2 R_M (where R_M is Mercury's radius, or 2440 km) (Alexeev et al. 2010; Anderson et al. 2011). Now we know that Mercury's magnetosphere is highly dynamic (e.g., Slavin et al. 2009, 2010; DiBraccio et al. 2013), so it cannot be considered as a stable structure where plasma distributes according to well-characterized populations, like in the Earth's magnetosphere. At Mercury, no stable ring current is observed and magnetic storms driven by adiabatic convection cannot develop. On the contrary, fast (few seconds) events like Flux Transfer Events (FTEs) (Slavin et al. 2012b), dipolarizations (Sundberg et al. 2012), plasmoids (Slavin et al. 2012a; DiBraccio et al. 2015a) are observed. Further, bursts of low- and moderate-energy (tens to hundreds keV) electrons (Ho et al. 2012) are often recorded.

Together with the protons and alpha particles (He^{2+}) of solar wind origin, MESSENGER's Fast Imaging Plasma Spectrometer (FIPS) detected ions of planetary origin like He^+, Na^+, and several other heavy ion species (Zurbuchen et al. 2008) while Mercury Atmospheric and Surface Composition Spectrometer (MASCS) UltraViolet and Visible Spectrometer (UVVS) (McClintock and Lankton 2007) mapped the Ca^+ tail on the nightside (Vervack et al. 2010). In particular, on the dayside, a solar wind-originating plasma population mixed with heavy planetary ions (Na^+ group) was observed in the region of the magnetospheric cusp. On the nightside, plasma ions were observed near the equator, in the central plasma sheet (Raines et al. 2013). Finally, increased plasma fluxes were observed in the magnetosheath as well as sparsely distributed planetary ions that span the magnetopause (MP) boundary (as identified in magnetic field measurements (Anderson et al. 2012; Winslow et al. 2013)). These features are observed on nearly every orbit, despite highly variable solar wind and IMF conditions (Gershman et al. 2012; Baker et al. 2013).

The solar wind and planetary ions interact with the surface to produce ion sputtering, backscattering, and internal structure alteration via chemical sputtering and/or enhanced diffusion (Mura et al. 2009; Sarantos et al. 2009). The surface-released material populates the neutral gas environment of Mercury as a tenuous and non-collisional regime, constituting the exosphere.

The presence of neutral atoms in Mercury's environment was also discovered during the Mariner 10 flybys; H, He, and O were detected in the atmosphere by the onboard UV spectrometer (Broadfoot et al. 1974). Later, Na, K, and Ca were detected through ground-based Earth observations (Potter and Morgan 1985, 1986; Bida et al. 2000). Doressoundiram et al. (2009) defined an upper limit for Al, Fe, and Si by ground-based observations. Finally, MESSENGER UVVS provided a systematic in situ detection of Na, Ca and Mg (Domingue et al.

2007; McClintock et al. 2009). New ground-based observations and new methods and technologies (e.g., Leblanc et al. 2008; Mangano et al. 2013), coupled with simulations (e.g., Schmidt 2013) permit the investigation of spatial and temporal variations in the exosphere, providing insight to magnetospheric and solar activity variation dependencies. UVVS measurements, surprisingly, have shown little exospheric response to magnetospheric activity.

The most globally attributed and systematically observed element at Mercury is Na, since its doublet is relatively easy to detect through the Earth's atmosphere and its abundance is high in Mercury's exosphere. A clear relation of Na distribution and its variability has been observed throughout the exosphere. The Na exosphere peaks frequently at mid-latitudes on the dayside, corresponding to the magnetic cusp regions where solar wind plasma is able to access the planetary surface (e.g., Killen et al. 2001). Nevertheless, experimental results exclude that the yield of direct sputtering can account for the observed Na intensity (McGrath et al. 1986; Johnson and Baragiola 1991). Modeling and recent Na temperature obtained by MESSENGER show that Photon Stimulated Desorption (PSD) is by far the most efficient process to release Na into the exosphere on the dayside of Mercury (Cassidy et al. 2015; Mura et al. 2009; Sarantos et al. 2009; Wurz et al. 2010), indicating that the processes are independent from each other (Leblanc and Johnson 2010; Mura et al. 2009). Also, the measurements by MESSENGER UVVS have shown evidence that variation of global intensity are well reproduced year by year (Cassidy et al. 2015), showing that solar wind action could account only for variation in the distribution, not in the global exosphere density.

Mercury's exosphere is continuously emptied and filled through a variety of chemical and physical processes acting on the planet's surface and environment (Killen et al. 2007; Leblanc et al. 2007). The neutral environment of the planet is not only generated by plasma-surface interactions, but it also interacts with the circulating plasma via charge exchange, and it also undergoes electron-impact and photo-ionization, creating a population of low-energy ions. These newly generated ions are accelerated (Delcourt et al. 2003; Seki et al. 2013) and contribute to the mini-magnetosphere. At a further step, the ions are either lost into the solar wind or impact again onto the surface.

Finally, we can conclude that the sources of the magnetospheric ions are mostly solar wind plasma entering the magnetosphere, ionization of exospheric species, and planetary ions from the surface. On the other hand, sinks of the ion populations are the surface, where plasma directly impacts, and the solar wind that picks up ions as it flows past the planet. To evaluate the source and sink balance in the Mercury environment, this global complex system should be investigated as a whole.

The forthcoming ESA—JAXA BepiColombo mission to Mercury (to be launched in 2017) (Benkhoff et al. 2010), consists of two Mercury-orbiting spacecraft to provide the opportunity for simultaneous two-point measurements. Thanks to this, the BepiColombo mission will offer an unprecedented opportunity to deeply investigate magnetospheric and exospheric dynamics at Mercury as well as their interactions with solar radiation and interplanetary dust (Milillo et al. 2010).

In the following sections of this chapter, the structure and dynamics of Mercury's magnetosphere are reviewed, with an emphasis on its local plasma environment. We examine both global and kinetic features that have been identified through magnetic field and plasma observations, organized into plasma sources and losses, as well as the exosphere and the surface processes that generate it. Finally, we discuss the contribution that modeling has made to our understanding of Mercury's magnetosphere and of the behavior of its plasma populations.

2 Magnetospheric Structure and Dynamics

2.1 Global Magnetosphere Configuration

The magnetosphere of Mercury is of interest in many respects. It is characterized by spatial and temporal scales much smaller than those at Earth (by a factor of 8 and 30, respectively (e.g., Russell and Walker 1985)). Boundary conditions also are quite different as compared to those at Earth with a dense solar wind and B_X-dominated IMF at the outer boundary as well as a tenuous atmosphere at the inner boundary. Mercury's intrinsic magnetic field has a dipole moment of 195 nT-R_M^3, that is aligned to within 3° of the planet's spin-axis but has a northward offset of 484 km (Anderson et al. 2011). As the supersonic, super-Alfvénic solar wind interacts with Mercury's intrinsic magnetic field, a planetary magnetosphere with an elongated magnetotail is formed. There are, however, notable differences between Mercury's magnetosphere and those of other planets with intrinsic magnetic field: Mercury possesses only a very tenuous exosphere consisting of planetary atoms, with some of them being ionized from the high solar radiation at Mercury's orbit (Zurbuchen et al. 2008, 2011; Raines et al. 2011, 2013). The lack of a conducting ionosphere implies that any field-aligned currents must close through the planet's regolith (Anderson et al. 2014). The solar wind is much more intense at Mercury's orbit than at any other planet of the solar system (Burlaga 2001). Although the solar wind velocity remains relatively constant throughout the heliosphere, its density at Mercury's orbit is 5–10 times larger than typical values at Earth. Additionally, the strength of the IMF is, on average, about 30 nT, increasing the solar wind Alfvén speed and enhancing the rate of reconnection with Mercury's magnetic field (Slavin and Holzer 1979).

The combination of Mercury's small dipole moment with the extreme solar parameters results in a small but dynamic magnetosphere (Fig. 1). In terms of planetary radii, the planet Mercury accounts for a much larger volume of its magnetosphere than Earth. At Mercury, the average subsolar magnetopause standoff distance is $\sim 1.45\ R_M$ (Winslow et al. 2013) where the typical standoff distance is $\sim 10\ R_E$ at Earth (Fairfield 1971). Upstream of the magnetosphere, Mercury's bow shock is located at an average distance of $\sim 1.96\ R_M$ away from the planet (Winslow et al. 2013). Due to the low Alfvénic Mach number (M_A) and low β, the ratio of plasma pressure to magnetic pressure, solar wind conditions at Mercury's orbit, the bow shock is weaker and exhibits smaller magnetic overshoots compared to the outer planets (Masters et al. 2013).

Like Earth, the open-closed field line boundaries of Mercury's magnetosphere map to high latitude, dayside regions defining magnetospheric cusps. The northern cusp is evident in both MESSENGER plasma and magnetic field data in the vast majority of orbits that cross this region. MESSENGER's passages over the southern cusp were at much larger altitudes and can only be indirectly inferred from measurements. The cusp appears as a strong enhancement in plasma flux, composed of solar wind and planetary ions (Zurbuchen et al. 2011; Raines et al. 2013) standing between two regions of much lower plasma density. These enhancements span Mercury latitudes of $\sim 30°$–80°N and local times of 6–14 h. The cusp is manifested in magnetic field data mainly as depressions in the field, attributed to the diamagnetic influence of the plasma present there. Winslow et al. (2012) performed a statistical analysis of these depressions. Their analysis showed that the cusp is a broad, highly variable region located around 56°–84°N magnetic latitude and 7–16 h local time, marking a similar region on Mercury's dayside as the plasma enhancements inferred from diamagnetic depressions. This spatial extent is more similar to the V-shaped outer cusp at Earth than the narrow cleft found at lower altitudes (Smith and Lockwood 1996; Lavraud et al. 2005).

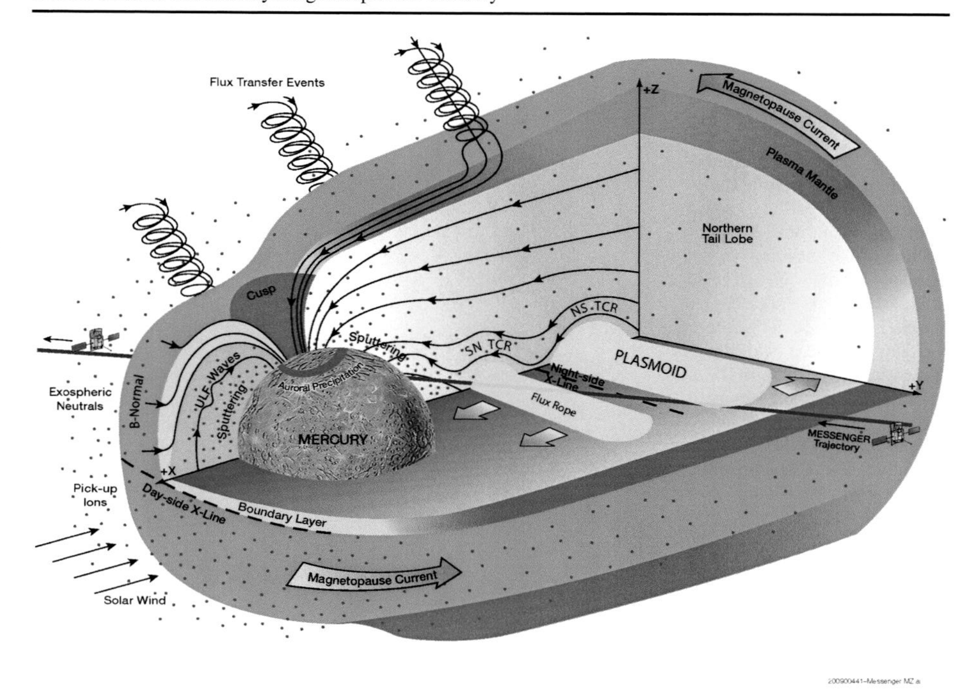

Fig. 1 Schematic of Mercury's Dungey cycle. This large-scale magnetospheric convection, responsible for the circulation of plasma and magnetic flux, is driven by steady and impulsive reconnection. Note the strong magnetic field normal to the dayside magnetopause, the large FTEs, and the reconnection line in the near-tail region. Figure from Slavin et al. (2009)

Ion measurements from MESSENGER's first Mercury flyby confirmed that Mercury's magnetosphere has an Earth-like central plasma sheet (Raines et al. 2011). The trajectory of this flyby was unique in that it passed nearby and almost parallel to Mercury's equatorial plane, providing an opportunity to observe across the plasma sheet not available in the orthogonal passes provided throughout the orbital phase. Those authors compared measurements at Mercury to a long baseline study of the plasma sheet at Earth (Baumjohann and Paschmann 1989). Accounting for the expected 5–10 fold higher solar wind densities at Mercury's orbit in the heliosphere, the measured proton density in Mercury's plasma sheet of 1–12 cm^{-3}, was comparable with those at Earth (0.2–0.5 cm^{-3}) during similarly quiet magnetospheric conditions. Proton temperature was much lower than the average at Earth, 2 MK versus 30–56 MK, respectively. Plasma β was also found to be lower and more steady at ~ 2 in Mercury's central plasma sheet. At the Earth, plasma β varies from ~ 0.3 near the edges of the plasma sheet, to ~ 30 at the center. More details concerning plasma sheet observations are included in Sect. 3 below.

2.1.1 Plasma Depletion Layers

The low-β conditions in Mercury's magnetosheath are further exacerbated by the frequent presence of plasma depletion layers (PDLs), caused by the draping and compression of the IMF as it encounters the magnetopause boundary (Fig. 2). This concept of PDLs was initially introduced by Zwan and Wolf (1976), who predicted that the natural draping of the

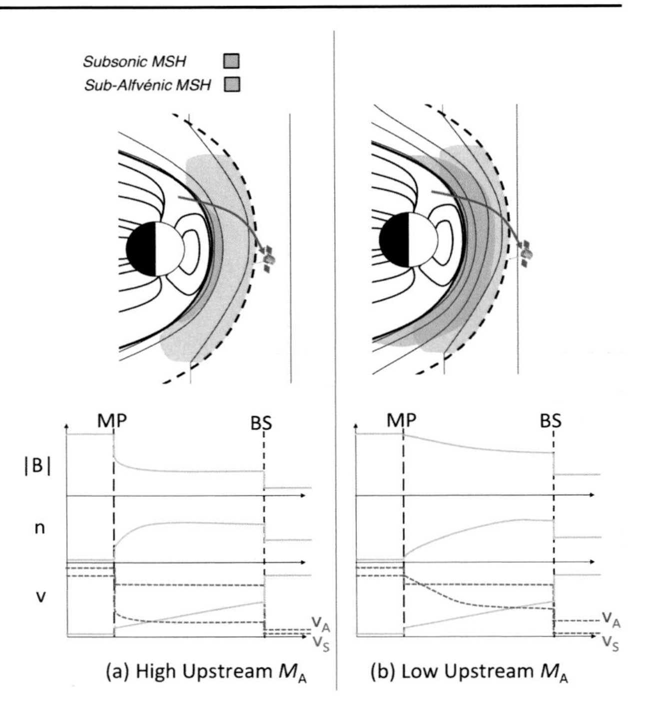

Fig. 2 Illustration of a spacecraft pass through the subsolar magnetosheath (MSH) for (**a**) high solar wind Mach number (M_A) and (**b**) low solar wind M_A. The MSH plasma is subsonic equatorward of the approximately ±45° latitude. With decreasing M_A, a larger fraction of the subsolar magnetosheath is sub-Alfvénic, as indicated by the *blue shaded region*. In addition, a thicker region of magnetic flux pileup is evident by an increase in $|\mathbf{B}|$ and a decrease in plasma density, n. The Alfvén speed (V_A) and sound speed (V_S) are also shown for both cases. Adapted from Gershman et al. (2013)

IMF would lead to the formation of low-β layers adjacent to the dayside magnetopause, which they termed plasma depletion layers. It was also predicted that the PDL thickness would be larger for low-M_A and low-β conditions, when magnetic pressure is dominating the magnetosheath, as is the case at Mercury (Zwan and Wolf 1976). Consistent with this prediction, Gershman et al. (2013) analyzed MESSENGER Magnetometer (MAG) and FIPS measurements to determine that lower upstream M_A ($M_A \sim 3$–5) values led to stronger depletion effects in the PDLs at Mercury. In this study, Gershman et al. (2013) identified 40 orbits where a PDL, adjacent to the dayside magnetopause, was observed as MESSENGER crossed through the magnetosheath. A typical PDL thickness was determined to be ~ 300 km, or $\sim 0.12\ R_M$. The PDLs were observed for both quasi-perpendicular and quasi-parallel shock geometries as well as for all IMF orientations. Despite the high frequency of reconnection occurring at Mercury's dayside magnetopause due to the low-β environment (DiBraccio et al. 2013), this substantial reconnection is not sufficient enough to transport all of the magnetic flux pileup and therefore the PDLs are a persistent feature of Mercury's magnetosheath. However, Gershman et al. (2013) also concluded that plasma depletion does not appear to exist during times of extended northward IMF.

2.1.2 Observations of Induction Effects

Given the mean subsolar magnetopause distance of only 1.45 R_M from the center of the planet (Winslow et al. 2013) and the high magnetopause reconnection rate (Slavin et al. 2009; DiBraccio et al. 2013), it seems reasonable to conclude that Mercury's surface may become directly exposed to the solar wind. Slavin and Holzer (1979) predicted that the low-M_A nature of Mercury's space environment, especially during periods of high solar wind pressure, would allow reconnection to erode the magnetopause down to the planetary surface. However, at the same time, Hood and Schubert (1979) and Suess and Goldstein (1979) predicted that induction effects at Mercury would cause the subsolar magnetopause to remain at or above 1.2 R_M.

Mercury's iron-rich, highly electrically conducting core, with a radius of 2000 km, makes up $\sim 80\ \%$ of the planet (Smith et al. 2012) and gives rise to an interaction that sets it apart from all other planetary magnetospheres. In the presence of this electrically conducting sphere, changes in upstream solar wind pressure will create changes in the magnetic field normal to the planetary surface. According to Faraday's law, these time-dependent changes will generate currents in the conducting core, which will serve to oppose this change in magnetic field and temporarily increase Mercury's magnetic moment, therefore limiting how far the magnetopause will be compressed (Hood and Schubert 1979; Suess and Goldstein 1979; Glassmeier et al. 2007).

To test these predictions and assess the roles of reconnection erosion and induction effects at Mercury, Slavin et al. (2014) analyzed three extreme solar wind dynamic pressure events using MESSENGER magnetic field and plasma measurements. Two of these events were due to coronal mass ejections (CMEs) and the third one was due to a high-speed stream. During these orbits, the magnetic field just inside the dayside magnetopause exceeded 300 nT with inferred solar wind pressures of $\sim$ 45–65 nPa. This field magnitude was double the typical strength of $\sim$ 150 nT just inside the magnetopause (DiBraccio et al. 2013; Winslow et al. 2013), which corresponds to solar wind ram pressures of $\sim$ 10 nPa. During these events, intense reconnection was observed in the form of frequent FTEs and steady reconnection rates derived from the normal magnetic field component to the magnetopause of 0.03–0.20.

In Fig. 3, the thin dashed curve illustrates the observed sixth-root relationship between solar wind dynamic pressure and magnetopause standoff distance determined by Winslow et al. (2013). The thick dashed line shows the predicted relationship between solar wind ram pressure and magnetopause standoff distance when induction effects are included. As evident in the figure, induction effect models predict that the magnetopause standoff distance will only be compressed below $\sim$ 1.2 R_M for solar wind pressures larger than $\sim$ 60 nT. The points on this plot indicate the magnetopause standoff distances, extrapolated to the subsolar point, for the boundary crossings observed during the three extreme solar wind events. The subsolar magnetopause was observed at much lower altitudes than predicted during these extreme solar wind intervals (Hood and Schubert 1979; Glassmeier et al. 2007) due to reconnection, which appears to be opposing the shielding effects of the induction currents. Therefore, during these days of extreme solar wind pressure, Mercury's magnetopause remained close to the surface due to the strong effect of dayside reconnection, which transfers magnetic flux into the magnetotail (Slavin and Holzer 1979). This result confirms that magnetic reconnection at Mercury is very intense and that both high-intensity reconnection as well as magnetosphere-core coupling must be included in global models of Mercury's magnetosphere during extreme solar wind pressure conditions.

2.2 Dungey Cycle at Mercury

Mercury's solar wind-driven magnetosphere experiences a circulation of plasma and magnetic flux similar to that of the Earth. This process is termed the Dungey cycle (Dungey 1961; Cowley 1982; see also Seki et al. 2015, this issue). The Dungey cycle begins with magnetic reconnection between the IMF and planetary magnetic field at the dayside magnetopause, resulting in open fields with one end rooted to the planet and the other in the solar wind. This open magnetic flux facilitates the exchange of solar wind and planetary plasma to and from

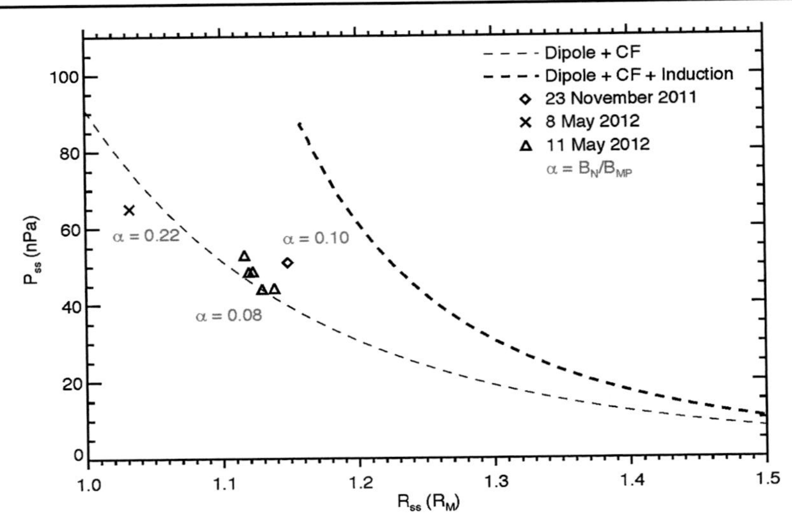

Fig. 3 Solar wind ram pressure, P_{SW}, versus extrapolated magnetopause standoff distance, R_{SS}, for the magnetopause crossings of Slavin et al. (2014). The magnetopause crossings on 23 November 2011, 8 May 2012, and 11 May 2012 are shown as a diamond, cross, and triangles, respectively. The dimensionless reconnection rate, α, averaged over the magnetopause crossings for each event, is also displayed. The sixth-root relationship (*thin dashed line*) determined from a large data set of MESSENGER magnetopause encounters at typical upstream pressures of ~ 5 to 15 nPa (Winslow et al. 2013) is compared with a theoretical model that includes the effects of induction in Mercury's interior (Glassmeier et al. 2007) (*thick dashed line*). Figure from Slavin et al. (2014)

the magnetosphere. The open fields are then carried downstream by the solar wind flow until they join the north and south lobes of the magnetotail. The oppositely directed fields of these tail lobes meet at the cross-tail current sheet where they reconnect. Tail reconnection creates two new magnetic field lines: a detached field line that rejoins the IMF and a closed field line with both ends attached to the planet. This closed field line convects sunward toward the planet, eventually moving toward the dayside and completing the cycle.

The Dungey cycle time is one of the keys for understanding the dynamical response of planetary magnetospheres to changes in the rates of magnetic reconnection at the magnetopause and in the magnetotail. It is determined by observing the rate of convection at various points in the cycle, as depicted in Fig. 1. For example, the cycle time may be deduced from the time for ionospheric plasma to $\mathbf{E} \times \mathbf{B}$ drift anti-sunward across the polar cap and return at lower latitudes to its point of initiation. Alternatively, the cross-magnetospheric electric field may be inferred from observations of the rate of magnetic flux being reconnected and transferred to/from the magnetotail or measured directly with electric field instrumentation. At Earth the time necessary for this cycle is in the range of 1–2 h (Siscoe et al. 1975; Cowley 1982). However at Mercury, Hill et al. (1976) noted that the lack of an ionosphere, and the expected resistive nature of the regolith, eliminates the need to take into account "line-tying" or "saturation" effects (see Kivelson and Ridley 2008) that reduce the cross-magnetospheric electric field at Earth from the maximum value, $-\mathbf{V}_{SW} \times \mathbf{B}_{SW}$, applied by the solar wind. Siscoe et al. (1975) then used scaling arguments and typical solar wind and IMF parameters to estimate that the Dungey cycle at Mercury would be of the order of 1 min.

MESSENGER's observations taken during its second flyby on 6 October 2008 provided the first opportunity to more directly infer the Dungey cycle time at Mercury. Slavin et al. (2009) used the magnetometer measurements (Anderson et al. 2007) to determine the magnetic field normal to the magnetopause and, with assumptions, calculated a cross-magnetospheric electric field of about 2 mV/m, which corresponds to a Dungey cycle time of 2 min. MESSENGER's third flyby on 29 September 2009 provided another opportunity to determine the Dungey cycle time when a series of loading–unloading events were observed as the magnetotail was traversed. At Earth magnetospheric substorms are often associated first with an interval of net magnetic flux transfer to the magnetotail, termed loading, which ends with the onset of magnetic reconnection in the cross-tail current layer and the dissipation of the magnetic flux stored in the tail (Baker et al. 1996). The duration of the tail loading and unloading intervals, sometimes referred to as the "growth" and "expansion" phases of the substorm because of the accompanying auroral signatures (McPherron et al. 1973), are typically on the order of the Dungey cycle time. Slavin et al. (2010) analyzed the magnetic field measurements during the third flyby and found a total of four loading–unloading events. In each case the duration of the event was $\sim$ 2–3 min and in reasonable agreement with the earlier estimate based upon the magnetic field normal to the magnetopause (i.e., dayside reconnection rate). We will show below that analogues to many aspects of the terrestrial substorm have been observed at Mercury, but on a time scale comparable to this miniature magnetosphere's Dungey cycle.

2.3 Magnetotail Loading & Unloading

As already discussed, dayside reconnection at Earth loads the tail lobes with magnetic flux and increases the tail's overall energy levels, which are later dissipated via tail reconnection and substorms. This enhanced loading of the tail lobes with magnetic flux causes the enhanced flaring of the flank magnetopause and increases the fraction of solar wind ram pressure applied directly to the magnetotail (Caan et al. 1973). In this manner, loading of the tail with magnetic flux is reflected in the magnetic field measurements both as an increase in the flaring of the magnetic field (i.e., $|B_Z|$ and/or $|B_Y|$) and in the total magnetic field magnitude. At Earth, the increase in the intensity of the lobe region during the substorm loading–unloading cycle is typically $\sim$ 10 to 30 % (Milan et al. 2004; Huang et al. 2009). However, the fractional enhancement in the lobe magnetic field observed at Mercury during the third flyby loading events appeared much larger, perhaps even reaching 100 % (Slavin et al. 2010).

MESSENGER observations since orbit insertion on 18 March 2011 have provided many opportunities to observe these loading–unloading events in the magnetotail. A comprehensive analysis has yet to be carried out, but Fig. 4 shows two examples of this phenomenon on 28 August 2011 where, between 19:45 and 19:55 UTC, two loading–unloading events are evident. MESSENGER was located in the south lobe of the tail at a distance of $\sim$ 3.3 R_M behind the planet. Each event begins with a total magnetic field intensity of $\sim$ 40 nT directed primarily in the $-X_{MSM}$ direction. The field then increases for $\sim$ 1 min until it reaches at peak value of $\sim$ 65 nT. This increase in total field is closely correlated with the B_Z component becoming more negative as the magnetic field flares away from the central axis of the tail. After the peak in total intensity the B_Z component becomes less negative as the intensity decreases back to its pre-substorm levels. The total increase in field magnitude during these events, $\sim$ 50 %, is significantly larger than observed at Earth, but below the larger values observed during the third flyby. The duration of the events, $\sim$ 2 min, is very close to the value determined from measurements of dayside magnetopause reconnection rate (Slavin et al. 2009; DiBraccio et al. 2013).

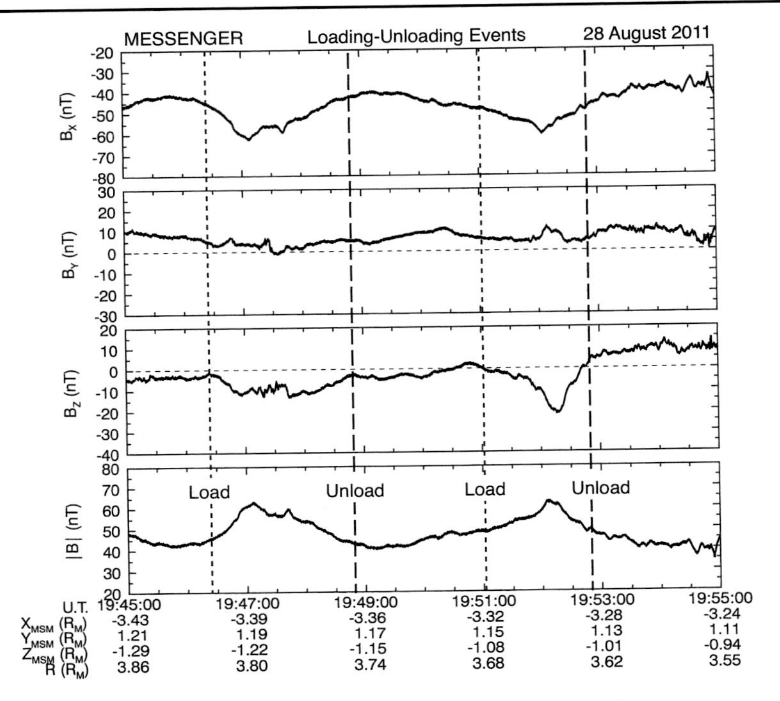

Fig. 4 Magnetic field measurements, displayed in Mercury Solar Magnetospheric coordinates. *Vertical dashed lines* identify the beginning and end of two loading–unloading cycles

3 Sources

Plasma in Mercury's magnetosphere either originates from the solar wind or the planet itself. The solar wind enters planetary magnetospheres mainly through five closely associated processes, all of which have been studied to some extent at Mercury: magnetopause reconnection followed by cusp precipitation, the plasma mantle, FTEs, direct impact of the solar wind on the surface due erosion and/or compression of the dayside magnetopause (Sect. 2.1.2), and Kelvin-Helmholtz driven reconnection on the flanks of the magnetotail. Planetary ions are formed at Mercury, either by ionization from exospheric neutral atoms or from processes that act directly on the surface. As a result, both the surface and the exosphere are significant plasma sources to Mercury's magnetosphere. These key sources of plasma to Mercury's magnetosphere are represented in Fig. 5.

3.1 Solar Wind Entry

3.1.1 Magnetopause Reconnection and the Plasma Mantle

Magnetopause magnetic reconnection is the dominant process for the transfer of mass, momentum, and energy between the solar wind and Mercury's magnetosphere. The resulting field topology exhibits a magnetic field component that is normal to the magnetopause, B_N. This was first observed at Mercury's magnetopause, indicating that magnetic reconnection had occurred, during the second MESSENGER flyby of Mercury on 6 October 2008 (Slavin et al. 2009). During this period, the IMF was oriented southward, a configuration that is

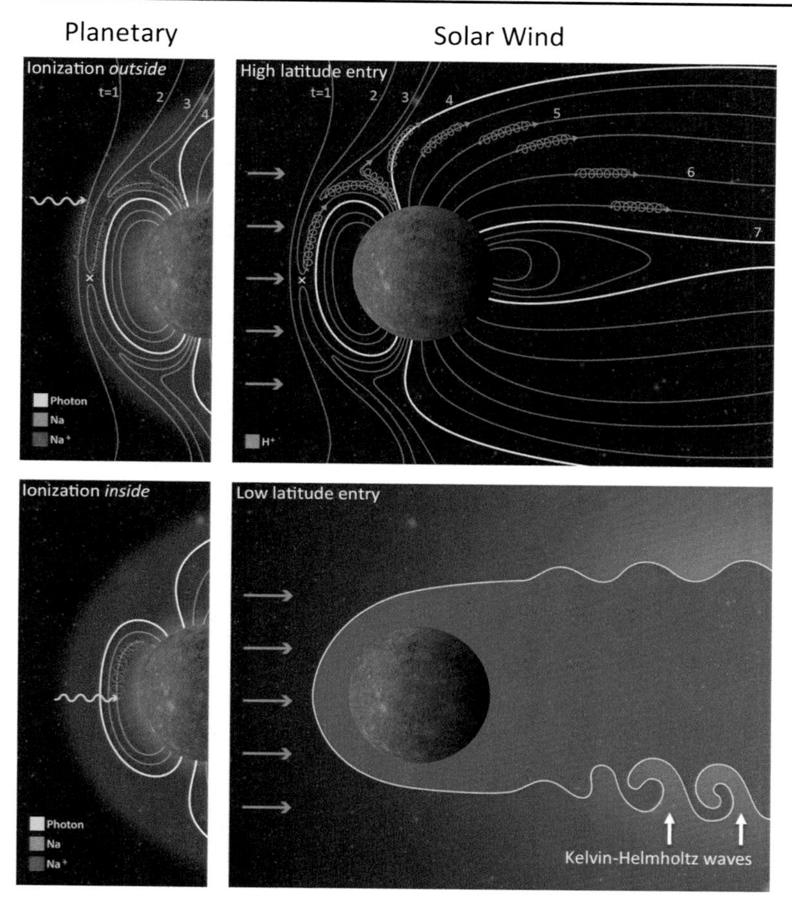

Fig. 5 Plasma sources to Mercury's magnetosphere. Energetic planetary ions, ionized upstream of the magnetopause, are transported into the magnetosphere on newly reconnected field lines after ionization outside the magnetopause (*upper left*, see Section 3.2.3), while lower energy planetary ions are created by ionization inside the magnetopause (*lower left*, see Section 3.2.2). Solar wind plasma can enter at via magnetopause reconnection (*upper right*, see Section 3.1.1) or at low latitudes via Kelvin-Helmholtz waves (*lower right*, see Section 3.1.3). The numbered field lines in the *top two panels* represent a time sequence of field line motion after magnetopause reconnection. The *lower right panel* is a slice through Mercury's equatorial plane, while the *other three panels* are slices through the noon-midnight plane

conducive to reconnection. Using a minimum variance analysis (MVA), Slavin et al. (2009) determined a significant, non-zero B_N, ~ 13 nT, at the outbound magnetopause crossing, indicating that the boundary was a rotational discontinuity. The dimensionless reconnection rate, α, is determined by:

$$\alpha = \frac{B_N}{B_{MP}}$$

where B_{MP} is the magnitude of the field just inside the magnetopause. During this second flyby, Slavin et al. (2009) calculated a reconnection rate of $\alpha = 0.13$.

In a statistical survey of magnetopause reconnection at Mercury, DiBraccio et al. (2013) identified 43 events with well-determined boundary normal vectors. The average B_N was

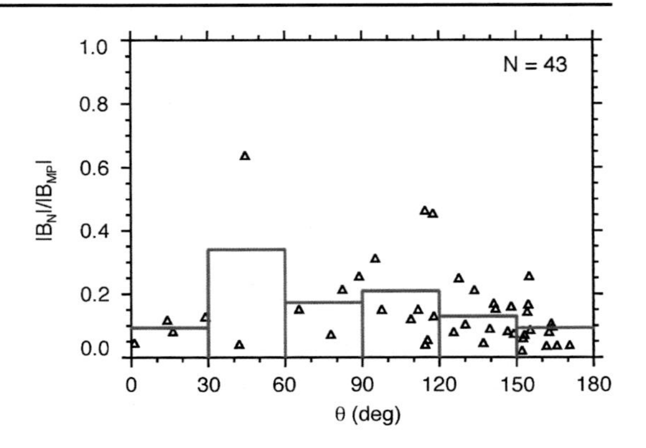

Fig. 6 Magnetopause shear angle θ compared with the rate of reconnection for the magnetopause crossings. The average reconnection rate was calculated in 30° bins, as indicated by the *red rectangles*. Little correlation between the two quantities is evident, indicating that the reconnection occurs at Mercury for a large range of shear angles

~ 20 nT, an order of magnitude larger than typical measurements at Earth. Additionally, the mean rate of reconnection resulting from this study was $\alpha = 0.15 \pm 0.02$, which is about a factor of three larger than the most extensive studies at Earth. However, more importantly, this study revealed that reconnection occurs at Mercury's magnetopause independent of the magnetic shear angle θ, the angle between the planetary field and the IMF (Fig. 6). In fact, DiBraccio et al. (2013) identified several reconnection events with $\theta < 30°$, including one event where $\theta \sim 1°$. Upon further inspection, the low-shear reconnection at Mercury appears to be a product of the low plasma β and decreased M_A of the solar wind in the inner heliosphere, as predicted by Slavin and Holzer (1979). The frequency of strong PDLs (Gershman et al. 2013) also appears to enhance the occurrence of reconnection for all IMF shear angles at Mercury (DiBraccio et al. 2013; Slavin et al. 2014).

Using MESSENGER FIPS and MAG data, DiBraccio et al. (2015b) presented the first observations of Mercury's plasma mantle, a main source for solar wind entry into the planet's magnetosphere, located in the high-latitude magnetotail. The plasma mantle is created as reconnected fields, populated with solar wind plasma, convect downstream of the planet and rejoin the magnetosphere as part of the Dungey cycle. The analysis of two successive orbits on 10 November 2012, revealed a dense population of solar wind protons present just inside the high-latitude tail magnetopause (DiBraccio et al. 2015b). These two events, with durations of 16 and 21 min, exhibited clear dispersions in the proton energy distributions observed by FIPS. This dispersion indicated that low-energy protons were transported much deeper into the magnetosphere than the higher energy particles, which escape to large downtail distances before they can $\mathbf{E} \times \mathbf{B}$ drift deeper toward the plasma sheet, where $\mathbf{E}$ and $\mathbf{B}$ is the cross-tail electric field and magnetic field magnitude, respectively. Frequent observations of FTEs throughout the magnetosheath, cusp, and into the magnetotail during these orbits are supportive of the high reconnection rates measured at Mercury and suggest that intense dayside reconnection is responsible for transporting solar wind plasma into Mercury's magnetosphere just as at Earth. Observations of Mercury's plasma mantle have provided direct evidence of one mechanism responsible for transporting solar wind plasma into the magnetosphere, which has consequences for surface space weathering especially through nightside plasma precipitation.

3.1.2 Flux Transfer Events

Reconnection is also observed at Mercury's magnetopause in the form of FTEs (Slavin et al. 2008, 2009, 2010, 2012b; Imber et al. 2014). FTEs are created as reconnection occurs

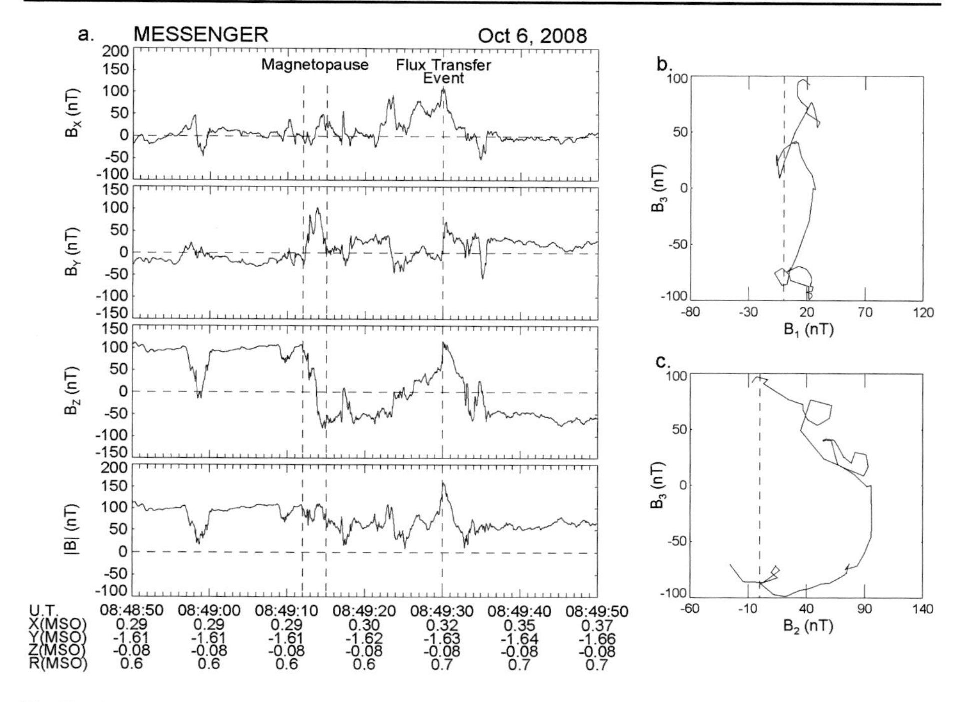

Fig. 7 (**A**) Magnetic field observations of the inner current sheet and MP boundary observed as MESSENGER exited the dawn-side magnetosphere. (**B**) Magnetic field measurements across the MP graphed in the plane of maximum and minimum variance. (**C**) Magnetic field measurements across the MP graphed in the plane of maximum and intermediate variance. Adapted from Slavin et al. (2009)

between the IMF and planetary magnetic field at multiple dayside X-lines. They are identified by their flux rope topology: a strong, axial-aligned core field with helical outer wraps increasing in pitch angle with radial distance from the center. In magnetic field data, the helical wraps are typically indicated by a bipolar signature, which also provides information about the direction that the flux rope is traveling. The core field is designated by a local field enhancement that is coincident with the inflection point of the bipolar signature. Flux ropes may also be remotely observed if the spacecraft does not directly pass through the FTE, but rather, encounters the draped and compressed fields surrounding the flux rope. These perturbations, called traveling compression regions (TCRs), are used to infer the dimensions of a flux rope.

During the first MESSENGER flyby of Mercury, Slavin et al. (2008) identified a ~ 4 s-duration FTE in the magnetosheath using magnetic field data implying a size of ~ 1200 km, or 0.5 R_M. The bipolar signature is evident in the B_Y component with an enhancement in both B_X and B_Z. During the second MESSENGER flyby, Slavin et al. (2009) reported a FTE with a core field strength of 160 nT and a duration of ~ 3 s (Fig. 7). The size of this flux rope was estimated to be ~ 900 km, or 0.4 R_M. After a more extensive review of the MESSENGER flybys, Slavin et al. (2010) reported on six FTEs encountered during the first and second Mercury flybys. The durations of these events ranged from 1–6 s and a flux rope modeling technique (Lepping et al. 1990, 1995, 1996) was implemented and determined the FTE diameters to range from 0.15–1.04 R_M. Additionally, the model results indicated that the magnetic flux content of these structures ranges from 0.001–0.2 MWb, or

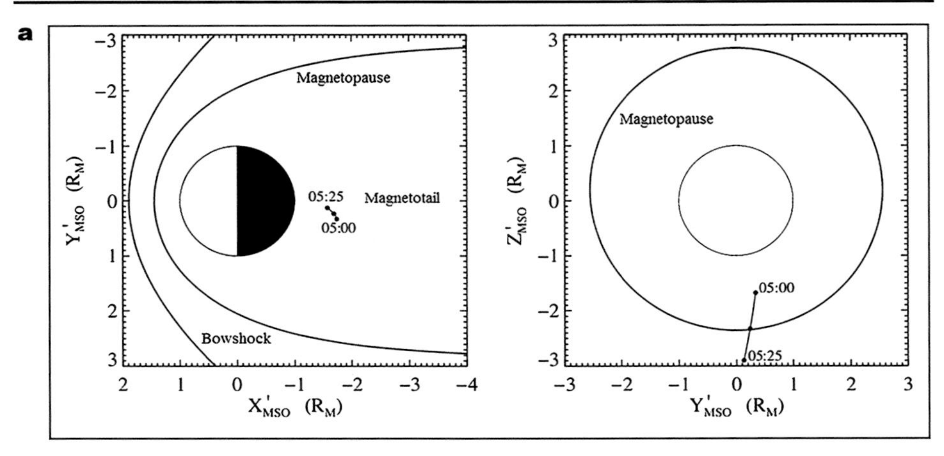

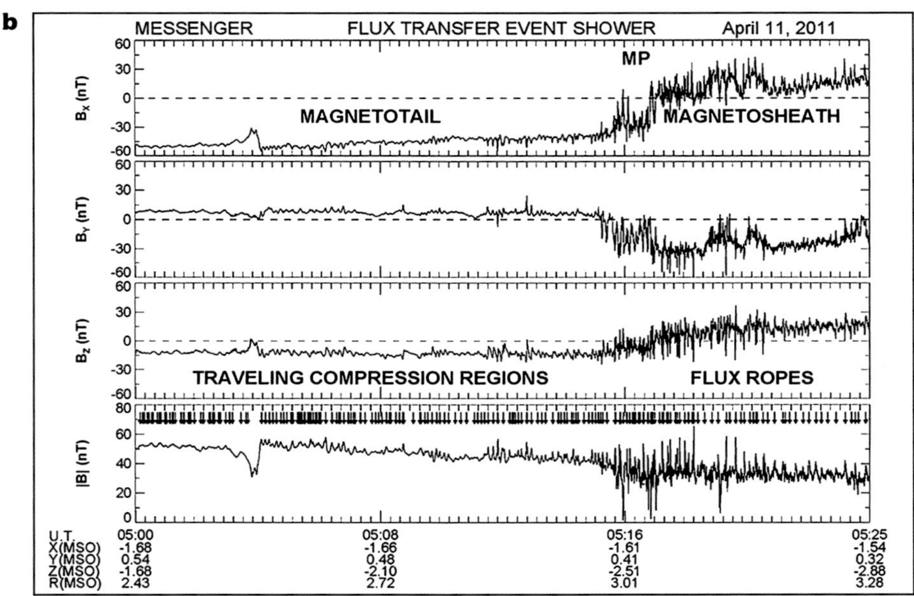

Fig. 8 MESSENGER trajectory from 05:00 to 05:25 UTC on 11 April 2011, projected on to the aberrated MSO *X-Y* and *Y-Z* planes. Note that the bow shock and magnetopause surfaces are shifted northward by 0.20 R_M to match the northeward offset in Mercury's internal magnetic dipole. (**b**) Magnetic field measurements taken during this interval span the outer portion of the southern lobe of Mercury's magnetotail, the magnetopause, and the nearby magnetosheath. *Vertical arrows in the fourth panel* mark 97 TCRs inside the magnetotail and 66 FTE-type flux ropes in the adjacent magnetosheath. Adapted from Slavin et al. (2012b)

about 5 % of the 4–6 MWb tail lobe flux (Slavin et al. 2010). Additionally, the largest of these events may contribute up to ~ 30 kV to the cross-magnetospheric electric potential.

At Mercury's magnetopause, flux ropes have been identified to occur as "FTE showers" (Slavin et al. 2012b). During a MESSENGER noon-midnight orbit on 11 April 2011, a combination of 163 FTEs and TCRs were observed over a 25 min interval as the spacecraft traversed the southern tail magnetopause (Fig. 8). During this orbit, the IMF was predominantly oriented northward. The average duration of the FTEs and TCRs was 1.7 s and 3.2 s, respectively, and all events were separated by periods of $\sim$ 8–10 s. By implementing the flux rope modeling technique of Hidalgo et al. (2002a, 2002b), the mean semimajor axis of the flux ropes was determined to be 0.15 R_M.

Most recently, Imber et al. (2014) performed a statistical study on FTEs observed in Mercury's subsolar magnetosheath. In this study, 58 large-amplitude FTEs, with core fields larger than the magnitude of the planetary field just inside the magnetopause, were selected. The average durations of these events were 2.5 s. MVA was used to determine their orientation and the force-free flux rope model of Lepping et al. (1990, 1995, 1996) was applied to estimate an average flux content of 0.06 MWb. Imber et al. (2014) concluded that unlike Earth, where FTEs contribute to $< 2\ \%$ of substorm flux transport, at least 30 % of the flux transport required to drive Mercury's 2–3 min substorms is contributed by FTEs.

3.1.3 *Kelvin-Helmholtz Waves*

Kelvin-Helmholtz (KH) instabilities are another well-known mechanism responsible for the transfer of mass, momentum, and energy from the solar wind into planetary magnetospheres. In situ observations of KH waves at a planetary magnetopause can be identified as surface waves creating a series of periodic magnetopause crossings. Indeed, the growth rate of KH waves relies on the velocity shear and finite Larmor radius effects. During the first MESSENGER flyby of Mercury, Slavin et al. (2008, 2009) reported possible KH wave activity after identifying three rotations along the dusk magnetopause while the IMF had a northward orientation. The durations of these field rotations were $\sim$ 5–25 s, implying spatial scales of $\sim$ 0.2–2 R_M. Sundberg et al. (2010) studied these events in further detail and concluded that the observed waves were not due to KH instabilities but might possibly indicate an initial perturbation leading to KH vortices further down the tail.

During the third MESSENGER flyby, Boardsen et al. (2010) identified magnetic field variations indicated by 15 dusk-side magnetopause crossings over a short 2-minute interval, likely suggesting the presence of KH instabilities. Additionally, a distinct sawtooth pattern present in B_Y and, to a lesser extent, B_X, supports the conclusion of highly steepened KH wave activity. Sundberg et al. (2011) revisited these observations and performed a reconstruction of the KH vortex, with the assumption that the wave pattern is quasi-stationary. This analysis concludes that the spatial reconstruction of a vortex pattern is in agreement with the field rotations located at the dusk-side magnetopause during the third MESSENGER flyby.

To understand the general characteristics of KH waves at Mercury's magnetopause, Sundberg et al. (2012) performed a survey of six KH wave trains by identifying the events based on sawtooth wave patterns (Fig. 9) and periodic magnetopause crossings in magnetic field data. The results provide clear evidence that KH waves are frequently observed at Mercury's dusk-side magnetopause with wave periods ranging from 10–40 s and large-amplitude oscillations ranging from 70–150 nT.

Gershman et al. (2015) showed that KH waves observed on Mercury's dusk-side ($\sim$ 18–21 h local time) magnetopause can be affected by the presence of heavy planetary ions. On the dusk-side, where Na^+-group ions (m/q 21–30) can dominate the pressure, KH waves appear at the Na^+ ion gyrofrequency. This kinetic-scale behavior is due to the large gyroradii of these planetary ions. This is contrasted with the fluid-scale behavior of other KH waves observed at Mercury, especially on the dayside region around the dusk terminator (12–18 h local time). This work constitutes the best evidence to date that Na^+-group ions can be dynamically important in the magnetosphere, an open question since the discovery of the Na-dominated exosphere at Mercury.

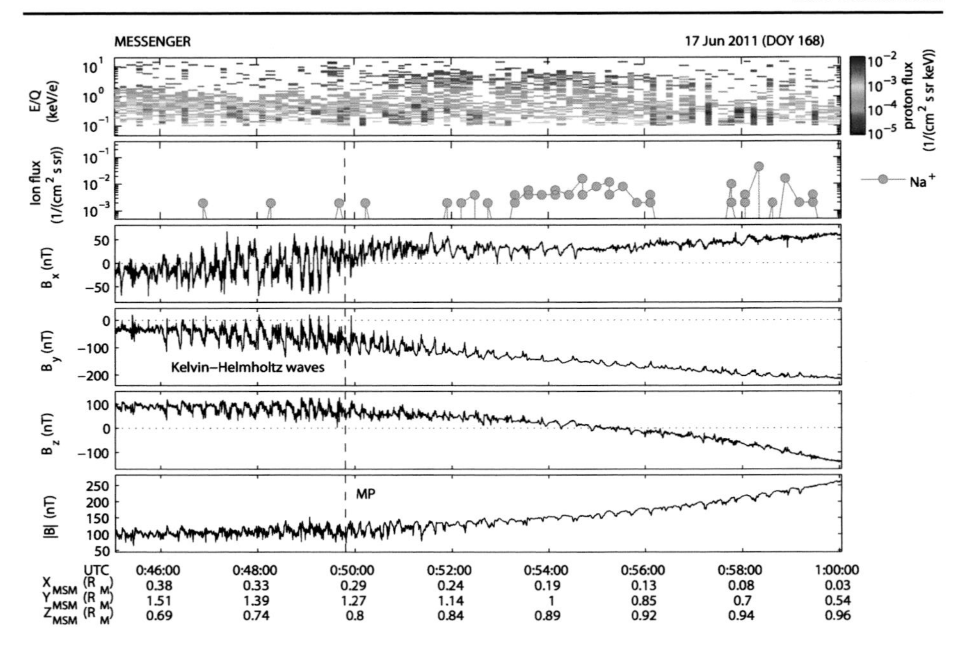

Fig. 9 KH observations on 17 June 2011. The *first and second panels* show the FIPS spectrogram of E/Q for the measured proton flux and the sodium ion count rate, respectively. The *third–fifth panels* are the magnetic field components and the *sixth panel* is magnetic field strength. Adapted from Sundberg et al. (2012)

3.2 Planetary Ions

3.2.1 Surface Processes

Exogenic processes acting on the surface, causing particle release, permanently populate Mercury's exosphere, the thin, collision-free, gaseous envelope around the planet. Ionization of these exospheric particles contributes significantly to the magnetospheric population of ions. Four processes have long been considered for particle release at Mercury: thermal desorption (TD), photon-stimulated desorption (PSD), micro-meteoritic impact vaporization (MIV), and ion-induced particle sputtering (IS). These particle-release processes have been reviewed several times (e.g., Wurz and Lammer 2003; Killen et al. 2007 and General Processes chapter of this issue) and have been extensively studied for Mercury (e.g., Mura et al. 2009; Wurz et al. 2010).

The intense solar irradiation of the surface is responsible for TD and PSD, i.e., these processes are confined to the dayside of Mercury. TD is restricted to volatile species, i.e., species that have an appreciable sublimation rate at the surface temperatures of Mercury, 100–700K. These volatiles (H_2, N_2, O_2, H_2O, CO_2, He, Ne, Ar, and molecular fragments thereof) are expected to constitute the major part of Mercury's dayside exosphere, but only He has been detected so far. Contributions by the other three processes are orders of magnitude lower (Wurz and Lammer 2003; Wurz et al. 2010). Since the evaporation rates for the dayside temperatures are large, volatile species falling onto the surface will be re-emitted almost immediately into the exosphere and will be thermally accommodated with the surface temperature. On the nightside, in contrast, some volatiles can condense and are thus removed from the exosphere. Thus, a day-night modulation in exospheric density of some volatiles is expected, as was observed for argon in the lunar exosphere (Stern 1999). Since TD-released particles have thermal energies, they all fall back onto the surface and escape (Jeans escape) is negligible. The contribution to the magnetosphere is via photoionization of exospheric gas. Since the scale heights of thermal particles are low, and thus the ballistic travel times are low, the flux of photoions from thermal species is moderate.

PSD, also driven by solar irradiation, is even more restricted than TD for species it can release from the surface: at Mercury only Na and K are released by this process. However, appreciable PSD yields of Na and K are only observed if the alkali metal is freed from the mineral bound in the crystal and is available as adsorbed atom on the surface (Yakshinskiy and Madey 1999, 2004). Impacting energetic plasma ions may cause the liberation of the alkali metal from the mineral, which was used in a recent 3D model to explain Na observations during Mercury transit of the Sun (Mura et al. 2009). Alternatively, a surface reservoir of Na was postulated to model the exospheric Na observations during a Mercury year (Leblanc and Johnson 2010). A part of these models is the consideration of the fate of alkali atoms when they fall back to the surface, which is discussed as sticking probability in surface physics. The sticking probability for atomic K is nearly constant over the surface temperature range of 100–500 K, whereas for Na it decreases with increasing temperature in this range (Yakshinskiy and Madey 2005), which influences the Na/K ratio to be observed in the exosphere. More recently, extensive UVVS observations of Na (K has not been observed by MESSENGER) have shown that TD is not a significant process for Na (Cassidy et al. 2015). It is also not seen in the other species regularly observed by UVVS: Ca (Burger et al. 2014) and Mg. The lack of TD is surprising for Na given that it is relatively volatile (Hunten et al. 1988) but may be explained by the relatively large binding energy seen for Na adsorbed on an ion-bombarded surface (Yakshinskiy et al. 2000).

MIV will take place everywhere on the surface of Mercury, on the day- and nightside. MIV fluxes at Mercury have been modeled by several authors (Cintala 1992; Müller et al. 2002; Cremonese et al. 2005; Bruno et al. 2007; Borin et al. 2010). These fluxes are usually considered omni-directional, though (Killen and Hahn 2015) showed that preferential dust bombardment on the dawn hemisphere could explain the concentration of Ca exosphere there (Burger et al. 2014). The impact of micro-meteorites and meteorites results in the release of surface material in form of gas and solid fragments (e.g., Cintala 1992) where the gas fraction is a hot thermal expanding cloud composed from all the material of the impact site. Most of the micro-meteorites are indeed very small particles, and thus a constant flux bombards Mercury's surface resulting in a constant contribution to the exosphere (Wurz et al. 2010). For typical solar wind conditions, MIV and IS give similar exospheric particle populations (Wurz et al. 2010). However, larger projectiles may sometimes hit the surface causing the exospheric density contribution from MIV to temporarily increase (for about 1 hour) by up to a factor of 100 for projectiles of 0.1 m (Mangano et al. 2007). Nevertheless, such episodic events have not been observed for the 15+ Mercury years that UVVS has been regularly observing Na, Ca, and Mg.

IS is the process of particle release upon the impact of an energetic ion on a solid surface. IS is a very well understood process because of its application in semi-conductor industry (Behrisch and Eckstein 2007). IS depends on the energy of the impacting ion, and the sputter yield, i.e., the number of surface atoms sputtered per incoming ion, is maximal for ions with energy of 1 keV/nuc. All atoms on the surface are released by IS more or less stoichiometrically causing a continuous erosion of the surface. IS arises either from solar wind ions at the locations where solar wind ions have access to the surface of Mercury or by magnetospheric ions, both given by the topology of Mercury's magnetosphere. For typical solar wind conditions the sputtering contribution to the exosphere is small (Wurz et al.

2010), but for CMEs with significantly higher plasma density and increased He^{2+} contents the ion sputtering contribution may increase dramatically for the duration of the CME passage, as was recently discussed for the Moon (Farrell et al. 2012). Sputtered particles have high kinetic energies, and a significant fraction of them can escape the gravitational field of the planet (Wurz et al. 2007, 2010). Because of their large exospheric scale height and the resulting long ballistic flight times, significant ionization of sputtered atoms occurs, which is species dependent, providing input to the ion population of Mercury's magnetosphere. In addition, about 0.1 to 10 % of the sputtered atoms are already ionized when sputtered from the surface (Benninghoven 1975), thus contributing directly to the magnetospheric ion population. Magnetospheric dynamics may cause some of these ions to return to the surface (Delcourt et al. 2003) and cause sputtering themselves, including on locations on the night side surface.

3.2.2 Neutral Observations

The components of Mercury's exosphere are sources of the magnetospheric ion population mostly through the photoionization process. For this reason, it is important to investigate the density, distribution, and variability of the neutral component to understand the plasma populations.

Generally, the observation of the exosphere can be performed by ground-based telescopes in the spectroscopic regions free of Earth atmospheric lines or by in situ measurements with ultraviolet-visible (UV-Vis) spectrometers and mass spectrometers. Particularly in the case of Mercury, ground-based observations can take advantage of both night telescopes and solar telescopes/towers and can provide global imaging of the extended exosphere (disk and tail). So far, only the elements Na, K, and Ca have been observed by ground-based telescopes. In situ measurements, instead, can provide high-resolution imaging of local density and allow the detection of lower intensity signals to extend the list of observable species. In both cases the exosphere brightness is calibrated using photometric models of Mercury's surface (Hapke 1981, 1984, 1986; Domingue et al. 1997).

Mercury's exosphere was discovered by a UV spectrometer onboard Mariner 10 (Broadfoot et al. 1977) that covered part of the extreme and far wavelength ranges (30–167 nm). It discovered atomic H and He, and made a possible detection of O. An occultation experiment on Mariner 10 also provided an upper limit of the total atmospheric abundance, which was higher than the sum of detected constituents (Fjelbo et al. 1976), meaning that some exospheric species remained still undetected. About a decade later, ground-based observations discovered Na, identified via the D1 and D2 emission lines (near 589 nm wavelength), which are caused by resonant scattering of sunlight (Potter and Morgan 1985). Later, also K and Ca have been detected by ground-based observations (Potter and Morgan 1986, 1997; Bida et al. 2000) and an upper limit for Al, Fe, and Si was defined (Doressoundiram et al. 2009). MESSENGER UVVS discovered Mg and Ca^+, and in its orbital phase regularly observed Na, Ca, Mg, and occasionally H. Its wavelength range (115 nm–600 nm) precluded observations of He and K.

The Broadfoot et al. (1976) Mariner 10 detection of atomic oxygen was 'very tentative', and it was not replicated by MESSENGER UVVS, which could have easily seen the claimed ~ 60 Rayleigh emission (Vervack et al. 2011). Wurz et al. (2010) predicted that ion sputtering and impact vaporization should produce large atomic oxygen column density (comparable in magnitude to the observed sodium), but it would be difficult to detect with UVVS given the poor efficiency with which atomic oxygen scatters sunlight (Killen et al. 2009). This hypothesized oxygen exosphere is a likely source for the abundant oxygen ions detected by FIPS (Zurbuchen et al. 2011; Raines et al. 2013).

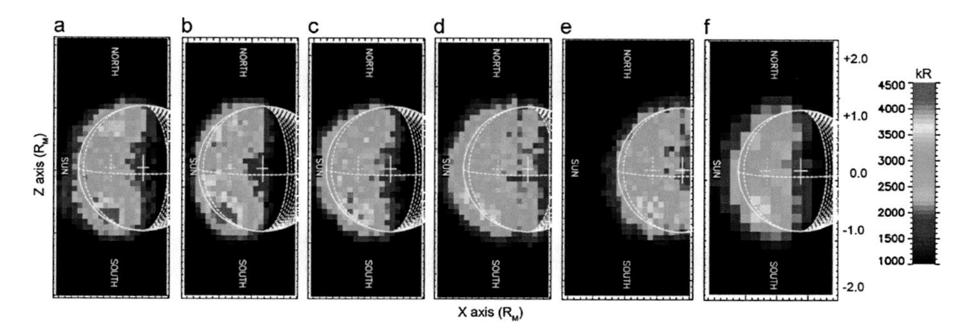

Fig. 10 Time sequence of scans of Na emission intensity (in kiloRayleigh) obtained in 13 July 2008 from 07:00 to 17:00 UT. The *X-Z* plane is the projection plane, with the *Z*-axis pointing northward; the *Y*-axis is along the direction Earth–Mercury (with the center being the sub-Earth point). The Sun is on the left. The *solid white line* denotes the disk of the planet and the cross indicating the center of the disk; in *white dashed region* of the disk not illuminated by the Sun, the sub-solar meridian, and the cross indicating the point of highest emission brightness due to solar reflection of the surface (Mangano et al. 2013)

Neutral observation of Na revealed, since their first detection, very distinctive features, such as recurrent peaks at mid latitudes (e.g., Potter et al. 1999) and a significant neutral tail in the anti-sunward direction (Kameda et al. 2009; Potter et al. 2002; Schmidt et al. 2012). Moreover, the variability of these features has been seen in almost three decades of Earth-based observations (Sprague et al. 1997; Potter et al. 2006; Leblanc et al. 2009). The average intensity and tail length modulate along the Mercury orbit in relation to the solar radiation pressure, which maximizes together with the velocity radial component (Leblanc et al. 2008). Kameda et al. (2009) related the average intensity modulation to the crossing of the interplanetary dust disk. The seasonal variation has been confirmed by MESSENGER UVVS (Cassidy et al. 2015). Most of the sodium exosphere is confined to low altitudes on the dayside; the scale height is only $\sim$ 100 km at low latitudes (Cassidy et al. 2015). This means that most of the ion source is deep within the magnetosphere, which has consequences for sodium ion kinetics (Raines et al. 2013, 2014; Gershman et al. 2014).

The improved spectral and temporal resolution of ground based observations allowed investigation of speed distributions (Leblanc et al. 2009) and detection of even more detailed features of the Na exosphere, which now range from time scales of days to hours. Daily variations are often due to changes in the position of Mercury around its orbit and to solar events (Killen et al. 2001; Potter et al. 2007). Hourly variations are attributed to normal solar wind fluctuations (mostly density and speed) and to rapidly changing IMF coupling with the planetary magnetic field (Mangano et al. 2013). Figure 10 shows an example of hourly variations of high latitude peaks in exospheric Na emission when observed from Earth. Similar double peaks at mid latitudes have been reported for the K exospheric distribution (Potter and Morgan 1986). This may indicate that both of these volatile species are linked to the solar wind impact onto the Mercury dayside surface below the cusps, even if it cannot be generated by direct ion sputtering (Mura et al. 2009). Observations of Ca (Burger et al. 2014) and UVVS observations of Na (Cassidy et al. 2015), instead, show different behavior apparently not related to solar wind impact but probably to MIV processes acting more efficiently in certain regions of the orbit due to higher MIV fluxes (Killen and Hahn 2015). In contrast to the rapid variability of the ground-based observations, UVVS observations of Na and Ca show little episodic variability as described below.

MESSENGER UVVS observations are quite different from, and complementary to, ground-based observations. UVVS provided unprecedented temporal coverage, observing

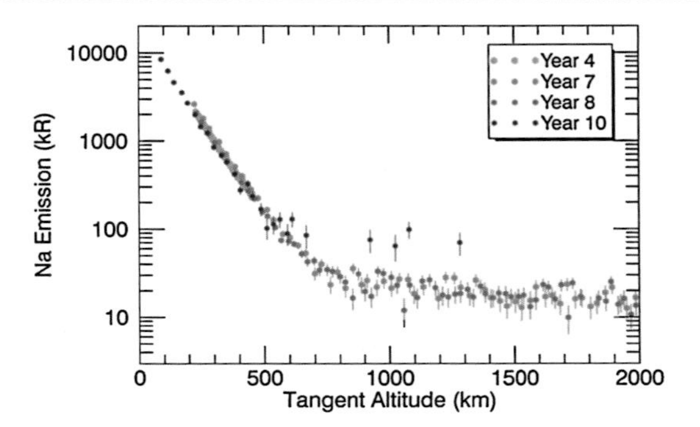

Fig. 11 Altitude profiles of sodium emission observed above Mercury's subsolar point by the MESSENGER MASCS UVVS instrument. These were taken over several Mercury years (as indicated by the legend), but all were taken near the same true anomaly angle, between 65°–70° in this example. Although the sodium exosphere varies temporally, this figure highlights the seasonal repeatability of MASCS observations

the exosphere almost daily for over 16 Mercury years. It also provided unprecedented spatial resolution: altitude profiles of exospheric emission resolve details down to the km scale (Fig. 11). UVVS had the advantage of *not* observing through Earth's atmosphere, but it had limitations, too. It was not an imaging spectrometer, and its field of view (FOV) and observation geometries were restricted by the many considerations of spacecraft operations in a challenging environment. It also had a relatively poor spectral resolution compared to the ground-based observations (~ 0.5 nm).

Some of the UVVS results are surprising in light of the decades of work published on Mercury's exosphere. In particular, ground observations (e.g. Mangano et al. 2013, above) and models show a Na exosphere that is highly variable on the time scale of hours. These sudden changes are thought to be in response to changing solar wind and IMF conditions. UVVS observations do not show this. The species that were regularly observed (sodium, calcium, magnesium) look quite similar from one Mercury year to the next, at least wherever consistent observing geometries were used over long periods of time (Burger et al. 2014; Cassidy et al. 2015). On the other hand, operational constraints have severely limited UVVS observations in the cusp, the most variable region. This may explain the differences, at least in part. Much of the UVVS data remains to be analyzed, so more progress on the variability of these exospheric species can be expected.

3.2.3 Plasma Observations

Overall Planetary Ion Composition and Distribution Plasma observations at Mercury began with the electron observations of Mariner 10 through three flybys in 1974–1975. Measurements from the first flyby convincingly showed Mercury to have an Earth-like interaction with the solar wind: There was a well-developed bow shock and a dense, hot plasma magnetosheath, surrounding a small magnetosphere (Ogilvie et al. 1974). Magnetometer measurements were compared with the plasma electron measurements and corroborated this interpretation (Ness et al. 1974). Within the magnetosphere, electrons were detected over the full energy range of the instrument, 13.4–687 eV, with a significant population in 200–680 eV range. These measurements were later interpreted as being from a hot plasma sheet (Ogilvie et al. 1977). Several energetic electron bursts were detected by the energetic particle instrument (Simpson et al. 1974; see also discussion in Wurz and Blomberg 2001), though they were later re-interpreted as being due to $>$ 36 keV electrons (Armstrong et al. 1975; Christon et al. 1987). Siscoe et al. (1975), Baker et al. (1986) and Christon et al. (1987) attributed these energetic bursts to substorms at Mercury. Fluxes and spectral shapes of plasma electrons were observed to be partially correlated with these energetic bursts. A hardware failure in the plasma ion instrument prevented any ion observations by Mariner 10 (Ogilvie et al. 1977). Measurements from both flybys were combined with neutral atom measurements from the ultraviolet spectrometer (Broadfoot et al. 1974, 1976) to infer that Mercury has no ionosphere, making the magnetosphere effectively bounded on the inside by the planet's surface.

The first plasma ion measurements at Mercury came with the first flyby of the MESSENGER spacecraft on 15 January 2008. The Fast Imaging Plasma Spectrometer (FIPS) (Andrews et al. 2007) detected ions throughout the entire Mercury space environment, confirming predictions of their presence (Zurbuchen et al. 2008). Protons and alpha particles (He^{2+}) from the solar wind were observed as the spacecraft traversed the magnetosphere, with highest abundance in the magnetosheath. Many heavy ions were also detected, ranging in mass per charge (m/q) from 6–40 amu/e. These ions were found with highest abundance within the magnetosphere, with Na^+ (or Mg^+) ions dominating the heavy ion population. As Na is one of the dominant atoms in the exosphere and is easily ionized (Wurz and Lammer 2003), these ions are generally taken to be Na^+, though the separation of Na^+ from Mg^+ ion has not yet been accomplished from FIPS data.

Once MESSENGER went into orbit around Mercury on 18 March 2011, the vast increase in the amount of data also necessitated a change in approach to a more automated approach of assigning counts to individual ion species that could be applied to the data in a largely automatic fashion. The main effect of this change was grouping of ions into ranges of m/q : O^+ group, m/q 14–20, including O^+ and any water group ions (e.g., H_2O^+, OH^+); Na^+ group, m/q 21–30, including Na^+, Mg^+ and Si^+. Substantially improved background removal was also accomplished in this new method, along with a much better estimation of signal to noise. The use of counts as measurement units was also replaced with a more physically relevant unit, the observed density (n_{obs}). This is the density computed from the counts measured, without any correction for those unobserved due to the limited FIPS field of view (FOV) on the three-axis stabilized MESSENGER spacecraft. These methods are explained in more detail in Raines et al. (2013).

A more complete picture of the distribution of ions in Mercury's space environment emerged from this much larger dataset. First, planetary ions were found throughout this space environment, both inside and outside of the magnetosphere. For the two most abundant species, Na^+-group and O^+-group ions, this distribution is not at all uniform. These ions show a very substantial abundance enhancement in the region of Mercury's northern magnetospheric cusp. Na^+-group and O^+-group ions are also very abundant in the nightside near-equatorial region, and often near high-latitude, dayside crossings of the magnetopause (Zurbuchen et al. 2011). Figure 12 shows this distribution, as a function of planetary latitude and local time, accumulated from 25 March 2011 through 31 December 2011. The different panels are accumulations over more than 500 orbits, indicating that these enhancements are very likely permanent features of Mercury's magnetosphere.

In contrast to Na^+-group and O^+-group ions, He^+ is much more evenly distributed throughout the space environment (Zurbuchen et al. 2011; Raines et al. 2013). This ion is present in the solar wind, but its enhanced abundance around the planet indicates that a significant fraction of its population comes from Mercury, either from the surface or exosphere. For example, He^+ has a distinct distribution from alpha particles (He^{2+}), not showing the magnetosheath enhancements very clearly observed in the doubly ionized He^{2+}. Very low

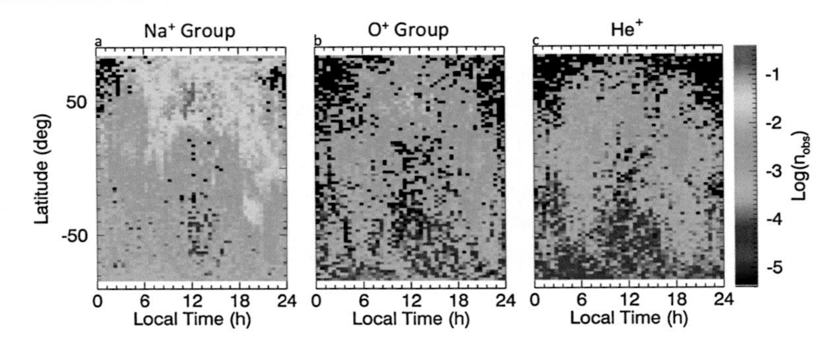

Fig. 12 Na^+-group (**a**), O^+-group (**b**), and He^+ (**c**) ion observed density as a function of local time and planetary latitude. Observed density is averaged within each 0.5 h by 2° local time–latitude bin. Unobserved regions are *colored white* while observed regions with zero counts are *colored black*. Updated from Zurbuchen et al. (2011)

plasma densities in Mercury's space environment make formation of He^+ from He^{2+} unlikely in any substantial quantities.

Of course, the distribution of planetary ions does not directly infer their sources. In the absence of a collisional atmosphere or ionosphere, the inner boundary of Mercury's magnetosphere is essentially the surface of the planet. Ions observed anywhere in this environment have been subject to the electromagnetic forces and processes of the magnetosphere, and their trajectories have been substantially affected. Furthermore, most of these processes are expected to be highly variable in time. One strategy employed for examining the relationship between observations and sources has been to look at the average behavior of ion distributions, hoping to find some commonality with the expected exosphere or surface sources. Raines et al. (2013) showed that the average observed density of Na^+-group and O^+-group ions varied substantially with true anomaly angle, the angle between Mercury and its orbital periapsis around its Keplarian orbit (Figure 2c of that work). He^+ ions showed a much less pronounced variation. These results were compared notionally with ground observations of the same variation of the exosphere. No clear correlation was apparent. In that same work *e*-folding heights of observed density versus altitude were computed for those same three planetary ion species, around three local times (dawn, noon, and dusk). These heights showed substantial differences across local time and species, with the smallest height always at noon and those of Na^+ group $\sim$ 2–6 times smaller than other ions. These ion *e*-folding heights are much larger, at least 5–10 times, than calculated scale heights for species of the neutral exosphere (Wurz and Lammer 2003; Wurz et al. 2010), likely confirming expectations that magnetospheric dynamics plays a substantial role.

Cusp Mercury's magnetospheric cusps have long been thought to be major sources of planetary ions for its magnetosphere, primarily through the process of solar wind sputtering (Lammer et al. 2003; Leblanc and Johnson 2003, 2010; Massetti et al. 2003). As discussed above, the abundance of planetary ions is largest there (Raines et al. 2013; Zurbuchen et al. 2011). The cusps, however, are very active, dynamic regions at Mercury, so a more detailed analysis was required to connect observed ions to cusp sources.

Raines et al. (2014) performed such a study of Mercury's Northern cusp region. Focusing on Na^+-group ions and protons, these authors selected 77 cusps with significant Na^+-group ion content from 518 orbits, spanning observations from September 2011 through May

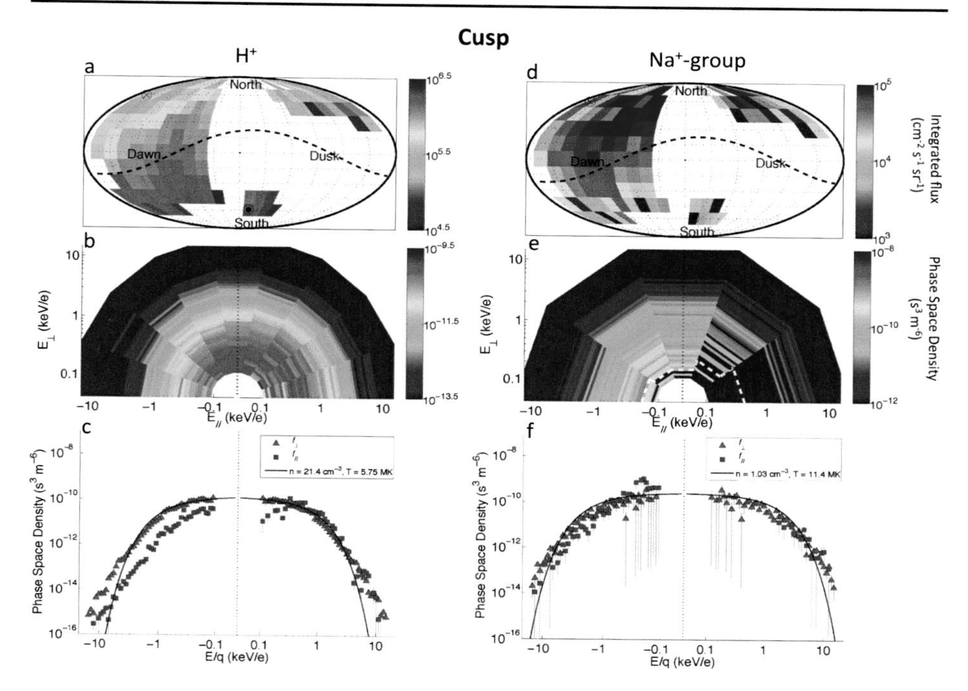

Fig. 13 Kinetic properties of protons and Na^+-group ions within the cusp, accumulated over 77 cusp crossings. *Top panels* (**a**), (**d**) show flow direction histograms for protons and Na^+-group ions. The *middle panels* (**b**), (**e**) are energy-resolved pitch angle distributions, which show the flow direction and energy of ions relative to the magnetic field in 20° (protons) and 36° (Na^+-group) bins. Slices through these distributions in the parallel, anti-parallel and perpendicular directions are shown in the *bottom panels* (**c**), (**f**). These figures show protons which are flowing down toward the surface, as well as loss cone of > 40° in width. Low energy (100–300 eV) Na^+-group ions appear to be upwelling from the surface, while those at energies up to 10 keV have large perpendicular energy components. Reproduced from Raines et al. (2014)

2012. They examined ion flow directions, energy-resolved pitch angle, energy and spatial distributions for these two species. Their main result was that Na^+-group ions in Mercury's cusp are too high in energy (2.7 keV on average) to be produced locally in the cusp. They also found a regular occurrence of keV-energy Na^+-group ions flowing northward in the dayside magnetosphere. From these measurements, the authors hypothesized that neutral Na atoms were ionized outside of the subsolar magnetopause and accelerated into the cusp by reconnection. This process may constitute a significant source of keV-energy planetary ions in Mercury's magnetosphere.

Two other interesting results emerged from this work, both of which can be more easily seen from energy-resolved pitch angle distributions. These plots (Fig. 13), which show the flow direction of ions relative to the magnetic field, are particularly interesting in the cusp because the magnetic field is largely radial there. This means that ions traveling in the anti-parallel magnetic field direction are effectively headed away from the surface, while those parallel ions are headed toward the surface. The energy-resolved pitch angle distribution for protons (Fig. 13b) shows a distinct depletion in flux coming up from the surface (anti-parallel, left side of figure), when compared with the flux going down toward the surface (parallel, right side of figure). This asymmetry likely results from the fact that a fraction of protons traveling toward the surface are lost to surface precipitation, rather than being

Table 1 Average kinetic properties in the central plasma sheet. From Gershman et al. (2014)

Species	Density (cm^{-3})	Temperature (MK)
H^+	7.81	9.29
He^{2+}	0.265	30.3
Na^+-group	0.663	15.7

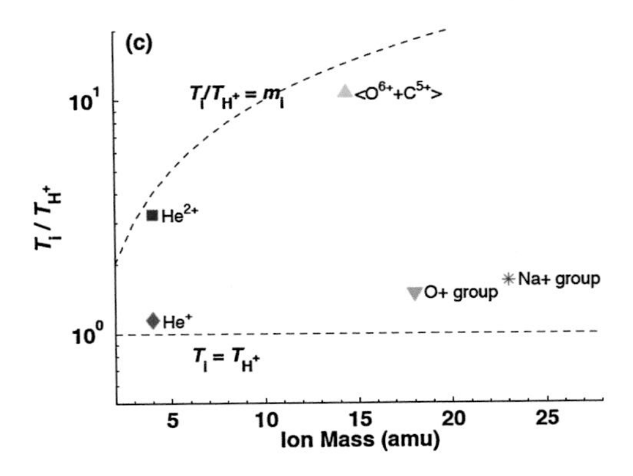

Fig. 14 (**a**) Average temperature T_i of each species relative to that of H^+. *Dashed lines* corresponding to $T_i/T_{H^+} = m_i$ are also shown

reflected in the increasing magnetic field there. This loss cone appears to be > 40° and constitutes a strong indication that protons are impacting Mercury's surface in the cusp. The opposite is observed for Na^+-group ions (Fig. 13e): At energies of 100–300 eV, they are enhanced in the anti-parallel direction and therefore appear to be streaming out of the cusp. This is especially visible as a small bump in the anti-parallel phase space density shown in the left half of Fig. 13f. Taken together, these two results may constitute a cause and effect observation of solar wind sputtering at Mercury, though some additional explanation of Na^+-group acceleration is required. Several studies provided a more quantitative look at proton precipitation at the cusp. Those are reviewed in Sect. 4.3.1.

Central Plasma Sheet A large collection of data from the orbital phase showed that average plasma sheet densities were in line with those observed in the first flyby, though average temperatures were higher (Gershman et al. 2014). In addition to values for protons, average density and temperature were also reported for alpha particles and Na^+-group ions (Table 1), giving a good average picture of plasma sheet ions for consideration by other studies. The estimated pressure contribution from plasma sheet protons was found to be in good agreement with the observed magnetic depressions there (Korth et al. 2011), providing an independent validation of these recovered plasma parameters.

One of the most interesting results from Gershman et al. (2014) comes from the relative temperatures of plasma sheet ions (Fig. 14). For solar wind ions, alpha particles and solar wind heavy ions (mostly O^{6+} and C^{5+}), the ratio of their temperature to that of protons is mass-proportional, i.e., $T_i/T_{H^+} = m_i$. This is expected for ions that are accelerated to the same speed, as is often the case in the solar wind and reconnection outflow. However, planetary Na^+-group and O^+-group ion temperatures show a roughly constant ratio to protons, as if they were accelerated through a potential. This may result from them having gyroradii which are large compared to plasma sheet magnetic field gradients, so that their motion in the plasma sheet is dominated by the cross-tail electric field. This result is consistent with findings by Raines et al. (2013) that Na^+-group ions are substantially enhanced in the pre-midnight plasma sheet when compared to the post-midnight side. These may both be observational evidence of the expected non-adiabatic behavior of heavy ions at Mercury, a point to which we return in some detail below.

4 Losses

There have yet to be any MESSENGER studies that focus on computing plasma loss rates from Mercury's magnetosphere. Estimating these rates from single spacecraft measurements of a highly dynamic system requires tightly coordinated and well-calibrated combination of models and data that has not yet been achieved. Work is heading in that direction, as described below in Sect. 5.1.2, so it is likely the plasma loss rates will be derived in the near future. Studies of several magnetospheric processes that contribute to plasma loss are described below.

4.1 Observations of Plasmoids and TCRs

Loading of the tail lobes and magnetopause flaring lead to thinning of the plasma sheet and its embedded cross-tail current layer for reasons that are still not well understood (Kuznetsova et al. 2007; Winglee et al. 2009; Raeder et al. 2010). When the current sheet thins, the normal magnetic field component is sufficiently reduced such that it becomes unstable to reconnection. A fundamental aspect of the reconnection process is the formation of magnetic islands with helical or quasi-loop-like topologies in the cross-tail current layer (Hesse and Kivelson 1998). These magnetic structures are called "plasmoids" (Hones et al. 1984). Similar to the FTEs at the magnetopause (Sect. 3.1.2), the lobe magnetic field becomes draped and locally compressed about the plasmoid, which can be observed as TCRs (Slavin et al. 1993). Because TCRs can be observed over a large fraction of the lobe region they are observed far more frequently than the underlying plasmoids that occupy a much smaller volume. Plasmoids and TCRs are highly correlated with the onset of magnetospheric substorms (Slavin et al. 1992; Moldwin and Hughes 1992). Many flux rope- or magnetic loop-like plasmoids can be formed during a given reconnection event, with some being carried sunward and others tailward by the fast Alfvénic jetting of plasma away from reconnection X-lines (Slavin et al. 2003). Indeed, initial analyses of the MESSENGER measurements have revealed the presence of sunward- and anti-sunward-moving plasmoids and TCRs (DiBraccio et al. 2015a; Slavin et al. 2009, 2012a).

Figure 15 displays a 90 sec-long interval on 29 August 2011 when MESSENGER had just entered the north lobe of the magnetotail $\sim 2.4\ R_M$ downstream of Mercury. The interval starts at 08:22:19 UTC with the spacecraft encountering a plasmoid. It is identified by the ~ 1.5 sec-long, large amplitude, north-then-south B_Z perturbation followed by a ~ 6 sec interval of weaker magnetic field with a southward orientation and higher frequency fluctuations. The plasmoid is then followed by a series of 9 traveling compression regions, which are similarly characterized by $\sim$ 1–2 sec north-then-south B_Z perturbations with a recovery period of ~ 5 sec. However, the TCRs differ in that they are strongly correlated with 10–15 % enhancements in the total magnetic field intensity. The absence of higher frequency fluctuations and the steady sunward orientation of the magnetic field indicate that all of these events take place in the northern lobe of the tail.

These observations are remarkably similar to the plasmoid and TCR events observed during the second flyby (Slavin et al. 2009). MESSENGER does not have the capability to measure the plasma flow during these events, but the mean ejection speed for plasmoids in

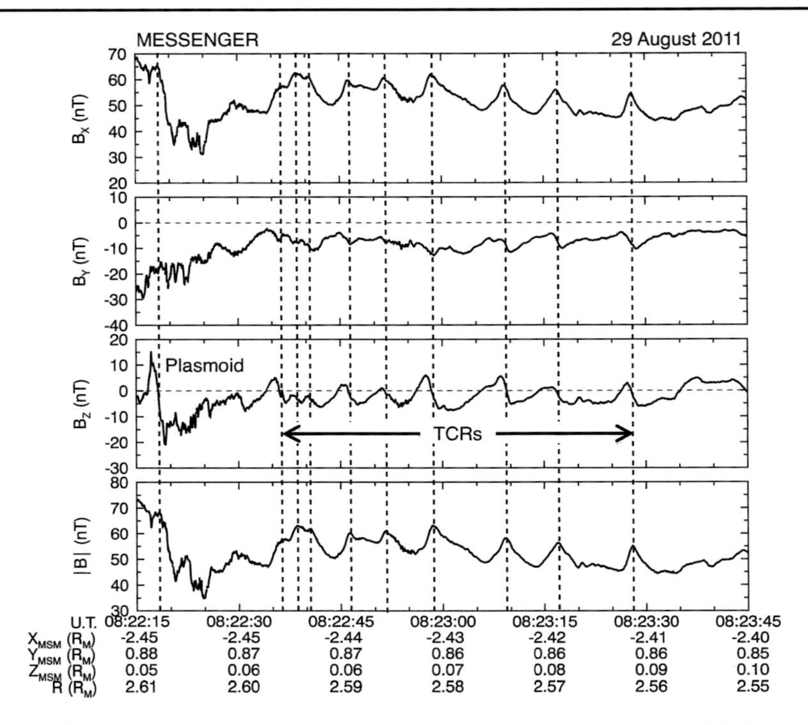

Fig. 15 Quasi-periodic plasmoids and TCRs observed during the orbital phase of the MESSENGER mission are marked with *vertical dashed lines*

the Earth's near-tail is $\sim$ 500–600 km/s (Ieda et al. 1998; Slavin et al. 2003). If we assume a speed of 500 km/s for these plasmoid and TCR events at Mercury, then the average diameters of these structures at Mercury are $\sim$ 500 km, or 0.2 R_M. This compares with $\sim$ 1 to 3 R_E plasmoid diameters in the near-tail of Earth (Slavin et al. 2003). Given the factor $\sim$ 8 scaling between the dimensions of these two magnetospheres, the diameters of plasmoids at these two planets appear to take up similar relative volumes at Mercury and Earth. It should also be noted that "chains" of plasmoids and TCRs, such as displayed in Fig. 15, are also common at Earth (Slavin et al. 1993, 2005; Imber et al. 2011). What is still not understood is whether these chains form simultaneously due to reconnection at multiple X-lines, as sometimes observed in simulations of ion tearing-mode reconnection (Schindler 1974; Tanaka et al. 2011) or to periodic episodes of reconnection at a smaller number of X-lines. Interestingly, the mean interval of 9 sec between the plasmoid and TCR events in Fig. 15 is very close the $\sim$ 8–10 sec spacing between flux transfer events observed at Mercury by Slavin et al. (2012b).

In a statistical survey of 49 flux rope-like plasmoids in Mercury's magnetotail, observed between 1.7 R_M and 2.8 R_M down the tail from the center of the planet, DiBraccio et al. (2015a) analyzed MESSENGER MAG and FIPS orbital data to determine the average characteristics of these structures. A superposed epoch analysis of the plasmoid-type flux rope events with north-then-south B_Z perturbations, consistent tailward motion, from DiBraccio et al. is displayed in Fig. 16. The magnetic field shows the characteristic variation expected for this type of flux rope (Slavin et al. 2003). In particular the strong core magnetic field in the $+/-$ Y direction centered on the bipolar B_Z variation associated with the outermost wraps of magnetic flux. This study concluded that the typical plasmoid diameter was

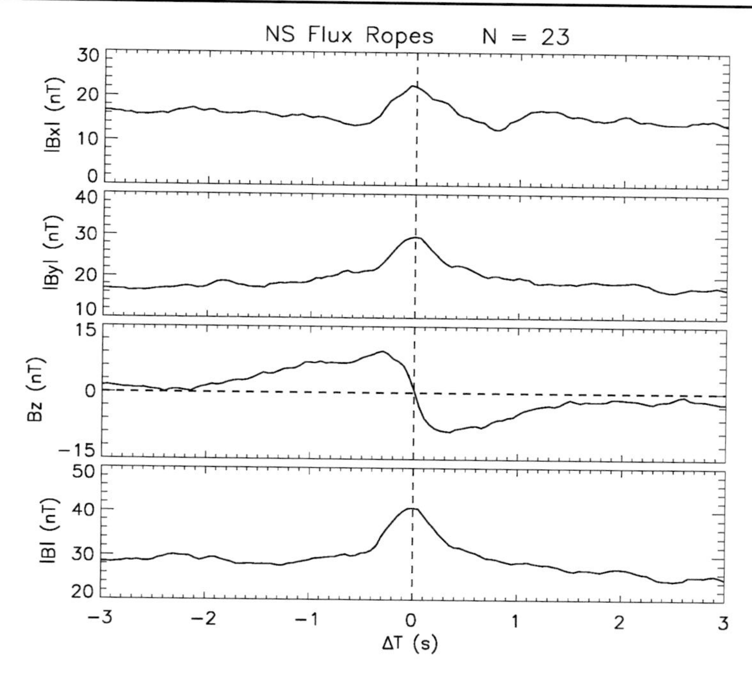

Fig. 16 Superposed epoch analysis of the magnetic fields measured during 23 plasmoid-type flux ropes in Mercury Solar Orbital coordinates (from DiBraccio et al. 2015a). Note the strong core magnetic field coincident with the bipolar variation in the B_Z perturbation. The continued negative B_Z following the plasmoid is due to continued reconnection involving lobe magnetic flux after the plasmoid is released as observed at Earth

~ 345 km, or $\sim 0.14\ R_M$, which is comparable to a proton gyroradius in the plasma sheet, or ~ 380 km. The events in this survey demonstrated that the magnetic variations of flux ropes at Mercury are similar to those observed at Earth but with timescales that are 40 times shorter at Mercury.

4.2 Observations of Dipolarization

An integral step in the substorm process is dipolarization of the fields in the near-tail (Baker et al. 1996). The transient increases in the northward, or dipolar, component of the equatorial magnetic field are closely associated with the braking of sunward-directed bursty-bulk flows originating at reconnection X-lines in the magnetotail (Angelopoulos et al. 1994). This is most readily understood as the result of the reconfiguration of the magnetotail into a lower energy state in which the stretched field lines created by the tail loading process quickly return to a more dipolar configuration (Shiokawa et al. 1994). At Earth these propagating dipolarization fronts are often accompanied by enhanced ion and electron fluxes up to hundreds of keV due to betatron acceleration as the magnetic field intensity increases (Ashour-Abdalla et al. 2011).

Earth-like dipolarization events were first observed at Mercury by Mariner 10 during it first flyby in 1974 (Baker et al. 1986; Christon et al. 1987). The observed magnetic field signatures were in good agreement with those expected from terrestrial dipolarization events, but with durations only of order 1 to 10 s as opposed to tens of min at Earth. Figure 17

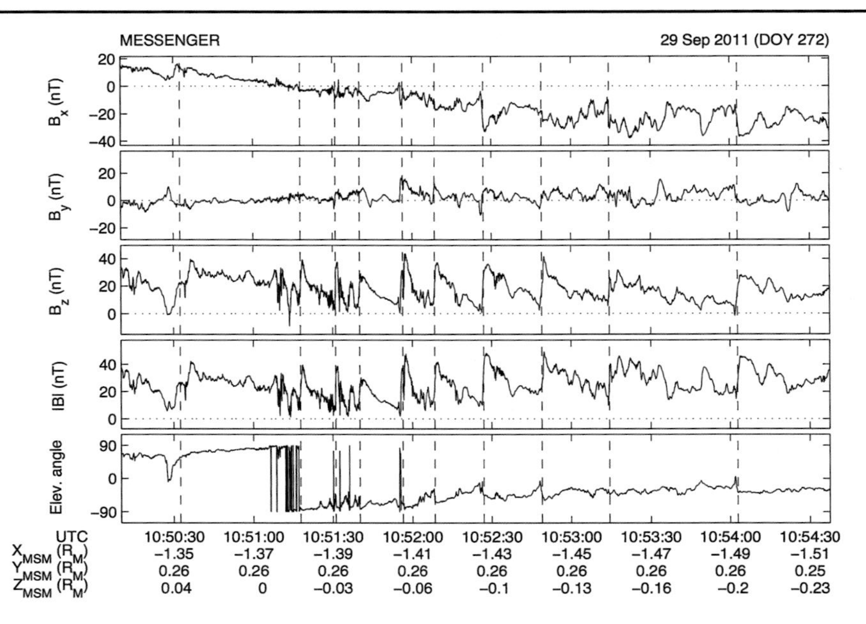

Fig. 17 A series of dipolarizations of the magnetic field in the plasma sheet observed by MESSENGER and analyzed by Sundberg et al. (2012). These brief, several second-long events are marked by *vertical dashed line*. Each has sudden, strong transitions from the magnetic field being highly stretched to a more dipolar configuration with a greatly enhanced B_Z component

displays an example of dipolarization events at Mercury on 29 September 2011 that have been analyzed by Sundberg et al. (2012). As shown, a series of 10 dipolarizations are seen to occur during a single plasma sheet encounter of several minutes at a distance of $\sim 1.4\ R_M$ downstream of Mercury's terminator plane. The dipolarization events, marked by dashed lines, are evident in the rapid (~ 1 s) increases and slow (~ 10 s) decays in the B_Z component of the magnetic field. The amplitudes of the magnetic field increases are 40–50 nT, similar to such events at Earth (Runov et al. 2011). The relatively short lifetime of the events is attributed to fast decay of the field-aligned currents that must accompany such dipolarizations. At Earth these currents close through an ionosphere with a conductance that is expected to be one or even two orders of magnitude larger than that of Mercury's regolith. The recurrence rate is generally in good agreement with those of plasmoids and traveling compression regions discussed previously.

4.3 Precipitation

4.3.1 Loss Cone Determination

Several studies provided a more quantitative look a proton precipitation at the cusp. Mapping of plasma pressures to invariant latitude (Korth et al. 2014) showed a clear north-south asymmetry on the nightside. This indicated increased particle loss through precipitation in the southern hemisphere, as anticipated from larger cusp that is created there by the northern offset of the planetary dipole. Winslow et al. (2014) used FIPS data to provide the first quantitative estimates of Mercury's loss cones and, from those, estimates of the surface fields in the cusp regions. In that work, self-normalized pitch angle distributions were summed over many cusp passages and then fit to an equation for pitch angle diffusion. The best-fit solutions gave loss cones of $59^\circ \pm 3^\circ$ for the northern cusp and $47^\circ{}^{+7}_{-13}$ in the southern cusp. The locations of the cusps were also mapped assuming an offset dipole field down from the spacecraft altitude to the surface. The northern cusp was found to be centered around 76.4°N latitude and noon local time, with a 15.6° extent in latitude and 7.5 h extent in local time. In the southern hemisphere, the cusp observations mapped to 23°–34°S latitude and 16–5.3 h local time. MESSENGER's orbit restricted the observation of the southern cusp to latitudes north of 30°S latitude, so uncertainties in the southern hemisphere are larger.

4.3.2 ULF Waves

Ultra-low frequency (ULF) waves were first detected in Mercury's magnetosphere by Mariner 10 (Russell 1989). Aside from acting as an important mechanism of energy transfer, these waves can increase plasma losses by scattering them into the loss cone. During the first MESSENGER flyby, Slavin et al. (2008) detected ULF waves in the magnetic field data between closest approach and the outbound magnetopause crossing with frequencies of ~ 0.5 to 1.5 Hz. Boardsen et al. (2009) performed a detailed analysis of these waves and found their fundamental mode was at frequencies between the He^+ and H^+ cyclotron frequencies (Fig. 18). Boardsen et al. (2009) concluded that wave frequency and amplitude increased from closest approach to the edge of a boundary layer located adjacent to the magnetopause; however, the frequency decreased by a factor of two and the amplitude increased by an order of magnitude inside the boundary layer.

Also inside Mercury's magnetosphere, Boardsen et al. (2012) surveyed coherent ULF waves at frequencies between 0.4–5 Hz. They were observed at the inner magnetosphere ($R < 0.2\ R_M$) at all MLTs. The waves are observed to be compressional and at maximum power near the equator on the nightside (Fig. 19), and become transverse with power decreasing for increasing magnetic latitudes. On average, the waves are strongly linear with wave-normal angles peaked around 90° and elliptical values < 0.3.

5 Modeling

5.1 Global Modeling of Mercury's Magnetosphere

5.1.1 MHD and Hybrid Models of Mercury's Magnetosphere

Global simulation models have been developed and applied to Mercury to understand the solar wind-magnetosphere interaction. These global models provide global context for interpreting and linking measurements obtained in various parts of the system, thereby extending our knowledge of Mercury's magnetospheric environment beyond that available from localized spacecraft observations. Two types of simulation models have been widely used in the global modeling of Mercury's magnetosphere, i.e., magnetohydrodynamic (MHD) simulation and hybrid simulation.

Global MHD simulation, in which both ions and electrons are treated as fluid, usually can provide a description of the global interaction over a reasonably large region around the obstacle and with relatively high resolution, at a feasible computational cost. MHD models have been used to characterize the large-scale structure of Mercury's magnetosphere under various solar wind and IMF conditions. For example, Kabin et al. (2000) employed a single-fluid MHD model to characterize the configuration of Mercury's magnetosphere

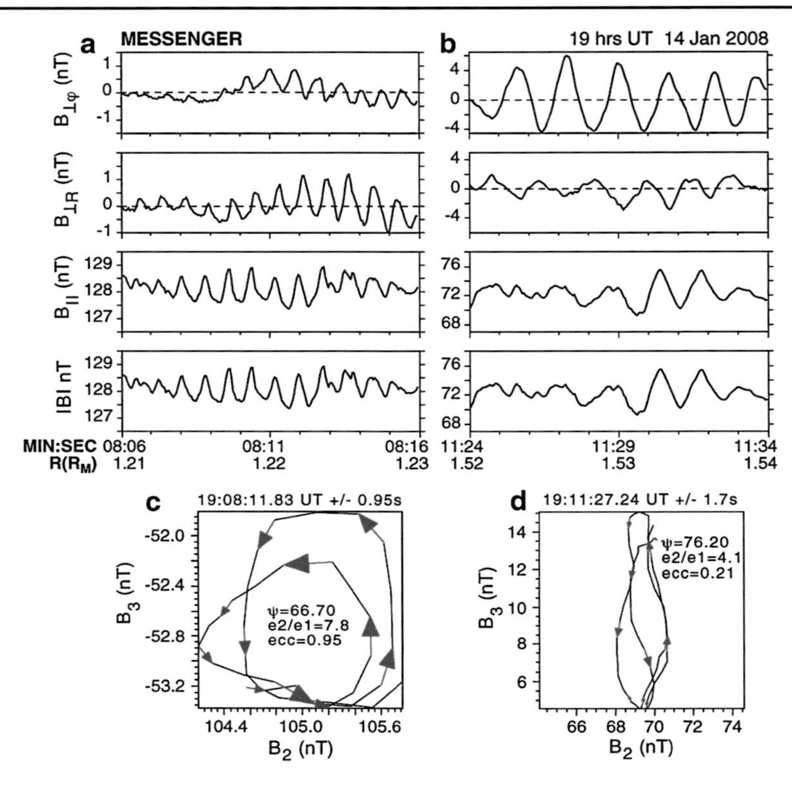

Fig. 18 (**a**) Time-series examples of ULF waves detected outbound from closest approach. (**b**) Example ULF waves. Hodograms of the time series shown in figures (**c**) 18a and (**d**) 18b. Axes B_2 and B_3 are the directions of intermediate and maximum variance, respectively. The wave-normal angle (Ψ), ratio of median to minimum eigenvalue ($e2/e1$) and ellipticity (ecc) are given for each hodogram. Adapted from Boardsen et al. (2009)

under extreme solar wind dynamic pressure conditions. Ip and Kopp (2002), also using a global MHD model, investigated the response of Mercury's magnetospheric configuration to different IMF orientations focusing particularly on the size of the polar caps, through which the solar wind particles can gain access to Mercury's surface. Recent MHD modeling efforts have been made to extend single-fluid MHD to multi-fluid MHD models. Kidder et al. (2008) adapted a multi-fluid model to Mercury that tracks the solar wind protons and planetary ions of Mercury origin as separate fluids, allowing for studying the effects of planetary heavy ions on the global magnetospheric structure. Benna et al. (2010) applied a two-fluid, Hall-MHD model in which the ion and electron fluids are treated separately with the inclusion of the Hall physics within the ideal MHD framework.

Another type of global simulations frequently used in the modeling of Mercury's magnetosphere is the hybrid model in which electrons are treated as a massless fluid while ions are represented as individual macro-particles. This allows for modeling ion kinetic effects, e.g., finite gyroradius effects and non-Maxwellian particle distributions. Compared to MHD simulation, hybrid simulation normally needs relatively expensive computational resources to achieve reasonably good resolution and to reduce system noise. With the rapid increase of computing power, it has recently become viable to apply a three-dimensional hybrid model to simulate Mercury's magnetosphere on a global scale (e.g.,

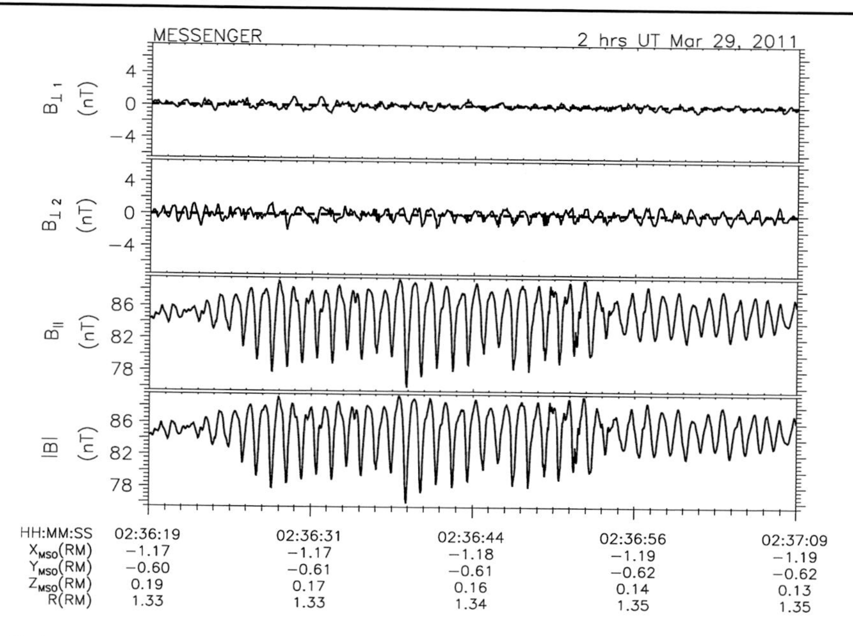

Fig. 19 Strongly compressional waves near Mercury's magnetic equator. Perpendicular components of the field from two-dimensional MVA are shown in the *first* (minimum perpendicular variance) and *second* (maximum perpendicular variance) *panels*. The *third panel* is the parallel field component and the *fourth panel* is the field magnitude. Peak-to-peak amplitudes are 10 nT. Adapted from Boardsen et al. (2012)

Kallio and Janhunen 2003; Trávníček et al. 2007, 2010; Wang et al. 2010; Müller et al. 2012; Richer et al. 2012). These hybrid simulations have provided significant insights into many of the fundamental plasma processes operating in Mercury's magnetosphere, especially those on the ion kinetic scale, such as energy-dependent particle drifts and wave generations resulting from ion temperature anisotropy.

Solar Wind Entry into the Magnetosphere Various modeling studies using global MHD and hybrid simulations have confirmed the picture of Mercury's magnetosphere derived from measurements: that reconnection is the dominant process transferring solar wind plasma and energy into Mercury's magnetosphere. Modeling has shown that other boundary processes, such as the Kelvin-Helmholtz instability (Paral and Rankin 2013) and ion kinetic motion across the magnetopause due to finite gyroradius effect (e.g., Müller et al. 2012), also contribute to the transfer of magnetosheath plasma into the magnetosphere as observed by MESSENGER. As an example, Fig. 20 shows the large-scale configuration of Mercury's magnetosphere from the hybrid model of Trávníček et al. (2010). Familiar magnetospheric structures can be readily identified in the figure, such as the bow shock, magnetosheath, cusps, tail lobes, and plasma sheet. Several modeling studies based on MHD and hybrid simulations have found that Mercury's magnetosphere changes its configuration considerably when the IMF orientation varies. In particular, as shown in Fig. 20, the dayside magnetopause is located closer to the planet during southward IMF compared to northward IMF (e.g., Ip and Kopp 2002; Kidder et al. 2008; Trávníček et al. 2010), consistent with the suggestion by Slavin and Holzer (1979) that enhanced low-latitude reconnection during periods of southward IMF can effectively erode the

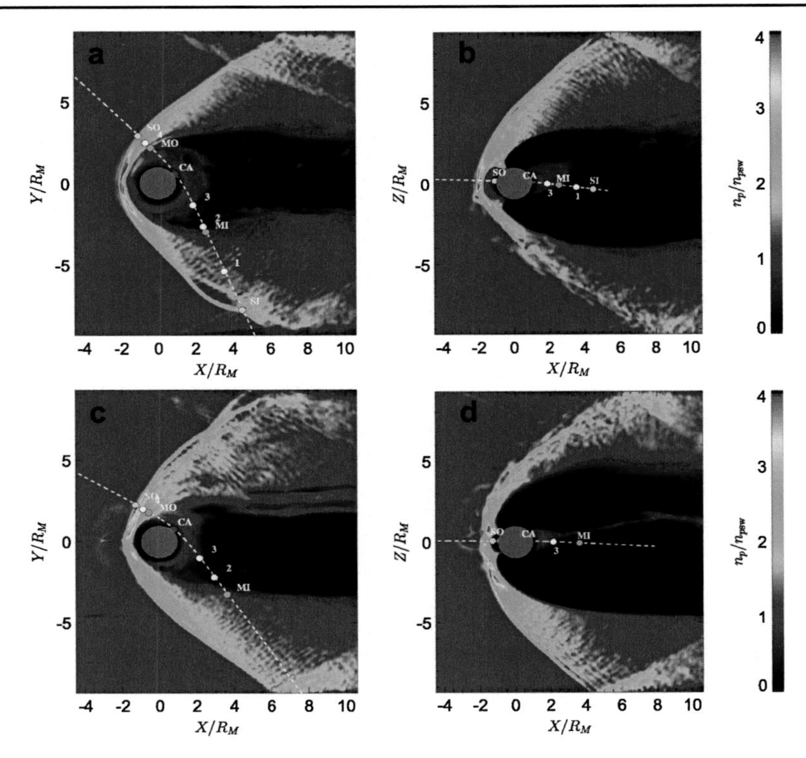

Fig. 20 Simulated magnetospheric configuration from the global hybrid model of Trávníček et al. (2010). Panels (**a**) and (**b**) show the simulated solar wind proton density in the equatorial and noon-midnight meridional planes, respectively, under northward IMF conditions. Panels (**c**) and (**d**) are the same but for southward IMF. *Colors in each panel* represent the density of solar wind protons normalized to the upstream value. MESSEGER trajectories of the M1 and M2 flybys are superimposed as *white dashed lines*

dayside magnetopause causing the boundary to move closer to the planet's surface. Correspondingly, the location and morphology of the cusps, through which the solar wind plasma can gain access to the low altitude region, also vary in response to solar wind and IMF changes. As described above, however, analysis of MESSENGER data (e.g., DiBraccio et al. 2013) does not support the strong bias of reconnection rate based on IMF direction alone. This behavior has not yet been captured in global models of Mercury's magnetosphere.

Aside from the IMF, solar wind dynamic pressure is another important factor that can significantly affect the size and configuration of Mercury's magnetosphere. Kabin et al. (2000) using an MHD model simulated Mercury's interaction with the solar wind for different upstream pressures. They showed that under extremely high solar wind dynamic pressure conditions Mercury's dayside magnetopause can be pushed all the way to the surface, a situation in which the solar wind plasma can directly impinge on the planet. Similar results have also been found in the hybrid simulation by Kallio and Janhunen (2003). However, results from MESSENGER observations paint a more nuanced picture: Slavin et al. (2014) found that increases in magnetic field due to induction in the planet's core act to resist this compression, where the resulting stand-off distance is the result of competition between these two processes. Some models have now included this induction (see paragraph "Simulation of the induction effect arising from the planetary core" below).

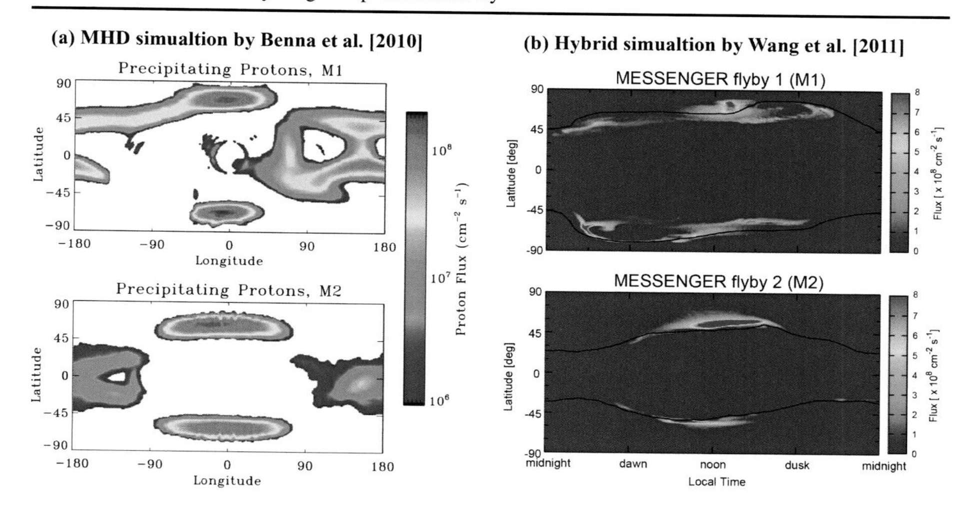

Fig. 21 Maps of precipitating solar wind proton fluxes (in the unit of particles per cm^2 per second) predicted by global magnetosphere simulations for the MESSENGER M1 and M2 flyby conditions. (**a**) From the MHD model by Benna et al. (2010) (figure adapted from Burger et al. 2010). (**b**) From the hybrid model by Wang and Ip (2011). For each model, the *top panel* shows the precipitation map for the M1 flyby conditions while the *bottom panel* is for the M2 flyby

Precipitation of Solar Wind Particles onto Mercury's Surface Once the solar wind enters the magnetosphere, the bulk of the plasma follows the large-scale Dungey cycle magnetospheric convection driven by the solar wind. Since precipitation of solar wind particles onto Mercury's surface is a major source of its exosphere and magnetosphere, it is of high interest to derive from global magnetosphere models quantitative information about this process as well as its dependence on the internal and external conditions.

Figure 21 shows maps of the solar wind precipitation onto Mercury's surface as predicted by MHD and hybrid simulations for the first two MESSENGER flybys. It should be noted that the two simulation models discussed here used different modeling approaches (MHD versus hybrid) and differ in many aspects, such as the input parameters and details of the numerical codes. Nevertheless, the two models give qualitatively similar results on the general features of ion precipitation. In general, there are two main regions on the surface to which solar wind particles can have access. One is the cusp region on the dayside and the other is the low-latitude region on the night side. As mentioned above, the location and size of the two cusps responds to IMF changes. This can be clearly seen by comparing the precipitation maps between the cases for the M1 (northward IMF) and M2 (southward IMF) flybys. The peak fluxes are centered above 70 degree latitude for the M1 flyby conditions whereas they move equatorward to about 65 degree latitude for the M2 flyby conditions, which is in accordance with the variation of the open-closed field line boundary as shown in the hybrid case (Fig. 21(b)). Both models predict a noticeable north-south asymmetry in the impact rate and the spatial distribution of the cusp precipitation. Such an asymmetry has been suggested to arise from the presence of a strong B_X component in the IMF (Sarantos et al. 2001), a typical feature of the Parker spiral at Mercury's orbit. Benna et al. (2010) attributed a north/south asymmetry in the sodium density observed during a MESSENGER flyby to this precipitation flux asymmetry. Others have also invoked ion flux to help explain sodium exosphere observations via the processes of ion sputtering, ion-enhanced diffusion, and chemical sputtering, as in the Monte Carlo models of Mura et al. (2009), Leblanc and

Johnson (2010), and Burger et al. (2010). In later work, the north/south asymmetry in precipitation was attributed mostly to Mercury's offset magnetic dipole (e.g., Winslow et al. 2014).

Ion precipitation is seen on the night side mainly at low latitudes where the magnetic field lines are closed field lines with both ends connected to the planet. The existence of such an ion impact region is consistent with the expectation that Mercury's plasma sheet ions have relatively large loss cone (Korth et al. 2014) because of the planet's weak intrinsic magnetic field. While the MHD and hybrid models show similar features of the cusp precipitation, the nightside precipitation appears to have different characteristics between the two models. The MHD model predicts a broad region of ion precipitation on the night side, which has been attributed to the absorption of particles in the drift belt formed in the equatorial region. Hybrid models, on the other hand, also predict the existence of such a drift belt near the planet (e.g., Trávníček et al. 2010). However, a surprising result of the hybrid simulations (e.g., Trávníček et al. 2010; Wang and Ip 2011) is that those ions precipitate onto the surface primarily at high latitudes, instead of near the equator as one might expect on the basis of finite gyroradius effect.

In addition to the external conditions, the internal conditions, such as the magnetic properties of the planet, may also affect the distribution of ion precipitation onto the surface. Richer et al. (2012) using a hybrid model explored the sensitivity of the global magnetospheric interaction to details of Mercury's intrinsic magnetic field. Two different internal field representations were used in their simulations: one contains a northward offset dipole (Anderson et al. 2011) and the other is a combination of a centered dipole and quadrupole fitted to the offset dipole derived from MESSENGER observations. They found that while the two internal field models yielded similar magnetic configuration in the northern hemisphere, the north-south asymmetry is more pronounced in the case with the a dipole plus a quadrupole field. This leads to very different precipitation patterns between the northern and southern hemispheres, an interesting result that needs to be checked against with observations of the low-altitude region of the southern hemisphere from future missions to Mercury, such as the BepiColombo mission.

Simulation of the Induction Effect Arising from the Planetary Core There is no doubt that the electromagnetic coupling between the planetary interior and the magnetosphere is an important element of Mercury's interaction system that needs to be included in global modeling, especially when considering the system response to time-varying external conditions. Most the global models applied to Mercury thus far excluded the planetary interior from the simulation domain. In those models, the electrical properties of the planet are mimicked through prescription of boundary conditions. To properly model the coupling between the magnetosphere and the core, it is desirable to explicitly include the planetary interior as part of the simulation domain. Such an attempt has been undertaken by Müller et al. (2012), who adapted a 3D hybrid model previously applied to planetary moons to Mercury and included the planetary interior with a specified conductivity distribution. The model has been applied to simulate MESSENGER flybys and shown to reproduce MESSENGER observations reasonably well. However, the induction effect arising from the core was not clearly demonstrated because the model employed steady solar wind conditions as input and focused on the steady-state behavior of the magnetosphere, as what has been done with most global models applied to Mercury. Jia et al. (2015) recently developed a global resistive MHD model that also explicitly includes the planetary interior with layers of different conductivities in their simulation. To characterize how the coupled system dynamically responds to the external forcing, they drive the simulation by using time-dependent solar wind conditions containing different types of disturbances typical of those seen at Mercury's orbit, such as CMEs and IMF rotations. Their results show that the reconfiguration of Mercury's magnetosphere indeed induces intense electric currents at the core where the electrical conductivity is high. Associated with those induced currents are strong magnetic perturbations present not only inside of the planet but also throughout the magnetosphere, clearly demonstrating that the induction effect plays an important role in determining the global magnetospheric structure.

While the modeling efforts discussed above represent a first step in characterizing Mercury's magnetosphere-core coupling in a self-consistent manner, future work is clearly needed to further quantify the induction effect. A particularly important question that should be addressed with self-consistent global simulations is how the strong magnetosphere-core coupling affects the extent to which the solar wind particles can have access to the planet's surface, which is of direct relevance to the plasma sources of Mercury's magnetosphere.

5.1.2 Exospheric Modeling

Global models of Mercury's neutral exosphere have made significant contributions to understanding of its origin from complex interactions between the Sun and the surface of the planet, as well as of seasonal variations due to Mercury's highly elliptical orbit (Leblanc and Johnson 2003, 2010; Mura et al. 2007; Burger et al. 2010, 2012; Wurz et al. 2010; Sarantos et al. 2011; Pfleger et al. 2015). Since the exosphere is collisionless, particle dynamics in these models are determined rather simply by gravity and radiation pressure; however, the sources and sinks of the exosphere add considerable complication and are the main area of active development. Global models typically include the source processes that have been described in Sect. 3.2.1: thermal desorption, ion sputtering, photon-stimulated desorption, and micrometeroid vaporization (TD, IS, PSD and MV). Additionally, the main loss processes included are photoionization, surface sticking, and gravitational escape. Of these three loss mechanisms, photoionization of exospheric neutral atoms is particularly important because it is also a significant *source* of planetary ions to Mercury's magnetosphere. The physics of these processes is well understood from laboratory measurements. Nevertheless, there is sufficient uncertainty in crucial parameters—such as Mercury's surface composition and the incident solar wind plasma—that the relative contributions of these processes are not well determined. Many researchers have sought to remedy this problem by using observations to constrain their models, either from Earth (ground-based) or MESSENGER. This synthesis of models and observations has been very effective in narrowing the parameter space, but the relative contributions of the various surface processes are still in dispute.

A subset of global exosphere models are able to simulate the dynamics of planetary ions in the system once they are created (Sarantos et al. 2009). This modeling capability is key to obtain a global understanding of plasma sources into Mercury's magnetosphere. It provides, thus far, the only quantitative connection to planetary ion sources from both the exosphere and surface, though their relative importance is also an open question. To model ion dynamics, exospheric models must also include the electric and magnetic fields of the magnetosphere. These models primarily focus on planetary ions, which are much lower in abundances than the solar wind ions that drive the magnetosphere. As a result, they incorporate static fields, typically from MHD models. Planetary ions are then *flown* through these fields by integrating their equation of motion directly. The ions are often treated as test particles, each representing a larger number of ions in the real system.

The ion component of global exospheric modeling can feedback into understanding the composition of the exosphere itself. From MESSENGER FIPS measurements, as well as from ion composition measurements around the Moon (e.g., Mall et al. 1998), we know that many more species exist that have not been observed. Many of these atmospheric species do not have emission lines in the MESSENGER UVVS spectral range; therefore FIPS measurements present the only way to update upper limits prior to BepiColombo orbit insertion. The observed seasonal variability of the exosphere (e.g., Leblanc and Johnson 2010, for Na; Burger et al. 2014, for Ca) has not yet been folded into ion model calculations.

Ions from the Exosphere Na^+ of exospheric origin is the only species that has been systematically studied with simulations. Trajectory tracings of Na ion test particles were performed in analytical (Delcourt et al. 2003), resistive MHD (Yagi et al. 2010; Seki et al. 2013), Hall MHD (Sarantos et al. 2009), and hybrid (Paral et al. 2010) simulated fields (Fig. 22). For the first three the exospheric model of Leblanc and Johnson (2003) was used, in which the finite Na reservoir is quickly depleted by thermal desorption leading to an exosphere with a dawn-dusk asymmetry, whereas the other two considered different mixes of photon stimulated desorption and sputtering, both spherically symmetric with respect to the Sun-Mercury line. Unfortunately, because of the small size of Mercury's magnetosphere, these tracings are very sensitive to the treatment of the inner boundary condition (Seki et al. 2013) and therefore differ between models.

A common feature of these simulations is an enhancement of Na^+ near dawn and in the morning sector. The estimated concentration peaks exceeded 10 cm^{-3} near the equator (Yagi et al. 2010). The pressure exerted by planetary ions in these simulations can locally surpass 10 % of the total, thus necessitating that sodium becomes one of the species of the MHD and hybrid simulations. Escape of planetary ions through a porous magnetopause, especially under southward IMF conditions, is evident in the simulations (Paral et al. 2009).

Magnetospheric ion recycling and its effect has been the subject of several works. "Self sputtering" is itself an inconsequential source for the exosphere (e.g., Delcourt et al. 2003; Poppe et al. 2013) as the recycled ion fluxes are a small portion of the inferred neutral efflux ($\sim 10^7\ cm^{-2}\ s^{-1}$). However, recycling could be important if ions neutralized in the soil increase the available reservoir for trace species (Killen et al. 2004). High ion recycling rates will obviously increase the reservoir for exogenous species of the exosphere that are in balance with the solar wind influx (e.g., He, Ne); but they could also increase the reservoir for exospheric Na and K, which are very nearly depleted on the dayside (Leblanc and Johnson 2003, 2010) to levels that can be supported by grain diffusion (Killen et al. 2004). Broad bands of nightside precipitation of Na^+ with fluxes $\sim 10^5\ cm^{-2}\ s^{-1}$ and extending up to $\pm 50°$ latitude form when realistic conditions about the surface conductance are adopted (Seki et al. 2013). These contain sub-keV ions which are deposited very near the top of the grains and should quickly diffuse to the grain surface. Schmidt (2013) proposed that ion precipitation to Mercury's nightside, which is shifted northward because the geomagnetic equator is displaced with respect to the geographic equator, is a mechanism for producing the north-south asymmetries of the dayside Na exosphere observed from ground-based telescopes (e.g., Potter et al. 2006).

Ions from the Surface Both precipitating protons and electrons can contribute to a surface ion source. McLain et al. (2011) suggested that electron stimulated desorption (ESD) could be an important source of Mercury's ions. Thresholds for such emission (~ 20 eV) are typically too high for solar wind electrons impinging onto the Moon but can clearly be exceeded at Mercury. While the typical yield for sputtered ions by proton impingement is in the range of 10^{-4} to 10^{-1} per impacting ion (Benninghoven 1975), the yields measured for ESD could be ten times higher, especially the more energetic the incident electrons (Wang et al. 1984). Ions and electrons from the solar wind should precipitate not only onto

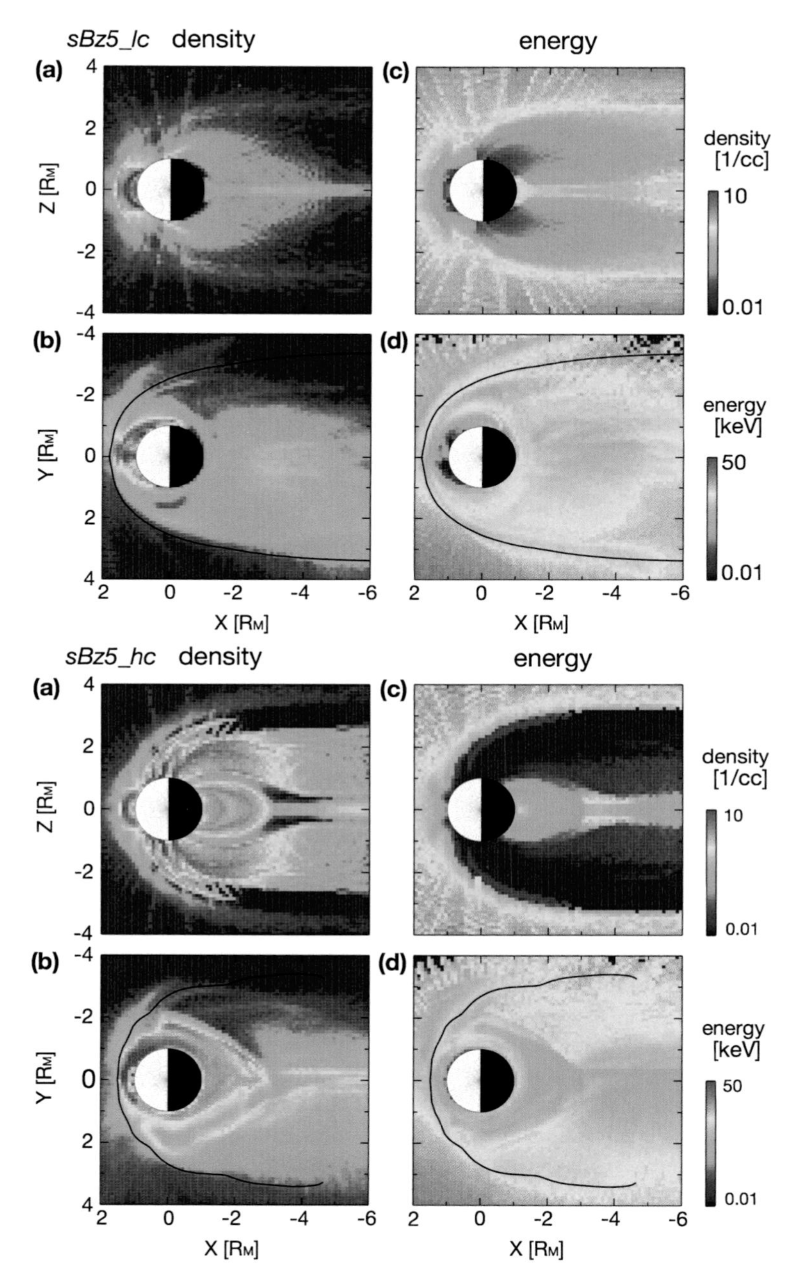

Fig. 22 Na ion distribution under the same southward IMF ($B_Z = -5$ nT) and solar wind conditions, initialized with the same exosphere model, but subject to different assumptions of the inner boundary condition. Case with high surface conductance shown in the *bottom panel*. The resulting ion distributions are markedly different as the formation of an X-line further from the planet inhibits escape in the second case (from Seki et al. 2013)

Mercury's cusp areas but also persistently onto the nightside in auroral regions as well as regions surrounding the geomagnetic equator (e.g., Benna et al. 2010). Thus, the Hermean surface at high latitudes of the dayside and low latitudes of the nightside are regions of planetary ion emission.

Outflow of ions released directly from the surface could be responsible for some of the cusp signatures observed by MESSENGER FIPS (Raines et al. 2014). Despite their sub-escape initial energies (~ 1 eV), such ions will be rapidly accelerated by centrifugal sources and escape into the magnetosphere (Delcourt et al. 2012). Their importance relative to photoions is uncertain. At the Moon, predictions from Sarantos et al. (2012) suggest that ions from exospheric neutrals dominate over surface ions for many metallic constituents such as Na^+, although for some species with more stringent exospheric limits (e.g., Ca^+) the surface should be the most important source. At Mercury such calculations are yet to be performed.

Estimating Total Ion Source Rates Models of the exosphere can provide a rough estimate for the sodium ion source rate in Mercury's magnetosphere, indirectly, via the commonly reported quantity of total exosphere content and the assumption, common to all models, that photoionization is the dominant ionization process. The answer varies from model to model, of course, but despite their major differences, all sodium models estimate the *content* to be on the order of 10^{28} sodium atoms (the exosphere content ranges 0.3–4×10^{28} in the following: Smyth and Marconi 1995; Killen et al. 2001; Mura et al. 2009; Leblanc and Johnson 2010; Mouawad et al. 2011).

These models do not use the same data sets and they even have different basic assumptions, yet they estimate the sodium content within the same order of magnitude. Consider the difference between the models described by Leblanc and Johnson (2003, 2010) and Burger et al. (2010): the two models have quite different mixtures of source processes and, even more fundamentally, differ in basic construction. The exosphere in Leblanc and Johnson is coupled to a large reservoir of adsorbed sodium atoms on Mercury's surface, while Burger et al. have no reservoir. Leblanc and Johnson (2010) ran their model for several simulated Mercury years and matched their results to several Earth years of ground-based observations; the Burger et al. model only simulated several hours and compared their result data taken during two of MESSENGER's Mercury flybys.

The ion source rate can be estimated from published results by multiplying this exosphere content by the photoionization frequency, which is on the order of 10^{-4} s^{-1} (Huebner et al. 1992). Assuming that most of the exosphere is exposed to sunlight, this gives a sodium ion source rate on the order of 10^{24} sodium ions s^{-1}, or a mass loss of several 10 s of g s^{-1}. This is comparable to the ion outflow from the other terrestrial planets (Strangeway et al. 2010), but some fraction of Mercury's ion production is lost to its surface. As discussed above, Na^+ is the most abundant planetary ion detected by FIPS. Ionized magnesium may contribute to the sodium ion signature owing to its similar mass, but it is much less abundant in the exosphere and has a longer lifetime against photoionization.

Sodium is the most abundant exospheric species identified so far, but there are several others. During one of MESSENGER's Mercury flybys, UVVS observed simultaneously neutral and ionized Ca (Vervack et al. 2010). Models applied to MASCS data support a strong localized source at dawn at high temperature (> 50000 K), probably related to micrometeoritic impact vaporization of Ca, in the form of CaO and CaOH, and subsequently dissociated (Burger et al. 2014). The model of UVVS calcium observations provides a Ca^+ photoion source on the order of 10^{23} calcium ions s^{-1} (Burger et al. 2014), though much of the calcium is ionized beyond the magnetosphere owing to its high-energy ejection process. Hydrogen and helium gases are thought to be neutralized solar wind plasma that are later (re-) ionized to contribute to the planetary ions detected by FIPS. The planetary helium is distinct from solar wind helium as it is singly, rather than doubly, ionized. Broadfoot et al. (1976) estimated that most of the planetary helium escapes Mercury's gravity before photoionization, although the energy distribution of the neutral helium is highly uncertain (Shemansky and Broadfoot 1977; Leblanc and Chaufray 2011). Broadfoot et al. (1976) estimated that the helium ion source rate from the helium exosphere is on the order of 10^{22} helium ions s^{-1}.

5.2 Ion Acceleration Processes and Non-adiabatic Behavior

5.2.1 Centrifugal Acceleration in Mercury's Lobes

To the first order, the large scale plasma circulation at Mercury resembles that at Earth, the coupling between the magnetosphere and the interplanetary magnetic field being responsible for a dawn-to-dusk convection electric field with antisunward transport at high latitudes (typically, above 50°) and return sunward flow at low latitudes. In this context, a process that readily affects planetary ions after their ejection into the magnetosphere is the centrifugal acceleration associated with the large scale $\mathbf{E} \times \mathbf{B}$ transport. Using a guiding center approach, Cladis (1986) showed that, during transport from high to low latitudes, ions expelled from the topside terrestrial ionosphere may be subjected to substantial acceleration in the parallel direction. Because of the small spatial scales of the Hermean environment, it was pointed out by Delcourt et al. (2002) that this acceleration is more pronounced at Mercury than at Earth, possibly leading to energization of heavy ions up to several hundreds of eVs or a few keVs in the lobes prior to their entry into the plasma sheet. This is at variance with the energy gain up to at most a few tens of eVs expected at Earth (e.g., Yau et al. 2012).

In particular, in contrast to Earth, the above centrifugal acceleration may play a specific role in the escape of planetary material at Mercury; hence, a prominent impact on the net plasma supply to the Hermean magnetosphere. Indeed, at Earth, unless a short-lived compression event affects the magnetosphere (e.g., Cladis et al. 2000), this acceleration is weak and operates over a long time as particles travel downtail in the lobes. Ions ejected from the terrestrial ionosphere with velocities smaller than the escape speed are not sufficiently accelerated by this mechanism to overcome gravity and return toward the ionosphere according to parabolic or hoping trajectories (e.g., Horwitz 1984). Because of the pronounced curvature of the magnetic field lines in the immediate vicinity of the planet surface, a different situation is obtained at Mercury with abrupt energization of the ions immediately after ejection into the magnetosphere (Delcourt 2013). In this latter study, it was found that the numerous populations that are released at very low energies such as those due to thermal desorption are rapidly accelerated up to $\sim 2V_{\mathbf{E}\times\mathbf{B}}$ ($V_{\mathbf{E}\times\mathbf{B}}$ being the $\mathbf{E} \times \mathbf{B}$ drift speed) in a like manner to the acceleration due to a moving magnetic mirror (Cowley 1984). Accordingly, instead of being trapped near the planet surface due to ejection velocities smaller than the escape speed, these ions readily overcome gravity and flow into the magnetosphere. Also, since the parallel velocity realized does not depend upon mass-to-charge ratio, all ion species are transported into a similar region of space in the pre-midnight sector of the inner magnetotail which may explain the density enhancements locally recorded by MESSENGER (Raines et al. 2013). Moreover, the study of Delcourt (2013) suggests that the centrifugal focusing of planetary material thus obtained depends little upon the convection rate, an increase of the convection electric field magnitude (and associated $\mathbf{E} \times \mathbf{B}$ drift speed) resulting into an increase of the particle parallel speed in the same proportion.

5.2.2 *Spatial Nonadiabaticity in Mercury's Magnetotail*

Upon reaching the field reversal in the magnetotail, particles may not conserve their magnetic moment (first adiabatic invariant) because of significant field variations on the length scale of the particle Larmor radius. A parameter that is often used to characterize this nonadiabatic behavior is the parameter κ defined as the square root of the minimum field line curvature radius-to-maximum Larmor radius ratio. For $\kappa > 3$, the particle motion is adiabatic and the guiding center approximation is valid while for κ of the order of unity or below, the motion is nonadiabatic with possibly large variations of the magnetic moment. (For more details see Seki et al. 2015, this issue.) At Earth, the transition from adiabatic to nonadiabatic regimes, viz. $\kappa \approx 1$, occurs in the mid-tail for plasma sheet ions. This region has been viewed either as the onset of prominent injections into the atmospheric loss cone and subsequent ion precipitation (leading to the Isotropy Boundary interpretation framework of Sergeev et al. 1993) or as a domain of enhanced trapping (hence, the "wall" picture put forward by Ashour-Abdalla et al. 1992), both pictures being valid since particles are subjected to either magnetic moment damping or enhancement (e.g., Delcourt et al. 1996). At Mercury, because of the weak intrinsic magnetic field and of the strong solar wind dynamical pressure, the magnetosphere is small and the planet occupies a much larger volume of it than at Earth. The nearly dipolar region of the inner terrestrial magnetosphere where the particle motion is essentially adiabatic is thus absent at Mercury, and it is expected that most ions behave nonadiabatically throughout the magnetotail. Computations of the adiabaticity parameter κ in model magnetospheres of Mercury actually suggest that the condition $\kappa \approx 3$ is met in the immediate vicinity of the planet. Hybrid simulations where a kinetic description is used for ions while electrons are treated as a massless fluid are thus most appropriate at Mercury (e.g., Kallio and Janhunen 2003; Trávníček et al. 2007; Richer et al. 2012).

The fact that ions behave nonadiabatically in most of the Hermean magnetotail is of importance for its structure and dynamics. In particular, be they of solar wind or planetary origin, ions at $\kappa < 1$ may display either quasi-trapped orbits with repeated crossings of the field reversal or Speiser-type orbits (Speiser 1965) with large energization along the dawn-dusk convection electric field during meandering motion about the midplane. Such nonadiabatic behaviors that are sometimes referred to as "quasi-adiabatic" because of possible conservation of the action integral I_Z (Büchner and Zelenyi 1989), are of paramount importance since they lead to the formation of thin current sheets embedded within a thick plasma sheet. Instability of these thin current sheets can lead to local current disruption and consequent reconfiguration of the magnetic field lines (e.g., Mitchell et al. 1990). Nonadiabatic particle behaviors also lead to the formation of nongyrotropic distribution functions; hence, significant off-diagonal terms in the plasma pressure tensor and a stress balance that does not rely on a large pressure gradient along the tail axis.

As planetary ions reach the magnetotail midplane after $\mathbf{E} \times \mathbf{B}$ transport over the polar cap, they are nearly aligned with the magnetic field owing to pitch angle folding from low to high altitudes. Would their motion be adiabatic (magnetic moment conserving), these ions would return to the planet vicinity after a single crossing of the magnetotail midplane and precipitate onto the surface. Far from such a behavior, planetary ions are subjected to prominent magnetic moment scattering upon interaction with the field reversal. As a result of this isotropization and temporary trapping, and without invoking other processes such as wave-particle interactions, these ions are found to substantially contribute to the plasma sheet populations. In a quantitative study of the Na^+ circulation at Mercury, Delcourt et al. (2003) considered a model exosphere of neutral sodium (Leblanc and Johnson 2003) and showed that this planetary material may contribute up to a few tenths of ions/cm^3 to the equatorial magnetotail, depending upon phase angle along Mercury orbit. Such densities are in qualitative agreement with those reported by Raines et al. (2013) in their analysis of MESSENGER data. Also, assuming a cross-polar cap potential drop of 20 kV, the simulations of Delcourt et al. (2003) put forward times of flight from the high-latitude dayside sector to the inner plasma sheet of the order of a few minutes on the average, together with a prominent asymmetry between dawn and dusk sectors due to westward drift of the ions.

During their nonadiabatic transport, Na^+ ions can be injected inside the loss cone which is much larger at Mercury than at Earth due to the weak planetary magnetic field; hence, their precipitation onto the planet surface. In the modified Luhmann-Friesen model considered by Delcourt et al. (2003), this ion precipitation is organized according to two narrow bands at mid-latitudes (between $\sim 30°$ and $\sim 40°$), the κ parameter varying from ~ 1 down to ~ 0.1 as the latitude increases. The poleward boundary of these precipitation bands is controlled by the width of the magnetotail, ions at higher latitudes (or, equivalently, at larger distances in the magnetotail) intercepting the magnetopause in the course of their duskward motion. Using results of MHD simulations, Seki et al. (2013) demonstrated that this overall precipitation pattern may significantly depend upon the planet surface conductivity as well as IMF orientation, the formation of a near-Mercury neutral line leading to significant downtail loss of planetary ions.

5.2.3 Temporal Nonadiabaticity in Mercury's Magnetotail

The nonadiabatic transport features described above in the magnetotail field reversal result from large magnetic field variations on the length scale of the particle Larmor radius. These features accordingly relate to spatial nonadiabaticity. In the case of explicit temporal variations of the field on the time scale of the particle gyroperiod, nonadiabatic features may appear as well. This latter temporal nonadiabaticity cannot be characterized with the help of the κ parameter, and it may actually occur in regions where $\kappa > 3$ (i.e., where the spatial adiabaticity condition is fulfilled). Such a temporal nonadiabaticity may emerge for instance during short-lived reconfigurations of the magnetospheric field lines. In this regard, it was shown that, at Earth, heavy ions originating from the ionosphere such as O^+ may be subjected to prominent nonadiabatic energization up to the hundred of keV range during substorm dipolarization (e.g., Delcourt 2002). This energization due to the electric field induced by the time-varying magnetic field preferentially affects O^+ ions that have cyclotron periods comparable to the dipolarization time scale. In contrast, protons that have smaller gyroperiods are transported adiabatically (provided that $\kappa > 3$) and subjected to Fermi-type or betatron energization. Because of the smaller characteristic time scales of the Mercury's environment (e.g., with a typical Dungey cycle time of ~ 2 min as opposed to ~ 1 hour at Earth), it may be anticipated that protons will be subjected to such a temporal nonadiabaticity during reconfigurations of the Hermean magnetotail.

Figure 23 shows the energy variations obtained for protons in the case of a 10-second model dipolarization of the magnetic field lines in the inner magnetotail of Mercury. In this figure, the H^+ post-dipolarization energy is shown as a function of initial energy and for different initial gyrophases. Because the particles considered here are equatorially trapped (i.e., 90° pitch angle at equator), no effect due to parallel motion and spatial nonadiabaticity is to be expected.

It is apparent from Fig. 23 that protons with low initial energies are systematically energized up to a level that gradually increases with initial distance (from left to right). In particular, in the right panel of Fig. 23, the low-energy protons initialized at 3 R_M are

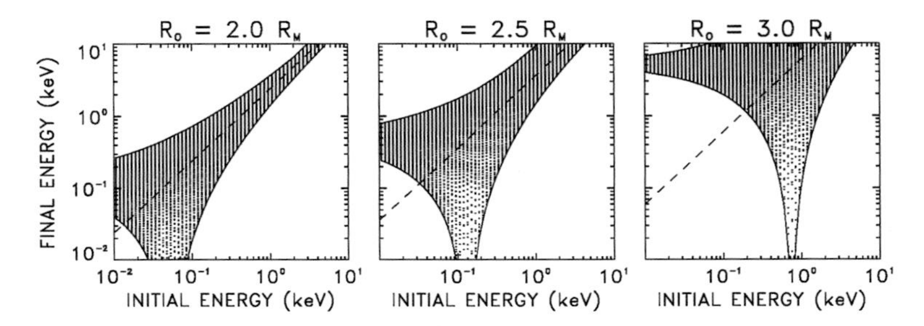

Fig. 23 Post-dipolarization energy versus initial energy for equatorially mirroring H^+ launched from different initial distances: (*from left to right*) 2, 2.5, and 3 R_M. In *each panel*, the various dots correspond to distinct initial gyrophases whereas the *dashed line* shows the final energy expected in the case of adiabatic (betatron-type) energization

systematically energized up to ~ 4 keV while being transported down to $\sim 1.8\ R_E$. This nonadiabatic behavior at low initial energies contrasts with that obtained at large initial energies where betatron-type energization (i.e., in proportion to the change in magnetic field magnitude) is obtained. Although short-lived fluctuations of the magnetic field that are not considered here may lead to deviations from these results, it is clearly apparent from Fig. 23 that, in a like manner to O^+ at Earth, protons may be transported in a nonadiabatic manner during dipolarization events at Mercury. Under the effect of the transient induced electric field, these ions may experience energy gains significantly above that expected from the large scale convection electric field alone.

Because temporal nonadiabaticity is to be expected whenever the magnetic field changes significantly on the time scale of the particle gyroperiod, it may be anticipated that ions will be transported nonadiabatically not only in the equatorial region but also at high latitudes. This follows from the short characteristic time scales at Mercury as well as from the weak intrinsic magnetic field that leads to large ion gyroperiods. An example of such behaviors is provided in Fig. 24 that shows the results of Na^+ simulations during a 20 s turning of the IMF from $B_X = 0$ to $B_X = 20$ nT (Delcourt et al. 2011).

The leftmost panels of Fig. 24 depict symmetrical Na^+ flows from the high latitude dayside sector above the polar cap as well as gradual centrifugal acceleration up to the keV range before reaching the nightside plasma sheet. On the other hand, during IMF turning (from left to right in Fig. 24), it is apparent that the Na^+ average energy (bottom panels) off equator rapidly increases up to several keVs. As discussed above, this energization occurs in a nonadiabatic manner and follows from resonance between the induced electric field and the particle gyromotion. At high latitudes, such a resonance is achieved for Na^+ and Fig. 24 thus suggests that IMF turning or short-lived magnetic transitions at Mercury may go together with the rapid production of heavy energetic material in the magnetospheric lobes.

6 Summary

Mercury's magnetosphere is dynamic and its environment is extreme. It is similar enough to allow application of terrestrial theory and approaches, yet it has differences sufficient to challenge some of them with the need for more sophistication. Mercury's intrinsic field is sufficient to stand off the solar wind, but creates a very small magnetosphere that responds dramatically to changing solar wind conditions. The main global dynamical behavior is

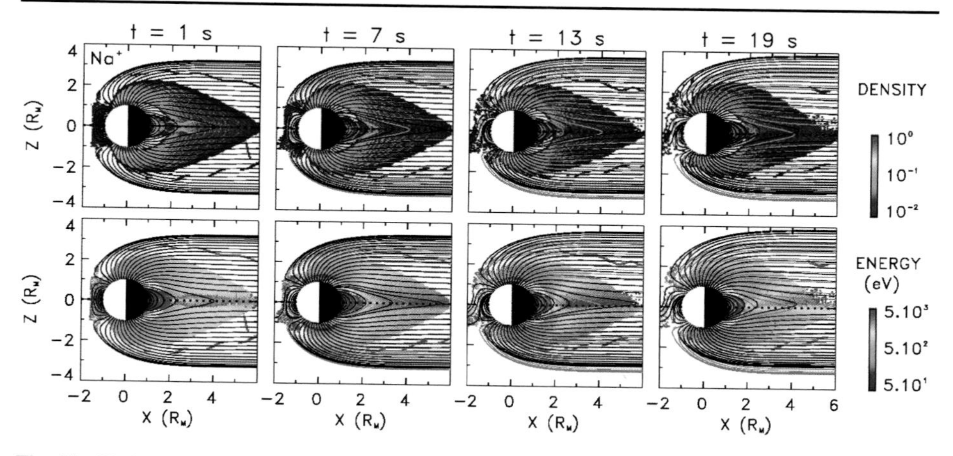

Fig. 24 (*Top*) model normalized density and (*bottom*) average energy of Na^+ ions in the noon-midnight meridian plane during a model IMF turning. *Panels from left to right* show snapshots at distinct times of the magnetic transition. *Black and grey lines* show the magnetic field lines at the corresponding time and in the initial configuration, respectively. Density and energy are coded according to the color scales at right. From Delcourt et al. (2011)

Dungey cycle circulation but at a rate about 30 times faster than at Earth, contributing to the magnetosphere's ability to reconfigure very quickly. This convection is driven by magnetic reconnection at the dayside magnetopause, but unlike other planetary magnetospheres, the reconnection rate is independent of magnetic shear angle. As a consequence of frequent magnetopause reconnection, Mercury's magnetosphere is dominated by Dungey cycle convection.

MESSENGER measurements have unambiguously proven that Mercury's magnetosphere is populated with plasma of both solar wind and planetary origins. The magnetospheric plasma distribution is similar to Earth's, concentrated at high-latitudes in the dayside magnetospheric cusp as well as in the equatorial plane of the central plasma sheet in the tail. Solar wind protons are the most abundant species in those regions, followed by solar wind alpha particles and Mercury-derived Na^+ ions (grouped with Mg^+ and Si^+ in observations). Two other planetary ion species have been studied, the O^+ group and He^+. Both are present in lower abundance than Na^+-group ions and have not been studied in much detail. Other planetary ions with $m/q > 30$ appear in FIPS data and work is underway to identify them. In the cusp, Na^+-group ions come from two sources: upwelling from the surface and swept in by reconnection from the vicinity of the dayside magnetopause. Plasma loss to the surface has been observed through the presence of a large loss cone of $\sim 59^\circ$ in a long-term average. In the central plasma sheet, protons dominate the number density by an order of magnitude, but Na^+-group ions can comprise up to 50 % of the mass density and 15 % of the thermal pressure. Observations of Kelvin-Helmholtz waves, along the magnetopause boundary, have provided the first conclusive evidence that these planetary ions are dynamically important in Mercury's magnetosphere.

The ultimate sources of planetary ions are certainly Mercury's tenuous exosphere and mineral surface, but their relative contributions have not yet been determined conclusively. Lacking a collisional ionosphere, Mercury's magnetosphere and exosphere are co-located, making their coupling via source and sink processes very direct. Global modeling of the exosphere is quite sophisticated, including all of these processes as well as effects related to surface reservoir and Mercury's extreme seasonal differences. Ground-based observations of the exosphere have been used extensively to constrain these models, though constraint with MESSENGER observations is just reaching maturity. Some of these models include generation and test particle tracing of planetary ions, but work to compare these quantitatively with MESSENGER plasma ion observations is only beginning.

7 Open Questions

Despite understanding many aspects of plasma sources in Mercury's magnetosphere, there are several top-level questions that still remain.

What is the role of the solar wind in generating the exosphere and populating the magnetosphere with planetary ions? The solar wind is expected to act as a highly variable exospheric (and likewise magnetospheric) source via precipitation and ion sputtering. The search for concrete observational evidence of this effect is still on going. Precipitation of solar wind protons and upwelling surface ions has been observed in situ by MESSENGER but a definitive cause and effect relationship has not been established. One complication is the fact that newly created ions, from all processes involved, have peak energies of just a few eV and to be observed they must be accelerated above FIPS energy minimum, which has been 46 eV for much of the MESSENGER mission. The limit was lowered in August 2014, down to effectively the spacecraft potential (10's of eV) but those measurements have not yet been analyzed in detail. Ground-based exospheric observations have shown variability on the timescale of hours, consistent with solar wind generation of exospheric neutrals, but have not been correlated in situ solar wind observations. Finally, confirmation with MESSENGER UVVS observations has proven elusive, due partly to operational constraints that have allowed for only very limited UVVS observations in the magnetospheric cusps where variability due to solar wind should be at a maximum. The frequency of these observations increased toward the end of the mission so the nature of this relationship may still be established from MESSENGER data. In any case, the BepiColombo Mercury Planetary Orbiter will measure these neutral atoms directly with the Strofio instrument, a low energy neutral gas mass spectrometer of the SERENA particle package (Orsini et al. 2010). Working with the particles and fields instrument on the Mercury Magnetospheric Orbiter, this correlation should be established in a straightforward way.

In what proportions do other exogenic processes contribute and how do they vary with season? Despite significant progress in tying observations to physical processes and the creation of realistic global models, no quantitative, consensus global picture of exosphere generation has emerged. One problem is that the system is under-constrained. The relative contributions to the exosphere of the many source and sink processes is likely different for each exospheric neutral species, and may be a function of Mercury season and even location on the surface. Exospheric composition depends on details of surface composition at significantly higher resolution than available. It may depend on other inputs such as micrometeriod impact rates or traversal of the interplanetary dust disk, which are at best difficult to measure. On-going work combining the available measurements with self-consistent exosphere/magnetosphere models will likely continue to improve understanding, as will additional measurements from the two-spacecraft BepiColombo mission. However, it is possible that there are just too many free parameters to adequately constrain with the relatively sparse measurements that are possible at Mercury.

Do heavy planetary ions make an important contribution to magnetospheric dynamics and if so, how? MESSENGER observations have shown that planetary ions change the scale of Kelvin-Helmholtz waves along Mercury's tail magnetopause. This dynamical contribution could have significant effect on plasma entry through this process. Observations in Mercury's central plasma sheet have shown that they can make up a significant portion of the pressure and dominate the mass density. This certainly sets the stage for participating in magnetospheric dynamics. That said, planetary ions have not yet been shown to influence plasmoids in the magnetotail or act as more than a tracer of plasma flowing through Mercury's northern magnetospheric cusp on newly reconnected field lines. Further analysis of MESSENGER data will likely shed more light on this question. However, the BepiColombo mission, with two spacecraft, more complete particles and fields instrumentation and the larger fields of view possible without sunshade obstructions, is well-poised to address this very complex question.

Does surface impact and sputtering by magnetospheric ions constitute a significant source to Mercury's exosphere and magnetosphere? Tracing protons on closed field lines in the magnetotail has shown losses that should be indicative of impact on the planet's surface. They have nearly the same energy as precipitating protons in the cusp and should, therefore, cause sputtering of ions and neutral atoms. Protons traveling toward the surface in closed field regions have not been observed, though MESSENGER's orbit and orientation is not well-suited to observing this precipitation. MHD and Hybrid simulations show this behavior, but as of yet, determining the contribution to the exosphere and magnetosphere of this process from simulations has not converged to clear values. In addition to protons, planetary ions such as Na^+ present in the magnetotail should behave similarly. Their higher energies, up to 10 keV, could make their precipitation contribute more than would be expected by relative number densities alone.

A lot has been learned about Mercury so far, from Mariner 10, ground observations, MESSENGER, and modeling, but there is much left to do. Analysis of MESSENGER orbital data is really just in the early stages; new results should continue to come out for years to come. In less than a decade after the end of the MESSENGER mission, BepiColombo will arrive to enable a new and potentially more detailed study of the closest planet to the Sun.

8 Compliance with Ethical Standards

This article is in compliance with the ethical standards laid out by the COPE guidelines. There are no conflicts of interest and no human or animal subjects were used in this work. All co-authors have contributed to this manuscript and have given their consent to its submission.

Acknowledgements This work was supported by the NASA Planetary Atmospheres program through grant NNX14AJ46G. Additional support was provided through the Solar System Workings program through grant NNX15AH28G, the Heliophysics Supporting Research program through grant NNX15AJ68G and the Discovery Data Analysis program through grants NNX15AK88G and NNX15AL01G. The MESSENGER project is supported by the NASA Discovery Program under contracts NASW-00002 to the Carnegie Institution of Washington and NAS5-97271 to the Johns Hopkins University Applied Physics Laboratory.

References

I.I. Alexeev, E.S. Belenkaya, J.A. Slavin, H. Korth, B.J. Anderson, D.N. Baker, S.A. Boardsen, C.L. Johnson, M.E. Purucker, M. Sarantos, S.C. Solomon, Mercury's magnetospheric magnetic field after the first two MESSENGER flybys. Icarus **209**, 23–39 (2010)

B.J. Anderson, M.H. Acuna, D.A. Lohr, J. Scheifele, A. Raval, H. Korth, J.A. Slavin, The magnetometer instrument on MESSENGER. Space Sci. Rev. **131**, 417–450 (2007)

B.J. Anderson et al., The global magnetic field of Mercury from MESSENGER orbital observations. Science **333**, 1859–1862 (2011). doi:10.1126/science.1211001

B.J. Anderson, C.L. Johnson, H. Korth, R.M. Winslow, J.E. Borovsky, M.E. Purucker, J.A. Slavin, S.C. Solomon, M.T. Zuber, R.L. McNutt Jr., Low-degree structure in Mercury's planetary magnetic field. J. Geophys. Res. **117**, E00L12 (2012). doi:10.1029/2012JE004159

B.J. Anderson, C.L. Johnson, H. Korth, J.A. Slavin, R.M. Winslow, R.J. Phillips, R.L. McNutt Jr., S.C. Solomon, Steady-state field-aligned currents at Mercury. Geophys. Res. Lett. **41**, 7444–7452 (2014). doi:10.1002/2014GL061677

G.B. Andrews et al., The energetic particle and plasma spectrometer instrument on the MESSENGER spacecraft. Space Sci. Rev. **131**, 523–556 (2007). doi:10.1007/s11214-007-9272-5

V. Angelopoulos, C.F. Kennel, F.V. Coroniti, R. Pellat, M.G. Kivelson, R.J. Walker, C.T. Russell, W. Baumjohann, W.C. Feldman, J.T. Gosling, Statistical characteristics of bursty bulk flow events. J. Geophys. Res. **99**, 21257–21280 (1994)

T.P. Armstrong, S.M. Krimigis, L.J. Lanzerotti, A reinterpretation of the reported energetic particle fluxes in the vicinity of Mercury. J. Geophys. Res. **80**, 4015–4017 (1975). doi:10.1029/JA080i028p04015

M. Ashour-Abdalla, L.M. Zelenyi, J.-M. Bosqued, V. Peroomian, Z. Whang, D. Schriver, R.L. Richard, The formation of the wall region: Consequences in the near-Earth magnetotail. Geophys. Res. Lett. **19**, 1739 (1992)

M. Ashour-Abdalla, M. El-Alaoui, M.L. Goldstein, M. Zhou, D. Schriver, R. Richard, R. Walker, M.G. Kivelson, K.J. Hwang, Observations and simulations of non-local acceleration of electrons in magnetotail magnetic reconnection events. Nat. Phys. **7**, 360–365 (2011)

D.N. Baker, J.A. Simpson, J.H. Eraker, A model of impulsive acceleration and transport of energetic particles in Mercury's magnetosphere. J. Geophys. Res. **91**, 8742–8748 (1986)

D.N. Baker, T.I. Pulkkinen, V. Angelopoulos, W. Baumjohann, R.L. McPherron, Neutral line model of substorms: Past results and present view. J. Geophys. Res. **101**, 12975–13010 (1996)

D.N. Baker, G. Poh, D. Odstrcil, C.N. Arge, M. Benna, C.L. Johnson, H. Korth, D.J. Gershman, G.C. Ho, W.E. McClintock, T.A. Cassidy, A. Merkel, J.M. Raines, D. Schriver, J.A. Slavin, S.C. Solomon, Solar wind forcing at Mercury: WSA-ENLIL model results. J. Geophys. Res. Space Phys. **118**, 45–57 (2013)

W. Baumjohann, G. Paschmann, Determination of the polytropic index in the plasma sheet. Geophys. Res. Lett. **16**, 295–298 (1989)

R. Behrisch, W. Eckstein, *Sputtering by Particle Bombardment: Experiments and Computer Calculations from Threshold to MeV Energies* (Springer, Berlin, 2007), p. 110

J. Benkhoff, J. van Casteren, H. Hayakawa, M. Fujimoto, H. Laakso, M. Novara, P. Ferri, H.R. Middleton, R. Ziethe, BepiColombo-comprehensive exploration of Mercury: Mission overview and science goals. Planet. Space Sci. **58**, 2–20 (2010)

M. Benna et al., Modeling of the magnetosphere of Mercury at the time of the first MESSENGER flyby. Icarus **209**, 3–10 (2010). doi:10.1016/j.icarus.2009.11.036

A. Benninghoven, Developments in secondary ion mass spectroscopy and applications to surface studies. Surf. Sci. **53**, 596–625 (1975). doi:10.1016/0039-6028(75)90158-2

T.A. Bida, R.M. Killen, T.H. Morgan, Discovery of calcium in Mercury's atmosphere. Nature **404**, 159–161 (2000)

S.A. Boardsen, B.J. Anderson, M.H. Acuña, J.A. Slavin, H. Korth, S.C. Solomon, Narrow-band ultra-low-frequency wave observations by MESSENGER during its January 2008 flyby through Mercury's magnetosphere. Geophys. Res. Lett. **36**, L01104 (2009). doi:10.1029/2008GL036034

S.A. Boardsen, T. Sundberg, J.A. Slavin, B.J. Anderson, H. Korth, S.C. Solomon, L.G. Blomberg, Observations of Kelvin-Helmholtz waves along the dusk-side boundary of Mercury's magnetosphere during MESSENGER's third flyby. Geophys. Res. Lett. **37**, L12101 (2010). doi:10.1029/2010GL043606

S.A. Boardsen, J.A. Slavin, B.J. Anderson, H. Korth, D. Schriver, S.C. Solomon, Survey of coherent $\sim$ 1 Hz waves in Mercury's inner magnetosphere from MESSENGER observations. J. Geophys. Res. **117**, A00M05 (2012). doi:10.1029/2012JA017822

P. Borin, M. Bruno, G. Cremonese, F. Marzari, Estimate of the neutral atoms' contribution to the Mercury exosphere caused by a new flux of micrometeoroids. Astron. Astrophys. **517**, A89 (2010)

A.L. Broadfoot, S. Kumar, M.J.S. Belton, Mercury's atmosphere from Mariner 10: Preliminary results. Science **185**, 166–169 (1974)

A.L. Broadfoot, D.E. Shemansky, S. Kumar, Mariner 10: Mercury atmosphere. Geophys. Res. Lett. **3**, 577–580 (1976)

A.L. Broadfoot, S.S. Clapp, F.E. Stuart, Mariner 10 ultraviolet spectrometer: Airglow experiment. Space Sci. Instrum. **3**, 199–208 (1977)

M. Bruno, G. Cremonese, S. Marchi, Neutral sodium atoms release from the surfaces of the Moon and Mercury induced by meteoroid impacts. Planet. Space Sci. **55**, 1494–1501 (2007)

J. Büchner, L.M. Zelenyi, Regular and chaotic charged particle motion in magnetotaillike field reversals: 1. Basic theory of trapped motion. J. Geophys. Res. **94**, 11821–11842 (1989). doi:10.1029/JA094iA09p11821

M.H. Burger, R.M. Killen, R.J. Vervack, E.T. Bradley, W.E. McClintock, M. Sarantos, M. Benna, N. Mouawad, Monte Carlo modeling of sodium in Mercury's exosphere during the first two MESSENGER flybys. Icarus **209**, 63–74 (2010). doi:10.1016/j.icarus.2010.05.007

M.H. Burger, R.M. Killen, W.E. McClintock, R.J. Vervack Jr., A.W. Merkel, A.L. Sprague, M. Sarantos, Modeling MESSENGER observations of calcium in Mercury's exosphere. J. Geophys. Res. **117**, E00L11 (2012). doi:10.1029/2012JE004158

M.H. Burger, R.M. Killen, W.E. McClintock, A.W. Merkel, R.J. Vervack Jr., T.A. Cassidy, M. Sarantos, Seasonal variations in Mercury's dayside calcium exosphere. Icarus **238**, 51–58 (2014)

L.F. Burlaga, Magnetic fields and plasmas in the inner heliosphere: Helios results. Planet. Space Sci. **49**, 1619–1627 (2001)

M.N. Caan, R.L. Mcpherron, C.T. Russell, Solar wind and substorm-related changes in lobes of geomagnetic tail. J. Geophys. Res. **78**, 8087–8096 (1973)

T.A. Cassidy, A.W. Merkel, M.H. Burger, M. Sarantos, R.M. Killen, W.E. McClintock, R.J. Vervack, Mercury's seasonal sodium exosphere: MESSENGER orbital observations. Icarus **248**, 547–559 (2015). doi:10.1016/j.icarus.2014.10.037

S.P. Christon, J. Feynman, J.A. Slavin, Dynamic substorm injections—Similar magnetospheric phenomena at Earth and Mercury, in *Magnetotail Physics*, ed. by A.T.Y. Lui (Johns Hopkins University Press, Baltimore, 1987), pp. 393–400

M.J. Cintala, Impact induced thermal effects in the lunar and Mercurian regoliths. J. Geophys. Res. **97**, 947–973 (1992)

J.B. Cladis, Parallel acceleration and transport of ions from polar ionosphere to plasma sheet. Geophys. Res. Lett. **13**, 893 (1986)

J.B. Cladis, H.L. Collin, O.W. Lennartsson, T.E. Moore, W.K. Peterson, C.T. Russell, Observations of centrifugal acceleration during compression of magnetosphere. Geophys. Res. Lett. **27**, 915 (2000)

S.W.H. Cowley, The causes of convection in the Earth's magnetosphere: A review of developments during the IMS. J. Geophys. Res. **20**, 531–565 (1982)

S.W.H. Cowley, The distant geomagnetic tail in theory and observation, in *AGU Monograph on "Magnetic Reconnection in Space and Laboratory Plasmas"*, vol. 30 (1984), p. 228

G. Cremonese, M. Bruno, V. Mangano, S. Marchi, A. Milillo, Release of neutral sodium atoms from the surface of Mercury induced by meteoroid impacts. Icarus **177**, 122–128 (2005)

D.C. Delcourt, Particle acceleration by inductive electric fields in the inner magnetosphere. J. Atmos. Sol.-Terr. Phys. **64**, 551 (2002)

D.C. Delcourt, On the supply of heavy planetary material to the magnetotail of Mercury. Ann. Geophys. **31**, 1673 (2013)

D.C. Delcourt, J.-A. Sauvaud, R.F. Martin Jr., T.E. Moore, On the nonadiabatic precipitation of ions from the near-Earth plasma sheet. J. Geophys. Res. **101**, 17409 (1996)

D.C. Delcourt, T.E. Moore, S. Orsini, A. Millilo, J.-A. Sauvaud, Centrifugal acceleration of ions near Mercury. Geophys. Res. Lett. **29**, 32 (2002). doi:10.1029/2001GL013829

D.C. Delcourt, S. Grimald, F. Leblanc, J.-J. Berthelier, A. Millilo, A. Mura, S. Orsini, T.E. Moore, A quantitative model of the planetary Na^+ contribution to Mercury's magnetosphere. Ann. Geophys. **21**, 1723–1736 (2003). doi:10.5194/angeo-21-1723-2003

D.C. Delcourt, T.E. Moore, M.-C. Fok, On the effect of IMF turning on ion dynamics at Mercury. Ann. Geophys. **29**, 987 (2011)

D.C. Delcourt, K. Seki, N. Terada, T.E. Moore, Centrifugally stimulated exospheric ion escape at Mercury. Geophys. Res. Lett. **39**, L22105 (2012). doi:10.1029/2012GL054085

G.A. DiBraccio, J.A. Slavin, S.A. Boardsen, B.J. Anderson, H. Korth, T.H. Zurbuchen, J.M. Raines, D.N. Baker, R.L. McNutt Jr., S.C. Solomon, MESSENGER observations of magnetopause structure and dynamics at Mercury. J. Geophys. Res. Space Phys. **118**, 997–1008 (2013). doi:10.1002/jgra.50123

G.A. DiBraccio, J.A. Slavin, S.M. Imber, D.J. Gershman, J.M. Raines, C.M. Jackman, S.A. Boardsen, B.J. Anderson, H. Korth, T.H. Zurbuchen, R.L. McNutt Jr., S.C. Solomon, MESSENGER observations of flux ropes in Mercury's magnetotail. Planet. Space Sci. **115**, 77–89 (2015a). doi:10.1016/j.pss.2014.12.016

G.A. DiBraccio, J.A. Slavin, J.M. Raines, D.J. Gershman, P.J. Tracy, S.A. Boardsen, T.H. Zurbuchen, B.J. Anderson, H. Korth, R.L. McNutt Jr., S.C. Solomon, First observations of Mercury's plasma mantle by MESSENGER. Geophys. Res. Lett. (2015b, accepted)

D.L. Domingue, A.L. Sprague, D.M. Hunten, Dependence of Mercurian atmospheric column abundance estimations on surface-reflectance modeling. Icarus **128**, 75–82 (1997)
D.L. Domingue, P.L. Koehn, R.M. Killen, A.L. Sprague, M. Sarantos, A.F. Cheng, E.T. Bradley, W.E. McClintock, Mercury's atmosphere: A surface-bounded exosphere. Space Sci. Rev. **131**, 161–186 (2007)
A. Doressoundiram, F. Leblanc, C. Foellmi, S. Erard, Metallic species in Mercury's exosphere: EMMI/New technology telescope observations. Astron. J. **137**, 3859–3863 (2009)
J.W. Dungey, Interplanetary magnetic field and the auroral zones. Phys. Rev. Lett. **6**, 47–48 (1961). doi:10.1103/PhysRevLett.6.47
R.C. Elphic, H.O. Funsten, B.L. Barraclough, D.J. McComas, M.T. Paffet, D.T. Vaniman, G. Heiken, Lunar surface composition and solar wind-induced secondary ion mass spectrometry. Geophys. Res. Lett. **18**, 2165–2168 (1991). doi:10.1029/91GL02669
J.H. Eraker, J.A. Simpson, Acceleration of charged particles in Mercury's magnetosphere. J. Geophys. Res. **91**, 9973–9993 (1986). doi:10.1029/JA091iA09p09973
D.H. Fairfield, Average and unusual locations for the Earth's magnetopause and bow shock. J. Geophys. Res. **76**, 6700–6716 (1971). doi:10.1029/JA076i028p06700
W.M. Farrell, J.S. Halekas, R.M. Killen, G.T. Delory, N. Gross, L.V. Bleacher, D. Krauss-Varben, P. Travnicek, D. Hurley, T.J. Stubbs, M.I. Zimmerman, T.L. Jackson, Solar-Storm/Lunar Atmosphere Model (SSLAM): An overview of the effort and description of the driving storm environment. J. Geophys. Res. **117**, E00K04 (2012). doi:10.1029/2012JE004070
G. Fjelbo, A. Kliore, D. Sweetnam, P. Esposito, B. Seidel, T. Howard, The occultation of Mariner 10 by Mercury. Icarus **29**, 407–415 (1976). doi:10.1016/0019-1035(76)90063-4
D.J. Gershman, T.H. Zurbuchen, L.A. Fisk, J.A. Gilbert, J.M. Raines, B.J. Anderson, C.W. Smith, H. Korth, S.C. Solomon, Solar wind alpha particles and heavy ions in the inner heliosphere observed with MESSENGER. J. Geophys. Res. **117**, A00M02 (2012). doi:10.1029/2012JA017829
D.J. Gershman, J.A. Slavin, J.M. Raines, T.H. Zurbuchen, B.J. Anderson, H. Korth, D.N. Baker, S.C. Solomon, Magnetic flux pileup and plasma depletion in Mercury's subsolar magnetosheath. J. Geophys. Res. **118**, 7181–7199 (2013). doi:10.1002/2013JA019244
D.J. Gershman, J.A. Slavin, J.M. Raines, T.H. Zurbuchen, B.J. Anderson, H. Korth, D.N. Baker, S.C. Solomon, Ion kinetic properties in Mercury's premidnight plasma sheet. Geophys. Res. Lett. **41**, 5740–5747 (2014). doi:10.1002/2014GL060468
D.J. Gershman, J.M. Raines, J.A. Slavin, T.H. Zurbuchen, T. Sundberg, S.A. Boardsen, B.J. Anderson, H. Korth, S.C. Solomon, MESSENGER observations of multi-scale Kelvin-Helmholtz vortices at Mercury. J. Geophys. Res. Space Phys. (2015, in revision)
K.-H. Glassmeier, J. Grosser, U. Auster, D. Constantinescu, Y. Narita, S. Stellmach, Electromagnetic induction effects and dynamo action in the Hermean system. Space Sci. Rev. **132**, 511–527 (2007). doi:10.1007/s11214-007-9244-9
B. Hapke, Bidirectional reflectance spectroscopy: 1. Theory. J. Geophys. Res. **86**, 3039–3054 (1981)
B. Hapke, Bidirectional reflectance spectroscopy: 3. Correction for macroscopic roughness. Icarus **59**, 41–59 (1984)
B. Hapke, Bidirectional reflectance spectroscopy: 4. The extinction coefficient and the opposition effect. Icarus **67**, 264–280 (1986)
M. Hesse, M.G. Kivelson, The formation and structure of flux ropes in the magnetotail, in *New Perspectives on the Earth's Magnetotail*, ed. by A. Nishida, D.N. Baker, S.W.H. Cowley (American Geophysical Union, Washington, 1998). doi:10.1029/GM105p0139
M.A. Hidalgo, C. Cid, A.F. Vinas, J. Sequeiros, A non-force-free approach to the topology of magnetic clouds in the solar wind. J. Geophys. Res. **107**, 1002 (2002a). doi:10.1029/2001JA900100
M.A. Hidalgo, T. Nieves-Chinchilla, C. Cid, Elliptical cross-section model for the magnetic topology of magnetic clouds. Geophys. Res. Lett. **29**, 1637 (2002b). doi:10.1029/2001GL013875
T.W. Hill, A.J. Dessler, R.A. Wolf, Mercury and Mars: The role of ionospheric conductivity in the acceleration of magnetospheric particles. Geophys. Res. Lett. **3**, 429–432 (1976). doi:10.1029/GL003i008p00429
G.C. Ho, S.M. Krimigis, R.E. Gold, D.N. Baker, B.J. Anderson, H. Korth, J.A. Slavin, R.L. McNutt Jr., R.M. Winslow, S.C. Solomon, Spatial distribution and spectral characteristics of energetic electrons in Mercury's magnetosphere. J. Geophys. Res. **117**, A00M04 (2012). doi:10.1029/2012JA017983
E.W. Hones, J. Birn, D.N. Baker, S.J. Bame, W.C. Feldman, D.J. Mccomas, R.D. Zwickl, J.A. Slavin, E.J. Smith, B.T. Tsurutani, Detailed examination of a plasmoid in the distant magnetotail with ISEE-3. Geophys. Res. Lett. **11**, 1046–1049 (1984)
L.L. Hood, G. Schubert, Inhibition of solar wind impingement on Mercury by planetary induction currents. J. Geophys. Res. **84**, 2641–2647 (1979)
J.L. Horwitz, Features of ion trajectories in the polar magnetosphere. Geophys. Res. Lett. **11**, 1111 (1984)
C.-S. Huang, A.D. DeJong, X. Cai, Magnetic flux in the magnetotail and polar cap during sawteeth, isolated substorms, and steady magnetospheric convection events. J. Geophys. Res. **114**, A07202 (2009). doi:10.1029/2009JA014232

W.F. Huebner, J.J. Keady, S.P. Lyon, Solar photo rates for planetary atmospheres and atmospheric pollutants. Astrophys. Space Sci. **195**, 1–289 (1992)
D.M. Hunten, T.H. Morgan, D.E. Shemansky, The Mercury atmosphere, in *Mercury*, ed. by F. Vilas, C.R. Chapman, M.S. Matthews (University of Arizona Press, Tucson, 1988), pp. 562–612
A. Ieda, S. Machida, T. Mukai, Y. Saito, T. Yamamoto, A. Nishida, T. Terasawa, S. Kokubun, Statistical analysis of the plasmoid evolution with Geotail observations. J. Geophys. Res. **103**, 4453–4465 (1998)
S.M. Imber, J.A. Slavin, H.U. Auster, V. Angelopoulos, A THEMIS survey of flux ropes and traveling compression regions: Location of the near-Earth reconnection site during solar minimum. J. Geophys. Res. **116**, A02201 (2011). doi:10.1029/2010JA016026
S.M. Imber, J.A. Slavin, S.A. Boardsen, B.J. Anderson, H. Korth, R.L. McNutt Jr., S.C. Solomon, MESSENGER observations of large dayside flux transfer events: Do they drive Mercury's substorm cycle? J. Geophys. Res. Space Phys. **119**, 5613–5623 (2014). doi:10.1002/2014JA019884
W.H. Ip, A. Kopp, MHD simulations of the solar wind interaction with Mercury. J. Geophys. Res. **107**, 1348 (2002). doi:10.1029/2001JA009171
X. Jia, J.A. Slavin, T.I. Gombosi, L. Daldorff, G. Toth, B. van de Holst, Global MHD simulations of Mercury's magnetosphere with coupled planetary interior: Induction effect of the planetary conducting core on the global interaction. J. Geophys. Res. Space Phys. (2015). doi:10.1002/2015JA021143
R.E. Johnson, R. Baragiola, Lunar surface: Sputtering and secondary ion mass spectrometry. Geophys. Res. Lett. **18**, 2169–2172 (1991)
K. Kabin, T.I. Gombosi, D.L. DeZeeuw, K.G. Powell, Interaction of Mercury with the solar wind. Icarus **143**, 397–406 (2000)
E. Kallio, P. Janhunen, Modelling the solar wind interaction with Mercury by a quasi-neutral hybrid model. Ann. Geophys. **21**, 2133 (2003)
S. Kameda, I. Yoshikawa, M. Kagitani, S. Okano, Interplanetary dust distribution and temporal variability of Mercury's atmospheric Na. Geophys. Res. Lett. **36**, L15201 (2009). doi:10.1029/2009GL039036
A. Kidder, R.M. Winglee, E.M. Harnett, Erosion of the dayside magnetosphere at Mercury in association with ion outflows and flux rope generation. J. Geophys. Res. **113**, A09223 (2008). doi:10.1029/2008JA013038
R.M. Killen, J.M. Hahn, Impact vaporization as a possible source of Mercury's calcium exosphere. Icarus **250**, 230–237 (2015). doi:10.1016/j.icarus.2014.11.035
R.M. Killen, W.H. Ip, The surface-bounded atmospheres of Mercury and the Moon. Rev. Geophys. **37**, 361–406 (1999)
R.M. Killen, A.E. Potter, P. Reiff, M. Sarantos, B.V. Jackson, P. Hick, B. Giles, Evidence for space weather at Mercury. J. Geophys. Res. **106**, 20509–20525 (2001)
R.M. Killen, M. Sarantos, A.E. Potter, P. Reiff, Source rates and ion recycling rates for Na and K in Mercury's atmosphere. Icarus **171**, 1–19 (2004)
R. Killen, G. Cremonese, H. Lammer, S. Orsini, A.E. Potter, A.L. Sprague, P. Wurz, M. Khodachenko, H.I.M. Lichtenegger, A. Milillo, A. Mura, Processes that promote and deplete the exosphere of Mercury. Space Sci. Rev. **132**, 433–509 (2007)
R. Killen, D. Shemansky, N. Mouawad, Expected emission from Mercury's exospheric species, and their ultraviolet-visible signatures. Astrophys. J. Suppl. Ser. **181**, 351–359 (2009)
M.G. Kivelson, A.J. Ridley, Saturation of the polar cap potential: Inference from Alfvén wing arguments. J. Geophys. Res. **113**, A05214 (2008). doi:10.1029/2007JA012302
H. Korth, B.J. Anderson, J.M. Raines, J.A. Slavin, T.H. Zurbuchen, C.L. Johnson, M.E. Purucker, R.M. Winslow, S.C. Solomon, R.L. McNutt Jr., Plasma pressure in Mercury's equatorial magnetosphere derived from MESSENGER magnetometer observations. Geophys. Res. Lett. **38**, L22201 (2011). doi:10.1029/2011GL049451
H. Korth, B.J. Anderson, D.J. Gershman, J.M. Raines, J.A. Slavin, T.H. Zurbuchen, S.C. Solomon, R.L. McNutt Jr., Plasma distribution in Mercury's magnetosphere derived from MESSENGER magnetometer and fast imaging plasma spectrometer observations. J. Geophys. Res. Space Phys. **119**, 2917–2932 (2014). doi:10.1002/2013JA019567
M.M. Kuznetsova, M. Hesse, L. Rastätter, A. Taktakishvili, G. Toth, D.L. DeZeeuw, A. Ridley, T.I. Gombosi, Multiscale modeling of magnetospheric reconnection. J. Geophys. Res. **112**, A10210 (2007). doi:10.1029/2007JA012316
H. Lammer, P. Wurz, M.R. Patel, R.M. Killen, C. Kolb, S. Massetti, S. Orsini, A. Milillo, The variability of Mercury's exosphere by particle and radiation induced surface release processes. Icarus **166**, 238–247 (2003)
B. Lavraud, H. Rème, M.W. Dunlop, J.-M. Bosqued, I. Dandouras, J.-A. Sauvaud, A. Keiling, T.D. Phan, R. Lundin, P.J. Cargill, C.P. Escoubet, C.W. Carlson, J.P. MacFadden, G.K. Parks, E. Moebius, L.M. Kistler, E. Amata, M.-B. Bavassano-Cattaneo, A. Korth, B. Klecker, A. Balogh, Cluster observes the high-altitude cusp region. Surv. Geophys. **26**, 135–175 (2005). doi:10.1007/s10712-005-1875-3

F. Leblanc, J.Y. Chaufray, Mercury and Moon He exospheres: Analysis and modeling. Icarus **216**, 551–559 (2011)

F. Leblanc, R.E. Johnson, Mercury's sodium exosphere. Icarus **164**, 261–281 (2003). doi:10.1016/S0019-1035(03)00147-7

F. Leblanc, R.E. Johnson, Mercury exosphere I. Global circulation model of its sodium component. Icarus **209**, 280–300 (2010)

F. Leblanc, E. Chassefiere, R.E. Johnson, D.M. Hunten, E. Kallio, D.C. Delcourt, R.M. Killen, J.G. Luhmann, A.E. Potter, A. Jambon, G. Crernonese, M. Mendillo, N. Yan, A.L. Sprague, Mercury's exosphere origins and relations to its magnetosphere and surface. Planet. Space Sci. **55**, 1069–1092 (2007)

F. Leblanc, A. Doressoundiram, N. Schneider, V. Mangano, A.L. Ariste, C. Lemen, B. Gelly, C. Barbieri, G. Cremonese, High latitude peaks in Mercury's sodium exosphere: Spectral signature using THEMIS solar telescope. Geophys. Res. Lett. **35**, L18204 (2008). doi:10.1029/2008GL035322

F. Leblanc, A. Doressoundiram, N. Schneider, S. Massetti, M. Wedlund, A. López Ariste, C. Barbieri, V. Mangano, G. Cremonese, Short-term variations of Mercury's Na exosphere observed with very high spectral resolution. Geophys. Res. Lett. **36**, L07201 (2009)

R.P. Lepping, J.A. Jones, L.F. Burlaga, Magnetic field structure of interplanetary magnetic clouds at 1 Au. J. Geophys. Res. **95**, 11957–11965 (1990)

R.P. Lepping, D.H. Fairfield, J. Jones, L.A. Frank, W.R. Paterson, S. Kokubun, T. Yamamoto, Cross-tail magnetic flux ropes as observed by the Geotail spacecraft. Geophys. Res. Lett. **22**(10), 1193–1196 (1995)

R.P. Lepping, J.A. Slavin, M. Hesse, J.A. Jones, A. Szabo, Analysis of magnetotail flux ropes with strong core fields: ISEE 3 observations. J. Geomagn. Geoelectr. **48**, 589–601 (1996)

E. Liljeblad, T. Sundberg, T. Karlsson, A. Kullen, Statistical investigation of Kelvin-Helmholtz waves at the magnetopause of Mercury. J. Geophys. Res. Space Phys. **119**, 9670–9683 (2014). doi:10.1002/2014JA020614

U. Mall, E. Kirsch, K. Cierpka, B. Wilken, A. Söding, F. Neubauer, G. Gloeckler, A. Galvin, Direct observation of lunar pick-up ions near the Moon. Geophys. Res. Lett. **25**, 3799–3802 (1998). doi:10.1029/1998GL900003

V. Mangano, A. Milillo, A. Mura, S. Orsini, E. De Angelis, A.M. Di Lellis, P. Wurz, The contribution of impulsive meteoritic impact vapourization to the Hermean exosphere. Planet. Space Sci. **55**, 1541–1556 (2007)

V. Mangano, S. Massetti, A. Milillo, A. Mura, S. Orsini, F. Leblanc, Dynamical evolution of sodium anisotropies in the exosphere of Mercury. Planet. Space Sci. **82–83**, 1–10 (2013)

S. Massetti, S. Orsini, A. Milillo, A. Mura, E. De Angelis, H. Lammer, P. Wurz, Mapping of the cusp plasma precipitation on the surface of Mercury. Icarus **166**, 229–237 (2003)

A. Masters, J.A. Slavin, G.A. DiBraccio, T. Sundberg, R.M. Winslow, C.L. Johnson, B.J. Anderson, H. Korth, A comparison of magnetic overshoots at the bow shocks of Mercury and Saturn. J. Geophys. Res. **118**, 4381–4390 (2013). doi:10.1002/jgra.50428

W.E. McClintock, M.R. Lankton, The Mercury atmospheric and surface composition spectrometer for the MESSENGER mission. Space Sci. Rev. **131**, 481–521 (2007). doi:10.1007/s11214-007-9264-5

W.E. McClintock, R.J. Vervack, E.T. Bradley, R.M. Killen, N. Mouawad, A.L. Sprague, M.H. Burger, S.C. Solomon, N.R. Izenberg, MESSENGER observations of Mercury's exosphere: Detection of magnesium and distribution of constituents. Science **324**, 610–613 (2009)

M.A. McGrath, R.E. Johnson, L.J. Lanzerotti, Sputtering of sodium on the planet Mercury. Nature **323**, 694–696 (1986)

J.L. McLain, A.L. Sprague, G.A. Grieves, D. Schriver, P. Travnicek, T.M. Orlando, Electron-stimulated desorption of silicates: A potential source for ions in Mercury's space environment. J. Geophys. Res. **116**, E03007 (2011). doi:10.1029/2010JE003714

R.L. McPherron, C.T. Russell, M.P. Aubry, Satellite studies of magnetospheric substorms on August 15, 1968: 9. Phenomenological model for substorms. J. Geophys. Res. **78**, 3131–3149 (1973). doi:10.1029/JA078i016p03131

S.E. Milan, S.W.H. Cowley, M. Lester, D.M. Wright, J.A. Slavin, M. Fillingim, C.W. Carlson, H.J. Singer, Response of the magnetotail to changes in the open flux content of the magnetosphere. J. Geophys. Res. **109**, A04220 (2004). doi:10.1029/2003JA010350

A. Milillo et al., The BepiColombo mission: An outstanding tool for investigating the Hermean environment. Planet. Space Sci. **58**, 40–60 (2010)

D.G. Mitchell, D.J. Williams, C.Y. Huang, L.A. Frank, C.T. Russell, Current carriers in the near-Earth cross-tail current sheet during substorm growth phase. Geophys. Res. Lett. **17**, 583 (1990)

M.B. Moldwin, W.J. Hughes, On the formation and evolution of plasmoids: A survey of isee-3 Geotail data. J. Geophys. Res. **97**, 19259–19282 (1992)

N. Mouawad, M.H. Burger, R.M. Killen, A.E. Potter, W.E. McClintock, R.J. Vervack, E.T. Bradley, M. Benna, S. Naidu, Constraints on Mercury's Na exosphere: Combined MESSENGER and ground-based data. Icarus **211**, 21–36 (2011)

M. Müller, S.F. Green, N. McBride, D. Koschny, J.C. Zarnecki, M.S. Bentley, Estimation of the dust flux near Mercury. Planet. Space Sci. **50**, 1101–1115 (2002)

J. Müller, S. Simon, Y.-C. Wang, U. Motschmann, D. Heyner, J. Schüle, W.-H. Ip, G. Kleindienst, G.J. Pringle, Origin of Mercury's double magnetopause: 3D hybrid simulation study with A.I.K.E.F. Icarus **218**, 666–687 (2012). doi:10.1016/j.icarus.2011.12.028

A. Mura, A. Milillo, S. Orsini, S. Massetti, Numerical and analytical model of Mercury's exosphere: Dependence on surface and external conditions. Planet. Space Sci. **55**, 1569–1583 (2007)

A. Mura, P. Wurz, H.I.M. Lichtenegger, H. Schleicher, H. Lammer, D. Delcourt, A. Milillo, S. Massetti, M.L. Khodachenko, S. Orsini, The sodium exosphere of Mercury: Comparison between observations during Mercury's transit and model results. Icarus **200**, 1–11 (2009)

N.F. Ness, K.W. Behannon, R.P. Lepping, Y.C. Whang, K.H. Schatten, Magnetic field observations near Mercury: Preliminary results from mariner 10. Science **185**, 151–160 (1974)

K.W. Ogilvie, J.D. Scudder, R.E. Hartle, G.L. Siscoe, H.S. Bridge, A.J. Lazarus, J.R. Asbridge, S.J. Bame, C.M. Yeates, Observations at Mercury encounter by plasma science experiment on mariner 10. Science **185**, 145–151 (1974)

K.W. Ogilvie, J.D. Scudder, V.M. Vasyliunas, R.E. Hartle, G.L. Siscoe, Observations at planet Mercury by plasma electron experiment: Mariner 10. J. Geophys. Res. **82**, 1807–1824 (1977)

S. Orsini, S. Livi, K. Torkar, S. Barabash, A. Milillo, P. Wurz, A.M. Di Lellis, E. Kallio, the SERENA team, SERENA: A suite of four instruments (ELENA, STROFIO, PICAM and MIPA) on board BepiColombo-MPO for particle detection in the Hermean environment. Planet. Space Sci. **58**, 166–181 (2010). doi:10.1016/j.pss.2008.09.012

S. Orsini, V. Mangano, A. Mura, D. Turrini, S. Massetti, A. Milillo, C. Plainaki, The influence of space environment on the evolution of Mercury. Icarus **239**, 281–290 (2014)

J. Paral, R. Rankin, Dawn-dusk asymmetry in the Kelvin-Helmholtz instability at Mercury. Nat. Commun. **4**, 1645 (2013). doi:10.1038/ncomms2676

J. Paral, P.M. Trávníček, K. Kabin, R. Rankin, T.H. Zurbuchen, Spatial distribution and energy spectrum of heavy ions in the Hermean magnetosphere with applications to MESSENGER flybys. Adv. Geosci. **15**, 1–16 (2009)

J. Paral, P.M. Trávníček, R. Rankin, D. Schriver, Sodium ion exosphere of Mercury during MESSENGER flybys. Geophys. Res. Lett. **37**, L19102 (2010). doi:10.1029/2010GL044413

M. Pfleger, H.I.M. Lichtenegger, P. Wurz, H. Lammer, E. Kallio, M. Alho, A. Mura, J.A. Martín-Fernández, M.L. Khodachenko, S. McKenna-Lawlor, 3D-modeling of Mercury's solar wind sputtered surface-exosphere environment. Planet. Space Sci. (2015). doi:10.1016/j.pss.2015.04.016

A.R. Poppe, J.S. Halekas, M. Sarantos, G.T. Delory, The self-sputtered contribution to the lunar exosphere. J. Geophys. Res. **118**, 1934–1944 (2013)

A. Potter, T.H. Morgan, Discovery of sodium in the atmosphere of Mercury. Science **229**, 651–653 (1985)

A. Potter, T.H. Morgan, Potassium in the atmosphere of Mercury. Icarus **67**, 336–340 (1986)

A. Potter, T.H. Morgan, Sodium and potassium atmospheres of Mercury. Planet. Space Sci. **45**, 95–100 (1997)

A. Potter, R.M. Killen, T.H. Morgan, Rapid changes in the sodium exosphere of Mercury. Planet. Space Sci. **47**, 1441–1448 (1999)

A. Potter, R.M. Killen, T.H. Morgan, The sodium tail of Mercury. Meteorit. Planet. Sci. **37**, 1165–1172 (2002)

A.E. Potter, R.M. Killen, M. Sarantos, Spatial distribution of sodium on Mercury. Icarus **181**, 1–12 (2006)

A.E. Potter, R.M. Killen, T.H. Morgan, Solar radiation acceleration effects on Mercury sodium emission. Icarus **186**, 571–580 (2007)

J. Raeder, P. Zhu, Y. Ge, G. Siscoe, Open geospace general circulation model simulation of a substorm: Axial tail instability and ballooning mode preceding substorm onset. J. Geophys. Res. **115**, A00I16 (2010). doi:10.1029/2010JA015876

J.M. Raines, J.A. Slavin, T.H. Zurbuchen, G. Gloeckler, B.J. Anderson, D.N. Baker, H. Korth, S.M. Krimigis, R.L. McNutt Jr., MESSENGER observations of the plasma environment near Mercury. Planet. Space Sci. **59**, 2004–2015 (2011). doi:10.1016/j.pss.2011.02.004

J.M. Raines, D.J. Gershman, T.H. Zurbuchen, M. Sarantos, J.A. Slavin, J.A. Gilbert, H. Korth, B.J. Anderson, G. Gloeckler, S.M. Krimigis, D.N. Baker, R.L. McNutt Jr., S.C. Solomon, Distribution and compositional variations of plasma ions in Mercury's space environment: The first three Mercury years of MESSENGER observations. J. Geophys. Res. Space Phys. **118**, 1604–1619 (2013). doi:10.1029/2012JA018073

J.M. Raines, D.J. Gershman, J.A. Slavin, T.H. Zurbuchen, H. Korth, B.J. Anderson, S.C. Solomon, Structure and dynamics of Mercury's magnetospheric cusp: MESSENGER measurements of protons and planetary ions. J. Geophys. Res. Space Phys. **119**, 6587–6602 (2014). doi:10.1002/2014JA020120

E. Richer, R. Modolo, C. Chanteur, S. Hess, F. Leblanc, A global hybrid model for Mercury's interaction with the solar wind: Case study of the dipole representation. J. Geophys. Res. **117** (2012). doi:10.1029/2012JA017898

A. Runov, V. Angelopoulos, X.-Z. Zhou, X.-J. Zhang, S. Li, F. Plaschke, J. Bonnell, A THEMIS multicase study of dipolarization fronts in the magnetotail plasma sheet. J. Geophys. Res. **116**, A05216 (2011). doi:10.1029/2010JA016316

C.T. Russell, ULF waves in the Mercury magnetosphere. Geophys. Res. Lett. **16**, 1253–1256 (1989). doi:10.1029/GL016i011p01253

C.T. Russell, R.J. Walker, Flux transfer events at Mercury. J. Geophys. Res. **90**, 11067 (1985)

C.T. Russell, D.N. Baker, J.A. Slavin, The magnetosphere of Mercury, in *Mercury*, ed. by F. Vilas, C.R. Chapman, M.S. Matthews (University of Arizona Press, Tucson, 1988), pp. 514–561

M. Sarantos, P.H. Reiff, T.W. Hill, R.M. Killen, A.L. Urquhart, A B_X-interconnected magnetosphere model for Mercury. Planet. Space Sci. **49**, 1629–1635 (2001)

M. Sarantos, J.A. Slavin, M. Benna, S.A. Boardsen, R.M. Killen, D. Schriver, P. Trávníček, Sodium-ion pickup observed above the magnetopause during MESSENGER's first Mercury flyby: Constraints on neutral exospheric models. Geophys. Res. Lett. **36**, L04106 (2009). doi:10.1029/2008GL036207

M. Sarantos, R.M. Killen, W.E. McClintock, E.T. Bradley, R.J. Vervack, M. Benna, J.A. Slavin, Limits to Mercury's magnesium exosphere from MESSENGER second flyby observations. Planet. Space Sci. **59**, 1992–2003 (2011)

M. Sarantos, R.E. Hartle, R.M. Killen, Y. Saito, J.A. Slavin, A. Glocer, Flux estimates of ions from the lunar exosphere. Geophys. Res. Lett. **39**, L13101 (2012). doi:10.1029/2012GL052001

K. Schindler, A theory of the substorm mechanism. J. Geophys. Res. **79**, 2803 (1974). doi:10.1029/JA079i019p02803

C.A. Schmidt, Monte Carlo modeling of north-south asymmetries in Mercury's sodium exosphere. J. Geophys. Res. **118**, 4564–4571 (2013). doi:10.1002/jgra.50396

C.A. Schmidt, J. Baumgardner, M. Mendillo, J.K. Wilson, Escape rates and variability constraints for high-energy sodium sources at Mercury. J. Geophys. Res. **117**, A03301 (2012). doi:10.1029/2011JA017217

K. Seki, N. Terada, M. Yagi, D.C. Delcourt, F. Leblanc, T. Ogino, Effects of the surface conductivity and IMF strength on the dynamics of planetary ions in Mercury's magnetosphere. J. Geophys. Res. **118**, 3233–3242 (2013). doi:10.1002/jgra.50181

K. Seki, et al., Space Sci. Rev. (2015, this issue). doi:10.1007/s11214-015-0170-y

V.A. Sergeev, M. Malkov, K. Mursula, Testing the isotropic boundary algorithm to evaluate the magnetic field configuration of the tail. J. Geophys. Res. **98**, 7609 (1993)

E.G. Shelley, R.G. Johnson, R.D. Sharp, Satellite observations of energetic heavy ions during a geomagnetic storm. J. Geophys. Res. **77**, 6104 (1972)

D.E. Shemansky, A.L. Broadfoot, Interaction of the surfaces of the Moon and Mercury with their exospheric atmospheres. Rev. Geophys. **15**, 491–499 (1977). doi:10.1029/RG015i004p00491

K. Shiokawa, K. Yumoto, Y. Tanaka, T. Oguti, Y. Kiyama, Low-latitude auroras observed at Moshiri and Rikubetsu ($L = 1.6$) during magnetic storms on February 26, 27, 29, and May 10, 1992. J. Geomagn. Geoelectr. **46**, 231–252 (1994)

J.A. Simpson, J.H. Eraker, J.E. Lamport, P.H. Walpole, Electrons and protons accelerated in Mercury's magnetic field. Science **185**, 160–166 (1974)

G.L. Siscoe, N.F. Ness, C.M. Yeates, Substorms on Mercury? J. Geophys. Res. **80**, 4359–4363 (1975). doi:10.1029/JA080i031p04359

J.A. Slavin, R.E. Holzer, The effect of erosion on the solar wind stand-off distance at Mercury. J. Geophys. Res. **84**, 2076–2082 (1979)

J.A. Slavin, M.F. Smith, E.L. Mazur, D.N. Baker, T. Iyemori, H.J. Singer, E.W. Greenstadt, ISEE-3 plasmoid and TCR observations during an extended interval of substorm activity. Geophys. Res. Lett. **19**, 825–828 (1992)

J.A. Slavin, M.F. Smith, E.L. Mazur, D.N. Baker, E.W. Hones, T. Iyemori, E.W. Greenstadt, ISEE-3 observations of traveling compression regions in the Earth's magnetotail. J. Geophys. Res. **98**, 15425–15446 (1993)

J.A. Slavin, R.P. Lepping, J. Gjerloev, D.H. Fairfield, M. Hesse, C.J. Owen, M.B. Moldwin, T. Nagai, A. Ieda, T. Mukai, Geotail observations of magnetic flux ropes in the plasma sheet. J. Geophys. Res. **108**, 1015 (2003). doi:10.1029/2002JA009557

J.A. Slavin, E.I. Tanskanen, M. Hesse, C.J. Owen, M.W. Dunlop, S. Imber, E.A. Lucek, A. Balogh, K.-H. Glassmeier, Cluster observations of traveling compression regions in the near-tail. J. Geophys. Res. **110**, A06207 (2005). doi:10.1029/2004JA010878

J.A. Slavin, R.P. Lepping, J. Gjerloev, D.H. Fairfield, M. Hesse, C.J. Owen, M.B. Moldwin, T. Nagai, A. Ieda, T. Mukai, MESSENGER: Exploring Mercury's magnetosphere. Space Sci. Rev. **131**, 133–160 (2007)

J.A. Slavin et al., Mercury's magnetosphere after MESSENGER's first flyby. Science **321**, 85–89 (2008). doi:10.1126/science.1159040

J.A. Slavin et al., MESSENGER observations of magnetic reconnection in Mercury's magnetosphere. Science **324**, 606–610 (2009)

J.A. Slavin et al., MESSENGER observations of extreme loading and unloading of Mercury's magnetic tail. Science **329**, 665–668 (2010)

J.A. Slavin et al., MESSENGER and mariner 10 flyby observations of magnetotail structure and dynamics at Mercury. J. Geophys. Res. **117**, A01215 (2012a). doi:10.1029/2011JA016900

J.A. Slavin et al., MESSENGER observations of a flux-transfer-event shower at Mercury. J. Geophys. Res. **117**, A00M06 (2012b). doi:10.1029/JA017926

J.A. Slavin, G.A. DiBraccio, D.J. Gershman, S.M. Imber, G.K. Poh, T.H. Zurbuchen, X. Jia, D.N. Baker, S.A. Boardsen, M. Sarantos, T. Sundberg, A. Masters, C.L. Johnson, R.M. Winslow, B.J. Anderson, H. Korth, R.L. McNutt Jr., S.C. Solomon, MESSENGER observations of Mercury's magnetosphere under extreme solar wind conditions. J. Geophys. Res. Space Phys. **119**, 8087–8116 (2014). doi:10.1002/2014JA020319

M.F. Smith, M. Lockwood, Earth's magnetospheric cusps. Rev. Geophys. **34**, 233–260 (1996). doi:10.1029/96RG00893

D.E. Smith et al., Gravity field and internal structure of Mercury from MESSENGER. Science **336**, 214–271 (2012). doi:10.1126/science.1218809

W.H. Smyth, M.L. Marconi, Theoretical overview and modeling of the sodium and potassium atmospheres of Mercury. Astrophys. J. **441**, 839–864 (1995)

C.S. Solomon, R.L. McNutt, R.E. Gold, D.L. Domingue, MESSENGER: Mission overview. Space Sci. Rev. **131**, 3–39 (2007)

T.W. Speiser, Particle trajectory in model current sheets: 1. Analytical solutions. J. Geophys. Res. **70**, 4219 (1965)

A.L. Sprague, R.W.H. Kozlowski, D.M. Hunten, N.M. Schneider, D.L. Domingue, W.K. Wells, W. Schmitt, U. Fink, Distribution and abundance of sodium in Mercury's atmosphere. Icarus **129**, 506–527 (1997)

S.A. Stern, The lunar atmosphere: History, status, current problems, and context. Rev. Geophys. **37**, 453–491 (1999)

R.J. Strangeway, C.T. Russell, J.G. Luhmann, T.E. Moore, J.C. Foster, S.V. Barabash, H. Nilsson, Does a planetary-scale magnetic field enhance or inhibit ionospheric plasma outflows? in *AGU Fall Meeting Abstracts* (2010), p. 1893

S.T. Suess, B.E. Goldstein, Compression of the Hermean magnetosphere by the solar wind. J. Geophys. Res. **84**, 3306–3312 (1979)

T. Sundberg, S.A. Boardsen, J.A. Slavin, L.G. Blomberg, H. Korth, The Kelvin-Helmholtz instability at Mercury: An assessment. Planet. Space Sci. **58**, 1434–1441 (2010). doi:10.1016/j.pss.2010.06.008

T. Sundberg, S.A. Boardsen, J.A. Slavin, L.G. Blomberg, J.A. Cumnock, S.C. Solomon, B.J. Anderson, H. Korth, Reconstruction of propagating Kelvin-Helmholtz vortices at Mercury's magnetopause. Planet. Space Sci. **59**, 2051–2057 (2011)

T. Sundberg, S.A. Boardsen, J.A. Slavin, B.J. Anderson, H. Korth, T.H. Zurbuchen, J.M. Raines, S.C. Solomon, MESSENGER orbital observations of large-amplitude Kelvin-Helmholtz waves at Mercury's magnetopause. J. Geophys. Res. **117**, A04216 (2012). doi:10.1029/2011JA017268

K.G. Tanaka, M. Fujimoto, I. Shinohara, On the peak level of tearing instability in an ion-scale current sheet: The effects of ion temperature anisotropy. Planet. Space Sci. **59**, 510–516 (2011). doi:10.1016/j.pss.2010.04.014

P. Trávníček, P. Hellinger, D. Schriver, Structure of Mercury's magnetosphere for different pressure of the solar wind: Three dimensional hybrid simulations. Geophys. Res. Lett. **34**, 5104 (2007). doi:10.1029/2006GL028518

P. Trávníček et al., Mercury's magnetosphere-solar wind interaction for northward and southward interplanetary magnetic field: Hybrid simulation results. Icarus **209**, 11–22 (2010). doi:10.1016/j.icarus.2010.01.008

R.J. Vervack, W.E. McClintock, R.M. Killen, A.L. Sprague, B.J. Anderson, M.H. Burger, E.T. Bradley, N. Mouawad, S.C. Solomon, N.R. Izenberg, Mercury's complex exosphere: Results from MESSENGER's third flyby. Science **329**, 672–675 (2010). doi:10.1126/science.1188572

R.J. Vervack, W.E. McClintock, R.M. Killen, A.L. Sprague, M.H. Burger, A.W. Merkel, M. Sarantos, MESSENGER searches for less abundant or weakly emitting species in Mercury's exosphere, in *AGU Fall Meeting Abstracts A2* (2011)

F. Vilas, C.R. Chapman, M.S. Mathews, *Mercury* (University of Arizona Press, Tucson, 1988)

Y.-C. Wang, W.-H. Ip, Source dependency of exospheric sodium on Mercury. Icarus **216**, 387–402 (2011). doi:10.1016/j.icarus.2011.09.023
Y.X. Wang, F. Ohuchi, P.H. Holloway, Mechanisms of electron stimulated desorption from soda-silica glass surfaces. J. Vac. Sci. Technol. A **2**(2), 732–737 (1984). doi:10.1116/1.572560
Y.-C. Wang, J. Mueller, U. Motschmann, W.-H. Ip, A hybrid simulation of Mercury's magnetosphere for the MESSENGER encounters in year 2008. Icarus **209**(pp. 46–52), 2010.05.020 (2010). doi:10.1016/j.icarus
R.M. Winglee, E. Harnett, A. Kidder, Relative timing of substorm processes as derived from multifluid/multiscale simulations: Internally driven substorms. J. Geophys. Res. **114**, A09213 (2009). doi:10.1029/2008JA013750
R.M. Winslow, C.L. Johnson, B.J. Anderson, H. Korth, J.A. Slavin, M.E. Purucker, S.C. Solomon, Observations of Mercury's northern cusp region with MESSENGER's magnetometer. Geophys. Res. Lett. **39**, L08112 (2012). doi:10.1029/2012GL051472
R.M. Winslow, B.J. Anderson, C.L. Johnson, J.A. Slavin, H. Korth, M.E. Purucker, D.N. Baker, S.C. Solomon, Mercury's magnetopause and bow shock from MESSENGER magnetometer observations. J. Geophys. Res. **118**, 2213–2227 (2013). doi:10.1002/jgra.50237
R.M. Winslow et al., Mercury's surface magnetic field determined from proton-reflection magnetometry. Geophys. Res. Lett. **41**, 4463–4470 (2014). doi:10.1002/2014GL060258
P. Wurz, L. Blomberg, Particle populations in Mercury's magnetosphere. Planet. Space Sci. **49**, 1643–1653 (2001)
P. Wurz, H. Lammer, Monte-Carlo simulation of Mercury's exosphere. Icarus **164**, 1–13 (2003)
P. Wurz, U. Rohner, J.A. Whitby, C. Kolb, H. Lammer, P. Dobnikar, J.A. Martín-Fernández, The lunar exosphere: The sputtering contribution. Icarus **191**, 486–496 (2007). doi:10.1016/j.icarus.2007.04.034
P. Wurz, J.A. Whitby, U. Rohner, J.A. Martín-Fernández, H. Lammer, C. Kolb, Self-consistent modelling of Mercury's exosphere by sputtering, micro-meteorite impact and photon-stimulated desorption. Planet. Space Sci. **58**, 1599–1616 (2010)
M. Yagi, K. Seki, Y. Matsumoto, D.C. Delcourt, F. Leblanc, Formation of a sodium ring in Mercury's magnetosphere. J. Geophys. Res. **115**, A10 (2010). doi:10.1029/2009JA015226
B.V. Yakshinksiy, T.E. Madey, Photon-stimulated desorption as a substantial source of sodium in the lunar atmosphere. Nature **400**, 642 (1999)
B.V. Yakshinskiy, T.E. Madey, Photon-stimulated desorption of Na from a lunar sample: Temperature-dependent effects. Icarus **168**, 53–59 (2004)
B.V. Yakshinskiy, T.E. Madey, Temperature-dependent DIET of alkalis from SiO_2 films: Comparison with a lunar sample. Surf. Sci. **593**, 202–209 (2005)
B.V. Yakshinskiy, T.E. Madey, V.N. Ageev, Thermal desorption of sodium atoms from thin SiO2 films. Surf. Rev. Lett. **7**, 75–87 (2000)
A.W. Yau, A. Howarth, W.K. Peterson, T. Abe, Transport of thermal-energy ionospheric oxygen (O^+) ions between the ionosphere and the plasma sheet and ring current at quiet times preceding magnetic storms. J. Geophys. Res. **117** (2012). doi:10.1029/2012JA017803
T.H. Zurbuchen, J.M. Raines, G. Gloeckler, S.M. Krimigis, J.A. Slavin, P.L. Koehn, R.M. Killen, A.L. Sprague, R.L. McNutt Jr., S.C. Solomon, MESSENGER observations of the composition of Mercury's ionized exosphere and plasma environment. Science **321**, 90–92 (2008). doi:10.1126/science.1159314
T.H. Zurbuchen et al., MESSENGER observations of the spatial distribution of planetary ions near Mercury. Science **333**, 1862 (2011)
B.J. Zwan, R.A. Wolf, Depletion of solar wind plasma near a planetary boundary. J. Geophys. Res. **81**, 1636–1648 (1976)

DOI 10.1007/978-1-4939-3544-4_5
Reprinted from *Space Science Reviews* Journal, DOI 10.1007/s11214-015-0187-2

The Earth: Plasma Sources, Losses, and Transport Processes

Daniel T. Welling[1] · Mats André[2] · Iannis Dandouras[3] · Dominique Delcourt[4] · Andrew Fazakerley[5] · Dominique Fontaine[4] · John Foster[6] · Raluca Ilie[1] · Lynn Kistler[7] · Justin H. Lee[8] · Michael W. Liemohn[1] · James A. Slavin[1] · Chih-Ping Wang[9] · Michael Wiltberger[10] · Andrew Yau[11]

Received: 7 March 2015 / Accepted: 13 July 2015 / Published online: 17 September 2015

Abstract This paper reviews the state of knowledge concerning the source of magnetospheric plasma at Earth. Source of plasma, its acceleration and transport throughout the system, its consequences on system dynamics, and its loss are all discussed. Both observational and modeling advances since the last time this subject was covered in detail (Hultqvist et al., Magnetospheric Plasma Sources and Losses, 1999) are addressed.

Keywords Magnetosphere · Plasma · Ionosphere · Solar wind

✉ D.T. Welling
dwelling@umich.edu

M. André
mats.andre@irfu.se

I. Dandouras
Iannis.Dandouras@irap.omp.eu

D. Delcourt
dominique.delcourt@lpp.polytechnique.fr

A. Fazakerley
a.fazakerley@ucl.ac.uk

D. Fontaine
dominique.fontaine@lpp.polytechnique.fr

J. Foster
jcf@haystack.mit.edu

R. Ilie
rilie@umich.edu

L. Kistler
lynn.kistler@unh.edu

J.H. Lee
Justin.H.Lee@aero.org

M.W. Liemohn
liemohn@umich.edu

J.A. Slavin
jaslavin@umich.edu

1 Introduction

Earth, being our most extensively explored solar system body, has decades of work dedicated to the sources, losses, and circulation of plasma within its magnetosphere. Indeed, a previous International Space Science Institute review book has already been dedicated to this topic (Hultqvist et al. 1999). This book painted a picture of the balance between ionospheric and solar wind plasma at every major magnetospheric region, from the high latitude ionosphere to the plasma sheet and inner magnetosphere. It is a comprehensive review of modeling and observational work performed up to the point of its publication.

Over the past decade and a half since the book's release, the community has continued to make significant strides in understanding the near-Earth plasma environment (see Chappell 2015 for history and current status). This review summarizes these advances. It will begin by focusing on recent advances in our knowledge of the entry mechanisms for each source. The transport and acceleration of the relevant populations from source to key magnetospheric regions will be reviewed, as well as the consequences each source has on magnetospheric dynamics. The review will conclude with loss mechanisms for magnetospheric plasma, then address the outstanding questions that remain in this broad subject area.

2 Sources

There are two important sources of plasma in Earth's magnetosphere: the solar wind, which provides almost exclusively hydrogen, and the Earth's ionosphere, which is capable of delivering considerable amounts of hydrogen as well as heavy ions, such as helium and oxygen. Other sources, important at other solar system bodies, are either not applicable (e.g., surface sputtering) or contribute so little as to be considered negligible (e.g., plasma from natural satellites). Here, we review progress in our understanding of the entry mechanisms for high latitude ionospheric plasma, low latitude ionospheric plasma, and solar wind plasma.

C.-P. Wang
cat@atmos.ucla.edu

M. Wiltberger
wiltbemj@ucar.edu

A. Yau
yau@ucalgary.ca

[1] University of Michigan, Ann Arbor, USA

[2] Swedish Institute of Space Physics, Uppsala, Sweden

[3] CNRS, IRAP, University of Toulouse, Toulouse, France

[4] LPP, Ecole Polytechnique-CNRS, Université Pierre et Marie Curie, Paris, France

[5] Mullard Space Science Laboratory, University College London, Holmbury St. Mary, UK

[6] Massachusetts Institute of Technology Haystack Observatory Westford, Massachusetts, USA

[7] University of New Hampshire, Durham, NH, USA

[8] The Aerospace Corporation, El Segundo, CA, USA

[9] University of California, Los Angeles, USA

[10] National Center for Atmospheric Research, High Altitude Observatory, Boulder, CO, USA

[11] University of Calgary, Calgary, Canada

2.1 High Latitude Ionospheric Plasma

The variety of observed ion outflows in the high-latitude ionosphere may be grouped into two categories: bulk ion flows with energies up to a few eV, in which all the ions acquire a bulk flow velocity, and suprathermal ion outflows in which in general a fraction of the ions are energized to much higher energies. The category of bulk ion flows includes the polar wind and auroral bulk O^+ up-flow from the topside auroral and polar-cap ionosphere. The category of suprathermal ion outflows includes ion beams, ion conics, transversely accelerated ions (TAI), and upwelling ions (UWI).

Observations of both thermal and superthermal ion outflows prior to the mid-1990 were the subject of the comprehensive review of Moore et al. (1999) under the ISSI Study Project on Source and Loss Processes. In this review, we shall focus on more recent outflow measurements from satellites and ground radar. These measurements were, in general, acquired in different phases in the 11-year solar cycle, and covered different ranges of both altitude and ion energy. It is important to take into account the relative phase in the solar cycle and the relative altitude and ion energy coverage between different measurements, as many ion outflow characteristics exhibit significant long-term variations as well as variability on the time scale of days within a solar rotation near solar maximum. For convenience in our discussions below, we will use the term "topside ionosphere" to refer to the altitude region below 1000 km, including the F-region, and the terms "low-", "mid-", and "high-altitude" to the regions between 1000–4000 km, between 4000–10,000 km, and above 10,000 km, respectively.

At both auroral and polar cap latitudes, a plasma flux tube undergoes a circulation cycle that begins with anti sunward flow and stretching in length, from $\sim$10 to $\sim$100 R_E. This occurs either as it disconnects from the conjugate hemisphere to connect into the solar wind during part of the Dungey cycle or as moves with the viscous flow in the low-latitude boundary layer. During the stretch part of the cycle, the ionospheric plasma can expand freely into the upper reaches of the flux tube because of the negligible plasma pressure there. This results in the formation of the polar wind: the spatial separation between the heavier ions and the electrons due to the Earth's gravitation produces a polarization electric field that acts to accelerate the ions in the upward direction. Additional acceleration mechanisms give rise to the so-called "non-classical" polar wind (Schunk 2007).

Polar wind ion observations have been made on a number of polar-orbiting satellites, including ISIS-2, DE-1, Akebono, and Polar; polar wind electron observations have also been made on DE-1 and Akebono. These observations spanned different phases of Solar Cycle 20 to 23, and a wide range of altitudes from 1000 km to $\sim$50,500 km (8 R_E) altitude (Yau et al. 2007). A composite picture of the polar wind emerges from these observations. The polar wind is regularly observed at all local times and polar latitudes, and is composed primarily of electrons and H^+, He^+ and O^+ ions; the ion composition varies with the solar cycle, and is dominated in density by O^+ ions up to 4000–7000 km. The dayside and the nightside velocity profiles are qualitatively similar for all three species, both having a monotonic increase in velocity with altitude, a similar mass dependence of the magnitude of the velocity, and the largest acceleration (increase of velocity with altitude) of the H^+ velocity below 4000 km.

Near solar maximum on the dayside, the altitude at which the ion reaches 1 km/s is near 2000 km for H^+, but near 3000 and 6000 km for He^+ and O^+, respectively; for all three species, the dayside velocity is significantly larger than on the nightside, being about 12, 6, and 4 km/s for H^+, He^+ and O^+ respectively, at 10,000 km, compared with $\sim$7, 4, and 3 km/s, respectively, on the nightside (Abe et al. 1993b). This is suggestive of

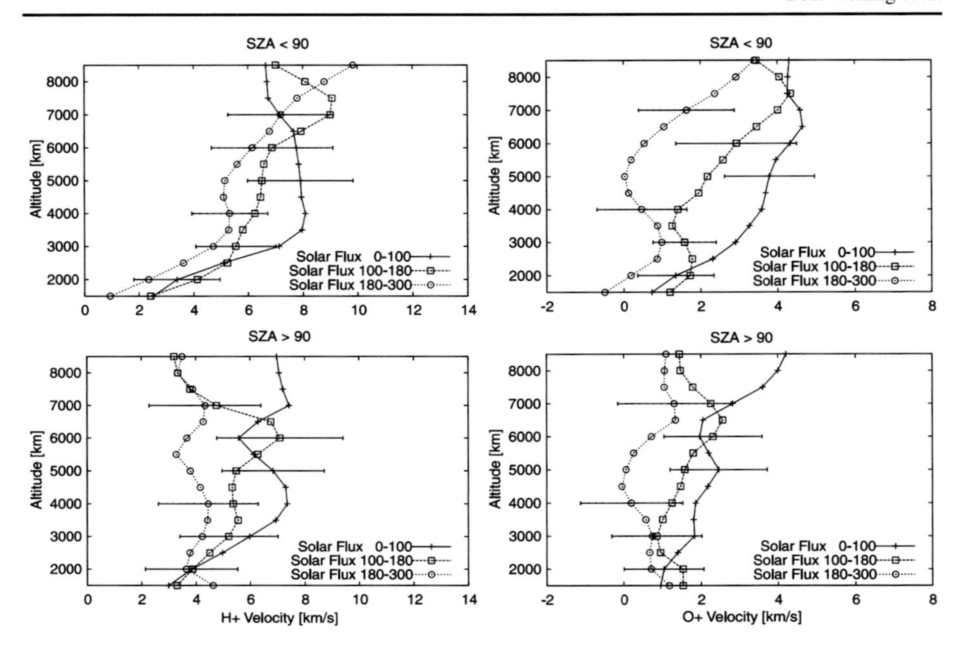

Fig. 1 Averaged H^+ (*left*) and O^+ (*right*) velocity observed on Akebono versus altitude for different solar flux levels: SZA $< 90°$ (*top row*) and SZA $> 90°$ (*bottom row*). From Abe (2004)

possible enhancement in the ambipolar electric field amplitude or presence of additional ion acceleration on the dayside due to escaping atmospheric photoelectrons (Tam et al. 2007). The averaged O^+ velocity begins to increase near 5000 km. This suggests that the O^+ ions above this altitude are predominantly upward; on the nightside, the averaged O^+ velocity starts to increase from zero at 7000 km. The magnitude of ion acceleration at a given altitude is found to correlate strongly with the electron temperature (Abe et al. 1993a). The ion velocity-to-electron temperature ratio also increases with altitude. This increase is consistent with the cumulative increase in ion velocity due to acceleration via ambipolar electric field along the field line. The variability (standard deviation) of the ion velocity is as much as 50 % of the mean during active times ($K_P \geq 3$), and larger during quiet times ($K_p \leq 2$). The mean velocity appears only weakly dependent on Kp for all three species.

Figure 1 shows the averaged H^+ and O^+ polar wind velocity at different solar flux levels (F10.7) as a function of altitude in the sunlit (SZA $< 90°$) and shadow (non-sunlit; SZA $> 90°$) regions, respectively. In the sunlit region, the H^+ velocity increases with altitude at all altitudes for all solar flux levels, except at low solar flux (F10.7 < 100), where it remains almost constant above 4000 km. However, the velocity gradient in different altitude regions varies with solar flux. At high solar flux (F10.7 > 180), the velocity increases continuously from 1500 km to 8500 km. In comparison, at low solar flux, the velocity increase with altitude is much larger below 3600 km and much smaller above 4000 km. As a result, the averaged velocity is about 50–60 % larger at 4000 km and comparable at $\sim$7000 km. The O^+ velocity in the sunlit region remains below 1 km/s below 6000 km, but increases with altitude above that height at high solar flux. A similar transition in the velocity is observed at 4000 km at medium solar flux. At low solar flux, the velocity increases gradually with altitude from 1500 to 7000 km, reaching 4 km/s at 5000 km. In other words, the altitudinal gradients of both H^+ and O^+ velocity have very similar solar flux dependence and altitude

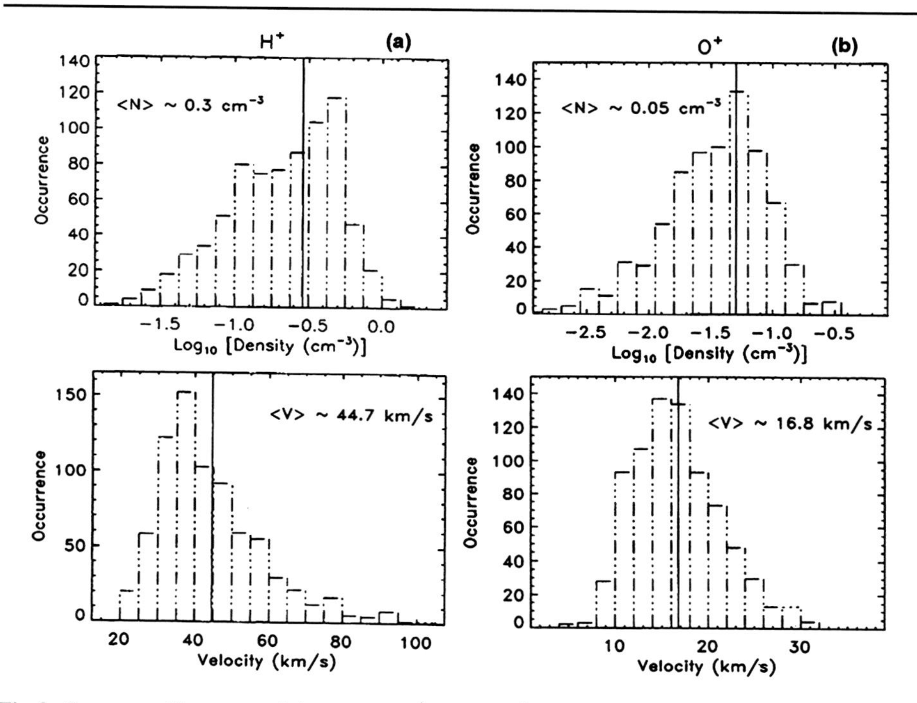

Fig. 2 Occurrence histograms of observed (**a**) H^+ and (**b**) O^+ polar wind density (*top*) and parallel velocity (*bottom*) on Polar at 50,500 km near solar minimum. From Su et al. (1998)

variations, namely, larger gradient below 5000 km and smaller gradient above 7000 km at low solar flux than at high solar flux. This results in generally higher H^+ and O^+ velocities below 7000 km and 8500 km, respectively, at low solar flux.

The observed ion outflow rate of H^+ and O^+ is also only weakly dependent on K_P, the O^+ rate at 6000–9000 km altitude increasing by a factor of 1.7 as K_P increases from 1 to 6 (Abe et al. 1996). The outflow rate of both species exhibits very similar IMF Bz dependence. It increases with B_Z under northward IMF conditions.

The magnetic local time (MLT) dependence of the polar wind ion flux strongly resembles that of the observed ion velocity: the ion flux is largest in the noon quadrant and smallest in the midnight quadrant. This is consistent with the larger ambipolar electric field in the sunlit polar wind. The polar wind H^+ flux (normalized to 2000 km altitude) in the noon quadrant is in the range of $1–20 \times 10^7$ $cm^{-2}\,s^{-1}$. The corresponding O^+ flux is typically a factor of 1.5–2.0 smaller. The fluxes of the different polar wind ion species have markedly different seasonal dependences in general. In the case of He^+, the flux has a winter-to-summer ratio of $\sim$20, which is attributed to the seasonal variations of neutral atmospheric helium and molecular nitrogen and the corresponding helium photo-ionization rate and He^+–N2 charge-exchange rate.

As the polar wind ions flow upward on open magnetic field lines to higher altitudes and undergo generally anti-sunward convection in the dayside cusp and the polar cap, they may be subject to a number of "non-classical" polar wind ion acceleration mechanisms (Yau et al. 2007). An example is centrifugal acceleration in the parallel direction due to strong $E \times B$ convection in regions of curved magnetic field at high altitudes above a few R_E. The result of this is that ions continue to increase in both drift speed and temperature. Figure 2 shows the occurrence distributions of the polar wind H^+ and O^+ ions near the apogee of the Polar satellite at ~50,050 km altitude, where the H^+ density averages ~0.3 cm^{-3} and the H^+ parallel velocity averages 45 km/s near solar minimum (Su et al. 1998). The corresponding O^+ density and velocity are about a factor of 6 and 2.7 smaller (i.e. ~0.05 cm^{-3} and ~17 km/s) respectively.

The observed velocity ratio between ion species on both Akebono and Polar spacecraft spans a wide range of values, and on average lies between unity and the inverse square root mass ratio of the species, i.e. $1 < V_{\parallel,H^+}/V_{\parallel,O^+} < \sqrt{m_{O^+}/m_{H^+}} = 4$. This suggests that a number of processes of comparable energy gain may be contributing to the overall ion acceleration. The temperature of polar wind ions is generally low. On Akebono, the temperature was found to be in the range of 0.05–0.35 eV below 10,000 km (Drakou et al. 1997), and the parallel-to-perpendicular temperature ratio was less than unity at 5000 km. At Polar apogee (~50,090 km), the averaged parallel H^+ and O^+ temperature is ~1.7 and ~7.5 eV, respectively, and the parallel-to-perpendicular temperature ratio is ~1.5 for H^+ and ~2.0 for O^+ (Su et al. 1998).

Interspersed with bulk polar flows is bulk auroral flow. Ion upflows at velocities exceeding 1 km/s have been observed in the topside ionosphere in both the nightside auroral zone and the dayside cleft on low-altitude polar-orbiting satellites, including the Dynamic Explorer 2 (DE-2) (Heelis et al. 1984) and the Hilat satellites, and from ground radars, including the Chatanika incoherent scatter radar and the European incoherent scatter radar (EISCAT) and EISCAT Svalbard radar (ESR). The term "upflow" is used instead of "outflow" to emphasize the very low (and below escape) energy nature of the flow. The observed ion upflow is highly variable in time and location, and generally confined to narrow latitude regions. Large upward ion flows often occur in regions of large ion convection velocities, and are dominated by O^+ and at times enhanced in molecular NO^+.

On DE-2 at 600–1000 km, the occurrence probability of upflow is generally larger than that of downflow in the auroral zone but smaller in the polar cap on both the dayside and the nightside. The peak probability spans the convection reversal on the dayside, and is more extended in latitude and located at lower latitude on the nightside. The probability for flows exceeding 100 m/s increases and moves equatorward with increasing K_P, from about 0.25 near 78° invariant at $K_P \leq 3-$ to about 0.35 near 70° at $K_P \geq 6$ on the dayside. In the polar cap (>78° invariant), the probability of upflow is several times larger during northward IMF than during southward IMF, and it is generally greater in the pre-noon sector than in the pre-midnight sector.

The observed upflow by EISCAT generally falls into two types. Type-1 upflow is associated with strong electric fields in regions of downward field-aligned currents and very low F-region electron densities adjacent to auroral arcs. It is characterized by ion temperature enhancements and perpendicular ion temperature anisotropy ($T_\perp > T_\parallel$). The latter is indicative of frictional heating of ions drifting through neutrals and production of strong pressure gradients, which push the ions upward. The type-2 upflow is typically observed above auroral arcs and is characterized by electron temperature enhancements, weak to moderate electric fields, and a stronger ion flux. All of these features are indicative of auroral electron precipitation and the resulting electron ionization occurs more frequently compared with type-1 upflow.

On average over the solar cycle, the field-aligned upflow occurrence probability at 500 km altitude is higher on the dusk side than on the dawn side, and peaks at ~23 % in the pre-midnight sector. The upflow velocity increases monotonically with altitude starting from about 300 km, to values exceeding 100 m/s at 500 km in the majority (~55 %) of cases (Foster et al. 1998). Roughly 50–60 % of the observed upflow events occurred during intervals of enhanced ion temperature. The observed dawn-dusk asymmetry and midnight-sector peak is believed to reflect the combined effects of both MLT and latitudinal variations of upflow at the location of the EISCAT radar at Tromsø, which at 66.2° invariant latitude, lies within the nightside auroral oval and equatorward of the dayside oval.

In contrast, at the EISCAT Svalbard radar (ESR), which at 75.4° invariant lies within the dayside oval and poleward of the nightside oval, the upflow on the dayside starts or reaches an observable velocity at higher altitudes, and has a larger occurrence frequency than on the nightside above 400 km (Liu et al. 2001), as well as a dawn-dusk asymmetry that increases with altitude in favor of the dawn side over the dusk side. The starting altitude of ion upflow increases with solar activity level, with approximately 25 % and 16 % of the dayside upflow events below 400 km (55 % and 34 % below 450 km) altitude in period of low and high solar activity (F10.7 < 140 and F10.7 > 140), respectively. The upflow occurrence frequency at 500 km altitude increases with K_P, and peaks around geomagnetic noon at $\sim$11–21 %, where the averaged ion flux reaches $2 \times 10^9\ \text{cm}^{-2}\text{s}^{-1}$ and is relatively independent of geomagnetic activity level (K_P). During quiet and moderately active periods, the downflow frequency peaks in the dawn sector (03–09 MLT) at $\sim$5–6 %. During disturbed periods, the downflow frequency peaks in the noon sector (10–15 MLT), and the peak frequency of $\sim$25 % exceeds the upflow frequency, and is consistent with the ESR being equatorward of most of the upflow events.

Approximately half of the dayside ion upflow events are accompanied by increases of both ion and electron temperatures, compared with only 10–20 % of events at other local times. About 20 % of the events are accompanied by electron temperature increases only, regardless of local time, and another 5–10 % of noon-sector events and 20–25 % of morning-sector events are accompanied by ion temperature increases. The remaining 15–40 % appear unaccompanied by any appreciable ion or electron heating.

The occurrence probability and morphological characteristics of the ion upflows observed at both EISCAT and ESR exhibit seasonal as well as diurnal and solar cycle variations (Foster et al. 1998). Above 300 km altitude at EISCAT, the occurrence frequency of upflow is greater during the winter months. Compared with the quieter phase of the solar cycle, the upflow during the active phase of the cycle has a larger ion flux, a smaller ion velocity, and its occurrence frequency has a more pronounced nightside maximum (Liu et al. 2001). The predominance of larger-flux events at solar maximum may be attributed to the higher prevailing ambient plasma density, and the smaller velocities in these events to a smaller per-capita amount of free energy available for acceleration and/or a larger energy loss to ion-neutral collisions. Compared with quiet times, the occurrence frequency of ion upflow is significantly larger at all altitudes during disturbed times ($K_P \geq 4$) (Liu et al. 2001). Furthermore, the starting altitude of upflow is lower (200–250 km), and the increase of occurrence frequency with geomagnetic activity is much more pronounced on the dawn side than on the dusk side, resulting in a higher frequency on the dawn side. The increase in frequency with altitude is also stronger.

The observed magnetic activity dependence of ion upflow is consistent with ion acceleration in the F-region and the topside ionosphere receiving important contributions from both $E \times B$-driven ion frictional heating and precipitating soft electron-driven electron heating. The effect of ion frictional heating is expected to increase with K_P and to be stronger on the dusk side and in the winter: this explains the higher occurrence frequency on the dusk side at EISCAT's latitude, and the increase in occurrence frequency with geomagnetic activity at both ESR and EISCAT. The effect of soft electron precipitation is expected to be stronger during disturbed times, particularly in the dusk quadrant, and to play a more dominant role on the dayside where the precipitating electrons tend to be softer: this explains the higher

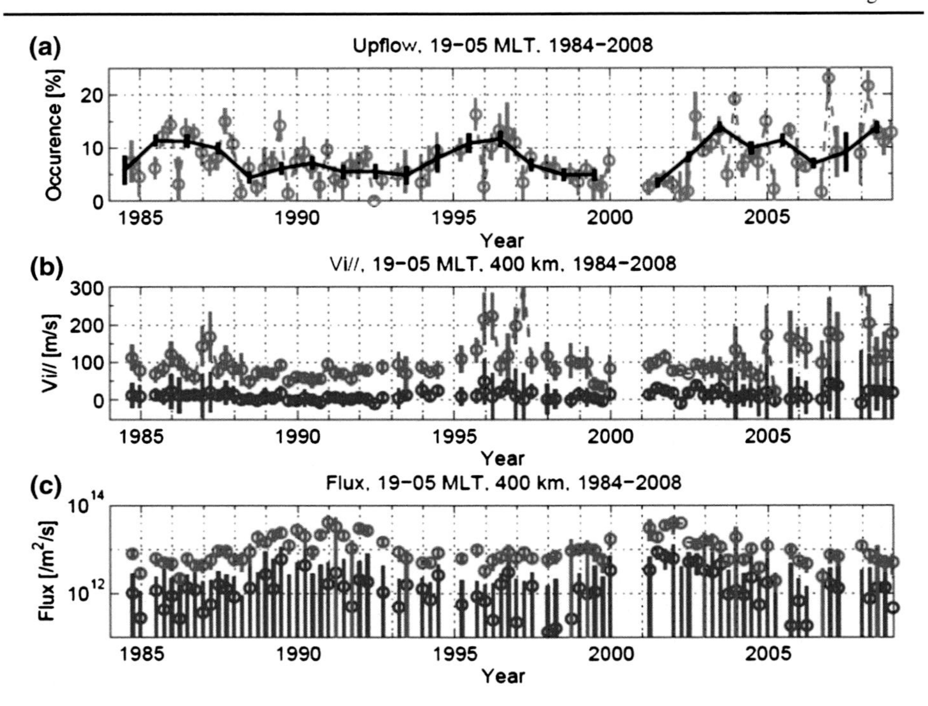

Fig. 3 (**a**) 12- (*black*) and 3-month (*grey*) averages of occurrence frequency of night side (19–05 MLT) ion up-flow at EISCAT starting between 200–550 km from 1984 to 2008; 3-month averages of (**b**) field-aligned ion velocity and (**c**) ion flux at 400 km over all up-flow events (*red*) and data samples (*blue*) From Ogawa et al. (2010)

dayside occurrence frequency at ESR compared with EISCAT at both quiet and disturbed times, and the higher frequency on the dawn side during disturbed times. It also suggests that soft electron-driven electron heating may be more efficient than convection-driven ion heating in driving ion upflow.

The long time coverage of the EISCAT data set makes it extremely valuable for studying the influence of solar activity on ion upflows. Figure 3 shows (a) the 12- (black) and 3-month (grey) averages of the observed occurrence frequency of nightside (19–05 MLT) between 200 and 550 km, and (b) the field-aligned ion velocity and (c) ion flux at 400 km at low ($F10.7 < 140$) and high ($F10.7 > 140$) solar flux, respectively, from 1984 to 2008. On average, the upward ion velocity in upflow events was a factor of 2 higher at low solar flux than at high solar flux ($F10.7 > 140$), when the upward ion flux was a factor of 4 higher. The larger flux at high solar flux (i.e. near solar maximum) is attributed to the stronger solar EUV flux and resulting ionization in the F-region, and the smaller velocity to the higher ion-neutral collision frequency due in turn to the higher exospheric temperature and neutral density in the thermosphere.

Ogawa et al. (2009) found the ion upflow occurrence frequency to increase with both solar wind density (above 30 cm^{-3}) and solar wind velocity (up to 700 km/s), and to peak in value inside the cusp, while the upward ion flux increases with solar wind density and decreases with solar wind velocity. Both IMF B_Y and B_Z are found to affect the upflow occurrence frequency, which increases with increasing magnitude of B_Y and peaks at B_Z ~ -5 nT. The apparent movement of the dayside ion upflow region may be understood in terms of the influence of solar wind velocity and density and the IMF B_Y and B_Z on the shape, size and location of the upflow region, since the location of the dayside cusp is known to move equatorward with decreasing IMF B_Z or increasing solar wind dynamic pressure.

Ogawa et al. (2010) found the average starting altitude of ion up-flow to track the measured electron density peak and to be typically 100–150 km higher than the latter. The distribution of starting altitude is quite different at low and high solar flux, respectively. At low solar flux, the distribution exhibits a broad peak starting at ~300 km; peaking near 450 km and extending to ~520 km. At high solar flux, the distribution shifted to higher altitude, starting near ~350 km; peaking more sharply near 450 km and extending to at least 540 km. The variation of the starting height with solar activity level can be attributed to the increased atmospheric density and ion-neutral collision frequency at a given altitude near solar maximum: the solar minimum neutral atomic oxygen density at the starting height of 300 km is $\sim 3 \times 10^8$ cm^{-3}, compared with the corresponding (solar maximum) density value of $\sim 3.3 \times 10^8$ cm^{-3} at the (increased, solar-maximum) starting height of ~450 km. This implies that the atmospheric density and ion-neutral collision frequency at the starting upflow altitude are comparable at solar minimum and maximum, respectively.

The DE-2 and the EISCAT/ESR radar observations demonstrate that both soft electron-driven electron heating and convection-driven ion heating play a significant role in auroral ion upflow production. Frictional heating of O^+ ions enhances the ion temperature in the F-region and increases the pre-existing parallel pressure gradient, and the ions respond by flowing to higher altitudes to attain a new equilibrium scale height distribution. Although the increase of the scale height is a transient feature, the upflow can remain if new plasma is horizontally convected into the heating region. Likewise, soft precipitating electrons deposit their energy in the F-region via electron impact ionization of the neutrals and collisional energy transfer with the neutrals, and thereby increase the average thermal electron energy (i.e. electron temperature) and enhance the ambipolar electric field.

The category of suprathermal ion outflows includes ion beams, ion conics, transversely accelerated ions (TAI), and upwelling ions (UWI). The occurrence and morphological characteristics of ion beams and conics in the different altitude regions were the subject of a number of statistical studies using S3-3, DE-1, Viking, Akebono, Freja, Fast and Polar satellite data, including several prior to 1997, which were reviewed in detail by Yau et al. (1997).

Ion beams are upflowing ions (UFI) that have a peak flux along the upward magnetic field direction. They are generally observed above 5000 km altitude, but are occasionally present down to about 2000 km during active aurora. The occurrence probability of both H^+ and O^+ ion beams increases with altitude at both quiet and active times. The increase is most prominent for the lower-energy (<1 keV) ions.

In contrast, ion conics have a peak flux at an angle to the upward magnetic field direction, and are observed down to sounding rocket altitudes (1000 km or below; Yau et al. 1983), and up to several Earth radii and beyond (Hultqvist 1983; Bouhram et al. 2004). At high altitude (above ~10,000 km), the occurrence probability of low-energy conics (<1 keV) decreases with increasing altitude. The motion of an ion conic is typically non-adiabatic as it evolves along the field line.

Transversely accelerated ions (TAI) have peak pitch angles at or close to 90°, and may be regarded as a special case of ion conics. They are regularly present down to about 3000 km (Whalen et al. 1991) on the dayside, and down to 1400–1700 km (Klumpar 1979; André et al. 1998) and to the active-time topside ionosphere above 400 km (Yau et al. 1983; Arnoldy et al. 1992) on the nightside.

Upwelling ions are observed exclusively in the morning sector of the auroral oval and the lower latitudes of the polar cap, and display the effects of both parallel (upward) and perpendicular energization to energies from one to tens of eV (Pollock et al. 1990). They are the most persistent suprathermal ion outflow feature in the cleft region, and are dominated by O^+ ions. Compared with ion conics with the same perpendicular energy, upwelling ions are more upward moving (have higher upward mean velocity). They often appear as field-aligned ion flows at other local times at higher altitudes in the presence of anti-sunward convection, hence the term "cleft ion fountain".

Both ion beams and ion conics are a common phenomenon, with occurrence frequencies sometimes higher than 50 % above 1 R_E altitude, and are dominated by H^+ and O^+ ions in the 10 eV to a few keV range. UFI's of a few tens of keV energy also occur occasionally. Distributions of UFI may evolve in different ways as they move upward. Ion conics often do not start as TAI distributions heated within a narrow altitude range and then move adiabatically up the geomagnetic field. Statistically (Miyake et al. 1993, 1996; Peterson et al. 1995), the energy of dayside ion conics increases with altitude, from $\sim$10 eV near 2000 km to $\leq$100 eV near 9000 km. The cone (apex) angle of ion conics decreases with altitude much more slowly than expected from adiabatic motion. In the so-called "restricted" ion conics, the ion distribution has a well-defined cone angle. However, in the so-called "extended" or "bimodal" conics (Klumpar et al. 1984), the cone angle increases with energy and the lower energy ions have a significant flux along the field line.

The occurrence probability of both H^+ and O^+ upflowing ions is fairly independent of magnetic activity (K_P index). However, the intensity distribution of O^+ UFI exhibits a marked dependence on magnetic activity that is absent in H^+. On DE-1 (Yau et al. 1984), the occurrence probability of intense ($>10^7$ $cm^{-2}\,s^{-1}\,sr^{-1}$) lower-energy (<1 keV) O^+ at active times ($K_P \geq 4-$) is a factor of 3 higher than at quiet times. A similar but smaller increase is also apparent in the occurrence probability of intense ($>10^6$ $cm^{-2}\,s^{-1}\,sr^{-1}$) higher-energy (>1 keV) ions. In contrast, the intensity distribution for H^+ remains fairly unchanged with K_P.

The observed O^+ UFI distributions exhibit significant seasonal and long-term variations, which are attributed to changes in the incident solar EUV flux on the atmosphere in different seasons of the year and at different phases of the 11-year solar cycle. The corresponding variations in the H^+ UFI distributions are much smaller. On DE-1, the probability of the O^+ UFI decreased by about a factor of 2 from near solar maximum in 1981 to the declining phase in 1984. The decrease in probability of intense UFI fluxes was even larger, by about a factor of 3–4. In contrast, there was no discernible change in the H^+ occurrence probability during the same period. Throughout the period, the occurrence probability of O^+ UFI was significantly higher in the summer than in the winter, the frequency of intense events being about a factor of 2 larger. The increase in occurrence probability, intensity, and conic abundance of O^+ UFI in periods of increased solar activity results in a large increase in the overall ion outflow rate.

Peterson et al. (2008) recast the observed ion outflow flux and energy distributions near Polar perigee in dynamic boundary-related coordinates. It was found that for all three ion species (H^+, O^+ and He^+), only a very small fraction ($\sim$2–3 %) of the observed energetic UFI was in the polar cap. However, their presence confirms that not only are energetic ions being transported by prevailing convection electric fields to the high-altitude polar cap, but they are also produced by ion acceleration events in the polar cap ionosphere. In the auroral zone, the flux in the midnight quadrant dominated, and consisted of $\sim$50 % of the total H^+ and He^+ flux and $\sim$30 % of the O^+ flux, compared with $\sim$37 % of O^+ flux in the noon quadrant where most of the flux was on cusp field lines (e.g. Zheng et al. 2005).

Figure 4 shows the net ion outflow rates of both H^+ and O^+, obtained by integrating the DE-1 ion flux measurements over all magnetic local times and all invariant latitudes

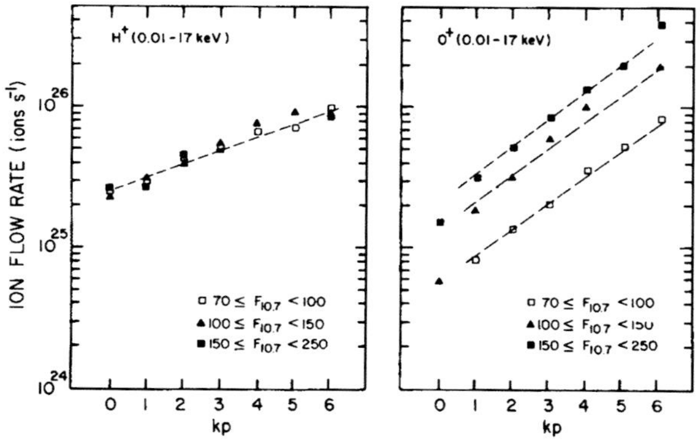

Fig. 4 H^+ and O^+ ion outflow rates at 0.01–17 keV observed at 16,000–24,000 km on DE-1, integrated over all MLT above 56° invariant latitude in both hemispheres as a function of K_P, for different ranges of F10.7. From Yau et al. (1988)

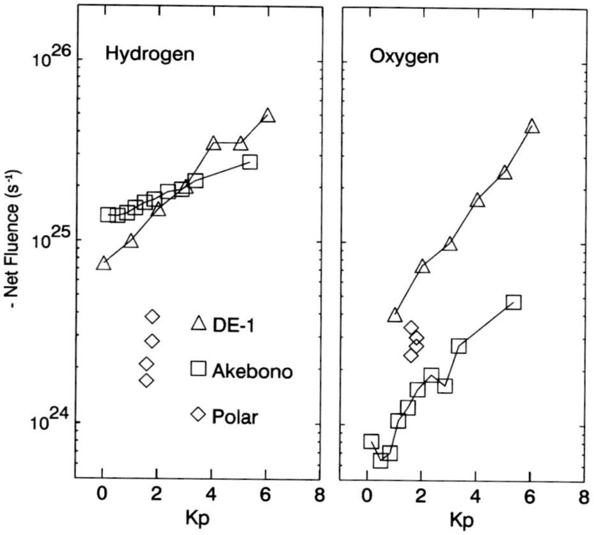

Fig. 5 H^+ and O^+ ion outflow rates near solar minimum as a function of K_P. *Squares* indicate low-energy rates on Akebono below 9000 km; *triangles* show suprathermal energy rates on DE-1 above 16,000 km; *diamonds* show suprathermal energy rates on POLAR below 9000 km. From Cully et al. (2003a)

above 56°, as a function of the magnetic K_P index for three F10.7 ranges (Yau et al. 1988). The O^+ rate increased exponentially with K_P, by a factor of 20 from $K_P = 0$ to 6, and exceeded 3×10^{26} ions s^{-1} at times of high solar and magnetic activity. The rate at low solar activity was about a factor of 4 smaller than that at high activity. In contrast, the H^+ rate was very similar across each of the three F10.7 ranges. In all three F10.7 ranges, the dependence of the O^+ rate on K_P was similar. In comparison, the H^+ rate increased with K_P more moderately, by a factor of 4 from $K_P = 0$ to 6.

Figure 5 compares the observed low-energy ion outflow rates observed on Akebono below 9000 km near solar minimum with the corresponding suprathermal energy rates on Polar at the same altitudes (15 eV–16 keV) and on DE-1 above 16,000 km (10 eV–16 keV), respectively. The rate of low-energy H^+ on Akebono is comparable with the suprathermal energy rate on DE-1 and a factor of 4-10 higher than the suprathermal energy rate on Polar. This indicates that significant acceleration of H^+ occurs above 9000 km in the high-latitude ionosphere. In contrast, the rate of low-energy O^+ below 9000 km is less than the corresponding suprathermal rate above this altitude, which is in turn less than the corresponding suprathermal rate above 16,000 km. This means that a significant fraction of O^+ is accelerated below 9,000 km, and that the acceleration continues between 9,000 and 16,000 km. In other words, a significant fraction of low-energy ions at low altitudes in the high-latitude ionosphere, including polar wind ions and auroral ion upflows, are accelerated to suprathermal energies at higher altitudes, where they lose their identity as thermal-energy ions. Thus, it is important to consider both thermal and suprathermal ion outflow in the high- latitude ionosphere as an integrated entity.

Additional suprathermal outflow arises from polar cap arcs of energetic electron precipitation. The difference with the auroral zone comes from the source region: the auroral zone is magnetically connected to the plasmasheet, while the polar cap is connected to the lobes. Therefore, the polar activity is expected to be closely related to the dynamics of distant magnetospheric regions or to the interaction between the solar wind and the magnetosphere, which vary between the different planets. For the Earth, the knowledge of polar cap arcs and of their plasma environment has benefited from flybys over the polar caps at different altitudes by numerous satellites with optical and in-situ instruments.

Polar cap arcs dominantly appear when the interplanetary magnetic field (IMF) is directed northward and during quiet geomagnetic conditions in the magnetosphere. A large variety of shapes, widths, lengths, and motions are reported from ground-based instruments and low-altitude satellites, which possibly suggests different driving mechanisms and different source regions for the electrons. Most studies focus on electrodynamics and in relation to the large-scale convection pattern (see reviews by Zhu et al. 1997; Kullen 2002). It is not yet fully understood whether polar cap arcs occur on open or closed field lines, and thus whether the source region is related to a highly distorted plasmasheet or to the magnetopause and boundary layers (Carlson 2005; Frey 2007). Recent work (Fear et al. 2014) connects arcs as observed by the IMAGE satellite with plasma observations from the Cluster constellation that are characteristic of popluations trapped on closed field lines. This supports the hypothesis that arcs occur on closed field lines.

Outflowing H^+ beams of polar arc source were first detected at low-altitude above the polar cap with characteristics significantly different from both the polar wind (Shelley et al. 1982) and from ionospheric ions escaping from the cusp/cleft ion fountain. Since this early detection, very few references exist in the literature about outflowing ions related to polar cap arcs before the Cluster observations. During periods of northward or weak IMF, Cluster flybys over the polar cap at relatively high altitudes (between 4 and 8 R_E) revealed that accelerated electron beams precipitating into the polar ionosphere were systematically accompanied by outflowing ion beams with a typical shape of inverted V (Nilsson et al. 2006). These observations are also correlated with the presence of convergent electric fields perpendicular to the magnetic field. As in the auroral zone, these characteristics are interpreted as the effect of a U-shape potential structure below the spacecraft, which accelerates ionospheric ions upwards. The simultaneous acceleration of the precipitating electron beams demonstrates that the potential structure must extend to altitudes higher than the spacecraft, whereas the acceleration region is assumed to be confined at the topside of the ionosphere in the auroral zone (Maggiolo et al. 2006; Teste et al. 2007). A case study with a good conjunction between Cluster observations of precipitating electrons and ion outflows at high altitude, and optical observations of an arc by the TIMED spacecraft, confirmed the relationship between polar cap arcs and accelerated ion outflows with typical shape of inverted "V"s (Maggiolo et al. 2012). A statistical study of ion outflows showed that they form elongated and sun-aligned structures with widths typically of the order of 30 km mapped to the ionospheric level. Their temperature is of the order of tens of eVs and they are accelerated to average energies of about 400–500 eV, with highest values up to 1–2 keV (Maggiolo et al. 2011). Periods of northward or weak IMF are not favorable for low-latitude reconnection processes at the magnetopause. During this period, substorm activity was limited and the auroral and geomagnetic activity in the magnetosphere was quite weak. The only signs of activity in the magnetosphere occurred in the polar cap with the presence of polar arcs. The associated outflowing ion beams represent a plasma source for the magnetosphere in such conditions of weak to northward IMF.

In addition, Teste et al. (2010) showed that these polar cap arc structures were surrounded by upwelling electron beams, accelerated from the ionosphere to low energies typically less than 100 eV. The return downward current carried by these outflowing electron beams was estimated and found to be comparable with the adjacent upward current carried by precipitating particles. This suggests that both upward and downward currents are part of the same current circuit closing through the ionosphere. Finally, during northward IMF periods, the polar cap exhibits successive sheets of outflowing ionospheric ions and electrons, the outflowing ion beams being associated with electron precipitations. These observations reinforce the role of the polar ionosphere as an alternative plasma (ions and electrons) source for the magnetosphere during periods of Northward IMF.

2.2 Low Latitude Ionospheric Plasma

At sub-auroral latitudes, ionospheric outflow slowly saturates closed flux tubes to create the plasmasphere. The plasmasphere is the torus of cold and dense plasma, which encircles the Earth at geomagnetic latitudes less than about 65°, occupying the inner magnetosphere out to a boundary known as the plasmapause (Carpenter 1962; Lemaire and Gringauz 1998; Kotova 2007). There, the density can drop by 1 to 2 orders of magnitude; the boundary is observed to be much more diffuse during prolonged quiet periods. The plasmasphere comprises the corotating region of the magnetosphere and is magnetically coupled to the ionosphere. Magnetic field lines are closed and approximately dipolar, permitting filling of the plasmasphere by plasma escaping from the Earth's ionosphere.

During quiet times the ionospheric plasma at mid-latitudes can expand upward along the magnetic field lines and fill them until the plasma gas pressure is equalized along the entire field line. In establishing the equilibrium between the plasmasphere and the ionosphere, plasma flows both to and from the plasmasphere. A net flow into the plasmasphere is often called "refilling" (Park 1970; Banks et al. 1971; Kotova 2007).

Early models of ionospheric plasma escape (Banks and Holzer 1969; Lemaire and Scherer 1970) predicted that the light ions H^+ and He^+ should flow out into the magnetosphere, while the heavy ions should remain gravitationally bound in the ionosphere and provide much smaller upwelling fluxes. H^+ is thus the principal plasmaspheric ion, while He^+ is the second most abundant species in the plasmasphere, accounting for approximately 5–10 % of the plasmasphere plasma. The ratio between He^+ and H^+ changes with geomagnetic activity, ranging from 3 % to about 50 % (Darrouzet et al. 2009, and references therein).

Heavy ion content in the plasmasphere is generally very low. Grew et al. (2007), by combining all measurements on a field line at $L = 2.5$, were able to solve simultaneous equations for the abundances of H^+, He^+ and heavier ions (taken to be O^+). For the $H^+ : He^+ : O^+$ ratio they found $\sim 82 : 15 : 3$ by number (~ 3 % O^+). An interesting deviation from this norm occurred just outside the plasmasphere, when the inferred O^+ proportion reached ~ 60 %. Dandouras et al. (2005) analysed ion composition measurements from the CIS experiment onboard the Cluster spacecraft (Rème et al. 2001) in the outer plasmasphere, at $L \sim 4$, and observed a quasi-absence of O^+ ions (O^+ less than 4 % by number). However, outside the main plasmasphere, a few low-energy O^+ observations occurred within detached plasma

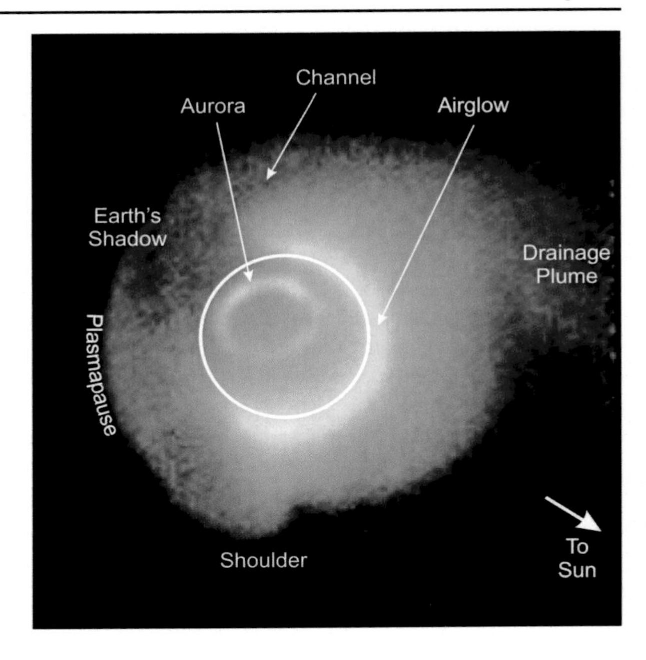

Fig. 6 Plasmasphere image obtained by the Extreme Ultraviolet Imager (EUV) onboard the IMAGE spacecraft, during a magnetic storm. From Sandel et al. (2003)

events, originating from deeper in the plasmasphere and having an outward expansion velocity towards higher L-shells. Chappell (1982) and Roberts et al. (1987), using the retarding potential ion mass spectrometer on board the Dynamics Explorer 1 satellite, also reported heavy ion observations in the region of the plasmasphere just inside the plasmapause. These observations allowed the separation of O^+, O^{++}, N^+ and N^{++} ions, all of which were observed in the plasmasphere.

The He^+ outside Earth's shadow resonantly scatters the solar 30.4 nm radiation, so that the plasmasphere glows (Fig. 6). The plasmaspheric He^+ emission is optically thin, therefore the integrated column density of He^+ along the line of sight through the plasmasphere is directly proportional to the intensity of the emission. The Extreme Ultraviolet Imager (EUV) onboard the IMAGE spacecraft allows for the study of the distribution of cold plasma in Earth's plasmasphere via imaging of the distribution of the He^+ ion through its emission at 30.4 nm (Sandel et al. 2000, 2003).

2.3 Solar Wind Plasma

Solar wind entry to the magnetosphere, due to magnetic reconnection with a southward oriented interplanetary magnetic field and the consequent occurrence of a solar wind driven convection cycle, was first proposed by Dungey (1961). The general concept is now widely accepted, but work is ongoing to develop a detailed understanding of when, where and how magnetopause reconnection proceeds. The recent reviews of Fuselier and Lewis (2011) and Paschmann et al. (2013) summarize our current understanding of this complex phenomena.

Reconnection at the magnetopause current sheet is a rather asymmetric situation, with the magnetic field strengths and orientations, as well as plasma properties differing on either side of the current sheet. As discussed in Hultqvist et al. (1999), early work (Sonnerup 1980) suggested that the magnetic fields either side of the current sheet should have equal components parallel to the reconnection line and that strongest rates of reconnection would occur

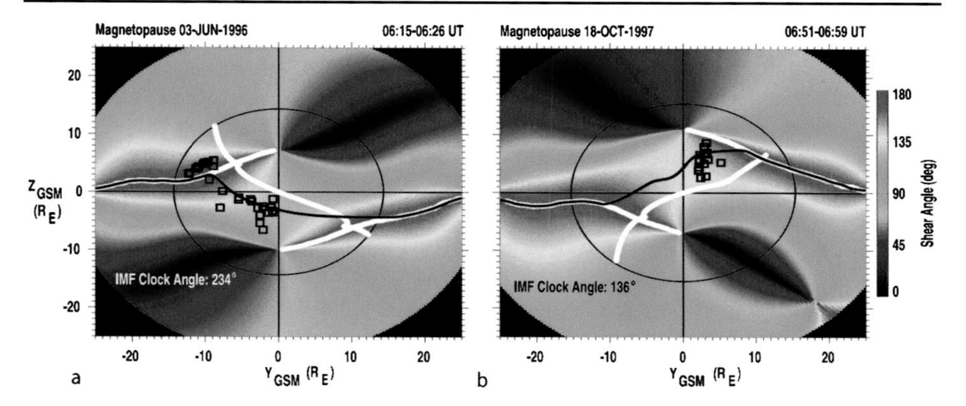

Fig. 7 This figure shows a projection of the magnetopause as seen from the Sun, in GSM coordinates, The *colour* represents the magnetic shear between draped magnetosheath magnetic field and magnetosphere boundary layer magnetic field, for southward interplanetary magnetic fields having significant dusk/dawn components. The *white region* superimposed *on red* shows where antiparallel reconnection would be favoured, while the remaining *white line* shows the locations where component reconnection as formulated in Moore (2002) is favoured. The *black trace* shows the locus of maximum magnetic shear, in the Trattner et al. (2007b) model which includes a treatment of dipole tilt, unlike Moore (2002), and thus shows a corresponding seasonal difference between the *left* and *right hand plots* (for northen hemisphere summer and winter respectively). The *black boxes* show inferred locations of reconnection, determined by analysis of dispersed ion signatures observed in the cusp regions with the Polar spacecraft. Adapted from Trattner et al. (2007a)

for anti-parallel fields. Cowley (1976) argued that the reconnection line forms in an orientation perpendicular to a line along which the magnetic fields, either side of the magnetopause current sheet, have opposite components (those being the "reconnecting components") with arbitrary components parallel to the line. This idea was developed by Cowley and Owen (1989) and, using more realistic magnetic field models, by Cooling et al. (2001) who explored the motion of newly reconnected flux across the magnetopause, and Moore (2002) who focused only on where reconnection is expected to occur. Cooling et al. (2001) introduced an assumption that reconnection can only proceed when the mean current density in the sheet exceeds a minimum level, or equivalently that the magnitude of the difference between the reconnecting magnetic field components exceeds a threshold. A given threshold can be exceeded at smaller shear angles where there are stronger magnetic fields, so we may expect reconnection at lower shear angles near the sub-solar magnetopause. At other locations with weaker reconnecting component fields, higher shear angles are needed, and in some locations the threshold may not met for any shear angle. Moore (2002) required that the reconnecting magnetic field components were equal and opposite. Trattner et al. (2007b) built on this earlier work by combining a magnetospheric magnetic field model with dipole tilt (Tsyganenko 1995) and a model of the draped magnetosheath magnetic field (Cooling et al. 2001) in order to determine both the magnetic shear and the locations most favourable for component reconnection at all points on the magnetopause for given solar wind conditions. Figure 7 illustrates typical predictions of the model showing, for southward IMF with a dominant dawn-dusk component, that the expected outcome is a component reconnection line in the subsolar regions that joins anti-parallel reconnection lines in the flank regions.

Trattner et al. (2007b, 2007a) used POLAR Toroidal Imaging Mass-Angle Spectrograph (TIMAS) observations of ion velocity dispersion observed in the high latitude magnetic cusp regions to infer the lower latitude locations of reconnection X-line from which the ions had travelled. This was done for a range of IMF conditions, and the results were compared with the predictions of anti-parallel and component reconnection models. It was concluded that both reconnection scenarios appear to occur, depending on IMF conditions. Furthermore, it was shown that the reconnection line occurs where the magnetic shear angle maximizes, giving reconnection X-line locations which may differ from the predictions of Moore (2002) due to seasonal non-zero dipole tilt (as illustrated in Fig. 7) and IMF B_X effects.

The "maximum magnetic shear" model has been tested (Trattner et al. 2012) by comparing the locations of 7 active low latitude reconnection lines directly observed by THEMIS spacecraft, with locations predicted by the model under the corresponding interplanetary magnetic field conditions. The study assumed that flow reversals are the signature of the spacecraft crossing an active reconnection line, rather than the signature of multiple reconnection lines. The model was shown to be very effective when IMF B_Y dominated. However, when the dominant IMF direction was southward or in the B_X direction, the model was less effective.

The Cluster spacecraft have crossed the dayside magnetopause across all latitudes, during more than a decade of operations, and provided observations which can be used to test the predictions of the maximum magnetic shear model. Fuselier et al. (2011) were able to identify 15 cases with clock angles between 105° and 228° (southward B_Z and varying B_Y components) where the antiparallel reconnection X-lines were predicted to lie polewards of the spacecraft and the component reconnection X-lines would lie equatorwards of the spacecraft. Careful analysis of spacecraft observations of ion and electron populations identified as being on newly reconnected field lines revealed whether the reconnection X-line in fact lay polewards or equatorwards of the spacecraft in each case. It was found that the observations were consistent with model predictions in 13 of the cases, while in the other two cases the observed flow direction in the magnetosheath boundary layer (MSBL) differed from the direction in the magnetospheric low latitude boundary layer (LLBL) preventing determination of the direction to the reconnection line. Counter-streaming electrons in the magnetosheath boundary layer were observed during 6 events. These were interpreted as indications of multiple reconnection, occurring perhaps initially at the equatorwards component reconnection line and later at an antiparallel reconnection line poleward of the spacecraft; this situation is not predicted by the maximum magnetic shear model.

Under northward interplanetary magnetic field conditions, as Dungey (1961) noted, magnetic reconnection may occur at high latitudes, poleward of the magnetospheric cusp, between magnetosheath and magnetotail lobe magnetic fields. Ongoing reconnection at a high-latitude site does not create or destroy closed magnetic flux, and thus does not provide an entry route for plasma into the magnetosphere. However, if the interplanetary magnetic field was to undergo reconnection with both the north and south magnetotail lobes, it could, in principle, form a newly closed dayside magnetic field line, containing trapped magnetosheath plasma. The idea that this process might occur under conditions of strongly northward IMF, and the suggestion that it could play a role in the formation of a low latitude boundary layer, was proposed by Song and Russell (1992) as a contribution to understanding why a well-defined structured LLBL occurs under strongly northward IMF.

A study by Twitty (2004) examined 3 years of Cluster high latitude magnetopause crossings tailward of the cusp, during intervals of relatively stable IMF, for evidence of reconnection associated plasma flows. Reversed energy-latitude ion dispersion signatures were used to confirm the interpretation that the observed plasma flows are signatures of reconnection. The survey showed that such flows were seen during ~90 % of the intervals when the IMF had a northward component, and almost never when the IMF had a southward component. The observations are predominantly from the northern hemisphere, and do not discriminate between single and dual-lobe reconnection.

Direct evidence of dual lobe reconnection has been sought by examining the properties of suprathermal electrons in the high latitude magnetosheath boundary layer (MSBL) under northward IMF conditions. Magnetosheath electrons flowing along the magnetic field towards a reconnection site are cooler than magnetosheath plasma returning from the reconnection site, as those electrons have been heated while crossing the magnetopause, reflecting at the ionosphere and heated again when recrossing the magnetopause to return to the magnetosheath boundary layer (Fuselier et al. 1997). A case study by Onsager et al. (2001) of a high latitude magnetopause crossing by the Polar spacecraft found evidence of open field lines connected to the northern hemisphere and others connected to the southern hemisphere. Bi-directional heated electrons in the MSBL were interpreted as evidence of high latitude reconnection in both hemispheres. The study also suggested that reconnection was occurring over a broad local time extent.

Statistical studies, which applied this technique to Cluster high latitude magnetopause crossing data, showed that the geomagnetic dipole tilt is the main influence on which hemisphere is more likely to show lobe reconnection, with the IMF tilt angle being less significant (Lavraud 2005; Lavraud et al. 2006). Observations of bi-directional heated MSBL electrons interpreted as evidence for dual-lobe reconnection were quite common and were shown to occur not only for strictly northward IMF, but across a range of clock angles smaller than 60°, in a sample of 56 magnetopause crossings.

Further work has demonstrated that the dual-lobe reconnection process must usually occur in two steps, which are separated in time. This interpretation reconciles contradictions between models proposed based on observations of bi-directional MSBL heated magnetosheath electron populations at high latitudes, which suggested that they should also be seen at low latitudes in the MSBL, with observations at lower latitudes in which heated MSBL electrons are typically uni-directional and bi-directional electrons are only seen inside the magnetopause current layer in the LLBL, as well as corresponding issues regarding observations of low latitude O^+ ions (Fuselier et al. 1995). Fuselier et al. (2012) revisited Cluster magnetopause observations, and analysed a large dataset covering 2001–2009, which included the high latitude crossings studied by Lavraud et al. (2006) and newer low latitude observations which became available due to the evolving orbit of Cluster. This statistical study confirmed the findings of Lavraud (2005) at high latitudes, while also confirming earlier works showing that the majority of low latitude events show uni-directional heated electrons in the MSBL. Figure 8 illustrates a way to reconcile these observations, in which time elapses between reconnections in the two hemispheres, during which the reconnected field line evolves and convects tailwards before the second reconnection occurs. As Fuselier et al. (2012) point out, since their study was confined to regions within 4 hours magnetic local time of noon, further work is needed to determine how far tailwards the reconnected field line typically convects before a second reconnection occurs.

In companion papers, Øieroset (2005) and Li (2005) presented complementary studies that indicate that magnetosheath plasma trapped by dual lobe reconnection might ultimately contribute to the formation of a cold dense magnetotail plasmasheet. A case study used Cluster observations to demonstrate the existence of a cold dense magnetotail plasmasheet (Øieroset et al. 2005) in association with a long duration interval of northward IMF. Low altitude observations from Defense Meteorological Satellite Program (DMSP) spacecraft demonstrated that the cold, dense plasmasheet was present across the span of the magnetotail from dawn to dusk, not only at the Cluster location. Furthermore, cusp ion dispersion signatures characteristic of high latitude lobe reconnection, poleward of the cusp, were observed by the low altitude FAST spacecraft, confirming lobe reconnection in at least one hemisphere. An MHD global magnetosphere simulation study (Li 2005) illustrated how

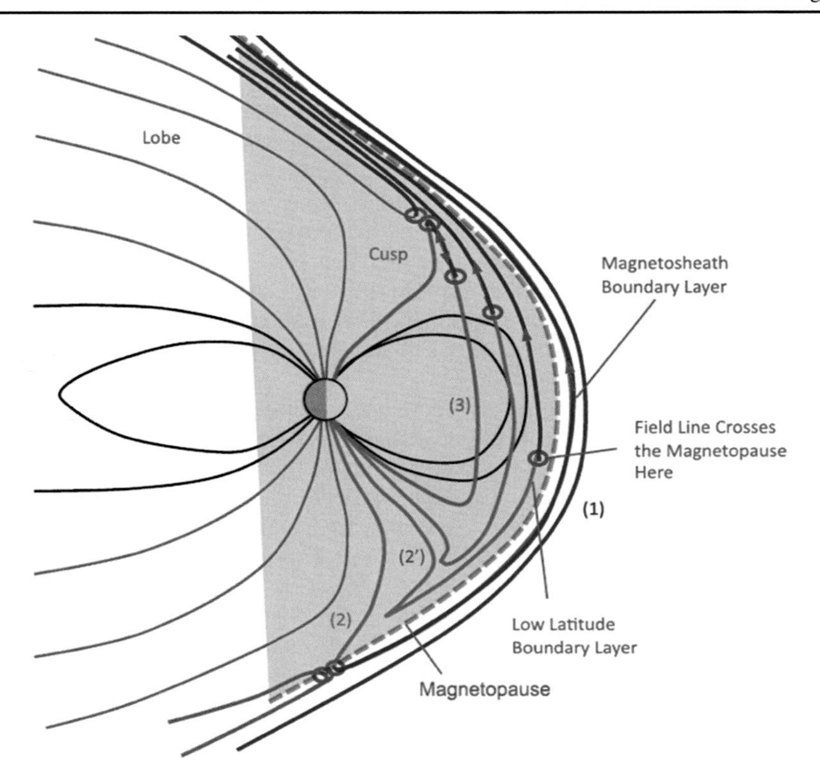

Fig. 8 The interpretation proposed by Fuselier et al. (2012) to explain magnetosheath boundary layer electron signatures at low and high latitudes under northward IMF conditions. A magnetosheath magnetic field line reconnects (in this sketch) poleward of the southern hemisphere cusp, and begins to contract to reduce curvature under the tension force. The part of the field line near the subsolar magnetopause meanwhile convects tailwards; the part of the field line outside the magnetopause carries uni-directional heated magnetosheath electrons travelling northwards. Only after enough time has elapsed for the magnetopause crossing point of the field line to reach quite high latitudes does the field line re-reconnect poleward of the northern hemisphere cusp, producing bidirectional heated magnetosheath electrons in a relatively localized high latitude magnetosheath boundary layer region

dual lobe reconnection might capture magnetosheath plasma and how this plasma may be transported to the magnetotail to be observed by Cluster, in a process similar to that envisaged by Song and Russell (1992).

Solar wind entry on the flanks of the magnetopause has been proposed to occur in the special context of rolled up Kelvin–Helmholtz vortices. This is of particular interest under northward IMF conditions, where it competes with the dual-lobe reconnection entry picture and the diffusive entry picture, as an explanation for the formation for the low latitude boundary layer and possibly the cold dense magnetotail plasmasheet, as described for example in Hultqvist et al. (1999).

Kelvin–Helmholtz waves on the magnetopause boundary have been recognised for many years (e.g., Otto and Fairfield 2000), and clearly represent a way to transfer energy and momentum from the magnetosheath flow to the magnetospheric boundary layer. The Cluster multi-spacecraft mission has enabled their properties to be measured more completely than by earlier missions. Owen et al. (2004) reported Cluster observations consistent with Kelvin–Helmholtz waves on the dawn flank magnetopause, during northward IMF, providing information on wavelength and propagation direction and noting a steepened leading (tailward) edge. This observation was consistent with predictions by Miura (1990) helping to resolve conflicting conclusions of earlier observational studies using one or two spacecraft datasets. Further studies of the conditions for formation of, and the development of magnetopause Kelvin–Helmholtz waves include Foullon et al. (2008) and Hwang et al. (2011, 2012) which demonstrate that, given suitable IMF conditions, such waves can grow at high latitudes under strong dawnward IMF, and under southern IMF, as well as the more commonly expected low latitude regions under northern IMF.

Hasegawa et al. (2004) examined a dusk flank magnetopause crossing and used multipoint Cluster data to identify specific plasma and magnetic signatures associated with Kelvin–Helmholtz vortices. They further argued that simultaneous observations of plasma at solar wind and magnetospheric energies, on the magnetosphere side of the magnetopause implied that plasma transport had occurred within the vortices, though they were unable to firmly identify the mechanism. Further work to confirm the result and to identify more examples of Kelvin–Helmholtz vortices followed, including studies by Hasegawa et al. (2006), which set out criteria with which to identify such vortices in single spacecraft datasets. The statistics of Hasegawa et al. (2006) and Taylor et al. (2012) show observations usually at low latitudes, and not only tailward but also sunward of the terminator (suggesting that at least sometimes they may develop very rapidly). Events have also been reported on the dusk flank during sourthward IMF conditions (Yan et al. 2014), as previously predicted.

It has been suggested for some time that these magnetopause disturbances are significantly contributing to solar wind plasma entry into the magnetosphere. The entangling of magnetospheric and magnetosheath magnetic field lines does not of itself enable plasma entry; it is necessary to also invoke a process such as magnetic reconnection or a cross-magnetic field diffusion process within the vortices. Nykyri et al. (2006) presented Cluster observations of reconnection inside Kelvin–Helmholtz vortices, but acknowledged that it was not clear that these reconnection events contributed to significant plasma transport into the magnetosphere.

Hasegawa et al. (2009) used simultaneous observations of the equatorial magnetopause about 15:00 MLT by Geotail and downstream about 19:00 MLT by Cluster to examine the formation of the LLBL during a prolonged interval of northward IMF. The Geotail observations show a LLBL for which high latitude reconnection was found to be the most plausible explanation. Cluster observes rolled up vortices and evidence is presented indicating that local reconnection at the edge of a rolled up vortex is seen by one of the spacecraft. However, it is suggested that this is a small scale event (other Cluster spacecraft did not see it) and in the absence of evidence for reconnection seen in relation to other vortices, it was concluded that vortex reconnection could not account for the significant plasma entry near Cluster. Based on data showing a larger plasma density in the LLBL at Cluster ($\sim$3 cm^{-3}) than Geotail ($\sim$2 cm^{-3}), it is argued that LLBL flux tubes must have gained material while convecting between Geotail and Cluster, as a reduced density ($\sim$1 cm^{-3}) might otherwise be expected due to expansion of the flux tubes. It should be noted that this conclusion relies on an assumption that Geotail and Cluster plasma instruments have good relative accuracy. Taylor et al. (2008) draw a similar conclusion from a study with Polar and Double Star TC-1, in a study of an interesting interval which has rolled up vortices, dual-lobe reconnection and a cold dense plasma sheet observed by Double Star TC-2.

Bavassano Cattaneo et al. (2010) performed a detailed examination of very large rolled up Kelvin Helmholtz vortices observed by Cluster during a long lasting interval of northward IMF in the dusk equatorial magnetopause region. The vortices are suggested to have been generated further upstream, due to their large size. Magnetospheric and magnetosheath

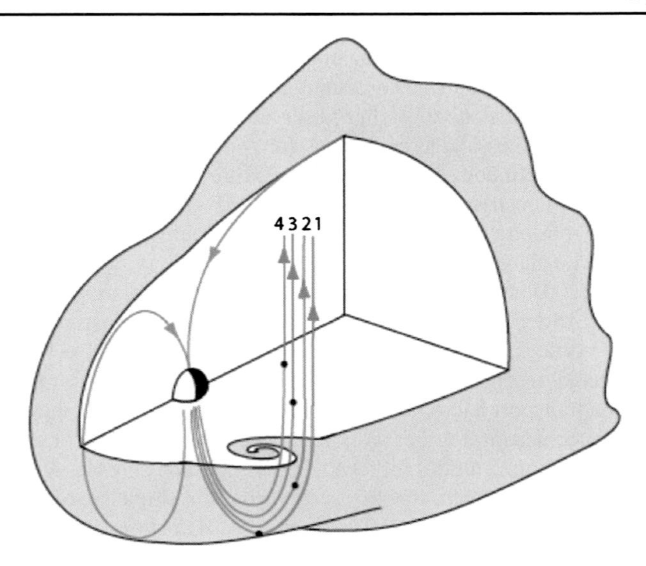

Fig. 9 The scenario described by Bavassano Cattaneo et al. (2010) in which magnetosheath magnetic field lines are reconnected at high latitudes under northward IMF conditions, and then convect tailwards to become entangled in Kelvin–Helmholtz vortices that form from Kelvin–Helmholtz boundary waves that also propagate tailwards along the magnetopause. The *blue parts* of these open field lines are within the magnetosphere while the *green parts* are outside. Near the equatorial plane, field lines 3 and 4 have crossed the magnetopause to become part of the low latitude boundary layer inside the magnetopause while field lines 1 and 2 are contributing to a magnetosheath boundary layer. The crossing point on a given field line effectively moves northwards as the field line moves tailwards

plasma coincide on the magnetosphere side of the magnetopause, similar to the findings of Hasegawa et al. (2004). Electron, proton and O^+ ion distributions observed in a succession of vortices show, in each case, a sequence of differing signatures consistent with crossing back and forward from the magnetosheath to the magnetosphere through a magnetosheath boundary layer (MSBL) and low latitude boundary layer (LLBL) of the kind that is expected for persistent lobe reconnection at high (southern) latitudes. In particular, parallel and antiparallel ion populations carry clear information about the different ages of the reconnected field lines in the inner and outer LLBL and the outer and inner MSBL. A key finding is that field aligned O^+ ions, while prevalent in the magnetotail plasmasheet, could not be found near the magnetopause current sheet, as might be expected for local reconnection allowing transport across the low latitude magnetopause. It is therefore suggested that, for this event at least, the reconnected field lines, while convecting tailwards became embedded in the developing vortices, as illustrated in Fig. 9.

Takagi et al. (2006) and Faganello et al. (2012) proposed that solar wind plasma entry to the magnetosphere can occur through so-called “double mid-latitude reconnection” occurring on the magnetospheric flanks as a consequence of rolling up of Kelvin–Helmholtz vortices. The concept is that there is a limited latitudinal extent over which conditions for vortex formation are favourable, and that magnetic flux that is separated by distances comparable to vortex scales sizes at low latitudes is able to reconnect at mid-latitudes, as shown in Fig. 10. In effect, this is similar to double-lobe (behind the cusp) reconnection, but the along-field distance between reconnection sites is smaller. Faganello et al. (2014) published a case study providing observational evidence for the occurrence of this scenario using THEMIS spacecraft data. There may not yet be a consensus on this scenario, as the same THEMIS

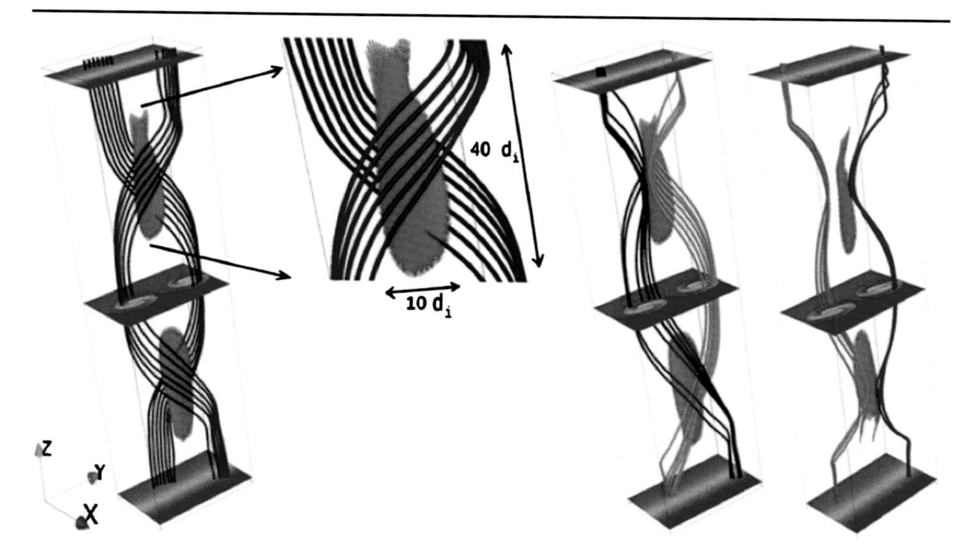

Fig. 10 Illustration from Faganello et al. (2012), of double mid-latitude reconnection associated with Kelvin Helmholtz vortices at the flank magnetopause. On the *left hand side* of the figure, *blue* (magnetospheric) and *red* (magnetosheath) field lines are shown becoming intertwined and twisted due to vortical flows at low latitude that do not occur at higher northerly or southerly latitudes. *Green shapes* show where mid-latitude current sheets form, which may become suspectible to magnetic reconnection. The *right hand pair* of figures show reconnection first at the upper current sheet, producing open field lines (*green* and *yellow*) and then at the lower current sheet producing newly closed pale *blue* field lines carrying a population of captured magnetosheath plasma

data were also used to provide support for an interpretation involving reconnection at low latitudes and sub-vortex scales, according to three-dimensional fully kinetic simulations by Nakamura et al. (2013).

It remains to be firmly demonstrated that reconnection-based processes can fully explain the observed magnetospheric boundary layers, particularly under northward IMF conditions, so it remains relevant to consider the relative effectiveness of other processes that may play a role. Alternative scenarios for plasma transport across the magnetic field at the magnetopause typically invoke a kinetic process that acts on scales smaller than those for which ideal magnetohydrodynamics is applicable.

For example, it has been proposed that cross-magnetic field diffusion may play a role in flank magnetopause plasma entry. Smets et al. (2007) used hybrid simulations to show that diffusion due to finite Larmor radius effects may occur at a tangential discontinuity magnetopause for southward IMF, and that its effectiveness is improved when Kelvin–Helmholtz waves activity occurs. Interestingly, this study predicts observations of "D-shaped" ion distributions, previously considered to be a unique indicator of magnetic reconnection.

When the wavelength of waves at the magnetopause is on the order of the ion gyroradius, the waves can lead to diffusive transport of transport of the magnetosheath ions across the magnetopause (Johnson and Cheng 1997; Chen 1999; Chaston et al. 2008). One of the likely types of waves is large-amplitude kinetic Alfvén waves (KAW). KAWs have been observed on the magnetospheric boundary (Tsurutani et al. 1982; Labelle and Treumann 1988; Anderson and Fuselier 1994). KAWs could result from mode conversion of magnetosheath compressions in the sharp magnetopause gradients at the magnetopause (Lee et al. 1994; Johnson and Cheng 1997, 2001). The mode conversion has been demonstrated with 2D hybrid simulations (Lin et al. 2010) and 3D simulations (Lin et al. 2012). Yao et al. (2011) surveyed wave power in the sheath and magnetopause and their results suggest that the wave power associated with transverse KAWs is enhanced along the dawn flank, which would provide enhanced transport. Particles can be heated nonlinearly by KAWs as they diffuse across the magnetopause (Johnson and Cheng 2001; Chaston et al. 2008). The parallel electric field of KAWs can heat electrons in the parallel direction (Hasegawa and Chen 1975; Hasegawa and Mima 1978). When the waves have large amplitudes, they can also heat ions in the perpendicular direction (Johnson and Cheng 2001). Chaston et al. (2008) showed observational evidence of stochastic heating of ions by KAWs as predicted by Johnson and Cheng (2001). The extent to which this source contributes plasma to the magnetosphere remains poorly understood.

In introducing magnetopause sources of plasma, Hultqvist et al. (1999) summarizes the total magnitude of this source with a single number that represents the sum of all above processes: 10^{26} ions/s. This number remains widely accepted today. Refinement of this number, its division amongst contributing processes, and its dependence on solar and magnetospheric conditions all remain open questions.

2.4 Other Sources

The Earth's magnetosphere is rarely considered to have any other sources beyond the solar wind and ionosphere. However, this is not strictly true. Other systems, especially those of the gas giants, can receive significant contributions from their satellites. Production can occur from surface sources, such as sputtering or volcanic activity, or from ionospheric processes on moons with sufficiently dense atmospheres. In a similar fashion, the Earth's moon can act as a third source of magnetospheric plasma.

The Moon crosses the Earth's magnetotail at $r \sim 60\ R_E$ for $\sim$5 days each month. The Moon does not have a significant atmosphere and only has a tenuous exosphere of neutral species. When in the magnetotail lobes, pickup ions can be produced on or above the lunar dayside by several mechanisms (Poppe et al. 2012):

1. photoionization of the neutral exosphere,
2. micrometeoroid bombardment of the surface,
3. photon- and electron-stimulated desorption on the surface, or
4. photo-ionized products of neutrals vented from a localized source in the lunar crust (see Seki 2015).

Lunar pickup ions are heavy ions, including He^+, C^+, O^+, Na^+, K^+, Ar^+, Al^+, and Si^+ (Tanaka et al. 2009; Saito et al. 2010). The Density of pickup ions can be in the order of 0.1 cm^{-3} and is several times higher than the density in the lobes (Harada et al. 2013; Zhou et al. 2013). Two electric fields can accelerate the freshly born ions: the photoelectric field from the existence of a high-energy tail of lunar-surface photoelectrons due to incident solar ultraviolet radiation and the convection electric field in the lobes. The ions can be accelerated to energies from several tens to several hundreds of eV (Poppe et al. 2012). Considering the pickup ions are only produced within the immediate neighborhood of the Moon, their contribution as a source for the Earth's tail plasma sheet is negligible in comparison with the mantle plasma.

3 Transport and Acceleration

3.1 Ionospheric Plasma Transport

Following the Dynamics Explorer 1 (DE-1) mission (1981) and before the Cluster mission, launched in 2000 into a 4 × 19 R_E elliptical polar orbit, the fate of ionospheric outflow at the nightside equatorial plane was studied using Akebono measurements in LEO (e.g., Cully et al. 2003a, 2003b), Polar measurements in HEO (apogee ~9 R_E) (e.g. Huddleston et al. 2005), and by employing particle trajectory modeling (e.g., Delcourt et al. 1989) to predict where the outflow observed by the spacecraft ended up; the modeling done using the Akebono and Polar measurements both showed the ionosphere to be capable of providing enough low energy plasma to fill the magnetosphere, lending support to an early prediction motivated by DE-1 observations (Chappell et al. 1987).

More recent studies on the occurrence of magnetospheric low energy plasma and its solar or terrestrial origin have drawn on the enhanced observational capabilities provided by the four-satellite Cluster mission and, beginning in 2007, the five-satellite Time History of Events and Macroscale Interactions during Substorms (THEMIS) mission. The Cluster spacecraft orbits allow for sampling of plasma directly above the polar caps and also at larger geocentric distance in the tail lobes and plasma sheet; the instrumentation was also designed to observe the dominant ion species for studying sources and transport and could combat some of the difficulties of observing low energy plasma with the ability to actively control the spacecraft potential (ASPOC). Together, these capabilities enable a cradle-to-grave inquiry into the transport of low energy plasma (O^+ in particular) from the ionosphere into the magnetosphere and near the equatorial plane as well as its energization (e.g., Kistler et al. 2005, 2010b; Liao et al. 2012, 2014). Such observations motivate recent particle trajectory modeling studies (Yau et al. 2012) and are also the subject of multiple global magnetospheric simulations (Lotko 2007; Glocer et al. 2009a, 2009b; Brambles et al. 2011; Yu and Ridley 2013). In the following, we review our current knowledge of ion transport and acceleration based on past and recent measurements.

The transport of low-energy plasma through the magnetosphere is, in its simplest form, a combination of the parallel motion along the field line and the convective $E \times B$ motion perpendicular to the field. Low-energy, in this case, refers to ions for which gradient and curvature drifts are not significant. As has been discussed above, there is essentially always ionospheric outflow at some level, due to the ambipolar electric field, and the fate of that outflow depends on the configuration (i.e. open or closed) and convective motion of the field-line. In the inner magnetosphere, for example, the ions that flow out on magnetic field lines that are corotating are able to accumulate, forming the dense plasmasphere, while ions that flow out at higher L-shells are continually convected towards the magnetopause, and so the density never reaches high levels.

At high latitudes, particularly in the cusp and auroral regions, there is further acceleration of the ions. The outflow in these regions covers a wide range of energies, from eV up to as high as 10 keV. In the case of the cusp, the combination of the parallel motion and $E \times B$ motion leads to what is known as the "velocity filter" effect, or the "tail lobe ion spectrometer" (Horwitz 1986). The $E \times B$ motion of the ions does not depend on energy, while the parallel velocity increases with energy. Thus, as a field line convects over the polar cap and into the lobe, the high energy ions are able to travel further down the tail than lower energy ions. Thus there is a separation of the ions by their velocity. Horwitz (1986) modeled this effect, providing maps that showed how the velocity of the ions entering the plasma sheet increased with distance downtail. Modeling by Delcourt et al. (1992) also showed the energy dependence of the transport paths. Because the process separates ions by velocity, not energy, lighter species go further down that tail than heavier species of the same energy, and at a particular location, different species with the same velocity, corresponding to different energies, are observed (Chappell et al. 1987). These populations are loosely referred to as "beams" as the flow outward along field lines.

Because O^+ has a higher energy than H^+ for the same velocity, it is easier to observe the O^+ ion beams in the lobes that result from the cusp outflow than the H^+ beams. The H^+ beams are often below the energy threshold of the plasma instruments, and/or below the energy of the positive spacecraft potential in the lobe. Thus, observations supporting the "ion spectro" picture have originated with (Candidi et al. 1982) and continue to come from O^+ ions. Liao et al. (2010) performed a statistical study of the occurrence of these O^+ beams using data from the Cluster/CODIF instrument ($\sim > 40$ eV). They found that the occurrence frequency of the ions increased with geomagnetic activity, although the beams could be observed for all levels of activity. They also found that the spatial distribution depended strongly on IMF B_Y, with O^+ from the northern cusp streaming towards the dawnside lobe when the IMF B_Y is positive, while O^+ from the south stream towards the duskside lobe. Liao et al. (2012) showed a positive correlation between solar activity and O^+ lobe beam observations. Although the beam occurrence frequency decreased with lower solar activity, their trends showed that in the lobes still occurred between 0–25 % of the time approaching solar minimum without accounting for ions below the instrument detection threshold. They noted seasonal as well as an orbital bias: the equatorial magnetosphere between 4 and 15 R_EE was not sampled often due to the Cluster orbit. Still, their results suggested ionospheric outflows contribute to the equatorial plasma content at all levels of solar activity but that this contribution should be greater near solar maximum, as shown in Fig. 11. Liao et al. (2015), compared the phase space density of the individual O^+ lobe beams with the phase space density of the outflowing cusp density and confirmed that the observed beam flux and the increase in energy of the beams down the tail are consistent with the velocity filter effect during quiet times, with no significant acceleration of the O^+ along this path. A small increase due to centripetal acceleration, however, as suggested by Nilsson et al. (2010), is not excluded. However, during active times, more acceleration during the transport is observed.

Because there is also significant cusp and polar cap outflow of H^+, H^+ should also be present in the lobes. However, studies of low-energy H^+ are plagued by electric shielding, as sunlit spacecraft are often charged positively from 10 to 100 V. Several methods have been employed to overcome this difficulty and obtain H^+ measurements. Relaxation sounders, which are antenna that obtain electron densities by emitting waves at characteristic plasma frequencies and observing the resulting plasma resonance (Harvey et al. 1978), do not suffer from electric shielding. These instruments were the first to measure cold, dense plasma in the plasma sheet boundary layer (Etcheto and Saint-Marc 1985). For particle detectors, one method to overcome shielding is to artificially lower the spacecraft potential by emitting positive ions (Moore et al. 1997; Torkar et al. 2008). Despite these efforts, a potential of typically a few Volts remains. During some periods, a spacecraft can be temporarily in eclipse and hence negatively charged, and low-energy ions can reach the onboard detectors (Seki et al. 2003). An alternative is to obtain the total plasma density from wave observations of the plasma frequency and then subtract the hot ion density observed by particle detectors (Sauvaud et al. 2001).

An alternative way to obtain the density is to use the fact that the spacecraft potential depends on the density and can, after calibration, be used to estimate the total density (Lybekk et al. 2012; Haaland et al. 2012). To also estimate the velocity of low-energy ions that can not reach a charged spacecraft, a recently developed technique has been used to analyze Cluster

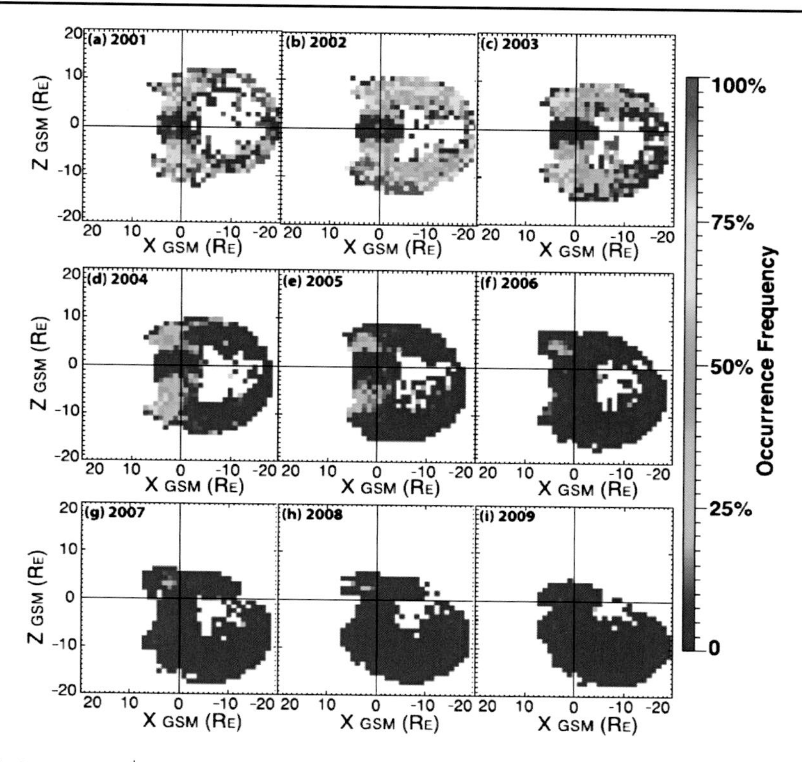

Fig. 11 Streaming O^+ occurrence from Liao et al. (2012). Each frame shows results from a different year and, therefore, different points in the solar cycle and phase of the Cluster II orbit

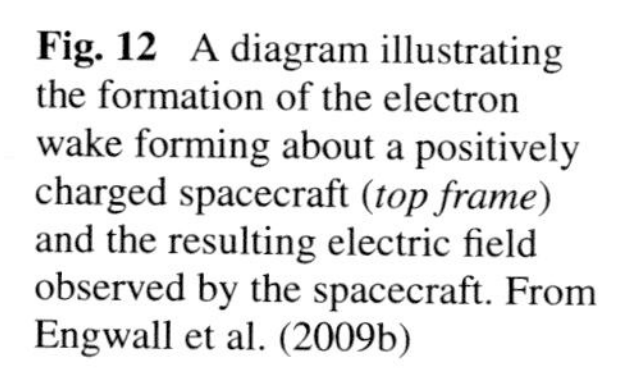

Fig. 12 A diagram illustrating the formation of the electron wake forming about a positively charged spacecraft (*top frame*) and the resulting electric field observed by the spacecraft. From Engwall et al. (2009b)

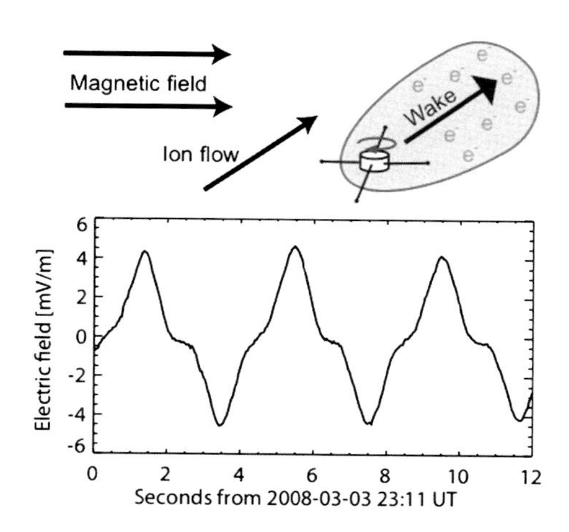

data. A supersonic flow of positive low-energy ions can create an enhanced wake behind a positively charged spacecraft. Here the ions are diverted by the potential structure and not by the much smaller spacecraft (Engwall et al. 2006, 2009b, 2009a; André and Cully 2012; André et al. 2015). The conditions for the enhanced wake formation sketched in Fig. 12

(top) are that the ion flow energy, $mv^2/2$, exceeds both the thermal energy, kT, and also is lower than the equivalent energy of the spacecraft potential, eV_{SC}:

$$kT < mv^2/2 < eV_{SC} \tag{1}$$

The ion wake will be filled with electrons, whose thermal energy is higher than the ram kinetic energy, in contrast to that of the ions. The negative space charge density can then create a local wake electric field close to the spacecraft. This electric field can be observed with electric field instruments using probes mounted on wire booms.

For a given velocity, lighter ions, such as H^+ will be more affected by the spacecraft and hence create a larger wake. Figure 12 (bottom) shows an example of a wake electric field observation close to the subsolar magnetopause. The non-sinusoidal repetitive pattern is due to the wake and indicates the presence of low-energy ions. Combining observations of the wake electric field with observations of the magnetic field and observations of the geophysical electric field using another method (the drift of keV electrons artificially emitted from the spacecraft) gives the ion velocity. Combing the velocity with the density from the spacecraft potential, the flux of low-energy ions can be determined (Engwall et al. 2009b, 2009a; André et al. 2010).

Using the Cluster dataset and this unique method to obtain low-energy ion fluxes in the night-side ($X_{GSM} < 0$) André et al. (2015) showed the occurrence rate of low-energy ions was 60–70 % in the lobes to out to X_{GSM} of about 15 R_E during all parts of the solar cycle, as indicated in Fig. 13. The ions' very low energy clearly identified them as ionospheric plasma and their high occurrence rate confirmed both their existence and their prominence in the lobes. Their statistics also showed a decrease in the occurrence rate to <20 % on approach to the plasma sheet ($|Z_{GSM}| < 2$ R_E) and they suggested that the low occurrence at small Z_{GSM} distances was due to the low-energy ions being energized above 10 eV upon approach and passage through the plasma sheet. They also noted, however, an observational limitation arising near the plasma sheet due to the detection method requiring both a steady magnetic field and the absence of ambient hot plasma.

A comparison of the statistical distribution of the O^+ beams from CODIF and the H^+ beams from Engwall et al. (2009b) (see Kronberg et al. 2014, Fig. 9) shows that the velocity distributions are very similar, which is again consistent with picture that the ions are distributed in the tail according to their velocity (not energy). However backtracing of the H^+ distributions observed indicated that at least some fraction of the lobe ions come from the polar cap, not from the cusp (Li et al. 2012). It is not surprising that the source of the lobe beams is mixed, as ions clearly flow from both the cusp and polar cap regions.

Geotail also measured beams of ions at higher energies (~keV) 100s of R_E down the magnetotail (Hirahara et al. 1997; Seki et al. 1999). These beams are observed in the plasma mantle, and consist of both ionospheric-source ions (e.g. O^+) as well as solar wind ions (H^+ and He^{++}). These beams were found to be too energetic to be consistent with just the velocity filter effect. Seki et al. (1996, 1998, 1999) investigated the possible transport routes and identified three possible sources: cusp outflow, recirculation of upward flowing ions from the nightside auroral region, and a dayside trapped population that enters the mantle through dayside reconnection. Recent measurements from Cluster (Nilsson et al. 2012, 2013) show that energetic O^+ is further accelerated in the high altitude cusp, and mixes with magnetosheath solar wind. This population moving tailward is likely the source of the deep tail beams.

The ions beams in the lobe move into the plasma sheet when the lobe field lines reconnect. Orsini et al. (1990) showed that the O^+ beams in the lobe accelerate and isotropize as they move into the plasma sheet. Kistler et al. (2010b) showed that during geomagnetic

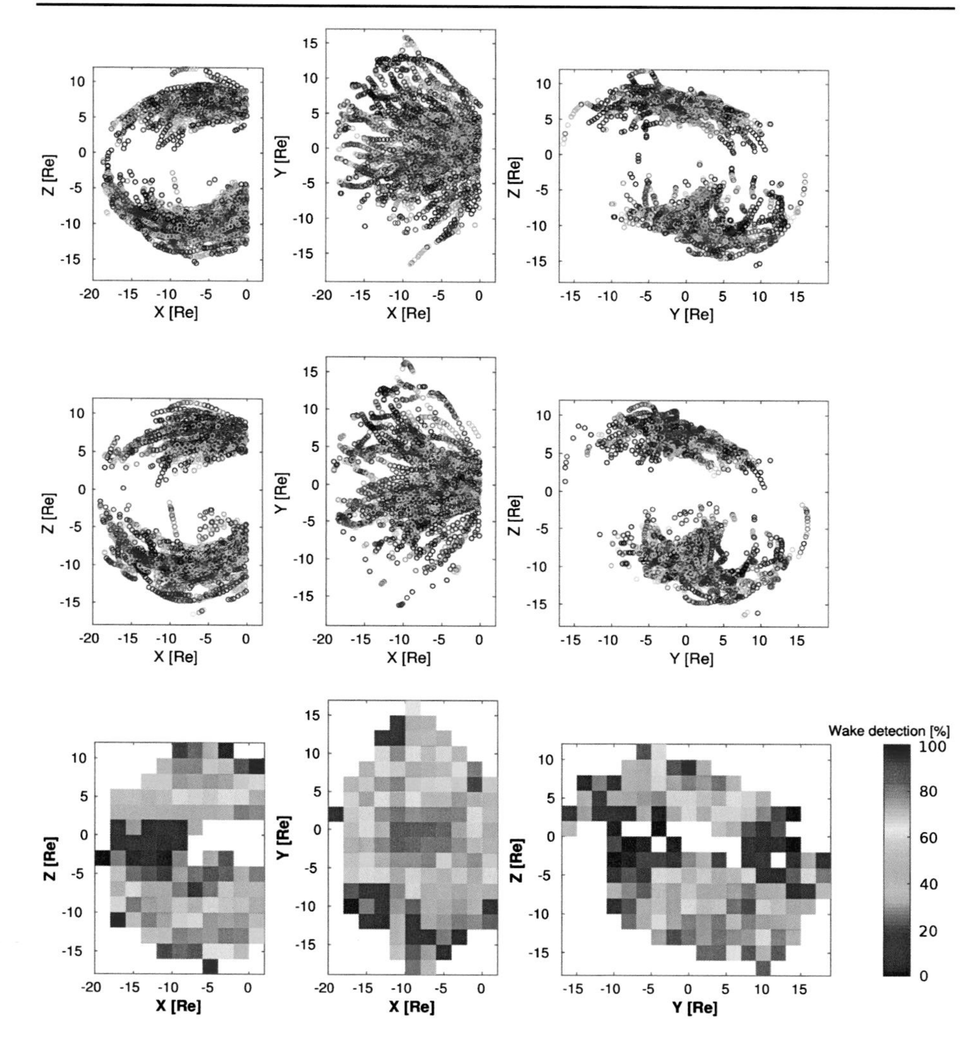

Fig. 13 Occurrence of low-energy ions. Detection of flowing low-energy ions in the GSM X-Z, X-Y, and Y-Z planes: (*top*) Cluster 1 2001–2009; (*middle*) Cluster 3 2001–2010; and (*bottom*) the sum of all data (*top* and *middle*) grouped in bins of 2 R_E by 2 R_E. From André et al. (2015)

storms, the lobe O^+ beams are observed crossing the plasma sheet boundary layer, and into the ~20 R_E plasma sheet. Once they cross the neutral sheet, the beams become isotropized. Hirahara et al. (1994), using Geotail Low Energy Particle (LEP) data, found that the lobe beams increase in energy as they move into the plasma sheet, due to enhanced $E \times B$ drift. Liao et al. (2015) also found that the beams increase in energy, and that on average, the perpendicular increase is consistent with an enhanced drift speed, due to a relatively constant average electric field, but a decreasing magnetic field towards the c enter of the plasma sheet. There is also an increase in the parallel direction, due to either wave heating, or non-adiabatic acceleration of the O^+. Kistler et al. (2010a), using STEREO/PLASTIC data from the deep tail pass of the STEREO-B spacecraft at 200–300 R_E, found that O^+ is also a constant presence in the deep tail plasma sheet during quiet times indicating that O^+ still has access to the plasma sheet downstream of the distant neutral line.

Liemohn (2005) performed a statistical survey of the <300 eV streaming ions in the nightside lobe and plasma sheet at closer radial distances (~9.5 R_E) using POLAR/TIDE measurements. They found that the cold ion streams were common, occurring >70 % of the time at these distances. They found that the tailward streaming lobe beams became bidirectional beams when they entered the plasma sheet. This shows that in contrast to further down the tail, at these closer radial distances, the field line curvature radius at the center of the plasma sheet is normally not small enough to scatter the ions, so the ions mirror and bounce. During active times, the bi-directional beams were less frequent, indicating that in these cases, the ions did scatter and isotropize.

The bouncing ion populations seen by Polar and other satellites were pursued by Chappell et al. (2008), who noticed a similarity between their persistent occurrence in Polar observations and in a compilation of particle observations from multiple past satellite missions (ATS, ISEE, SCATHA, DE-1). As this population convects inward, it drifts eastward due to the corotation electric field, remaining outside the closed drift paths of the plasmapause. Chappell et al. (2008) named this population the warm plasma cloak, due to its observed features that showed it to be a bi-directional streaming population of warm (~10 eV to few keV) plasma draped over the plasmasphere that was being blown sunwards by convection. This population co-exists with the more energetic ring current. The authors also performed particle trajectory modeling to explain its formation, showing that a polar wind proton that is centrifugally accelerated and crosses the plasma sheet at a smaller geocentric distance would pick up less energy, would not get deflected around dusk by magnetic drifts on earthward approach, and instead flow around the dawnside due to combined convection and co-rotation drifts that transport it towards the dayside magnetopause.

In addition to ions from the lobe, ions can also enter the plasma sheet directly from the nightside aurora region, which can also lead to bidirectional ion beams. Daglis and Axford (1996) suggested that the auroral outflow provides a fast feeding of the inner plasma sheet with O^+ during the substorm expansion phase. This was based on observations of the increase in the O^+ energy density at substorm onset, using AMPTE/CHEM observations (Daglis et al. 1994) at distances close to the AMPTE/CCE apogee, 8.8 R_E. This study showed that the O^+ energy density has a strong correlation with Auroral Envelope (AE) index. AMPTE/CHEM covered the energy range 1 keV to 300 keV, but its sensitivity to O^+ below ~30 keV was very low. Thus, while it could measure the accelerated O^+, it could not verify the auroral source. Gazey et al. (1996) reported an example where the EISCAT radar observed a discrete auroral arc associated with considerable upflow of ionospheric plasma. At the same time, the MICS instrument on the CRRES satellite observed two substorm injections, the second of which was O^+ dominated. MICS also measures energetic ions, from 50–300 keV. They concluded that the auroral outflow could be the source of the O^+, although they found they could not exclude a cleft source. Sauvaud et al. (2004) showed an example where an injection of O^+ from the nightside aurora accounted for 80 % of the O^+ in the mid-tail region during a geomagnetic storm.

Finally, coexisting with these warmer plasma sheet populations are a significant cold population in the equatorial nightside magnetosphere. As in the case with the global ion beams, these ions are difficult to measure because they are often below the lower energy threshold of the plasma instruments, and they are also often below the spacecraft potential, which tends to charge positive when the spacecraft is exposed to sunlight. Seki et al. (2003) used a time period when the Geotail spacecraft was in eclipse at a distance of 9 R_E down the tail, to show that there existed a cold population that had a density (~0.2 cm^{-3}) equal to the hot population occurring at the same time. Hirahara (2004) used a different technique to find cold ions in the same region. During time periods when Pc5 ULF waves occurred, multiple cold ion species (H^+, He^+, and O^+) otherwise invisible to particle detectors were accelerated into the energy range of the particle instrument. Hirahara (2004) showed that these cold ions were present 40–70 % of the time that the Pc5 waves were observed. These observations confirmed that the cold ions were observed simultaneously with a hot ion component at the inner edge of the plasma sheet, indicating that ionospheric cold plasma could cross the plasma sheet without being significantly energized. These cold ions, with partial densities comparable to the energetic ion component, were observed more frequently during the rising phase of the solar cycle. As a result, the authors suggested that the observed cold ion signatures were due to direct feeding of ionospheric outflow into the plasma sheet that was dependent on solar activity.

Nightside equatorial cold ions were sampled using the ULF wave technique by Lee and Angelopoulos (2014) out to ~13 R_E during predominantly quiet times (observation interval between 2008 and 2013). They used the THEMIS satellites to sample cold ions during intervals of enhanced bulk plasma flows (convection or ULF waves) that accelerated ambient cold ions above the spacecraft potential so they could be detected by the particle instruments carried by the three inner THEMIS spacecraft (low inclination, 1.5 by 13 R_E). They estimated the partial densities and temperatures of the three dominant ion species (H^+, He^+ and O^+) during such flow intervals and showed that all three occurred around 1–20 % of the time on the nightside, but that the heavier ions were more abundant and also warmer than the protons (H^+: few to 10 eV, He^+: 10s eV, O^+: 100s eV), as illustrated in Fig. 14. These nightside equatorial observations support the interpretation by Engwall et al. (2009b) that the outflowing LEP observed with Cluster II were likely energized above the energy needed to form a wake at locations near the plasma sheet. Lee and Angelopoulos (2014) used the heavy ion density ratios and higher temperatures on the nightside to also infer a major ionospheric source of LEP at $L < 13\ R_E$. They noticed another trend: the median temperatures of all three species were quite warm (10–100s eV) and traced out a path from pre-midnight through the dawn side, consistent with particles in the warm plasma cloak, with evidence of another path of the nightside warm ions along the dusk side. The dawnside trend implied that the heavy ions, likely to originate from the nightside ionosphere, could make it to the equator, gain moderate energy from injections or waves, and then become part of the cloak, which was discussed but not directly observed by Chappell et al. (2008).

During geomagnetic storms, enhanced convection brings the hot plasma sheet population into the inner magnetosphere. The inward motion to a stronger magnetic field increases the energy of the ions through conservation of the first adiabatic invariant. The first measurements of the ring current population (Krimigis et al. 1985; Gloeckler and Hamilton 1987) showed that during moderately active times, the ionospheric contribution was about equal to the solar wind contribution. Hamilton et al. (1988) showed that during a very large storm, ionospheric O^+ became the dominant contributor to the plasma pressure at the peak of the main phase. Greenspan and Hamilton (2002) performed a statistical study of the O^+/H^+ ratio during storms, using 68 storms that covered the rising phase of solar cycle 22. They found that both Dst and F10.7, a measure of solar EUV, are important and nearly independent predictors of the O^+/H^+ energy density ratio. Thus a large storm, at any time, will have high O^+, while even a small storm at solar maximum can have a high O^+ contribution.

Modeling of ion transport during storm times (e.g., Kistler et al. 1989, 1999; Jordanova et al. 2001, 2003, 2010) have shown that particle drift from the near-earth plasma sheet, with a large convection electric field bringing the ions into the inner magnetosphere, and then a reduced convection electric trapping the ions in the inner magnetosphere, is able to explain

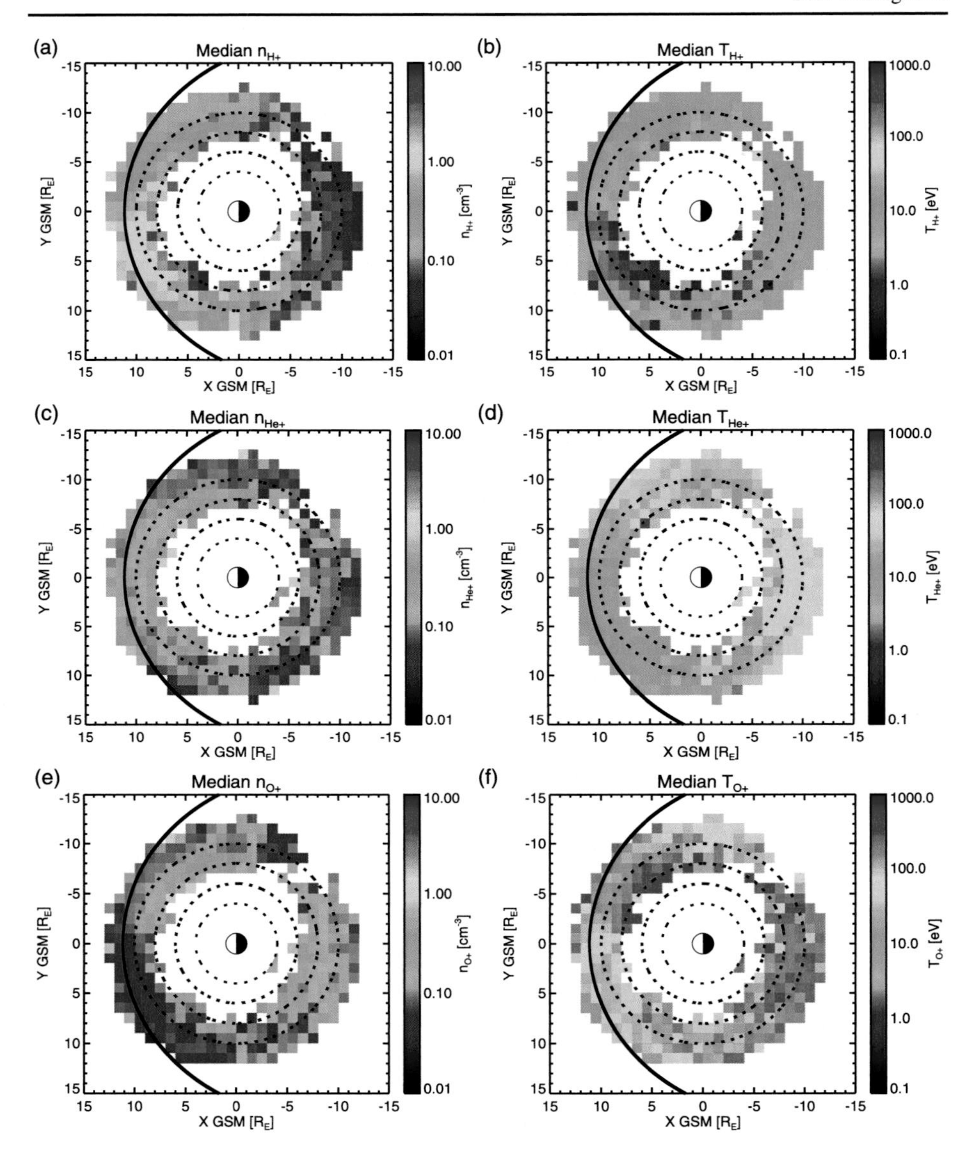

Fig. 14 Global cold ion species properties from Lee and Angelopoulos (2014)

the observed ring current spectra. The complex, non-maxwellian energy spectra observed in the ring current result from the competition between gradient curvature and $E \times B$ drifts in the inner magnetosphere, combined with loss processes along the drift path. An open question, however, is how the O^+ gets accelerated to become the dominant species in the ring current. The observations and possible mechanisms on this question have recently been reviewed by Keika et al. (2013). They address whether this is mainly due to the enhanced O^+ density in the plasma sheet from the increased entry from the lobe and night side aurora, or whether O^+ is also preferentially accelerated in the plasma sheet. As discussed above, it is clear that the O^+ is enhanced in the plasma sheet during storm times, and so that is certainly part of the answer. The question of whether O^+ is preferentially accelerated in the plasma sheet, and the role of substorms in generating the ring current is reviewed in Sect. 3.4

3.2 Plasmaphere Transport

Much of the most exciting new work concerning transport of plasmaspheric material concerns investigation of the plasmapause, or the outer boundary of the plasmasphere. The position of the plasmapause is determined by the interplay between the corotation and the convection electric fields. The magnetospheric convection electric field, controlled by the solar wind conditions and the level of geomagnetic activity, is a key factor in all existing theories for the formation of the plasmapause (Pierrard et al. 2008, and references therein).

The configuration and dynamics of the plasmapause are highly sensitive to geomagnetic disturbances. During extended periods of relatively quiet geomagnetic conditions the plasmasphere expands and the plasmapause can become diffuse, with a gradual fall-off of plasma density. Inversely, during increasing magnetospheric activity, the plasmasphere gets compressed and the plasmapause is eroded. Plasmaspheric ions can then be peeled off and escape toward the outer magnetosphere. Observations and modelling efforts have demonstrated that, for instance, plasma tongues can be wrapped around the plasmasphere, shoulders can be formed, or that plasma irregularities can be detached from the main body of the plasmasphere and form plumes (Lemaire 2001; Goldstein 2003; Sandel et al. 2003; Dandouras et al. 2005; Pierrard et al. 2008).

The plasmaspheric plumes are especially relevant because they constitute a cold plasma outflow mechanism, from the plasmasphere to the outer magnetosphere. They are associated with active periods, and during these periods they contribute typically $\sim 2 \times 10^{26}$ ions/s to the magnetospheric populations (Borovsky and Denton 2008). Recent studies have also demonstrated that plumes may affect dayside merging conditions (e.g., Walsh et al. 2014), discussed further in Sect. 4. The remote sensing observations of the plasmasphere by the IMAGE spacecraft and the in situ observations obtained by the Cluster constellation provide some novel views of this region.

Figure 15 gives an example of a plasmaspheric plume development during a magnetic substorm on the June 10, 2001, following a steady increase of the K_P activity index in the two preceding days. The plasmapause formation is simulated (Pierrard and Cabrera 2005), based on the instability mechanism for the plasmapause formation (Lemaire 2000, 2001; Pierrard and Lemaire 2004) and depending on the time history of the values of K_P. The development of a plume is clearly visible in the dusk LT sector at 7 UT. Figure 15(b) shows the EUV/IMAGE observation at 07:00 UT. A plume is indeed observed in the same LT sector, as predicted by the simulations.

Another way to observe large scale plume dynamics is to observe their connection with the ionosphere. For cold plasmas originating in the ionosphere and outer plasmasphere, $E \times B$ drift redistribution keeps both low altitude (F region O^+) and high altitude (topside H^+) ions on the same flux tube as they are convected from the plasmasphere boundary layer (PBL) to higher latitude field lines. Incoherent scatter radar observations reveal plumes of ionospheric storm enhanced density (SED; primarily O^+) extending from the dusk sector PBL to the vicinity of the noontime cusp (Foster 1993). These radar observations of SED have been projected into the equatorial plane by Su et al. (2001) and compared with geosynchronous orbit observations of a sunward-streaming plume of plasmaspheric material. That study concluded that the eroded plasmaspheric/ionospheric material is extended along the magnetic field and that SED is an ionospheric signature of the erosion of the outer plasmasphere.

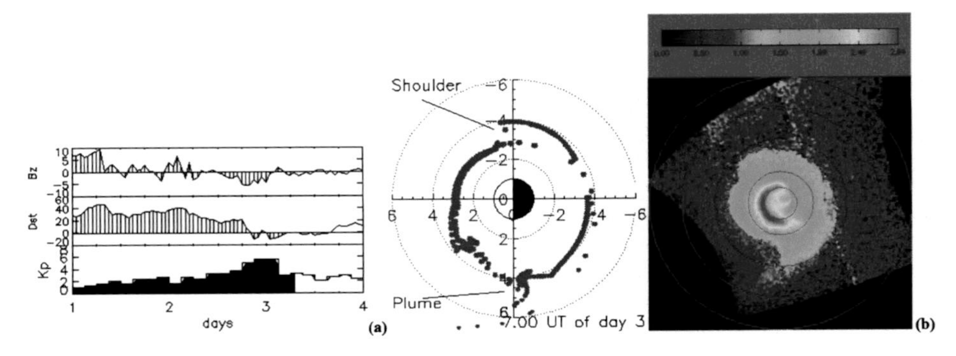

Fig. 15 Plasmaspheric plume development on the June 10, 2001, 07:00 UT. The center frame shows simulation results from Pierrard and Cabrera (2005), based on the instability mechanism (Lemaire 2000, 2001), the E5D electric field model (McIlwain 1986) and the value of K_P. The plasmapause in the geomagnetic equatorial plane corresponds to the *blue line*. The indexes B_Z, D_{ST} and K_P, observed during the previous and following days, are shown in the *left frame*. The *dotted circles* correspond to $L = 1, 2, 4$ and 6. The *right frame* shows EUV observations for this event projected in the geomagnetic equatorial plane. The *red line* corresponds to 40 % of the maximum intensity of the image and permits one to visualize the plasmapause. The *red circles* correspond to $L = 1, 2, 4, 6$ and 8. From Pierrard and Cabrera (2005)

In conjunction with radar observations, the spatial-temporal evolution of the plasmasphere/ionosphere plume can be measured through observations of Total Electron Content (TEC) from ground-based Global Positioning Satellite (GPS) receivers. The satellites of the GPS constellation are in 12-hr circular orbits (~20,000 km altitude) with orbital inclination ~55°. The GPS satellites have apogee near 20,000 km ($L \sim 4$) and the integrated total electron content (TEC), determined from analysis of their transmissions, is the combined contribution of the ionosphere and the overlying plasmasphere. An example of an SED plume originating from the duskside ionosphere and traversing poleward, illustrated by GPS TEC, is shown in Fig. 16. The narrow band of elevated TEC extending anti-sunward from the cusp across polar latitudes to the nightside auroral oval reveals how these plumes earned their other common title: the polar tongue of ionization (TOI) (Foster 2005; Thomas et al. 2013).

These GPS TEC observations have enabled new studies that further connect SED plumes with the plasmasphere. Figure 17 shows the plasmasphere erosion plume on October 8, 2013 (GPS TEC mapped to GSM equatorial plane) and the intersecting orbits of Van Allen Probes RBSP-A and Themis E_{SC}. The Van Allen Probes satellites, with their 5.5 R_E apogee, were well positioned to observe the plume in-situ, and found good agreement with the TEC maps. Additionally, it was found that the plume was oxygen rich: the O^+/H^+ density ratio increased threefold within the plume (Foster et al. 2014a).

An analysis of the April 11, 2001 event by Foster (2004) indicates that at F-region heights a plume of storm enhanced density stretched continuously from the ionospheric projection of the dusk plasmapause to the dayside cusp. Separate calculations using observations from the Millstone Hill radar, DMSP overflights, and ground-based GPS total electron content (TEC) indicate that the Storm Enhanced Density (SED) plume carried a flux of $>10^{26}$ ions/s into the cusp ionosphere during the peak of the event. At magnetospheric heights, they calculated that the associated plasmasphere drainage plume transported a flux of $>10^{27}$ ions/s to the dayside magnetopause. For comparison, Elphic et al. (1997) have estimated the flux of plasmaspheric ions, which are injected into the magnetotail and convected up and over the polar cap during strong disturbances, to be $\sim 10^{26}$ ions/s.

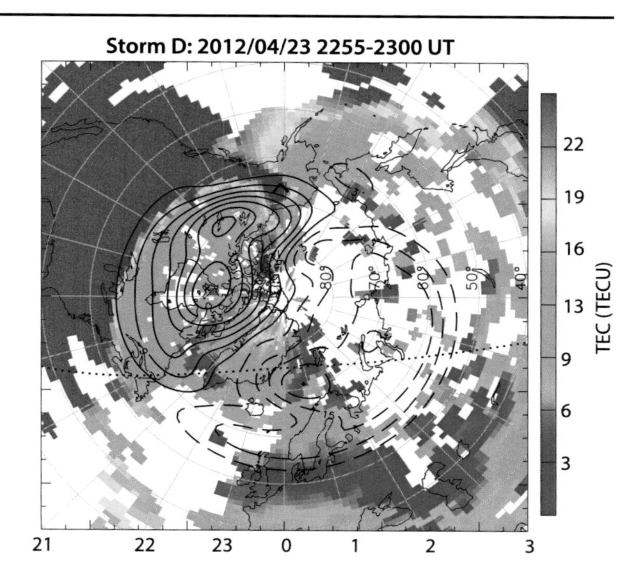

Fig. 16 An example of an SED plume traversing from the duskside ionosphere over the pole (*colored contours*). Units are total electron content unit (TECU), where 1 TECU $= 10^{16}$ e/m^2. *Black solid/dashed lines* are contours of electric potential as observed via radar. From Zou et al. (2014)

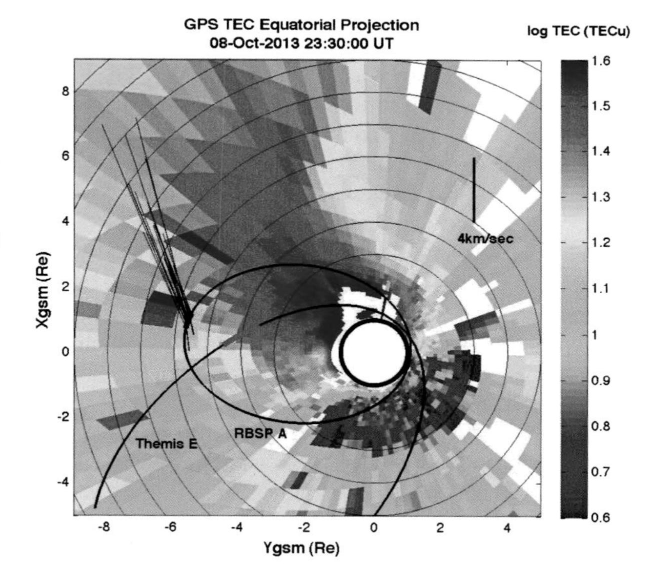

Fig. 17 Plasmasphere erosion plume on October 8, 2013 (GPS TEC mapped to GSM equatorial plane with the sun at the *top*). Orbits of the Van Allen Probes RBSP-A and Themis E are shown. 10 km/s sunward velocity (*vectors* shown) was observed along the outer portion of the plume. From Foster et al. (2014a)

Foster et al. (2014b) investigated geospace cold plasma redistribution combining GPS TEC and incoherent scatter radar ionospheric observations with in situ data in the outer plasmasphere from the Van Allen Probes spacecraft and in the topside ionospheric heights with DMSP. For a moderately disturbed event, they estimate the total fluence of eroded ionospheric/plasmaspheric ions carried antisunward at polar latitudes in the TOI channel to be $\sim 5 \times 10^{25}$ ions/s. A similar calculation of the ion fluence in the SED/erosion plume that carries the eroded plasmasphere material toward the cusp found the sunward fluence across a 5 degree span latitude to be $\sim 7 \times 10^{25}$ ions/s, which compares well with the 5×10^{25} ions/s antisunward fluence observed at that time in the TOI.

Using ground-based TEC maps and measurements from the THEMIS spacecraft, Walsh et al. (2014) investigated simultaneous, magnetically interconnected ionosphere—magnetosphere observations of the plasmaspheric plume and its involvement in unsteady magnetic reconnection. The observations show the full circulation pattern of the plasmaspheric plume and validate the connection between signatures of variability in the dense plume and reconnection at the magnetopause as measured in-situ and through TEC measurements in the ionosphere. The location of THEMIS at the reconnecting magnetopause mapped to the point in the ionosphere, where the TOI is formed, and enhancements in TEC stream tailward over the pole on open field lines. That study confirmed that the formation of the TOI in the ionosphere is spatially linked to the presence of the plume and reconnection at the magnetopause. The dense plasma on newly opened magnetic field lines convected tailward over the pole as observed in the motion of TOI patches in the ionosphere and in-situ at the magnetopause. Foster et al. (2014a) observed such plume/TOI plasma at 5.5 R_E in the midnight sector and its role in substorm injection and particle acceleration to energetic ($\sim$100 keV) and highly relativistic ($\sim$5 MeV) energies. These multi-instrument observational studies demonstrate the extent of plasmaspheric recirculation through the magnetosphere and the effect it has on global dynamics.

Are plasmaspheric plumes the only mode for plasmaspheric material release to the magnetosphere? As indicated above, plasmaspheric plumes are associated with active periods and with fluctuations of the convective large-scale electric field, governed by solar wind conditions. In 1992, however, an additional way for plasmaspheric material release to the magnetosphere was proposed: the existence of a plasmaspheric wind, steadily transporting cold plasmaspheric plasma outwards across the geomagnetic field lines, even during prolonged periods of quiet geomagnetic conditions (Lemaire and Schunk 1992). This wind is expected to be a slow radial flow pattern, providing a continual loss of plasma from the plasmasphere, for all local times and for $L > \sim 2$. It is thus similar, but on a completely different scale, to that of the subsonic expansion of the equatorial solar corona.

The existence of this wind has been proposed on a theoretical basis: it is considered to be the result of plasma interchange motion driven by an imbalance between gravitational, centrifugal, and pressure gradient forces (André and Lemaire 2006; Pierrard et al. 2009). Such a radial plasma transport implies that the plasma streamlines are not closed, and therefore the cold plasma elements slowly drift outward from the inner plasmasphere to the plasmapause, along wound up spiral drift paths. Figure 18 shows the displacements of the plasma elements (the blue $\times$ symbols) from their initial positions, i.e. the black dots initially aligned along the dipole magnetic field lines, which are represented by the solid lines. The innermost arc of blue $\times$ symbols was thus initially along the innermost magnetic field solid line shown in Fig. 18 (see Pierrard et al. 2009). As shown in this figure, this outward radial transport effect is strongest at the geomagnetic equator.

Indirect evidence suggesting the presence of a plasmaspheric wind has been provided from the plasma refilling timing. Following the erosion of the plasmasphere after a severe geomagnetic storm, the plasma refilling time at $L > 3$ can be 4 days or even as long as 8 days (Park 1970; Banks et al. 1971; Kotova 2007; Obana et al. 2010). Considering a simple refilling scenario, with an ionization flux varying with time as the equatorial density increases, Lemaire and Schunk (1992) estimated the equatorial densities in drifting and refilling flux tubes and noted that a flux tube located at $R = 4$ R_E would take only 2.5 days to completely refill and reach a state of diffusive equilibrium. This refilling timing difference, between calculated and observed times, suggests a continuous plasma leak from the plasmasphere, even during quiet conditions, consistent with the plasmaspheric wind. Evidence for such a continuous plasma leak, outside the plasmapause, has been also provided by global EUV

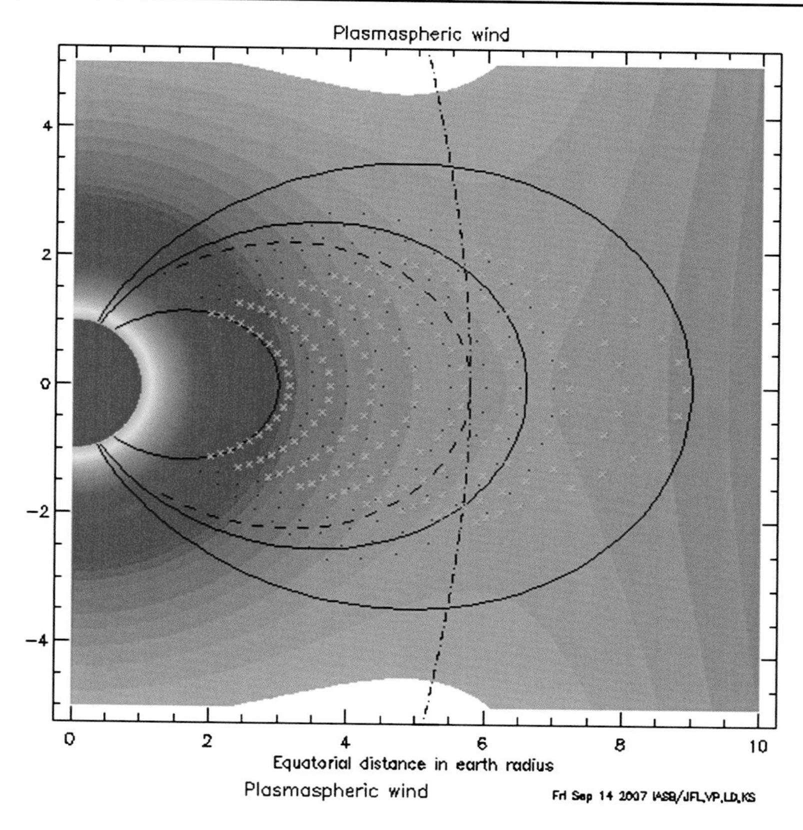

Fig. 18 Plasmaspheric wind formation simulation, as the result from a plasma interchange motion driven by an imbalance between gravitational, centrifugal and pressure gradient forces. It shows the displacements of the plasma elements (the *blue* × symbols) from their initial positions, i.e. the *black dots* initially aligned along the dipole magnetic field lines which are represented by the *solid lines* (Pierrard et al. 2009). Courtesy of Joseph F. Lemaire, Nicolas André and Viviane Pierrard, from a numerical simulation available at http://plasmasphere.aeronomie.be/plasmaspherewindsimulation.html

imaging of the plasmasphere (Yoshikawa 2003). Indirect evidence for the plasmaspheric wind has been also provided from the smooth density transitions from the plasmasphere to the subauroral region, observed during quiet conditions and at various magnetic local times (Tu et al. 2007).

Experimental direct evidence for the plasmaspheric wind has been provided recently (Dandouras 2013) based on the analysis of the ion distribution functions, acquired in the outer plasmasphere by the Cluster Ion Spectrometry (CIS) experiment onboard the Cluster spacecraft. As shown in the example presented in Fig. 19, the ion distribution functions obtained close to the magnetic equator reveal an imbalance between the outward and inward moving ions, both for H^+ and for He^+ ions, corresponding to a net outward flow. This outflow has been observed during all quiet or moderately active magnetospheric conditions events analysed, in all MLT sectors, and is consistent with the plasmaspheric wind proposed on a theoretical basis by Lemaire and Schunk (1992). Calculations show that the observed radial outflow corresponds to a 5×10^{26} ions/s plasma loss rate from the plasmasphere, which at the same time constitutes a cold plasma supply to the outer magnetosphere.

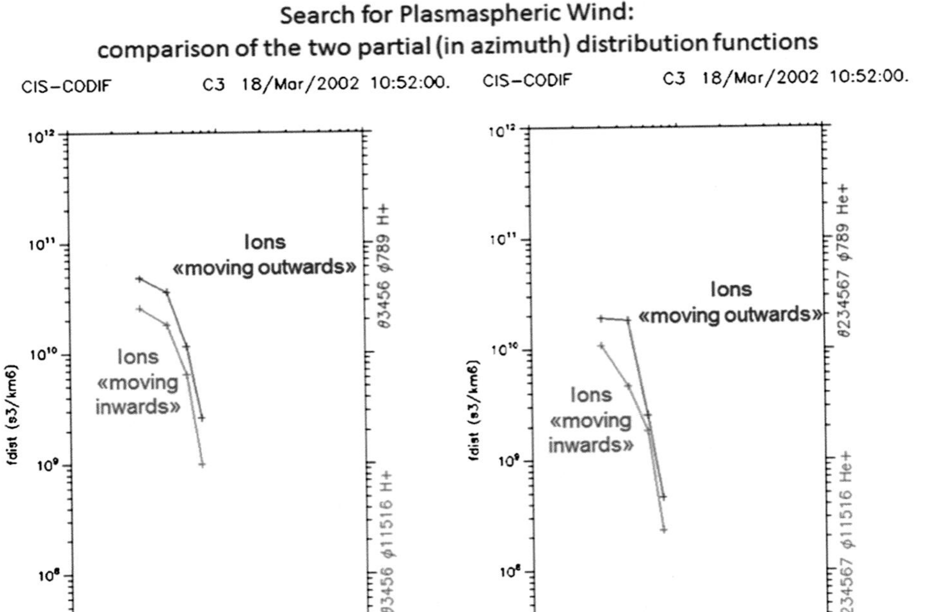

Fig. 19 Partial distribution functions in the outer plasmasphere and close to the magnetic equator, corresponding to ions flowing radially outwards (*blue plots*) and to ions flowing radially inwards (*red plots*). *Left panel* is for H^+ ions and *right panel* is for He^+ ions. Ordinate axis is in phase space density units (ions s^3 km^{-6}). The systematic imbalance between the outwards and inwards propagating ions reveals a net outward flow. From Dandouras (2013)

These plasmaspheric transport mechanisms, i.e., plumes (localised plasma releases, mainly during active periods) and the plasmaspheric wind (continuous outflow, even during prolonged periods of quiet geomagnetic conditions), appear to contribute strongly to other magnetospheric regions. The solar wind source is of the order of 10^{27} ions/s and the high-latitude ionospheric source is of the order of 10^{26} ions/s, varying by a factor of $\sim$3, as a function of the activity level and particularly dependent on the IMF orientation (Moore 2005; Haaland et al. 2009; Li et al. 2012). It appears thus that plasmaspheric recirculation constitutes a substantial plasma source for the outer magnetosphere and cusp, and it is comparable to the other sources as the solar wind and the high-latitude ionosphere.

3.3 Solar Plasma Transport

Particles originating from the solar wind can enter the open field line region of the magnetosphere (the lobes) through upward flow out of the cusp or via reconnection just tailward of the cusp. The plasma entering the lobes is called the mantle plasma. The mantle plasma is magnetosheath-like with reduced density and velocity. The mantle plasma is relatively much denser and colder than the plasma sheet plasma and with substantial tailward field-aligned bulk flow. The mantle plasma spreads across the full width of the lobes and reaches the plasma sheet via $E \times B$ drift under the influence of the dawn-to-dusk magnetospheric electric field (Pilipp and Morfill 1978). Conversely, as the solar wind gains access to dayside and terminator regions of the magnetosphere at low latitudes, it forms the Low Latitude Boundary Layer (LLBL). This is a region of northward, closed magnetic flux just inside the magnetopause on the dawn and dusk sides of the magnetotail (Fairfield 1979; Slavin et al. 1985; Kaymaz et al. 1994). Understanding the transport of mantle and LLBL plasma to the plasma sheet and inner magnetosphere is critical for determining the influence of the solar plasma source on magnetospheric dynamics.

Indication of the existence of the mantle plasma was first reported by Hones et al. (1972) and was later confirmed and termed "plasma mantle" by Rosenbauer et al. (1975). The mantle plasma can be seen in the near-Earth region (Haerendel and Paschmann 1975; Taguchi et al. 2001), at the lunar distance (Hardy et al. 1975; Wang et al. 2014), and in the distant tail (Gosling et al. 1984; Slavin et al. 1985; Maezawa and Hori 1998). The plasma mantle is confined to high latitudes closer to the Earth, but spreads to lower latitudes farther down stream as the mantle plasma $E \times B$ drifts down to the plasma sheet (Slavin et al. 1985; Siscoe and Kaymaz 1999).

Two main sources have been suggested for the mantle plasma. For the cusp source, plasma enters the cusp first and then mirror back to nightside high latitude tail. For the magnetopause source, plasma can enter the lobe through open field lines at any downtail location of the magnetopause. From the MHD point of view (Siscoe and Sanchez 1987; Siscoe et al. 2001) mantle plasma can be described as a slow-mode expansion fan of the plasma from the magnetosheath entering through merging lines along the magnetopause. The mantle source is often found to be mixed with plasma from the ionosphere (Seki et al. 1996).

As the mantle plasma flows tailward along the magnetic field lines, it $E \times B$ drifts toward the equator, thus providing particles into the tail plasma sheet (Speiser 1968; Sckopke et al. 1976). The mantle plasma at low latitudes is often found to be adjacent to the plasma sheet and is often mixed with plasma from the plasma sheet boundary layer (PSBL) and plasma sheet (Akinrimisi et al. 1989; Maezawa and Hori 1998). The particle supply depends on the spatial distribution of the mantle plasma. Pilipp and Morfill (1978) theoretically predicted the cross-magnetosphere and down-tail profiles for the mantle plasma resulting from the parallel and perpendicular transport of particles coming from either the cusp or magnetopause source. The model predicted quite different cross-tail profiles corresponding to the source. With the magnetopause source, density, temperature, and bulk velocity are the highest at the magnetopause and decrease with increasing distance away from the magnetopause, while with the cusp source there are almost no cross-tail variations at large downtail distances.

Once the mantle particles reach low-latitudes and become incorporated into the plasma sheet through tail reconnection, they are either transported Earthward in Bursty Bulk Flows (BBFs) (Baumjohann et al. 1990; Angelopoulos et al. 1992) or lost to flow down the tail where they will eventually join the solar wind. Therefore, the location of the reconnection X-lines, from which the earthward and tailward flows emanate, regulates the transport and fate of mantle plasma. At substorm onset, these X-lines form closer to the Earth, $X \sim -20$ to 30 R_E, and they are termed the "near-Earth neutral line" (NENL) (Nagai 2005; Imber et al. 2011). In fact, the frequent observations of flux ropes in this region with diameters of several Earth radii suggests the simultaneous existence of multiple X-lines near the time of onset (Slavin et al. 2003), complicating mantle transport. Observations in the distant magnetotail have shown the persistent presence of a "distant neutral line" (DNL) at $\sim X = -120$ to -140 R_E (Zwickl et al. 1984; Slavin et al. 1985). Earthward of the DNL,

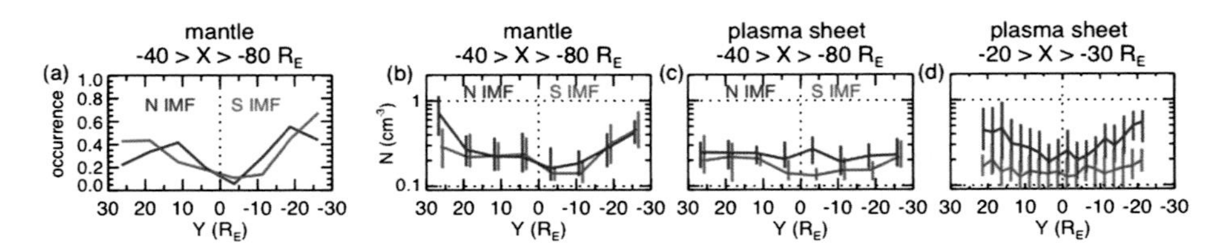

Fig. 20 The occurrence rates (**a**) and density (**b**) for the mantle plasma observed by ARTEMIS. The ion number density for the plasma sheet at $X < -40\ \mathrm{R}_E$ observed by ARTEMIS (**c**) and at $X > -30\ \mathrm{R}_E$ observed by Geotail (**d**). The *blue* (*red*) *curves* show the profiles corresponding to the 4-hr averaged IMF $\mathrm{B}_Z > 0$ (<0). The *curves* indicate median values and *vertical lines* indicate the 25 % and 75 % quartiles. Adapted from Wang et al. (2014)

the plasma sheet flow is sunward except during the expansion phase of substorms when fast flows carry flux ropes, also called plasmoids, tailward. It is unclear whether the DNL is a single stable reconnection X-line or the statistical aggregation of the tailward retreating NENLs from successive substorms (Slavin et al. 1987). Plasma mantle particles reaching the plasma sheet beyond the DNL are all lost down the tail at all times.

The mantle occurrence rate is higher and mantle thickness (the distance from the magnetopause) is larger during southward IMF than northward IMF (Paschmann et al. 1976; Sckopke et al. 1976). The mantle plasma at the lunar distance (60 R_E) has been studied using the instruments on the surface of the moon (Hardy et al. 1975, 1976, 1979). Their results showed that mantle plasma can appear at all Y, but with lower occurrence rate at smaller $|Y|$. The occurrence has a strong dawn-dusk asymmetry depending on the IMF B_Y direction (Hardy et al. 1975; Gosling et al. 1984).

Figure 20 shows the occurrence rates and plasma density for the mantle plasma as a function of Y under north and southward IMF in the magnetotail from $X = -40$ to $-80\ \mathrm{R}_E$, observed by the two ARTEMIS spacecraft from August 2010 to December 2012 (Wang et al. 2014). Both the occurrence rates and densities are highest near the flanks and decrease with decreasing $|Y|$, suggesting that the particle supply to the plasma sheet becomes smaller at smaller $|Y|$. There are no significant differences in these cross-tail profiles between north and south IMF conditions, suggesting that the particle supply is independent of the IMF B_Z direction. Figures 20(c) and 20(d) show the plasma sheet density in the tail ($-40 > X > -80\ \mathrm{R}_E$) and in the near-Earth tail ($-20 > X > -30\ \mathrm{R}_E$) respectively. The magnitude of the plasma sheet density is slightly less than the mantle density during southward IMF. However, during northward IMF the density in the tail at smaller $|Y|$ and in the near-Earth tail is substantially higher than during south IMF. The comparisons suggest that during southward IMF the mantle plasma supply is likely important to the plasma sheet, while during northward IMF cross-tail transport may be needed to bring particles from the flanks toward midnight.

The plasma sheet gradually becomes colder and denser as northward IMF proceeds (Terasawa et al. 1997; Øieroset 2005; Wing et al. 2005, 2006; Wang et al. 2010). The cold-dense plasma is often a mixture of one cool and one warm population (Wang et al. 2012). The cool population can be seen extending from the flanks to midnight during prolonged northward IMF. Both the particle supplies from the low-latitude boundary layer (LLBL) and plasma mantles are strongest near the flanks, thus insufficient to account for the increase of cool particles deep inside the magnetosphere. Therefore, there are likely cross-tail transport processes allowing the cold particles to have access from the flanks to midnight. The gradual increase of cool population during northward IMF suggests that the cross-tail transport is a slow process.

Analysis of transport paths shows that $E \times B$ drift delivers particles toward the earth and the flanks, and thus cannot bring the flank source particles across the tail to midnight (Wang et al. 2007, 2009). Magnetic drift can bring particles from the dawn flank into the midnight plasma sheet (Spence and Kivelson 1993; Wang 2004), however, magnetic drift is too small to move cold particles into the plasma sheet from the dawn flank. Despite that, large-scale $E \times B$ drift transport particles mainly earthward and toward the flanks when closer to the Earth (due to shielding of the convection $E \times B$ field), the plasma sheet flow is constantly fluctuating in both its magnitude and direction even during quiet times (Angelopoulos et al. 1993). The magnitudes of flow fluctuation are significantly larger than the average flow speed. Ionospheric velocity measurements inferred from the SuperDARN radar also suggest that even under steady driven conditions, there are significant ionospheric velocity fluctuations (Bristow 2008). The convection velocity fluctuations are also observed in the lobes by Cluster (Förster et al. 2007).

The flow fluctuation can result in diffusive particle transport if the particle number density has a spatial gradient. It has been proposed (Terasawa et al. 1997; Antonova 2005; Borovsky 2003; Weygand 2005) that diffusion may transport cold particles from the flanks deep into the plasma sheet. The diffusion coefficient associated with flow fluctuations in the plasma sheet has been estimated (Borovsky et al. 1997, 1998; Ovchinnikov et al. 2000; Nagata et al. 2007; Stepanova et al. 2011).

Efficiency of diffusive transport of particles depends on both the distributions of the diffusion coefficients and particle spatial gradients. To evaluate whether diffusion is capable of bringing particles across the tail within the typical observed time scale, (Wang et al. 2010) estimated diffusion coefficients associated with turbulent flows from Geotail observations. They performed a simulation of density evolution due to diffusive and drift transport of particles with the sources at the flanks. In the simulation, the flank sources, drift velocities, and diffusion coefficient are IMF and time-dependent and are established using Geotail data. The simulation results show that diffusive transport due to turbulence can move cold particles from the flank to the midnight meridian during northward IMF to form cold dense plasma sheet with density increase rates consistent with the statistical Geotail results.

However, using the THEMIS observations, Stepanova et al. (2011) showed that diffusion coefficients decrease quickly with decreasing distances from the Earth. This suggests that diffusive transport may become too weak to account for the formation of cold-dense plasma sheet in the near-Earth region ($r < \sim 15\ R_E$). It has been suggested that interchange motion may be another transport mechanism (Johnson and Wing 2009). The reconnection within a rolled-up K–H vortex should create cold-dense plasma with relatively lower- entropy (i.e., the entropy parameter, $PV^{5/3}$, where P is plasma pressure and V is flux tube volume per unit magnetic flux) than the surrounding hot plasma sheet plasma. This can lead to interchange instability that transports the cold plasma inward. Wang et al. (2014) used the Rice Convection Model (RCM) to simulate the evolution of colder, denser, and lower-entropy ions and electrons that are presumably created locally along the flanks by the Kelvin–Helmholtz vortices and subsequent reconnection. The RCM simulation quantitatively reproduces many prominent features of the formation of cold-dense plasma sheet simultaneously observed by five THEMIS probes near and away from the flank, indicating that interchange motion is a plausible inward transport mechanism for cold particles in the near-Earth plasma sheet.

3.4 Substorm Acceleration

A notable feature of the expansion phase of substorms in the inner terrestrial magnetosphere is the relaxation of magnetic field lines from a stretched configuration to a more dipolar one.

During such events, a variety of in situ measurements reveal that heavy ions (O^+) may be subjected to prominent energization up to the hundred of keV range (Ipavich et al. 1984; Möbius et al. 1986; Nosé et al. 2000). In a number of instances, this effect seems to depend upon mass-to-charge ratio since no similar energization is noticeable for protons. As an example, the observations of energetic neutral atoms reported by Mitchell et al. (2003) show evidence of energetic O^+ injections in conjunction with auroral break-ups, but no significant change in the energetic proton flux. Some energization process thus appears to be at work during such events that preferentially affects O^+ as compared to H^+. A possible mechanism to explain this mass selective ion energization is an impulsive energization under the effect of the electric field, induced by the magnetic field line relaxation.

Indeed, as magnetic field lines rapidly evolve from tail-like to dipole-like configurations, the electric field induced by the magnetic transition is responsible for a convection surge that rapidly injects particles into the inner magnetosphere (Mauk 1986). If the time scale of this reconfiguration is large compared to the gyroperiod of the particles, their magnetic moment (first adiabatic invariant) is conserved and the adiabatic (guiding center) approximation is valid. In contrast, if the time scale of the reconfiguration is comparable to (or smaller than) the particle gyroperiod, the magnetic field varies significantly within a cyclotron turn, the guiding center approximation is not valid, and the particle magnetic moment may not be conserved during transport. This temporal nonadiabaticity (i.e., due to explicit time variations of the magnetic field) differs from spatial nonadiabaticity (i.e., due to field variations on the length scale of the particle Larmor radius like in the magnetotail current sheet) and it may actually occur in regions of the magnetotail where the ion motion would otherwise be adiabatic (i.e., $\kappa > 3$, where κ is the adiabaticity parameter defined as the square root of the minimum curvature radius-to-maximum Larmor radius ratio). In the inner terrestrial magnetosphere, dipolarization of the magnetic fied lines typically occurs on a time scale of a few minutes, which is on the order of the cyclotron period of O^+ in this region of space. Accordingly, while protons may be transported in an adiabatic manner and experience betatron or Fermi-type energization, heavy ions may experience prominent nonadiabatic energization during such events (see Seki et al. 2015, for more detailed discussion of these processes).

Unlike the energy gain due to the large-scale convection electric field that is constrained by the magnitude of the cross-polar cap potential drop, there is no well defined limit for the energization that can be achieved from the induced electric field (Heikkila and Pellinen 1977; Pellinen and Heikkila 1978). As a matter of fact, single-particle trajectory calculations in model reconfigurations of the magnetic field lines reveal that O^+ energization up to the 100 keV range may readily be achieved in the inner magnetosphere (Delcourt et al. 1990). Since this energization occurs in a nonadiabatic manner and goes together with prominent enhancement of the magnetic moment, it radically changes the long-term behavior of the particles. This is illustrated in Fig. 21, from Delcourt (2002), which shows model O^+ trajectories in two distinct cases; steady state (top panels) and assuming a one-minute dipolarization of the field lines at some point during transport (bottom panels). In this figure, the test O^+ ion is considered to originate from the nightside auroral zone and it can be seen that, in steady state, this ion intercepts the mid-tail where it is subjected to magnetic moment scattering upon crossing of the field reversal. As a result, the O^+ subsequently bounces back and forth between high-altitude mirror points, while drifting westward. This ion is ultimately lost into the dusk magnetopause, the net energy gain realized being of the order of 20 keV. In the bottom panels of Fig. 21, a drastically different behavior can be seen as a result of substorm dipolarization. The magnetic field line reconfiguration is here assumed to occur 40 minutes after ejection of the test O^+ from the topside ionosphere and it is apparent that, as a result of this reconfiguration, the ion is rapidly transported from the mid-tail down

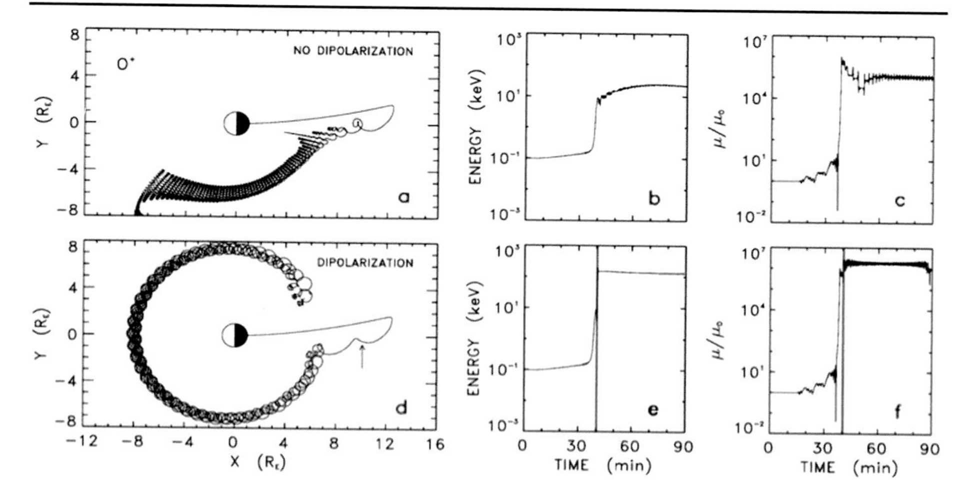

Fig. 21 Trajectories of test O^+ launched from the nightside auroral zone (0000 MLT) with an initial energy of 100 eV and considering either (*top panels*) steady state or (*bottom panels*) 1-min dipolarization during transport. *Left panels* show the trajectory projection in the X-Y plane, whereas center and *right panels* show the kinetic energy and magnetic moment (normalized to the initial value) versus time, respectively. In the *bottom panels*, dipolarization occurs after 40-min time-of-flight (*shaded area* in (**e**) and (**f**)). The *arrow* in (**d**) indicates the O^+ position at the dipolarization onset. From Delcourt (2002)

to the geosynchronous vicinity. During this convection surge, the O^+ experiences a prominent energization that exceeds 100 keV. The nonadiabatic character of this energization is apparent from the rightmost panel that shows further magnetic moment enhancement on the time scale of the dipolarization. Given this large post-dipolarization energy realized, the O^+ motion subsequently is dominated by gradient drift around the planet and, instead of being lost at the magnetopause (top panels), it rapidly encircles the Earth with a drift period of about 50 minutes. It is clearly apparent from Fig. 21 that a short-lived convection surge with prominent nonadiabatic energization is an efficient process to populate the outer ring current with heavy ions of ionospheric origin.

However, a more recent statistical analysis of all the substorm events from 10 years of Geotail data (Ono et al. 2009) indicated that while the greater enhancement of O^+ over H^+ was observed over the energy range 9–36 keV, at higher energies the picture was more mixed. Some events showed the O^+ spectrum becoming harder than H^+, as had been reported before, but other events showed the H^+ spectrum becoming harder than O^+. To explain the new observations, Ono et al. (2009) have suggested that the acceleration is due to the magnetic field fluctuations during the dipolarization, not due to the dipolarization itself. They found that the biggest increases did not occur when the time scales of the dipolarization and the gyrofrequencies were matched. Instead, it was found that the most significant acceleration occurred when the power in the shorter time scale fluctuations was close to the ion gyrofrequencies. In some cases, this power favored the O^+, but in other cases it favored the H^+. Nosé et al. (2014) examined magnetic fluctuations that occurred during dipolarization for 7 events inside geosynchronous orbit. They modeled the ion acceleration in the electromagnetic fields, and found that the O^+ was accelerated in the energy 0.5–5 keV by these fluctuations, while the H^+ was not significantly affected, consistent with the observations. In light of the Delcourt (2002) work, and these recent simulations, it seems likely that both the dipolarization itself, and the smaller scale fluctuations associated with it play a role in the ion acceleration.

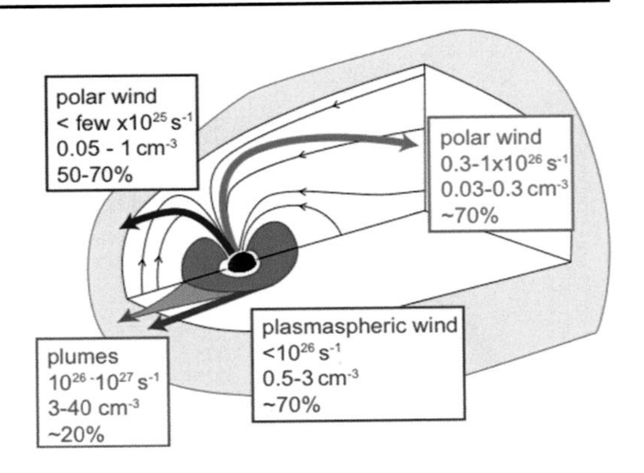

Fig. 22 Estimations of low energy particle densities from André and Cully (2012)

4 Consequences

The immediate and inescapable consequence of the various sources of plasma in the terrestrial system is mass and energy loading of different regions of the magnetosphere, chronicled in detail by Hultqvist et al. (1999). Since the book's release, subsequent work has been driven by a growing awareness of the magnitude of the ionospheric source of light and heavy ions and the potential effects this population has on a vast number of magnetospheric regions. Low-energy ions of ionospheric origin with energies below tens of eV dominate most of the volume of the terrestrial magnetosphere at least 50–70 % of the time. Orders of magnitude estimates for low-energy ion density, outflow and the percentage of time these ions dominate the density are given in Fig. 22 (André and Cully 2012). The nightside outflow is often dominated by low-energy ions. The H^+ outflow is estimated to be about 10^{26} ions/s (Engwall et al. 2009a), which is larger than the previously observed energetic outflow at high altitudes, and consistent with observations at low altitudes (Peterson et al. 2008). On the dayside, the outflow of low-energy ions is very variable. When plasmaspheric plumes are not present, the outflow is typically a few times 10^{26} ions/s, while in plumes (occurring about 20 % of the time) the outflow can be up to 10^{27} ions/s. New results show that low-energy ions can dominate 50–70 % of the time just inside the magnetopause, even when there are no plasmaspheric plumes (André and Cully 2012). The large amount of low-energy plasma detected puts strong limits on heating and acceleration mechanisms. The low-energy plasma will also lower the Alfvén velocity and the dayside reconnection rate, and will also change the micro-physics of the reconnection separatrix region (André et al. 2010). Indeed, a recent review is dedicated to the role of heavy ion outflow in global dynamics (Kronberg et al. 2014). These effects create a new paradigm in which solar wind control of magnetospheric dynamics must compete with internal feedback from ionospheric mass.

A clear example of this is the potential of the plasmasphere population to affect dayside reconnection rates. Borovsky and Steinberg (2006), as part of a larger study of magnetospheric preconditioning before Corotating Interacting Region (CIR) driven storms, initially suggested the possibility of magnetopause mass loading via plasmaspheric plumes. Borovsky and Denton (2008) provided empirical evidence of this effect by examining the AU, AL, and Polar Cap Index (PCI) activity indices. They found that, for a given solar wind electric field ($-vB_Z$), all three indices were statistically lower during periods when a plume was observed at geosynchronous locations versus periods when no plume was observed. Subsequently, this effect was shown to manifest in global, resistive MHD models (Borovsky et al. 2008). Reconnection electric field about the magnetopause was calculated using the

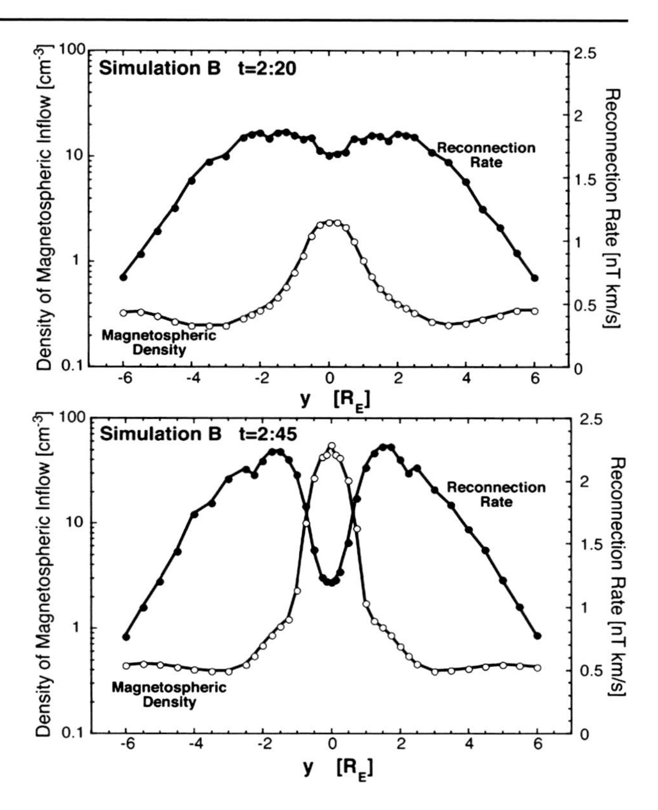

Fig. 23 Reconnection rate and magnetospheric density about the dayside magnetopause before a plasmaspheric plume arrives at the subsolar point (*top frame*) and after (*bottom frame*). From Borovsky and Denton (2008)

formula derived by Cassak and Shay (2007). Figure 23 shows the reconnection electric field (black dotted line) and density along the magnetopause (white dotted line) as a function of distance along the dayside magnetopause, both before (top frame) and after (bottom frame) plume arrival. Once the plume arrives, the local reconnection rate drops dramatically, yielding an overall reduced reconnection rate. Early THEMIS observations used a combination of techniques to obtain cold plasma densities near the magnetopause (McFadden et al. 2008), establishing plume presence in the region. Further observational work has connected in-situ observations of plume arrival at the magnetopause with the onset of bursty reconnection (Walsh et al. 2014). This connection still requires further investigation; indeed, it has been suggested that any plasmasphere impact on the magnetopause would be local and that the magnetosphere shape would adjust to compensate for the mass loading effect (Lopez et al. 2010).

Because the plasmasphere dominates the mass content of the inner magnetosphere, it therefore plays an essential role in governing the radiation belt dynamics (Horne and Thorne 1998; Thorne 2010; Chen et al. 2012). During prolonged geomagnetically quiet periods the plasmapause coincides mostly with the outer edge of the outer radiation belt of energetic electrons (>2 MeV). However, during higher geomagnetic activity time periods, the plasmapause is located closer to the inner boundary of the outer radiation belt (Darrouzet et al. 2013).

The inclusion of the high-latitude ionospheric plasma source in global models has resulted in a set of surprising large-scale effects (recently reviewed in detail by Wiltberger 2015). Initially, global fluid models relied on simple inner boundary conditions (i.e., uniform mass density) to passively include this source (e.g., Winglee 1998; Walker et al. 2003; Zhang et al. 2007). Though simple, this outflow specification can form time and space dependent outflows into the magnetosphere (Welling and Liemohn 2014) and dominate the central plasma sheet (Welling and Ridley 2010). Winglee (2002) found that if a heavy ion component was included in a simple, passive outflow source, the modeled cross polar cap potential was reduced significantly compared to an identical simulation that used an all-hydrogen inner boundary. Similar results were obtained when more realistic outflow specifications were applied. Glocer et al. (2009a, 2009b) and Welling et al. (2011), using a first-principles-based outflow model to drive heavy and light ion, polar-wind-like outflow in global MHD, found a similar reduction in CPCP. Brambles et al. (2010) used an empirical formula (Strangeway et al. 2005) that drove outflow of O^+ as a function of joule heating and AC Poynting flux calculated by the Lyon–Fedder–Mobarry model. It was found that the CPCP reduction was produced if the outflow was slow and dense. Though each study provides a unique hypothesis as to why such an effect manifests, not one has been verified to date (Welling and Zaharia 2012).

Other global simulations continue to show that global dynamics dependend on ionospheric outflow. Because the ring current can be fed significant mass from ionospheric sources (Welling and Ridley 2010; Welling et al. 2011), magnetospheric shape appears beholden to the strength of the ionospheric source (Brambles et al. 2010; Garcia et al. 2010). The source location, density, and outflow velocity all appear to be factors in driving this affect (Garcia et al. 2010). Yu and Ridley (2013) noted that as heavy ion outflow populations arrive at the plasma sheet, they can affect the location of reconnection. More dramatically, heavy ion populations that arrive near the magnetic X-line can alter reconnection rates enough to trigger a magnetospheric substorm (Wiltberger et al. 2010; Winglee and Harnett 2011). When causal outflow (i.e., outflow that is a function of magnetospheric dynamics) is employed for periods of strong driving, outflow-triggered substorms can drive additional heavy ion outflows, triggering further substorms (Brambles et al. 2011, 2013; Ouellette et al. 2013). These periodic substorms resemble global sawteeth oscillations (Huang 2003; Henderson 2004), a mode of magnetospheric activity previously unachievable with ideal MHD. A statistical investigation of the O^+/H^+ ratio during sawteeth, substorm, and non-substorm storm periods by Liao et al. (2014) suggests that heavy ion outflow plays a role in sawteeth triggering, but that high O^+ concentrations are neither a necessary or sufficient condition. All of these studies support the view that ionospheric outflow is an integral part in magnetosphere-ionosphere coupling.

5 Losses

5.1 Charge Exchange

In the inner magnetosphere, charge exchange of ions with the neutral hydrogen geocorona is a slow but persistent loss process. The charge exchange cross sections depend on species and energy, so the effects of charge exchange can be clearly identified by the associated composition changes. Kistler et al. (1998) showed an example from the FAST satellite in which the composition of the plasma sheet population changed from a hydrogen dominated population to a helium dominated population as the spacecraft moved into the inner magnetosphere. Comparison with simulations showed that this was expected because of the shorter charge exchange lifetime of H^+ at these energies (1–10 keV). Hamilton et al. (1988) examined the role of charge exchange in explaining the two-phase decay that is often observed for very large storms. They found that the fast initial decay is consistent with the charge exchange lifetime of the energetic (75–100 keV) O^+ that dominated the main phase ring current. Many examples comparing modeled ring current spectra with observations (e.g. Kistler et al. 1989, 1999; Jordanova et al. 1996, 2001) have shown the importance of charge exchange in explaining the composition changes in the energy ranges where the ion drifts are slow and go deep into the inner magnetosphere.

5.2 Advective Loss

At times of enhanced convection, the outflow of ring current ions on open drift paths to the dayside magnetopause dictates the decay of ring current (Takahashi et al. 1990; Ebihara and Ejiri 1998; Liemohn et al. 1999) and the dawn-dusk component of the solar wind electric field is the parameter that sets up the time scale for ion loss. Also, the energy of the particle along with the timescale of recovery of the cross polar cap potential controls the amount of plasma trapped on the closed field lines (Takahashi et al. 1990).

Due to the long duration of a geomagnetic storm, the particles that are injected on the nightside are able to drift completely through the inner magnetosphere. This energy and convection dependent drift can move the energetic particles from the nightside to the magnetopause in only few hours. The dayside outflow usually takes place during the main and early recovery phase of a storm (Takahashi et al. 1990; Liemohn et al. 1999, 2001; Kozyra 2002), when the ring current is highly asymmetric and most of its energy is flowing along open drift paths (Liemohn et al. 2001; Kozyra 2002). The formation of the symmetric ring current is inhibited by these losses from convection to the dayside magnetopause (Liemohn et al. 1999).

Observations of energetic O^+ ions in the magnetosheath and upstream of the bow shock during times of elevated convection confirms not only the loss of ring current ions to the magnetopause (Möbius et al. 1986; Christon et al. 2000; Zong et al. 2001; Posner 2002), but it is estimated that the loss rate of O^+ ions to the magnetopause can be as high as 6.1×10^{23} ions/s (Zong et al. 2001).

Based on in-situ observations by Geotail/EPIC, Keika (2005) estimate that a minimum of 23 % of the total ring current fast decay is due to dayside ion outflow, even in the case of a sudden northward turning of the interplanetary magnetic field, which causes a sudden decrease in the convection electric field. However, increased convection will push particles closer to the Earth where charge exchange processes can contribute to the rapid decay of the ring current (Ilie et al. 2013). The spatial configuration of the open drift paths and how deep the particles penetrate into the inner magnetosphere determines whether charge exchange makes a significant contribution to the ring current losses as the ions drift through the inner magnetosphere to the dayside magnetopause region.

An example of a high convection event is presented in Fig. 24, from Kozyra and Liemohn (2003), clearly showing that the ion outflow losses dominate the main phase of the storm, while the charge exchange processes contribute significantly to the ring current decay during the recovery phase. The convection strength controls this loss process, i.e. increasing convection will increase the outflow loss and vice versa.

Particles of the same energy but with different pitch angle may follow different drift paths due to MLT-asymmetry of the magnetospheric electric field (Roederer and Schulz 1971) or magnetic field (Roederer 1967; Roederer et al. 1973), which is the so-called drift shell splitting. For high-energy ions dominated by magnetic drift, magnetic drift shell splitting is important. For example, Takahashi et al. (1997) showed that in a realistic magnetic field configuration (compressed on the dayside and stretched on the nightside) for particles of the

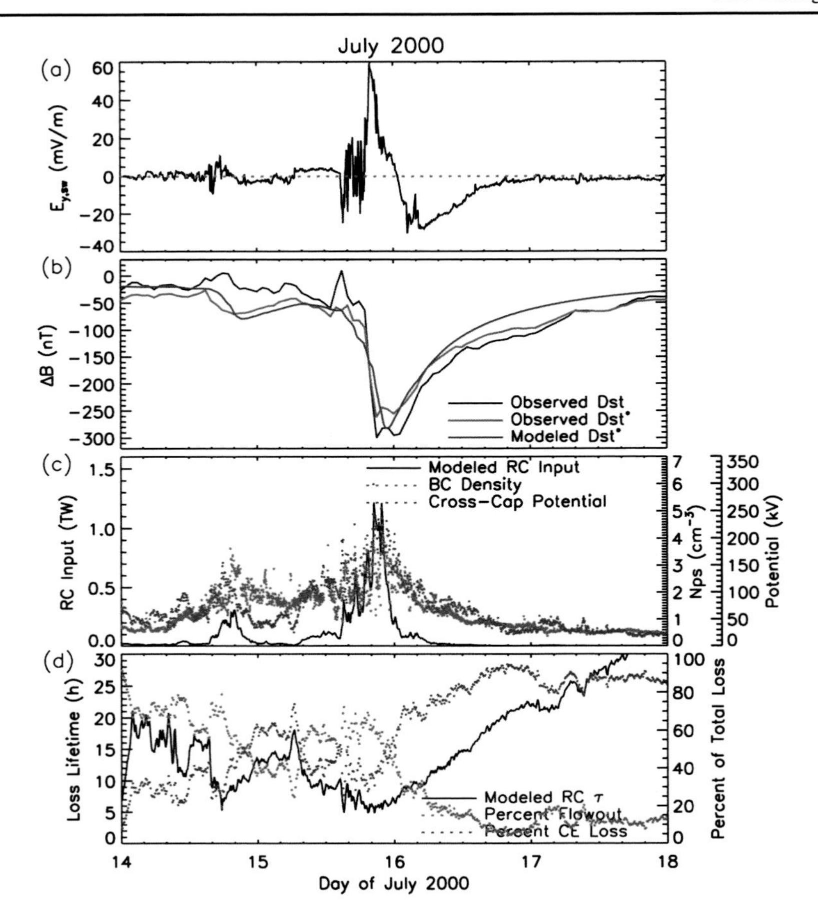

Fig. 24 Simulation results for the July 14, 2000 magnetic storm. *Top panel*: eastward component of the solar wind electric field (Ey, sw (mV/m)). *Second panel*: modeled Dsts nT (*blue line*), the observed Dst (*black line*) and observed Dsts (*red line*). *Third panel*: energy input through the nightside outer boundary ($L = 6.75$) of the model (*black line*), plasma density at geosynchronous orbit (*red dotted line*) and the cross polar cap potential (*blue dotted line*). *Bottom panel*: the globally-averaged loss lifetime for the ring current is presented in the *bottom panel* (*black line*) along with percentage of loss due to charge exchange (*blue dotted line*) and flow-out (*red dotted line*). From Kozyra and Liemohn (2003)

same energy at $r = r_0$ at noon, 90° ions come from $r < r_0$ at the midnight MLT while 30° ions come from $r > r_0$.

Given this magnetic drift shell splitting, 90° ions from the nightside at larger r are more likely to hit the duskside magnetopause than are ions of other pitch-angles. These particles are lost to the magnetopause and thus cannot complete the drift circle and return back to the nightside. As a result, there are relatively fewer ions near 90° than ions of other pitch-angles in the post-midnight sector. This process is known as magnetopause shadowing. Magnetopause shadowing produces butterfly pitch angle distributions (PADs) with negative anisotropy (more particles in the parallel than perpendicular directions) in the post-midnight sector (Sibeck et al. 1987; Fritz et al. 2003). Figure 25 shows the statistical spatial distributions of the pitch-angle anisotropy for 10, 20, 45, and 100 keV ions observed by THEMIS (Wang et al. 2012). It can be seen for ions above ~40 keV, anisotropy is negative is the

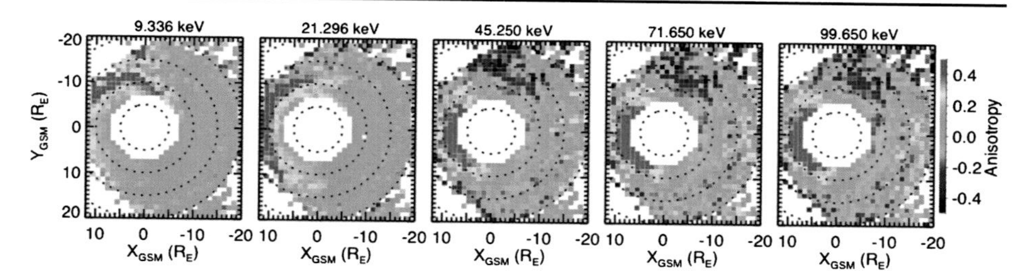

Fig. 25 Equatorial distributions of the ion pitch-angle anisotropy for different energies for $-5 > D_{ST} > -20$ nT. Anisotropy $= 0$ indicates an isotropic distribution. Anisotropy >0 (<0) indicates higher particle fluxes in the perpendicular (*parallel*) direction. Adapted from Wang et al. (2012)

post-midnight MLTs at $r \sim 10$–15 R_E. The shape of pitch-angle distribution in the negative anisotropy region is dominantly a butterfly distribution. Thus the magnetopause shadowing is an important loss mechanism for high-energy ions with near 90° pitch-angle.

Opposite of magnetopause loss is plasma exhaust downtail, which has an estimated loss rate of 10^{28-29} ions/s (Hultqvist et al. 1999). Tailward flow is frequently punctuated by transients, such as plasmoids resulting from magnetospheric substorm events (see the recent reviews of McPherron 2015 and Eastwood and Kiehas 2015). New observations of these flows in the deep tail (Opitz et al. 2014) has been afforded by the Solar Terrestrial Relations Observatory (STEREO). During January through May of 2007, STEREO Ahead (STA) was 200–800 R_E upstream of the Earth and observing the undisturbed solar wind flow. Simultaneously, STEREO Behind (STB) was 200–800 R_E down stream of the Earth, passing in and out of the magnetotail. The Solar Electron Proton Telescopes (SEPT, Luhmann et al. 2007) on both STA and STB observed 110–2200 keV ion enhancements corresponding to corotating interacting region (CIR) events. During periods when STB was in the tail, the ion fluxes were both of higher magnitude and more anti-sunward aligned, indicating an additional magnetospheric source. Additionally, the enhanced STB-observed fluxes were impulsive in nature and correlated with increases in the Auroral Electrojet (AE) index, indicating a substorm source. These new observations confirm previous observations of escaping energetic magnetosphere populations and ties them to substorm activity.

5.3 Atmospheric Precipitation

Particles with a pitch angle contained in the so-called atmospheric loss-cone (i.e. with their mirror points located below the topside ionosphere) precipitate into the atmosphere and are lost from the magnetosphere. It is generally accepted that wave-particle interaction processes develop during magnetospheric transport and cause a persistent pitch-angle scattering into the loss cone. Electron cyclotron waves are known to cause pitch-angle scattering, but recent works show that whistler-mode chorus waves could play a dominant role (Ni et al. 2011a, 2011b). Plasmasheet particles are continuously lost during their transport from their injection in the tail on the night side until the plasmasheet inner edge and then until they reach the dayside. Diffuse aurorae develop at the ionospheric footprint of their magnetic field lines and form two belts permanently surrounding the magnetic North and South poles. The polar boundary of the auroral oval corresponds to the field-aligned mapping of the injection region in the distant tail on the night side and of the exit region near the magnetopause on the dayside. The equatorward boundary corresponds to the mapping the plasmasheet inner edge.

Discrete and intense auroral arcs appear at smaller scale inside the auroral oval mainly during disturbed magnetic conditions (see Frey 2007, for a review). They are brighter than diffuse aurorae and can be observed from the ground with different sizes, and shapes and fast motions. As diffuse aurorae, they are also caused by precipitating plasmasheet particles, but usually with higher energies. Acceleration processes develop under various conditions, during transient events in the plasmasheet, or depending on solar-wind magnetosphere interactions, or due to wave-particle interaction processes. As noticed quite some time ago by Akasofu (1964), the appearance of discrete auroral arcs is often related to disturbed conditions, as during substorms, impulsive events of great magnitude, responsible for global changes in the magnetosphere: magnetic reconfiguration, particle acceleration, electric current and field enhancements.

Resonant pitch angle scattering also has the potential to remove resonant ions on timescales of under one hour. This timescale is therefore much shorter than the loss rate associated with collisional processes (Feldstein et al. 1994). The wave particle interaction mechanism is primarily important during the main phase of the storm (Gonzalez et al. 1989), possibly contributing to the geomagnetic trapping and acceleration of ionospheric ions that are injected during the main phase of a storm. Nevertheless, due to their localized nature (Jordanova et al. 1998), their contribution to the decay of the ring current is small relative to outflow, charge exchange and Coulomb collision losses.

Coulomb collisions between charged particles can also cause losses from the magnetosphere. While energy degradation from hot to cold particles occurs during these interactions (e.g., Fok et al. 1995; Jordanova et al. 1999; Liemohn et al. 2000) and in fact the energy deposition from this process is the cause of stable auroral red arcs (Kozyra et al. 1997), the primary contribution to mass loss from Coulomb collisions is via pitch angle scattering. Several studies have shown that this term is smaller than scattering due to wave-particle interactions and much smaller than either dayside flow out or charge exchange (e.g., Fok et al. 1993; Kozyra et al. 1998; Jordanova et al. 1998; Liemohn et al. 1997, 1999).

6 Open Questions

A great deal of work has been performed since the release of the Hultqvist et al. (1999) book. However, a great deal of questions remain unanswered. The balance of the contribution of solar and ionospheric plasma to the magnetosphere has shifted to the ionospheric source, especially in light of the expanded observations of the cold, "invisible" source. However, this topic is far from settled. The contribution of solar plasma is still only tenuously understood, especially as it is often difficult to separate from ionosphere populations. The magnitude of the flank-entering solar source is still undetermined, with further research required to determine what mechanisms efficiently allow mass entry into the magnetosphere.

Our understanding of the plasmasphere has also drastically transformed from a passive population to a critical reservoir of cold ions that has far-reaching implications. Further research into the effectiveness of the plasmaspheric wind in supplying cold ions to the outer magnetosphere is ongoing, as is work to determine the importance of plume material recirculating into other regions. The possibility of this material affecting solar-magnetosphere coupling by altering reconnection rates is also only tenuously understood.

Finally, with the advent of global models that better capture the different ionospheric sources of plasma, the self-consistent effects of all plasma sources on global dynamics are being rapidly explored. Recent work demonstrates that ionospheric outflow may regulate many global features, such as the development of substorms and sawteeth oscillations. A plethora of studies to scrutinize these potential relationships and demonstrate—or refute—their existence remain to be performed.

Conflict of interest The authors declare that they have no conflict of interests. This work did not include human or animal subjects.

References

T. Abe, B.A. Whalen, A.W. Yau, S. Watanabe, E. Sagawa, K.I. Oyama, Altitude profile of the polar wind velocity and its relationship to ionospheric conditions. Geophys. Res. Lett. **20**(24), 2825–2828 (1993a). doi:10.1029/93GL02837

T. Abe, B.A. Whalen, A.W. Yau, R.E. Horita, S. Watanabe, E. Sagawa, EXOS D (Akebono) suprathermal mass spectrometer observations of the polar wind. J. Geophys. Res. **98**(A7), 11191 (1993b). doi:10.1029/92JA01971

T. Abe, S. Watanabe, B.A. Whalen, A.W. Yau, E. Sagawa, Observations of polar wind and thermal ion outflow by Akebono/SMS. J. Geomagn. Geoelectr. **48**(3), 319–325 (1996). doi:10.5636/jgg.48.319. https://www.jstage.jst.go.jp/article/jgg1949/48/3/48_3_319/_article

T. Abe, Long-term variation of the polar wind velocity and its implication for the ion acceleration process: Akebono/suprathermal ion mass spectrometer observations. J. Geophys. Res. **109**(A9), 09305 (2004). doi:10.1029/2003JA010223

S.-I. Akasofu, The development of the auroral substorm. Planet. Space Sci. **12**(4), 273–282 (1964). doi:10.1016/0032-0633(64)90151-5. http://www.sciencedirect.com/science/article/pii/0032063364901515

J. Akinrimisi, S. Orsini, M. Candidi, H. Balsiger, *Ion Dynamics in the Plasma Mantle* (1989). http://www.researchgate.net/publication/234190759_Ion_dynamics_in_the_plasma_mantle

B.J. Anderson, S.A. Fuselier, Response of thermal ions to electromagnetic ion cyclotron waves. J. Geophys. Res. **99**(A10), 19413 (1994). doi:10.1029/94JA01235

M. André, C.M. Cully, Low-energy ions: a previously hidden solar system particle population. Geophys. Res. Lett. **39**(3) (2012). doi:10.1029/2011GL050242

M. André, K. Li, A.I. Eriksson, Outflow of low-energy ions and the solar cycle. J. Geophys. Res. Space Phys. (2015). doi:10.1002/2014JA020714

M. André, A. Vaivads, Y.V. Khotyaintsev, T. Laitinen, H. Nilsson, G. Stenberg, A. Fazakerley, J.G. Trotignon, Magnetic reconnection and cold plasma at the magnetopause. Geophys. Res. Lett. **37**(22) (2010). doi:10.1029/2010GL044611

M. André, P. Norqvist, L. Andersson, L. Eliasson, A.I. Eriksson, L. Blomberg, R.E. Erlandson, J. Waldemark, Ion energization mechanisms at 1700 km in the auroral region. J. Geophys. Res. **103**(A3), 4199 (1998). doi:10.1029/97JA00855. http://adsabs.harvard.edu/abs/1998JGR...103.4199A

N. André, J.F. Lemaire, Convective instabilities in the plasmasphere. J. Atmos. Sol.-Terr. Phys. **68**(2), 213–227 (2006). doi:10.1016/j.jastp.2005.10.013. http://www.sciencedirect.com/science/article/pii/S1364682605002932

V. Angelopoulos, W. Baumjohann, C.F. Kennel, F.V. Coroniti, M.G. Kivelson, R. Pellat, R.J. Walker, H. Lühr, G. Paschmann, Bursty bulk flows in the inner central plasma sheet. J. Geophys. Res. **97**(A4), 4027 (1992). doi:10.1029/91JA02701

V. Angelopoulos, C.F. Kennel, F.V. Coroniti, R. Pellat, H.E. Spence, M.G. Kivelson, R.J. Walker, W. Baumjohann, W.C. Feldman, J.T. Gosling, C.T. Russell, Characteristics of ion flow in the quiet state of the inner plasma sheet. Geophys. Res. Lett. **20**(16), 1711–1714 (1993). doi:10.1029/93GL00847

E.E. Antonova, The structure of the magnetospheric boundary layers and the magnetospheric turbulence. Planet. Space Sci. **53**, 161–168 (2005). doi:10.1016/j.pss.2004.09.041. http://www.sciencedirect.com/science/article/pii/S0032063304001813

R.L. Arnoldy, K.A. Lynch, P.M. Kintner, J. Vago, S. Chesney, T.E. Moore, C.J. Pollock, Bursts of transverse ion acceleration at rocket altitudes. Geophys. Res. Lett. **19**(4), 413–416 (1992). doi:10.1029/92GL00091

P.M. Banks, T.E. Holzer, High-latitude plasma transport: the polar wind. J. Geophys. Res. **74**(26), 6317–6332 (1969). doi:10.1029/JA074i026p06317

P.M. Banks, A.F. Nagy, W.I. Axford, Dynamical behavior of thermal protons in the mid-latitude ionosphere and magnetosphere. Planet. Space Sci. **19**(9), 1053–1067 (1971). doi:10.1016/0032-0633(71)90104-8. http://www.sciencedirect.com/science/article/pii/0032063371901048

W. Baumjohann, G. Paschmann, H. Lühr, Characteristics of high-speed ion flows in the plasma sheet. J. Geophys. Res. **95**(A4), 3801 (1990). doi:10.1029/JA095iA04p03801

M.B. Bavassano Cattaneo, M.F. Marcucci, Y.V. Bogdanova, H. Rème, I. Dandouras, L.M. Kistler, E. Lucek, Global reconnection topology as inferred from plasma observations inside Kelvin–Helmholtz vortices. Ann. Geophys. **28**(4), 893–906 (2010). doi:10.5194/angeo-28-893-2010. http://www.ann-geophys.net/28/893/2010/angeo-28-893-2010.html

J. Borovsky, M.F. Thomsen, R.C. Elphic, The driving of the plasma sheet by the solar wind. J. Geophys. Res. **103**(A8), 17617–17639 (1998)

J.E. Borovsky, MHD turbulence in the Earth's plasma sheet: dynamics, dissipation, and driving. J. Geophys. Res. **108**(A7), 1284 (2003). doi:10.1029/2002JA009625

J.E. Borovsky, M.H. Denton, A statistical look at plasmaspheric drainage plumes. J. Geophys. Res. **113**(A9), 09221 (2008). doi:10.1029/2007JA012994

J.E. Borovsky, J.T. Steinberg, The "calm before the storm" in CIR/magnetosphere interactions: occurrence statistics, solar wind statistics, and magnetospheric preconditioning. J. Geophys. Res. **111**(A7), 7–10 (2006). doi:10.1029/2005JA011397

J.E. Borovsky, R.C. Elphic, H.O. Funsten, M.F. Thomsen, *The Earth's Plasma Sheet as a Laboratory for Flow Turbulence in High-β MHD* (Cambridge University Press, Cambridge, 1997). doi:10.1017/S0022377896005259. http://journals.cambridge.org/abstract_S0022377896005259

J.E. Borovsky, M. Hesse, J. Birn, M.M. Kuznetsova, What determines the reconnection rate at the dayside magnetosphere? J. Geophys. Res. **113**(A7), 07210 (2008). doi:10.1029/2007JA012645

M. Bouhram, B. Klecker, W. Miyake, H. Rème, J.-A. Sauvaud, M. Malingre, L. Kistler, A. Blăgău, On the altitude dependence of transversely heated O⁺ distributions in the cusp/cleft. Ann. Geophys. **22**(5), 1787–1798 (2004). doi:10.5194/angeo-22-1787-2004. http://www.ann-geophys.net/22/1787/2004/angeo-22-1787-2004.html

O.J. Brambles, W. Lotko, P.a. Damiano, B. Zhang, M. Wiltberger, J. Lyon, Effects of causally driven cusp O+ outflow on the storm time magnetosphere-ionosphere system using a multifluid global simulation. J. Geophys. Res. **115**, 1–4 (2010). doi:10.1029/2010JA015469

O.J. Brambles, W. Lotko, B. Zhang, M. Wiltberger, J. Lyon, R.J. Strangeway, Magnetosphere sawtooth oscillations induced by ionospheric outflow. Science **332**(6034), 1183–1186 (2011). doi:10.1126/science.1202869

O.J. Brambles, W. Lotko, B. Zhang, J. Ouellette, J. Lyon, M. Wiltberger, The effects of ionospheric outflow on ICME and SIR driven sawtooth-events. J. Geophys. Res. Space Phys. **118**(10), 6026–6041 (2013). doi:10.1002/jgra.50522

W. Bristow, Statistics of velocity fluctuations observed by SuperDARN under steady interplanetary magnetic field conditions. J. Geophys. Res. Space Phys. **113**(A11), 11202 (2008). doi:10.1029/2008JA013203

M. Candidi, S. Orsini, V. Formisano, The properties of ionospheric O/+/ ions as observed in the magnetotail boundary layer and northern plasma lobe. J. Geophys. Res. **87**, 9097–9106 (1982)

H.C. Carlson, Accelerated polar rain electrons as the source of Sun-aligned arcs in the polar cap during northward interplanetary magnetic field conditions. J. Geophys. Res. **110**(A5), 05302 (2005). doi:10.1029/2004JA010669

D.L. Carpenter, Electron-density variations in the magnetosphere deduced from whistler data. J. Geophys. Res. **67**(9), 3345–3360 (1962). doi:10.1029/JZ067i009p03345

P.A. Cassak, M.A. Shay, Scaling of asymmetric magnetic reconnection: general theory and collisional simulations. Phys. Plasmas **14**(10), 102114 (2007). doi:10.1063/1.2795630. http://scitation.aip.org/content/aip/journal/pop/14/10/10.1063/1.2795630

C.R. Chappell, Initial observations of thermal plasma composition and energetics from dynamics explorer-1. Geophys. Res. Lett. **9**(9), 929–932 (1982). doi:10.1029/GL009i009p00929

C.R. Chappell, The role of the ionosphere in providing plasma to the terrestrial magnetosphere—an historical overview. Space Sci. Rev. (2015, submitted)

C.R. Chappell, T.E. Moore, J.H. Waite, The ionosphere as a fully adequate source of plasma for the Earth's magnetosphere. J. Geophys. Res. **92**(A6), 5896 (1987). doi:10.1029/JA092iA06p05896

C.R. Chappell, M.M. Huddleston, T.E. Moore, B.L. Giles, D.C. Delcourt, Observations of the warm plasma cloak and an explanation of its formation in the magnetosphere. J. Geophys. Res. **113**(A9), 09206 (2008). doi:10.1029/2007JA012945

C. Chaston, J. Bonnell, J.P. McFadden, C.W. Carlson, C. Cully, O. Le Contel, A. Roux, H.U. Auster, K.H. Glassmeier, V. Angelopoulos, C.T. Russell, Turbulent heating and cross-field transport near the magnetopause from THEMIS. Geophys. Res. Lett. **35**(17), L17S08 (2008). doi:10.1029/2008GL033601

L. Chen, Theory of plasma transport induced by low-frequency hydromagnetic waves. J. Geophys. Res. **104**(A2), 2421 (1999). doi:10.1029/1998JA900051

L. Chen, W. Li, J. Bortnik, R.M. Thorne, Amplification of whistler-mode hiss inside the plasmasphere. Geophys. Res. Lett. **39**(8) (2012). doi:10.1029/2012GL051488

S.P. Christon, M.I. Desai, T.E. Eastman, G. Gloeckler, S. Kokubun, A.T.Y. Lui, R.W. McEntire, E.C. Roelof, D.J. Williams, Low-charge-state heavy ions upstream of Earth's bow shock and sunward flux of ionospheric O+1, N+1, and O+2 ions: Geotail observations. Geophys. Res. Lett. **27**(16), 2433–2436 (2000). doi:10.1029/2000GL000039

B.M.A. Cooling, C.J. Owen, S.J. Schwartz, Role of the magnetosheath flow in determining the motion of open flux tubes. J. Geophys. Res. **106**(A9), 18763 (2001). doi:10.1029/2000JA000455

S.W.H. Cowley, Comments on the merging of non-antiparallel magnetic fields. J. Geophys. Res. **81**, 3455–3458 (1976)

S.W.H. Cowley, C.J. Owen, A simple illustrative model of open flux tube motion over the dayside magnetopause. Planet. Space Sci. **37**(11), 1461–1475 (1989). doi:10.1016/0032-0633(89)90116-5. http://www.sciencedirect.com/science/article/pii/0032063389901165

C.M. Cully, E. Donovan, A.W. Yau, G.G. Arkos, Akebono/Suprathermal mass spectrometer observations of low-energy ion outflow: dependence on magnetic activity and solar wind conditions. J. Geophys. Res. **108**(A2), 1093 (2003a). doi:10.1029/2001JA009200

C.M. Cully, E.F. Donovan, A.W. Yau, H.J. Opgenoorth, Supply of thermal ionospheric ions to the central plasma sheet. J. Geophys. Res. **108**(A2), 1092 (2003b). doi:10.1029/2002JA009457

I.A. Daglis, W.I. Axford, Fast ionospheric response to enhanced activity in geospace: ion feeding of the inner magnetotail. J. Geophys. Res. **101**(A), 5047–5066 (1996)

L.A. Daglis, S. Livi, E.T. Sarris, B. Wilken, Energy density of ionospheric and solar wind origin ions in the near-Earth magnetotail during substorms. J. Geophys. Res. **99**, 5691–5703 (1994)

I. Dandouras, Detection of a plasmaspheric wind in the Earth's magnetosphere by the Cluster spacecraft. Ann. Geophys. **31**(7), 1143–1153 (2013). doi:10.5194/angeo-31-1143-2013. http://www.ann-geophys.net/31/1143/2013/angeo-31-1143-2013.html

I. Dandouras, V. Pierrard, J. Goldstein, C. Vallat, G.K. Parks, H. Rème, C. Gouillart, F. Sevestre, M. Mccarthy, L.M. Kistler, B. Klecker, A. Korth, P. Bavassano-Cattaneo, M.B. Escoubet, A. Masson, No TitleMultipoint observations of ionic structures in the plasmasphere by CLUSTER'CIS and comparisons with IMAGE-EUV observations and with model simulations, in *Inner Magnetosphere Interactions: New Perspectives from Imaging*, ed. by J. Burch, M. Schulz, H. Spence (Am. Geophys. Union, Washington, 2005). doi:10.1029/159GM03

F. Darrouzet, V. Pierrard, S. Benck, G. Lointier, J. Cabrera, K. Borremans, N.Y. Ganushkina, J.D. Keyser, Links between the plasmapause and the radiation belt boundaries as observed by the instruments CIS, RAPID, and WHISPER onboard Cluster. J. Geophys. Res. Space Phys. **118**(7), 4176–4188 (2013). doi:10.1002/jgra.50239

F. Darrouzet, D.L. Gallagher, N. André, D.L. Carpenter, I. Dandouras, P.M. Décréau, J.D. Keyser, R.E. Denton, J.C. Foster, J. Goldstein, M.B. Moldwin, B.W. Reinisch, B.R. Sandel, J. Tu, Plasmaspheric density structures and dynamics: properties observed by the CLUSTER and IMAGE missions, in *The Earth's Plasmasphere*, ed. by F. Darrouzet, J. De Keyser, V. Pierrard (Springer, New York, 2009), pp. 55–106. 978-1-4419-1322-7. doi:10.1007/978-1-4419-1323-4. http://link.springer.com/10.1007/978-1-4419-1323-4

D.C. Delcourt, J.A. Sauvaud, A. Pedersen, Dynamics of single-particle orbits during substorm expansion phase. J. Geophys. Res. **95**(A12), 20853 (1990). doi:10.1029/JA095iA12p20853

D.C. Delcourt, C.R. Chappell, T.E. Moore, J.H. Waite, A three-dimensional numerical model of ionospheric plasma in the magnetosphere. J. Geophys. Res. **94**(A9), 11893 (1989). doi:10.1029/JA094iA09p11893. http://adsabs.harvard.edu/abs/1989JGR....9411893D

D.C. Delcourt, T.E. Moore, J.A. Sauvaud, C.R. Chappell, Nonadiabatic transport features in the outer cusp region. J. Geophys. Res. **97**, 16833 (1992)

D.C. Delcourt, Particle acceleration by inductive electric fields in the inner magnetosphere. J. Atmos. Sol.-Terr. Phys. **64**(5-6), 551–559 (2002). doi:10.1016/S1364-6826(02)00012-3. http://www.sciencedirect.com/science/article/pii/S1364682602000123

E. Drakou, A.W. Yau, T. Abe, Ion temperature measurements from the Akebono suprathermal mass spectrometer: application to the polar wind. J. Geophys. Res. **102**(A8), 17523 (1997). doi:10.1029/97JA00099

J.W. Dungey, The steady state of the Chapman–Ferraro problem in two dimensions. J. Geophys. Res. **66**(4), 1043–1047 (1961). doi:10.1029/JZ066i004p01043. http://adsabs.harvard.edu/abs/1961JGR....66.1043D

J.P. Eastwood, S.A. Kiehas, in *Magnetotails in the Solar System*, ed. by A. Keiling, C.M. Jackman, P. Delmare (Wiley, Hoboken, 2015), pp. 269–287. Chap. 16

Y. Ebihara, M. Ejiri, Modeling of solar wind control of the ring current buildup: a case study of the magnetic storms in April 1997. Geophys. Res. Lett. **25**(20), 3751–3754 (1998). doi:10.1029/1998GL900006

R.C. Elphic, M.F. Thomsen, J.E. Borovsky, The fate of the outer plasmasphere. Geophys. Res. Lett. **24**(4), 365–368 (1997). doi:10.1029/97GL00141

E. Engwall, A.I. Eriksson, M. André, I. Dandouras, G. Paschmann, J. Quinn, K. Torkar, Low-energy (order 10 eV) ion flow in the magnetotail lobes inferred from spacecraft wake observations. Geophys. Res. Lett. **33**(6), 06110 (2006). doi:10.1029/2005GL025179

E. Engwall, A.I. Eriksson, C.M. Cully, M. André, R. Torbert, H. Vaith, *Earth's Ionospheric Outflow Dominated by Hidden Cold Plasma* (2009a). doi:10.1038/ngeo387. http://dx.doi.org/10.1038/ngeo387

E. Engwall, A.I. Eriksson, C.M. Cully, M. André, P.A. Puhl-Quinn, H. Vaith, R. Torbert, Survey of cold ionospheric outflows in the magnetotail. Ann. Geophys. **27**(8), 3185–3201 (2009b). doi:10.5194/angeo-27-3185-2009. http://www.ann-geophys.net/27/3185/2009/

J. Etcheto, A. Saint-Marc, Anomalously high plasma densities in the plasma sheet boundary layer. J. Geophys. Res. **90**(A6), 5338 (1985). doi:10.1029/JA090iA06p05338

M. Faganello, F. Califano, F. Pegoraro, T. Andreussi, S. Benkadda, Magnetic reconnection and Kelvin–Helmholtz instabilities at the Earth's magnetopause. Plasma Phys. Control. Fusion **54**(12), 124037 (2012). doi:10.1088/0741-3335/54/12/124037. http://stacks.iop.org/0741-3335/54/i=12/a=124037

M. Faganello, F. Califano, F. Pegoraro, A. Retinò, Kelvin–Helmholtz vortices and double mid-latitude reconnection at the Earth's magnetopause: comparison between observations and simulations. Europhys. Lett. **107**(1), 19001 (2014). doi:10.1209/0295-5075/107/19001. http://stacks.iop.org/0295-5075/107/i=1/a=19001

D.H. Fairfield, On the average configuration of the geomagnetic tail. J. Geophys. Res. **84**(A5), 1950 (1979). doi:10.1029/JA084iA05p01950

R.C. Fear, S.E. Milan, R. Maggiolo, A.N. Fazakerley, I. Dandouras, S.B. Mende, Direct observation of closed magnetic flux trapped in the high-latitude magnetosphere. Science **346**(6216), 1506–1510 (2014). doi:10.1126/science.1257377. http://www.sciencemag.org/content/346/6216/1506.short

Y.I. Feldstein, A.E. Levitin, S.A. Golyshev, L.A. Dremukhina, U.B. Vestchezerova, T.E. Valchuk, A. Grafe, Ring current and auroral electrojets in connection with interplanetary medium parameters during magnetic storm. Ann. Geophys. **12**(7), 602–611 (1994). doi:10.1007/s00585-994-0602-6. http://www.ann-geophys.net/12/602/1994/

M.-C. Fok, J.U. Kozyra, A.F. Nagy, C.E. Rasmussen, G.V. Khazanov, Decay of equatorial ring current ions and associated aeronomical consequences. J. Geophys. Res. **98**(A11), 19381 (1993). doi:10.1029/93JA01848

M.-C. Fok, P.D. Craven, T.E. Moore, P.G. Richards, Ring current-plasmasphere coupling through Coulomb collisions, in *Cross-Scale Coupling in Space Plasmas*, ed. by J.L. Horwitz, N. Singh, J.L. Burch (Am. Geophys. Union, Washington, 1995). doi:10.1029/GM093p0161

M. Förster, G. Paschmann, S.E. Haaland, J.M. Quinn, R.B. Torbert, H. Vaith, C.A. Kletzing, High-latitude plasma convection from Cluster EDI: variances and solar wind correlations. Ann. Geophys. **25**(7), 1691–1707 (2007). https://hal.archives-ouvertes.fr/hal-00318357/

C. Foster, M. Lester, J.A. Davies, A statistical study of diurnal, seasonal and solar cycle variations of F-region and topside auroral upflows observed by EISCAT between 1984 and 1996. Ann. Geophys. **16**(10), 1144–1158 (1998). doi:10.1007/s00585-998-1144-0. http://www.ann-geophys.net/16/1144/1998/

J.C. Foster, Stormtime observations of the flux of plasmaspheric ions to the dayside cusp/magnetopause. Geophys. Res. Lett. **31**(8), 08809 (2004). doi:10.1029/2004GL020082

J.C. Foster, Multiradar observations of the polar tongue of ionization. J. Geophys. Res. **110**(A9), 9–31 (2005). doi:10.1029/2004JA010928

J.C. Foster, P.J. Erickson, D.N. Baker, S.G. Claudepierre, C.A. Kletzing, W. Kurth, G.D. Reeves, S.A. Thaller, H.E. Spence, Y.Y. Shprits, J.R. Wygant, Prompt energization of relativistic and highly relativistic electrons during a substorm interval: Van Allen Probes observations. Geophys. Res. Lett. **41**(1), 20–25 (2014a). doi:10.1002/2013GL058438

J.C. Foster, P.J. Erickson, A.J. Coster, S. Thaller, J. Tao, J.R. Wygant, J.W. Bonnell, Storm time observations of plasmasphere erosion flux in the magnetosphere and ionosphere. Geophys. Res. Lett. **41**(3), 762–768 (2014b). doi:10.1002/2013GL059124

J.C. Foster, Storm time plasma transport at middle and high latitudes. J. Geophys. Res. **98**(A2), 1675 (1993). doi:10.1029/92JA02032

C. Foullon, C.J. Farrugia, A.N. Fazakerley, C.J. Owen, F.T. Gratton, R.B. Torbert, Evolution of Kelvin–Helmholtz activity on the dusk flank magnetopause. J. Geophys. Res. **113**(A11), 11203 (2008). doi:10.1029/2008JA013175

H.U. Frey, Localized aurora beyond the auroral oval. Rev. Geophys. **45**(1), 1003 (2007). doi:10.1029/2005RG000174

T.A. Fritz, M. Alothman, J. Bhattacharjya, D.L. Matthews, J. Chen, Butterfly pitch-angle distributions observed by ISEE-1. Planet. Space Sci. **51**(3), 205–219 (2003). doi:10.1016/S0032-0633(02)00202-7. http://www.sciencedirect.com/science/article/pii/S0032063302002027

S.A. Fuselier, W.S. Lewis, Properties of near-Earth magnetic reconnection from in-situ observations. Space Sci. Rev. **160**(1-4), 95–121 (2011). doi:10.1007/s11214-011-9820-x. http://link.springer.com/10.1007/s11214-011-9820-x

S.A. Fuselier, B.J. Anderson, T.G. Onsager, Particle signatures of magnetic topology at the magnetopause: AMPTE/CCE observations. J. Geophys. Res. **100**(A7), 11805 (1995). doi:10.1029/94JA02811

S.A. Fuselier, B.J. Anderson, T.G. Onsager, Electron and ion signatures of field line topology at the low-shear magnetopause. J. Geophys. Res. **102**(A3), 4847 (1997). doi:10.1029/96JA03635

S.A. Fuselier, K.J. Trattner, S.M. Petrinec, Antiparallel and component reconnection at the dayside magnetopause. J. Geophys. Res. **116**(A10), 10227 (2011). doi:10.1029/2011JA016888

S.A. Fuselier, K.J. Trattner, S.M. Petrinec, B. Lavraud, Dayside magnetic topology at the Earth's magnetopause for northward IMF. J. Geophys. Res. **117**(A8), 08235 (2012). doi:10.1029/2012JA017852

K.S. Garcia, V.G. Merkin, W.J. Hughes, Effects of nightside O+ outflow on magnetospheric dynamics: results of multifluid MHD modeling. J. Geophys. Res. **115**(May), 1–9 (2010). doi:10.1029/2010JA015730

N.G.J. Gazey, M. Lockwood, M. Grande, C.H. Perry, P.N. Smith, S. Coles, A.D. Aylward, R.J. Bunting, H. Opgenoorth, B. Wilken, EISCAT/CRRES observations: nightside ionospheric ion outflow and oxygen-rich substorm injections. Ann. Geophys. **14**(1), 1032–1043 (1996)

A. Glocer, G. Tóth, T. Gombosi, D. Welling, Modeling ionospheric outflows and their impact on the magnetosphere, initial results. J. Geophys. Res. **114**(A5), 1–16 (2009a). doi:10.1029/2009JA014053

a. Glocer, G. Tóth, Y. Ma, T. Gombosi, J.-C. Zhang, L.M. Kistler, Multifluid block-adaptive-tree solar wind roe-type upwind scheme: magnetospheric composition and dynamics during geomagnetic storms initial results. J. Geophys. Res. **114**(A12), 12203 (2009b). doi:10.1029/2009JA014418

G. Gloeckler, D.C. Hamilton, AMPTE ion composition results. Phys. Scr. T **18**, 73–84 (1987)

J. Goldstein, Control of plasmaspheric dynamics by both convection and sub-auroral polarization stream. Geophys. Res. Lett. **30**(24), 2243 (2003). doi:10.1029/2003GL018390

W.D. Gonzalez, B.T. Tsurutani, A.L.C. Gonzalez, E.J. Smith, F. Tang, S.-I. Akasofu, Solar wind-magnetosphere coupling during intense magnetic storms (1978–1979). J. Geophys. Res. **94**(A7), 8835 (1989). doi:10.1029/JA094iA07p08835

J.T. Gosling, D.N. Baker, S.J. Bame, E.W. Hones, D.J. McComas, R.D. Zwickl, J.A. Slavin, E.J. Smith, B.T. Tsurutani, Plasma entry into the distant tail lobes: ISEE-3. Geophys. Res. Lett. **11**(10), 1078–1081 (1984). doi:10.1029/GL011i010p01078

M.E. Greenspan, D.C. Hamilton, Relative contributions of H+ and O+ to the ring current energy near magnetic storm maximum. J. Geophys. Res. Space Phys. **107**(A), 1043 (2002)

R.S. Grew, F.W. Menk, M.A. Clilverd, B.R. Sandel, Mass and electron densities in the inner magnetosphere during a prolonged disturbed interval. Geophys. Res. Lett. **34**(2), 02108 (2007). doi:10.1029/2006GL028254

S. Haaland, B. Lybekk, K. Svenes, A. Pedersen, M. Förster, H. Vaith, R. Torbert, Plasma transport in the magnetotail lobes. Ann. Geophys. **27**(9), 3577–3590 (2009). doi:10.5194/angeo-27-3577-2009. http://www.ann-geophys.net/27/3577/2009/angeo-27-3577-2009.html

S. Haaland, A. Eriksson, E. Engwall, B. Lybekk, H. Nilsson, A. Pedersen, K. Svenes, M. André, M. Förster, K. Li, C. Johnsen, N. Østgaard, Estimating the capture and loss of cold plasma from ionospheric outflow. J. Geophys. Res. **117**(A7), 07311 (2012). doi:10.1029/2012JA017679

G. Haerendel, G. Paschmann, Entry of solar wind plasma into the magnetosphere, in *Physics of the Hot Plasma in the Magnetosphere*, ed. by B. Hultqvist, L. Stenflo (Springer, New York, 1975), pp. 23–43. doi:10.1007/978-1-4613-4437-7_2

D.C. Hamilton, G. Gloeckler, F.M. Ipavich, W. Stüdemann, B. Wilken, G. Kremser, Ring current development during the great geomagnetic storm of February 1986. J. Geophys. Res. **93**(A12), 14343 (1988). doi:10.1029/JA093iA12p14343

Y. Harada, S. Machida, J.S. Halekas, A.R. Poppe, J.P. McFadden, ARTEMIS observations of lunar dayside plasma in the terrestrial magnetotail lobe. J. Geophys. Res. Space Phys. **118**(6), 3042–3054 (2013). doi:10.1002/jgra.50296

D.A. Hardy, J.W. Freeman, H.K. Hills, Plasma observations in the magnetotail, in *Magnetospheric Particles and Fields*, ed. by B.M. McCormac, Hingham, Mass., USA (1976), p. 89

D.A. Hardy, H.K. Hills, J.W. Freeman, A new plasma regime in the distant geomagnetic tail. Geophys. Res. Lett. **2**(5), 169–172 (1975). doi:10.1029/GL002i005p00169

D.A. Hardy, H.K. Hills, J.W. Freeman, Occurrence of the lobe plasma at lunar distance. J. Geophys. Res. **84**(A1), 72 (1979). doi:10.1029/JA084iA01p00072

C. Harvey, J. Etcheto, Y. Javel, R. Manning, M. Petit, The ISEE electron density experiment. IEEE Trans. Geosci. Electron. **16**(3), 231–238 (1978). doi:10.1109/TGE.1978.294553. http://ieeexplore.ieee.org/lpdocs/epic03/wrapper.htm?arnumber=4071924

A. Hasegawa, L. Chen, Kinetic process of plasma heating due to Alfvén wave excitation. Phys. Rev. Lett. **35**(6), 370–373 (1975). doi:10.1103/PhysRevLett.35.370. http://link.aps.org/doi/10.1103/PhysRevLett.35.370

A. Hasegawa, K. Mima, Anomalous transport produced by kinetic Alfvén wave turbulence. J. Geophys. Res. **83**(A3), 1117 (1978). doi:10.1029/JA083iA03p01117

H. Hasegawa, M. Fujimoto, T.-D. Phan, H. Rème, A. Balogh, M.W. Dunlop, C. Hashimoto, R. Tandokoro, Transport of solar wind into Earth's magnetosphere through rolled-up Kelvin–Helmholtz vortices. Nature **430**(7001), 755–758 (2004). doi:10.1038/nature02799. http://dx.doi.org/10.1038/nature02799

H. Hasegawa, M. Fujimoto, K. Takagi, Y. Saito, T. Mukai, H. Rème, Single-spacecraft detection of rolled-up Kelvin–Helmholtz vortices at the flank magnetopause. J. Geophys. Res. **111**(A9), 09203 (2006). doi:10.1029/2006JA011728

H. Hasegawa, A. Retinò, A. Vaivads, Y. Khotyaintsev, M. André, T.K.M. Nakamura, W.-L. Teh, B.U.O. Sonnerup, S.J. Schwartz, Y. Seki, M. Fujimoto, Y. Saito, H. Rème, P. Canu, Kelvin–Helmholtz waves at the Earth's magnetopause: multiscale development and associated reconnection. J. Geophys. Res. **114**(A12), 12207 (2009). doi:10.1029/2009JA014042. http://adsabs.harvard.edu/abs/2009JGRA..11412207H

R.A. Heelis, J.D. Winningham, M. Sugiura, N.C. Maynard, Particle acceleration parallel and perpendicular to the magnetic field observed by DE-2. J. Geophys. Res. **89**(A6), 3893 (1984). doi:10.1029/JA089iA06p03893

W.J. Heikkila, R.J. Pellinen, Localized induced electric field within the magnetotail. J. Geophys. Res. **82**(10), 1610–1614 (1977). doi:10.1029/JA082i010p01610

M.G. Henderson, The may 2–3, 1986 CDAW-9C interval: a sawtooth event. Geophys. Res. Lett. **31**(11), 11804 (2004). doi:10.1029/2004GL019941

M. Hirahara, Periodic emergence of multicomposition cold ions modulated by geomagnetic field line oscillations in the near-Earth magnetosphere. J. Geophys. Res. **109**(A3), 03211 (2004). doi:10.1029/2003JA010141

M. Hirahara, M. Nakamura, T. Terasawa, T. Mukai, Y. Saito, T. Yamamoto, A. Nishida, S. Machida, S. Kokubun, Acceleration and heating of cold ion beams in the plasma sheet boundary layer observed with GEOTAIL. Geophys. Res. Lett. **21**, 3003–3006 (1994) (ISSN 0094-8276)

M. Hirahara, T. Terasawa, T. Mukai, M. Hoshino, Y. Saito, S. Machida, T. Yamamoto, S. Kokubun, Cold ion streams consisting of double proton populations and singly charged oxygen observed at the distant magnetopause by Geotail: a case study. J. Geophys. Res. **102**(A), 2359–2372 (1997)

E.W. Hones, J.R. Asbridge, S.J. Bame, M.D. Montgomery, S. Singer, S.-I. Akasofu, Measurements of magnetotail plasma flow made with Vela 4B. J. Geophys. Res. **77**(28), 5503–5522 (1972). doi:10.1029/JA077i028p05503

R.B. Horne, R.M. Thorne, Potential waves for relativistic electron scattering and stochastic acceleration during magnetic storms. Geophys. Res. Lett. **25**(15), 3011–3014 (1998). doi:10.1029/98GL01002

J.L. Horwitz, The tail lobe ion spectrometer. J. Geophys. Res. **91**, 5689–5699 (1986)

C.-S. Huang, Periodic magnetospheric substorms and their relationship with solar wind variations. J. Geophys. Res. **108**(A6), 1255 (2003). doi:10.1029/2002JA009704

M.M. Huddleston, C.R. Chappell, D.C. Delcourt, T.E. Moore, B.L. Giles, M.O. Chandler, An examination of the process and magnitude of ionospheric plasma supply to the magnetosphere. J. Geophys. Res. Space Phys. **110** (2005). doi:10.1029/2004JA010401

B. Hultqvist, On the origin of the hot ions in the disturbed dayside magnetosphere. Planet. Space Sci. **31**(2), 173–184 (1983). doi:10.1016/0032-0633(83)90052-1. http://www.sciencedirect.com/science/article/pii/0032063383900521

B. Hultqvist, M. Øieroset, G. Paschmann, R.A. Treumann (eds.), *Magnetospheric Plasma Sources and Losses*. Space Sciences Series of ISSI, vol. 6 (Springer, Dordrecht, 1999). 978-94-010-5918-3. doi:10.1007/978-94-011-4477-3. http://www.springerlink.com/index/10.1007/978-94-011-4477-3

K.-J. Hwang, M.M. Kuznetsova, F. Sahraoui, M.L. Goldstein, E. Lee, G.K. Parks, Kelvin–Helmholtz waves under southward interplanetary magnetic field. J. Geophys. Res. **116**(A8), 08210 (2011). doi:10.1029/2011JA016596

K.-J. Hwang, M.L. Goldstein, M.M. Kuznetsova, Y. Wang, A.F. Viñas, D.G. Sibeck, The first in situ observation of Kelvin–Helmholtz waves at high-latitude magnetopause during strongly dawnward interplanetary magnetic field conditions. J. Geophys. Res. **117**(A8), 08233 (2012). doi:10.1029/2011JA017256

R. Ilie, R.M. Skoug, H.O. Funsten, M.W. Liemohn, J.J. Bailey, M. Gruntman, The impact of geocoronal density on ring current development. J. Atmos. Sol.-Terr. Phys. **99**, 92–103 (2013). doi:10.1016/j.jastp.2012.03.010. http://www.sciencedirect.com/science/article/pii/S1364682612000946

S.M. Imber, J.A. Slavin, H.U. Auster, V. Angelopoulos, A THEMIS survey of flux ropes and traveling compression regions: location of the near-Earth reconnection site during solar minimum. J. Geophys. Res. **116**(A2), 02201 (2011). doi:10.1029/2010JA016026

F.M. Ipavich, A.B. Galvin, G. Gloeckler, D. Hovestadt, B. Klecker, M. Scholer, Energetic (>100 keV) O+ ions in the plasma sheet. Geophys. Res. Lett. **11**(5), 504–507 (1984). doi:10.1029/GL011i005p00504

J.R. Johnson, C.Z. Cheng, Kinetic Alfvén waves and plasma transport at the magnetopause. Geophys. Res. Lett. **24**(11), 1423–1426 (1997). doi:10.1029/97GL01333

J.R. Johnson, C.Z. Cheng, Stochastic ion heating at the magnetopause due to kinetic Alfvén waves. Geophys. Res. Lett. **28**(23), 4421–4424 (2001). doi:10.1029/2001GL013509
J.R. Johnson, S. Wing, Northward interplanetary magnetic field plasma sheet entropies. J. Geophys. Res. Space Phys. **114** (2009). doi:10.1029/2008JA014017
V.K. Jordanova, L.M. Kistler, J.U. Kozyra, G.V. Khazanov, A.F. Nagy, Collisional losses of ring current ions. J. Geophys. Res. **101**(A1), 111 (1996). doi:10.1029/95JA02000
V.K. Jordanova, C.J. Farrugia, J.M. Quinn, R.M. Thorne, K.E. Ogilvie, R.P. Lepping, G. Lu, A.J. Lazarus, M.F. Thomsen, R.D. Belian, Effect of wave-particle interactions on ring current evolution for January 10-11, 1997: initial results. Geophys. Res. Lett. **25**(15), 2971–2974 (1998). doi:10.1029/98GL00649
V.K. Jordanova, R.B. Torbert, R.M. Thorne, H.L. Collin, J.L. Roeder, J.C. Foster, Ring current activity during the early B z < 0 phase of the January 1997 magnetic cloud. J. Geophys. Res. **104**(A11), 24895 (1999). doi:10.1029/1999JA900339
V.K. Jordanova, L.M. Kistler, C.J. Farrugia, R.B. Torbert, Effects of inner magnetospheric convection on ring current dynamics: March 10–12, 1998. J. Geophys. Res. **106**(A), 29705–29720 (2001)
V.K. Jordanova, L.M. Kistler, M.F. Thomsen, C.G. Mouikis, Effects of plasma sheet variability on the fast initial ring current decay. Geophys. Res. Lett. **30**(6), 41–44 (2003)
V.K. Jordanova, R.M. Thorne, W. Li, Y. Miyoshi, Excitation of whistler mode chorus from global ring current simulations. J. Geophys. Res. **115**, 1–10 (2010). doi:10.1029/2009JA014810
Z. Kaymaz, G.L. Siscoe, N.A. Tsyganenko, R.P. Lepping, Magnetotail views at 33 R E : IMP 8 magnetometer observations. J. Geophys. Res. **99**(A5), 8705 (1994). doi:10.1029/93JA03564
K. Keika, Outflow of energetic ions from the magnetosphere and its contribution to the decay of the storm time ring current. J. Geophys. Res. **110**(A9), 09210 (2005). doi:10.1029/2004JA010970
K. Keika, L.M. Kistler, P.C. Brandt, Energization of O+ ions in the Earth's inner magnetosphere and the effects on ring current buildup: a review of previous observations and possible mechanisms. J. Geophys. Res. Space Phys. **118**(7), 4441–4464 (2013)
L.M. Kistler, F.M. Ipavich, D.C. Hamilton, G. Gloeckler, B. Wilken, Energy spectra of the major ion species in the ring current during geomagnetic storms. J. Geophys. Res. **94**, 3579–3599 (1989)
L.M. Kistler, E. Möbius, D.M. Klumpar, M.A. Popecki, L. Tang, V. Jordanova, B. Klecker, W.K. Peterson, E.G. Shelley, D. Hovestadt, M. Temerin, R.E. Ergun, J.P. McFadden, C.W. Carlson, F.S. Mozer, R.C. Elphic, R.J. Strangeway, C.A. Cattell, R.F. Pfaff, FAST/TEAMS observations of charge exchange signatures in ions mirroring at low altitudes. Geophys. Res. Lett. **25**(1), 2085–2088 (1998)
L.M. Kistler, B. Klecker, V.K. Jordanova, E. Möbius, M.A. Popecki, D. Patel, J.A. Sauvaud, H. Rème, A.M. Di Lellis, A. Korth, M. McCarthy, R. Cerulli, M.B. Bavassano-Cattaneo, L. Eliasson, C.W. Carlson, G.K. Parks, G. Paschmann, W. Baumjohann, G. Haerendel, Testing electric field models using ring current ion energy spectra from the Equator-S ion composition (ESIC) instrument. Ann. Geophys. **17**(1), 1611–1621 (1999)
L.M. Kistler, C. Mouikis, E. Möbius, B. Klecker, J.A. Sauvaud, H. Réme, A. Korth, M.F. Marcucci, R. Lundin, G.K. Parks, A. Balogh, Contribution of nonadiabatic ions to the cross-tail current in an O+ dominated thin current sheet. J. Geophys. Res. Space Phys. **110**(A6), 06213 (2005). doi:10.1029/2004JA010653
L.M. Kistler, C.G. Mouikis, B. Klecker, I. Dandouras, Cusp as a source for oxygen in the plasma sheet during geomagnetic storms. J. Geophys. Res. **115**(A3), 03209 (2010a). doi:10.1029/2009JA014838
L.M. Kistler, a.B. Galvin, M.a. Popecki, K.D.C. Simunac, C. Farrugia, E. Moebius, M.a. Lee, L.M. Blush, P. Bochsler, P. Wurz, B. Klecker, R.F. Wimmer-Schweingruber, A. Opitz, J.-a. Sauvaud, B. Thompson, C.T. Russell, Escape of O+ through the distant tail plasma sheet. Geophys. Res. Lett. **37**(21) (2010b). doi:10.1029/2010GL045075
D.M. Klumpar, Transversely accelerated ions: an ionospheric source of hot magnetospheric ions. J. Geophys. Res. **84**(A8), 4229 (1979). doi:10.1029/JA084iA08p04229
D.M. Klumpar, W.K. Peterson, E.G. Shelley, Direct evidence for two-stage (bimodal) acceleration of ionospheric ions. J. Geophys. Res. **89**(A12), 10779 (1984). doi:10.1029/JA089iA12p10779
G.A. Kotova, The Earth's plasmasphere: state of studies (a review). Geomagn. Aeron. **47**(4), 409–422 (2007). doi:10.1134/S0016793207040019. http://link.springer.com/10.1134/S0016793207040019
J.U. Kozyra, Multistep Dst development and ring current composition changes during the 4–6 June 1991 magnetic storm. J. Geophys. Res. **107**(A8), 1224 (2002). doi:10.1029/2001JA000023
J.U. Kozyra, A.F. Nagy, D.W. Slater, High-altitude energy source(s) for stable auroral red arcs. Rev. Geophys. **35**(2), 155 (1997). doi:10.1029/96RG03194
J.U. Kozyra, M.-C. Fok, E.R. Sanchez, D.S. Evans, D.C. Hamilton, A.F. Nagy, The role of precipitation losses in producing the rapid early recovery phase of the Great Magnetic Storm of February 1986. J. Geophys. Res. **103**(A4), 6801 (1998). doi:10.1029/97JA03330
J.U. Kozyra, M.W. Liemohn, Ring current energy input and decay, in *Magnetospheric Imaging—the Image Prime Mission*, ed. by J.L. Burch (Springer, Berlin, 2003), pp. 105–131. doi:10.1007/978-94-010-0027-7_6

S.M. Krimigis, R.W. McEntire, T.A. Potemra, G. Gloeckler, F.L. Scarf, E.G. Shelley, A synthesis of ring current spectra and energy densities measured with AMPTE/CCE. Geophys. Res. Lett. **12**, 329–332 (1985). Magnetic storm of September 4, 1984 (ISSN 0094-8276)
E.A. Kronberg, M. Ashour-Abdalla, I. Dandouras, D.C. Delcourt, E.E. Grigorenko, L.M. Kistler, I.V. Kuzichev, J. Liao, R. Maggiolo, H.V. Malova, K.G. Orlova, V. Peroomian, D.R. Shklyar, Y.Y. Shprits, D.T. Welling, L.M. Zelenyi, Circulation of heavy ions and their dynamical effects in the magnetosphere: recent observations and models. Space Sci. Rev. **184**(1-4), 173–235 (2014). doi:10.1007/s11214-014-0104-0. http://link.springer.com/10.1007/s11214-014-0104-0
A. Kullen, Solar wind dependence of the occurrence and motion of polar auroral arcs: a statistical study. J. Geophys. Res. **107**(A11), 1362 (2002). doi:10.1029/2002JA009245
J. Labelle, R.A. Treumann, Plasma waves at the dayside magnetopause. Space Sci. Rev. **47**(1–2) (1988). doi:10.1007/BF00223240. http://link.springer.com/10.1007/BF00223240
B. Lavraud, Characteristics of the magnetosheath electron boundary layer under northward interplanetary magnetic field: implications for high-latitude reconnection. J. Geophys. Res. **110**(A6), 06209 (2005). doi:10.1029/2004JA010808
B. Lavraud, M.F. Thomsen, B. Lefebvre, S.J. Schwartz, K. Seki, T.D. Phan, Y.L. Wang, A. Fazakerley, H. Rème, A. Balogh, Evidence for newly closed magnetosheath field lines at the dayside magnetopause under northward IMF. J. Geophys. Res. **111**(A5), 05211 (2006). doi:10.1029/2005JA011266
J.H. Lee, V. Angelopoulos, On the presence and properties of cold ions near Earth's equatorial magnetosphere. J. Geophys. Res. Space Phys. **119**(3), 1749–1770 (2014). doi:10.1002/2013JA019305
L.C. Lee, J.R. Johnson, Z.W. Ma, Kinetic Alfvén waves as a source of plasma transport at the dayside magnetopause. J. Geophys. Res. **99**(A9), 17405 (1994). doi:10.1029/94JA01095
J. Lemaire, M. Scherer, Model of the polar ion-exosphere. Planet. Space Sci. **18**(1), 103–120 (1970). doi:10.1016/0032-0633(70)90070-X. http://www.sciencedirect.com/science/article/pii/003206337090070X
J. Lemaire, R.W. Schunk, Plasmaspheric wind. J. Atmos. Terr. Phys. **54**(3-4), 467–477 (1992). doi:10.1016/0021-9169(92)90026-H. http://www.sciencedirect.com/science/article/pii/002191699290026H
J.F. Lemaire, K.I. Gringauz, *The Earth's Plasmasphere* (Cambridge University Press, Cambridge, 1998), p. 376. 0521675553. http://books.google.com/books?hl=en&lr=&id=xdnWRdPEvdQC&pgis=1
J.F. Lemaire, The formation plasmaspheric tails. Phys. Chem. Earth, Part C, Sol.-Terr. Planet. Sci. **25**(1-2), 9–17 (2000). doi:10.1016/S1464-1917(99)00026-4. http://www.sciencedirect.com/science/article/pii/S1464191799000264
J.F. Lemaire, The formation of the light-ion trough and peeling off the plasmasphere. J. Atmos. Sol.-Terr. Phys. **63**(11), 1285–1291 (2001). doi:10.1016/S1364-6826(00)00232-7. http://www.sciencedirect.com/science/article/pii/S1364682600002327
K. Li, S. Haaland, A. Eriksson, M. André, E. Engwall, Y. Wei, E.A. Kronberg, M. Fränz, P.W. Daly, H. Zhao, Q.Y. Ren, On the ionospheric source region of cold ion outflow. Geophys. Res. Lett. **39**(18) (2012). doi:10.1029/2012GL053297
W. Li, Plasma sheet formation during long period of northward IMF. Geophys. Res. Lett. **32**(12), L12S08 (2005). doi:10.1029/2004GL021524
J. Liao, L.M. Kistler, C.G. Mouikis, B. Klecker, I. Dandouras, J.-C. Zhang, Statistical study of O+ transport from the cusp to the lobes with Cluster CODIF data. J. Geophys. Res. **115**, 1–15 (2010). doi:10.1029/2010JA015613
J. Liao, L.M. Kistler, C.G. Mouikis, B. Klecker, I. Dandouras, Solar cycle dependence of the cusp O+ access to the near-Earth magnetotail. J. Geophys. Res. **117**(A10), 10220 (2012). doi:10.1029/2012JA017819
J. Liao, X. Cai, L.M. Kistler, C.R. Clauer, C.G. Mouikis, B. Klecker, I. Dandouras, The relationship between sawtooth events and O+ in the plasma sheet. J. Geophys. Res. Space Phys. **119**(3), 1572–1586 (2014). doi:10.1002/2013JA019084
J. Liao, L.M. Kistler, C.G. Mouikis, B. Klecker, I. Dandouras, Acceleration of O+ from the cusp to the plasma sheet. J. Geophys. Res. Space Phys. (2015)
M.W. Liemohn, Occurrence statistics of cold, streaming ions in the near-Earth magnetotail: survey of polar-TIDE observations. J. Geophys. Res. **110**(A7), 07211 (2005). doi:10.1029/2004JA010801
M.W. Liemohn, G.V. Khazanov, J.U. Kozyra, Guided plasmaspheric hiss interactions with superthermal electrons: 1. Resonance curves and timescales. J. Geophys. Res. **102**(A6), 11619 (1997). doi:10.1029/97JA00825
M.W. Liemohn, J.U. Kozyra, V.K. Jordanova, G.V. Khazanov, M.F. Thomsen, T.E. Cayton, Analysis of early phase ring current recovery mechanisms during geomagnetic storms. Geophys. Res. Lett. **26**(18), 2845–2848 (1999). doi:10.1029/1999GL900611
M.W. Liemohn, J.U. Kozyra, P.G. Richards, G.V. Khazanov, M.J. Buonsanto, V.K. Jordanova, Ring current heating of the thermal electrons at solar maximum. J. Geophys. Res. **105**(A12), 27767 (2000). doi:10.1029/2000JA000088

M.W. Liemohn, J.U. Kozyra, M.F. Thomsen, J.L. Roeder, G. Lu, J.E. Borovsky, T.E. Cayton, Dominant role of the asymmetric ring current in producing the stormtime Dst*. J. Geophys. Res. **106**(A6), 10883 (2001). doi:10.1029/2000JA000326

Y. Lin, J.R. Johnson, X.Y. Wang, Hybrid simulation of mode conversion at the magnetopause. J. Geophys. Res. **115**(A4), 04208 (2010). doi:10.1029/2009JA014524

Y. Lin, J.R. Johnson, X. Wang, Three-dimensional mode conversion associated with kinetic Alfvén waves. Phys. Rev. Lett. **109**(12), 125003 (2012). doi:10.1103/PhysRevLett.109.125003. http://link.aps.org/doi/10.1103/PhysRevLett.109.125003

H. Liu, S.-Y. Ma, K. Schlegel, Diurnal, seasonal, and geomagnetic variations of large field-aligned ion upflows in the high-latitude ionospheric F region. J. Geophys. Res. **106**(A11), 24651 (2001). doi:10.1029/2001JA900047

R.E. Lopez, R. Bruntz, E.J. Mitchell, M. Wiltberger, J.G. Lyon, V.G. Merkin, Role of magnetosheath force balance in regulating the dayside reconnection potential. J. Geophys. Res. **115**(A12), 12216 (2010). doi:10.1029/2009JA014597

W. Lotko, The magnetosphere–ionosphere system from the perspective of plasma circulation: a tutorial. J. Atmos. Sol.-Terr. Phys. **69**(3), 191–211 (2007). doi:10.1016/j.jastp.2006.08.011. http://www.sciencedirect.com/science/article/pii/S1364682606002604

J.G. Luhmann, D.W. Curtis, P. Schroeder, J. McCauley, R.P. Lin, D.E. Larson, S.D. Bale, J.-A. Sauvaud, C. Aoustin, R.A. Mewaldt, A.C. Cummings, E.C. Stone, A.J. Davis, W.R. Cook, B. Kecman, M.E. Wiedenbeck, T. von Rosenvinge, M.H. Acuna, L.S. Reichenthal, S. Shuman, K.A. Wortman, D.V. Reames, R. Mueller-Mellin, H. Kunow, G.M. Mason, P. Walpole, A. Korth, T.R. Sanderson, C.T. Russell, J.T. Gosling, STEREO IMPACT investigation goals, measurements, and data products overview. Space Sci. Rev. **136**(1-4), 117–184 (2007). doi:10.1007/s11214-007-9170-x. http://adsabs.harvard.edu/abs/2008SSRv..136..117L

B. Lybekk, A. Pedersen, S. Haaland, K. Svenes, A.N. Fazakerley, A. Masson, M.G.G.T. Taylor, J.-G. Trotignon, Solar cycle variations of the Cluster spacecraft potential and its use for electron density estimations. J. Geophys. Res. Space Phys. **117**(A1) (2012). doi:10.1029/2011JA016969

K. Maezawa, T. Hori, The distant magnetotail: its structure, IMF dependence, and thermal properties, in *New Perspectives on the Earth's Magnetotail*, ed. by A. Nishida, D.N. Baker, S.W.H. Cowley (Am. Geophys. Union, Washington, 1998). doi:10.1029/GM105p0001

R. Maggiolo, J.A. Sauvaud, D. Fontaine, A. Teste, E. Grigorenko, A. Balogh, A. Fazakerley, G. Paschmann, D. Delcourt, H. Rème, A multi-satellite study of accelerated ionospheric ion beams above the polar cap. Ann. Geophys. **24**(6), 1665–1684 (2006). doi:10.5194/angeo-24-1665-2006. http://www.ann-geophys.net/24/1665/2006/angeo-24-1665-2006.html

R. Maggiolo, M. Echim, J. De Keyser, D. Fontaine, C. Jacquey, I. Dandouras, Polar cap ion beams during periods of northward IMF: Cluster statistical results. Ann. Geophys. **29**(5), 771–787 (2011). doi:10.5194/angeo-29-771-2011. http://www.ann-geophys.net/29/771/2011/angeo-29-771-2011.html

R. Maggiolo, M. Echim, C. Simon Wedlund, Y. Zhang, D. Fontaine, G. Lointier, J.-G. Trotignon, Polar cap arcs from the magnetosphere to the ionosphere: kinetic modelling and observations by Cluster and TIMED. Ann. Geophys. **30**(2), 283–302 (2012). doi:10.5194/angeo-30-283-2012. http://www.ann-geophys.net/30/283/2012/angeo-30-283-2012.html

B.H. Mauk, Quantitative modeling of the "convection surge" mechanism of ion acceleration. J. Geophys. Res. **91**(A12), 13423 (1986). doi:10.1029/JA091iA12p13423

J.P. McFadden, C.W. Carlson, D. Larson, J. Bonnell, F.S. Mozer, V. Angelopoulos, K.-H. Glassmeier, U. Auster, Structure of plasmaspheric plumes and their participation in magnetopause reconnection: first results from THEMIS. Geophys. Res. Lett. **35**(17), L17S10 (2008). doi:10.1029/2008GL033677

C.E. McIlwain, A Kp dependent equatorial electric field model. Adv. Space Res. **6**(3), 187–197 (1986). doi:10.1016/0273-1177(86)90331-5. http://www.sciencedirect.com/science/article/pii/0273117786903315

R.L. McPherron, Earth's magnetotail, in *Magnetotails in the Solar System*, ed. by A. Keiling, C.M. Jackman, P. Delmare (Wiley, Hoboken, 2015), pp. 61–84. Chap. 3

D.G. Mitchell, P. C:son Brandt, E.C. Roelof, D.C. Hamilton, K.C. Retterer, S. Mende, Global imaging of O+ from IMAGE/HENA. Space Sci. Rev. **109**(1-4), 63–75 (2003). doi:10.1023/B:SPAC.0000007513.55076.00. http://link.springer.com/10.1023/B:SPAC.0000007513.55076.00

A. Miura, Kelvin–Helmholtz instability for supersonic shear flow at the magnetospheric boundary. Geophys. Res. Lett. **17**(6), 749–752 (1990). doi:10.1029/GL017i006p00749

W. Miyake, T. Mukai, N. Kaya, On the evolution of ion conics along the field line from EXOS D observations. J. Geophys. Res. **98**(A7), 11127 (1993). doi:10.1029/92JA00716

W. Miyake, T. Mukai, N. Kaya, On the origins of the upward shift of elevated (bimodal) ion conics in velocity space. J. Geophys. Res. **101**(A12), 26961 (1996). doi:10.1029/96JA02601

E. Möbius, D. Hovestadt, B. Klecker, M. Scholer, F.M. Ipavich, C.W. Carlson, R.P. Lin, A burst of energetic O+ ions during an upstream particle event. Geophys. Res. Lett. **13**(13), 1372–1375 (1986). doi:10.1029/GL013i013p01372

T.E. Moore, The dayside reconnection X line. J. Geophys. Res. **107**(A10), 1332 (2002). doi:10.1029/2002JA009381

T.E. Moore, Plasma sheet and (nonstorm) ring current formation from solar and polar wind sources. J. Geophys. Res. **110**(A2), 02210 (2005). doi:10.1029/2004JA010563

T.E. Moore, C.R. Chappell, M.O. Chandler, P.D. Craven, B.L. Giles, C.J. Pollock, J.L. Burch, D.T. Young, J.H. Waite, J.E. Nordholt, M.F. Thomsen, D.J. McComas, J.J. Berthelier, W.S. Williamson, R. Robson, F.S. Mozer, High-altitude observations of the polar wind. Science **277**(5324), 349–351 (1997). doi:10.1126/science.277.5324.349. http://www.sciencemag.org/cgi/doi/10.1126/science.277.5324.349

T.E. Moore, W.K. Peterson, C.T. Russell, M.O. Chandler, M.R. Collier, H.L. Collin, P.D. Craven, R. Fitzenreiter, B.L. Giles, C.J. Pollock, Ionospheric mass ejection in response to a CME. Geophys. Res. Lett. **26**(15), 2339–2342 (1999). doi:10.1029/1999GL900456

T. Nagai, Solar wind control of the radial distance of the magnetic reconnection site in the magnetotail. J. Geophys. Res. **110**(A9), 09208 (2005). doi:10.1029/2005JA011207

D. Nagata, S. Machida, S. Ohtani, Y. Saito, T. Mukai, Solar wind control of plasma number density in the near-Earth plasma sheet. J. Geophys. Res. Space Phys. **112**(A9), 09204 (2007). doi:10.1029/2007JA012284

T.K.M. Nakamura, W. Daughton, H. Karimabadi, S. Eriksson, Three-dimensional dynamics of vortex-induced reconnection and comparison with THEMIS observations. J. Geophys. Res. Space Phys. **118**(9), 5742–5757 (2013). doi:10.1002/jgra.50547

B. Ni, R.M. Thorne, R.B. Horne, N.P. Meredith, Y.Y. Shprits, L. Chen, W. Li, Resonant scattering of plasma sheet electrons leading to diffuse auroral precipitation: 1. Evaluation for electrostatic electron cyclotron harmonic waves. J. Geophys. Res. **116**(A4), 04218 (2011a). doi:10.1029/2010JA016232

B. Ni, R.M. Thorne, N.P. Meredith, R.B. Horne, Y.Y. Shprits, Resonant scattering of plasma sheet electrons leading to diffuse auroral precipitation: 2. Evaluation for whistler mode chorus waves. J. Geophys. Res. **116**(A4), 04219 (2011b). doi:10.1029/2010JA016233

H. Nilsson, M. Waara, S. Arvelius, O. Marghitu, M. Bouhram, Y. Hobara, M. Yamauchi, R. Lundin, H. Rème, J.-A. Sauvaud, I. Dandouras, A. Balogh, L.M. Kistler, B. Klecker, C.W. Carlson, M.B. Bavassano-Cattaneo, A. Korth, Characteristics of high altitude oxygen ion energization and outflow as observed by Cluster: a statistical study. Ann. Geophys. **24**(3), 1099–1112 (2006). doi:10.5194/angeo-24-1099-2006. http://www.ann-geophys.net/24/1099/2006/angeo-24-1099-2006.html

H. Nilsson, E. Engwall, A. Eriksson, P.A. Puhl-Quinn, S. Arvelius, Centrifugal acceleration in the magnetotail lobes. Ann. Geophys. **28**(2), 569–576 (2010). doi:10.5194/angeo-28-569-2010. http://www.ann-geophys.net/28/569/2010/angeo-28-569-2010.html

H. Nilsson, I.A. Barghouthi, R. Slapak, A.I. Eriksson, M. André, Hot and cold ion outflow: spatial distribution of ion heating. J. Geophys. Res. **117**(A11), 11201 (2012)

H. Nilsson, I.A. Barghouthi, R. Slapak, A.I. Eriksson, M. André, Hot and cold ion outflow: observations and implications for numerical models. J. Geophys. Res. Space Phys. **118**(1), 105–117 (2013)

M. Nosé, A.T.Y. Lui, S. Ohtani, B.H. Mauk, R.W. McEntire, D.J. Williams, T. Mukai, K. Yumoto, Acceleration of oxygen ions of ionospheric origin in the near-Earth magnetotail during substorms. J. Geophys. Res. **105**(A4), 7669 (2000). doi:10.1029/1999JA000318

M. Nosé, K. Takahashi, K. Keika, L.M. Kistler, K. Koga, H. Koshiishi, H. Matsumoto, M. Shoji, Y. Miyashita, R. Nomura, Magnetic fluctuations embedded in dipolarization inside geosynchronous orbit and their associated selective acceleration of O+ ions. J. Geophys. Res. Space Phys. **119**(6), 4639–4655 (2014). doi:10.1002/2014JA019806

K. Nykyri, A. Otto, B. Lavraud, C. Mouikis, L.M. Kistler, A. Balogh, H. Rème, Cluster observations of reconnection due to the Kelvin–Helmholtz instability at the dawnside magnetospheric flank. Ann. Geophys. **24**(10), 2619–2643 (2006). doi:10.5194/angeo-24-2619-2006. http://www.ann-geophys.net/24/2619/2006/angeo-24-2619-2006.html

M. Øieroset, Global cooling and densification of the plasma sheet during an extended period of purely northward IMF on October 22–24, 2003. Geophys. Res. Lett. **32**(12), L12S07 (2005). doi:10.1029/2004GL021523

M. Øieroset, J. Raeder, T.D. Phan, S. Wing, J.P. McFadden, W. Li, M. Fujimoto, H. Rème, A. Balogh, Global cooling and densification of the plasma sheet during an extended period of purely northward IMF on October 22-24, 2003. Geophys. Res. Lett. **32**, 1–4 (2005). doi:10.1029/2004GL021523

Y. Obana, F.W. Menk, I. Yoshikawa, Plasma refilling rates for $L = 2.3$–3.8 flux tubes. J. Geophys. Res. **115**(A3), 03204 (2010). doi:10.1029/2009JA014191

Y. Ogawa, S.C. Buchert, R. Fujii, S. Nozawa, A.P. van Eyken, Characteristics of ion upflow and downflow observed with the European Incoherent Scatter Svalbard radar. J. Geophys. Res. **114**(A5), 05305 (2009). doi:10.1029/2008JA013817

Y. Ogawa, S.C. Buchert, A. Sakurai, S. Nozawa, R. Fujii, Solar activity dependence of ion upflow in the polar ionosphere observed with the European Incoherent Scatter (EISCAT) TromsøUHF radar. J. Geophys. Res. **115**(A7), 07310 (2010). doi:10.1029/2009JA014766
Y. Ono, M. Nosé, S.P. Christon, A.T.Y. Lui, The role of magnetic field fluctuations in nonadiabatic acceleration of ions during dipolarization. J. Geophys. Res. **114**(A5), 05209 (2009). doi:10.1029/2008JA013918
T.G. Onsager, J.D. Scudder, M. Lockwood, C.T. Russell, Reconnection at the high-latitude magnetopause during northward interplanetary magnetic field conditions. J. Geophys. Res. **106**(A11), 25467 (2001). doi:10.1029/2000JA000444
A. Opitz, J.-A. Sauvaud, A. Klassen, R. Gomez-Herrero, R. Bucik, L.M. Kistler, C. Jacquey, J. Luhmann, G. Mason, P. Kajdic, B. Lavraud, Solar wind control of the terrestrial magnetotail as seen by STEREO. J. Geophys. Res. Space Phys. **119**(8), 6342–6355 (2014). doi:10.1002/2014JA019988
S. Orsini, M. Candidi, M. Stockholm, H. Balsiger, Injection of ionospheric ions into the plasma sheet. J. Geophys. Res. **95**, 7915–7928 (1990)
A. Otto, D.H. Fairfield, Kelvin–Helmholtz instability at the magnetotail boundary: MHD simulation and comparison with Geotail observations. J. Geophys. Res. **105**(A9), 21175 (2000). doi:10.1029/1999JA000312
J.E. Ouellette, O.J. Brambles, J.G. Lyon, W. Lotko, B.N. Rogers, Properties of outflow-driven sawtooth substorms. J. Geophys. Res. Space Phys. **118**(6), 3223–3232 (2013). doi:10.1002/jgra.50309
I.L. Ovchinnikov, E.E. Antonova, Y.I. Yermolaev, Determination of the turbulent diffusion coefficient in the plasma sheet using the project INTERBALL data. Cosm. Res. **38**(6), 557–561 (2000). doi:10.1023/A:1026686600686. http://link.springer.com/article/10.1023/A%3A1026686600686
C.J. Owen, M.G.G.T. Taylor, I.C. Krauklis, A.N. Fazakerley, M.W. Dunlop, J.M. Bosqued, Cluster observations of surface waves on the dawn flank magnetopause. Ann. Geophys. **22**(3), 971–983 (2004). doi:10.5194/angeo-22-971-2004. http://www.ann-geophys.net/22/971/2004/angeo-22-971-2004.html
C.G. Park, Whistler observations of the interchange of ionization between the ionosphere and the protonosphere. J. Geophys. Res. **75**(22), 4249–4260 (1970). doi:10.1029/JA075i022p04249
G. Paschmann, G. Haerendel, N. Sckopke, H. Rosenbauer, P.C. Hedgecock, Plasma and magnetic field characteristics of the distant polar cusp near local noon: the entry layer. J. Geophys. Res. **81**(16), 2883–2899 (1976). doi:10.1029/JA081i016p02883
G. Paschmann, M. Øieroset, T. Phan, In-situ observations of reconnection in space. Space Sci. Rev. **178**(2-4), 385–417 (2013). doi:10.1007/s11214-012-9957-2. http://adsabs.harvard.edu/abs/2013SSRv..178..385P
R.J. Pellinen, W.J. Heikkila, Energization of charged particles to high energies by an induced substorm electric field within the magnetotail. J. Geophys. Res. **83**(A4), 1544 (1978). doi:10.1029/JA083iA04p01544
W.K. Peterson, L. Andersson, B.C. Callahan, H.L. Collin, J.D. Scudder, A.W. Yau, Solar-minimum quiet time ion energization and outflow in dynamic boundary related coordinates. J. Geophys. Res. **113**(A7), 07222 (2008). doi:10.1029/2008JA013059
W.K. Peterson, H.L. Collin, M.F. Doherty, C.M. Bjorklund, Extended (Bi-modal) ion conics at high altitudes, in *Space Plasmas: Coupling Between Small and Medium Scale Processes*. Geophysical Monograph, vol. 86 (1995). http://adsabs.harvard.edu/abs/1995GMS....86..105P
V. Pierrard, J. Cabrera, Comparisons between EUV/IMAGE observations and numerical simulations of the plasmapause formation. Ann. Geophys. **23**(7), 2635–2646 (2005). doi:10.5194/angeo-23-2635-2005. http://www.ann-geophys.net/23/2635/2005/angeo-23-2635-2005.html
V. Pierrard, J.F. Lemaire, Development of shoulders and plumes in the frame of the interchange instability mechanism for plasmapause formation. Geophys. Res. Lett. **31**(5) (2004). doi:10.1029/2003GL018919
V. Pierrard, G.V. Khazanov, J. Cabrera, J. Lemaire, Influence of the convection electric field models on predicted plasmapause positions during magnetic storms. J. Geophys. Res. **113**(A8), 08212 (2008). doi:10.1029/2007JA012612
V. Pierrard, J. Goldstein, N. André, V.K. Jordanova, G.A. Kotova, J.F. Lemaire, M.W. Liemohn, H. Matsui, Recent progress in physics-based models of the plasmasphere. Space Sci. Rev. **145**(1-2), 193–229 (2009). doi:10.1007/s11214-008-9480-7. http://link.springer.com/10.1007/s11214-008-9480-7
W.G. Pilipp, G. Morfill, The formation of the plasma sheet resulting from plasma mantle dynamics. J. Geophys. Res. **83**(A12), 5670 (1978). doi:10.1029/JA083iA12p05670
C.J. Pollock, M.O. Chandler, T.E. Moore, J.H. Waite, C.R. Chappell, D.A. Gurnett, A survey of upwelling ion event characteristics. J. Geophys. Res. **95**(A11), 18969 (1990). doi:10.1029/JA095iA11p18969
A.R. Poppe, R. Samad, J.S. Halekas, M. Sarantos, G.T. Delory, W.M. Farrell, V. Angelopoulos, J.P. McFadden, ARTEMIS observations of lunar pick-up ions in the terrestrial magnetotail lobes. Geophys. Res. Lett. **39**(17) (2012). doi:10.1029/2012GL052909
A. Posner, Association of low-charge-state heavy ions up to 200 R e upstream of the Earth's bow shock with geomagnetic disturbances. Geophys. Res. Lett. **29**(7), 1099 (2002). doi:10.1029/2001GL013449
H. Rème, C. Aoustin, J.M. Bosqued, I. Dandouras, B. Lavraud, J.A. Sauvaud, A. Barthe, J. Bouyssou, T. Camus, O. Coeur-Joly, A. Cros, J. Cuvilo, F. Ducay, Y. Garbarowitz, J.L. Medale, E. Penou, H. Perrier, D. Romefort, J. Rouzaud, C. Vallat, D. Alcaydé, C. Jacquey, C. Mazelle, C. D'Uston, E. Möbius,

L.M. Kistler, K. Crocker, M. Granoff, C. Mouikis, M. Popecki, M. Vosbury, B. Klecker, D. Hovestadt, H. Kucharek, E. Kuenneth, G. Paschmann, M. Scholer, N. Sckopke, E. Seidenschwang, C.W. Carlson, D.W. Curtis, C. Ingraham, R.P. Lin, J.P. McFadden, G.K. Parks, T. Phan, V. Formisano, E. Amata, M.B. Bavassano-Cattaneo, P. Baldetti, R. Bruno, G. Chionchio, A.D. Lellis, M.F. Marcucci, G. Pallocchia, A. Korth, P.W. Daly, B. Graeve, H. Rosenbauer, V. Vasyliunas, M. Mccarthy, M. Wilber, L. Eliasson, R. Lundin, S. Olsen, E.G. Shelley, S. Fuselier, A.G. Ghielmetti, W. Lennartsson, C.P. Escoubet, H. Balsiger, R. Friedel, J.-B. Cao, R.A. Kovrazhkin, I. Papamastorakis, R. Pellat, J. Scudder, B. Sonnerup, First multispacecraft ion measurements in and near the Earth's magnetosphere with the identical Cluster ion spectrometry (CIS) experiment. Ann. Geophys. **19**(10/12), 1303–1354 (2001). https://hal.archives-ouvertes.fr/hal-00329192/

W.T. Roberts, J.L. Horwitz, R.H. Comfort, C.R. Chappell, J.H. Waite, J.L. Green, Heavy ion density enhancements in the outer plasmasphere. J. Geophys. Res. **92**(A12), 13499 (1987). doi:10.1029/JA092iA12p13499

J.G. Roederer, On the adiabatic motion of energetic particles in a model magnetosphere. J. Geophys. Res. **72**(3), 981–992 (1967). doi:10.1029/JZ072i003p00981

J.G. Roederer, M. Schulz, Splitting of drift shells by the magnetospheric electric field. J. Geophys. Res. **76**(4), 1055–1059 (1971). doi:10.1029/JA076i004p01055

J.G. Roederer, H.H. Hilton, M. Schulz, Drift shell splitting by internal geomagnetic multipoles. J. Geophys. Res. **78**(1), 133–144 (1973). doi:10.1029/JA078i001p00133

H. Rosenbauer, H. Grünwaldt, M.D. Montgomery, G. Paschmann, N. Sckopke, Heos 2 plasma observations in the distant polar magnetosphere: the plasma mantle. J. Geophys. Res. **80**(19), 2723–2737 (1975). doi:10.1029/JA080i019p02723

Y. Saito, S. Yokota, K. Asamura, T. Tanaka, M.N. Nishino, T. Yamamoto, Y. Terakawa, M. Fujimoto, H. Hasegawa, H. Hayakawa, M. Hirahara, M. Hoshino, S. Machida, T. Mukai, T. Nagai, T. Nagatsuma, T. Nakagawa, M. Nakamura, K.-i. Oyama, E. Sagawa, S. Sasaki, K. Seki, I. Shinohara, T. Terasawa, H. Tsunakawa, H. Shibuya, M. Matsushima, H. Shimizu, F. Takahashi, In-flight performance and initial results of plasma energy angle and composition experiment (PACE) on SELENE (Kaguya). Space Sci. Rev. **154**(1-4), 265–303 (2010). doi:10.1007/s11214-010-9647-x. http://link.springer.com/10.1007/s11214-010-9647-x

B.R. Sandel, A.L. Broadfoot, C.C. Curtis, R.A. King, T.C. Stone, R.H. Hill, J. Chen, O.H.W. Siegmund, R. Raffanti, D.D. Allred, R.S. Turley, D.L. Gallagher, The extreme ultraviolet imager investigation for the image mission, in *The IMAGE Mission*, ed. by J.L. Burch (Springer, Berlin, 2000), pp. 197–242. doi:10.1007/978-94-011-4233-5_7

B.R. Sandel, J. Goldstein, D.L. Gallagher, M. Spasojevic, Extreme ultraviolet imager observations of the structure and dynamics of the plasmasphere. Space Sci. Rev. **109**(1-4), 25–46 (2003). doi:10.1023/B:SPAC.0000007511.47727.5b. http://link.springer.com/10.1023/B:SPAC.0000007511.47727.5b

J.-A. Sauvaud, R. Lundin, H. Rème, J.P. McFadden, C. Carlson, G.K. Parks, E. Möbius, L.M. Kistler, B. Klecker, E. Amata, A.M. Dilellis, V. Formisano, J.M. Bosqued, I. Dandouras, P. Décréau, M. Dunlop, L. Eliasson, A. Korth, B. Lavraud, M. Mccarthy, Intermittent thermal plasma acceleration linked to sporadic motions of the magnetopause, first Cluster results. Ann. Geophys. **19**(10/12), 1523–1532 (2001). https://hal.archives-ouvertes.fr/hal-00329207/

J.A. Sauvaud, P. Louarn, G. Fruit, H. Stenuit, C. Vallat, J. Dandouras, H. Rème, M. André, A. Balogh, M. Dunlop, L. Kistler, E. Möbius, C. Mouikis, B. Klecker, G.K. Parks, J. McFadden, C. Carlson, F. Marcucci, G. Pallocchia, R. Lundin, A. Korth, M. McCarthy, Case studies of the dynamics of ionospheric ions in the Earth's magnetotail. J. Geophys. Res. **109**(A), 1212 (2004)

R.W. Schunk, Time-dependent simulations of the global polar wind. J. Atmos. Sol.-Terr. Phys. **69**(16), 2028–2047 (2007). doi:10.1016/j.jastp.2007.08.009. http://www.sciencedirect.com/science/article/pii/S136468260700243X

N. Sckopke, G. Paschmann, H. Rosenbauer, D.H. Fairfield, Influence of the interplanetary magnetic field on the occurrence and thickness of the plasma mantle. J. Geophys. Res. **81**(16), 2687–2691 (1976). doi:10.1029/JA081i016p02687

K. Seki, M. Hirahara, T. Terasawa, I. Shinohara, T. Mukai, Y. Saito, S. Machida, T. Yamamoto, S. Kokubun, Coexistence of Earth-origin O+ and solar wind-origin H+/He++ in the distant magnetotail. Geophys. Res. Lett. **23**(9), 985–988 (1996). doi:10.1029/96GL00768

K. Seki, M. Hirahara, T. Terasawa, T. Mukai, S. Kokubun, Properties of He+ beams observed by Geotail in the lobe/mantle regions: comparison with O+ beams. J. Geophys. Res. **104**(A4), 6973 (1999). doi:10.1029/1998JA900142

K. Seki, General processes, in *Solar System Sources and Losses of Plasma* (2015)

K. Seki, T. Terasawa, M. Hirahara, T. Mukai, Quantification of tailward cold O+ beams in the lobe/mantle regions with Geotail data: constraints on polar O+ outflows. J. Geophys. Res. **103**(A), 29371–29382 (1998)

K. Seki, M. Hirahara, M. Hoshino, T. Terasawa, R.C. Elphic, Y. Saito, T. Mukai, H. Hayakawa, H. Kojima, H. Matsumoto, Cold ions in the hot plasma sheet of Earth's magnetotail. Nature **422**(6932), 589–592 (2003). doi:10.1038/nature01502. http://dx.doi.org/10.1038/nature01502

K. Seki, A. Nagy, C.M. Jackman, F. Crary, D. Fontaine, P. Zarka et al., A review of general physical and chemical processes related to plasma sources and losses for solar system magnetospheres. Space Sci. Rev. (2015). doi:10.1007/s11214-015-0170-y

E.G. Shelley, W.K. Peterson, A.G. Ghielmetti, J. Geiss, The polar ionosphere as a source of energetic magnetospheric plasma. Geophys. Res. Lett. **9**(9), 941–944 (1982). doi:10.1029/GL009i009p00941

D.G. Sibeck, R.W. McEntire, A.T.Y. Lui, R.E. Lopez, S.M. Krimigis, Magnetic field drift shell splitting: cause of unusual dayside particle pitch angle distributions during storms and substorms. J. Geophys. Res. **92**(A12), 13485 (1987). doi:10.1029/JA092iA12p13485

G.L. Siscoe, E. Sanchez, An MHD model for the complete open magnetotail boundary. J. Geophys. Res. **92**(A7), 7405 (1987). doi:10.1029/JA092iA07p07405

G.L. Siscoe, G.M. Erickson, B.U.O. Sonnerup, N.C. Maynard, K.D. Siebert, D.R. Weimer, W.W. White, Relation between cusp and mantle in MHD simulation. J. Geophys. Res. **106**(A6), 10743 (2001). doi:10.1029/2000JA000385

G. Siscoe, Z. Kaymaz, Spatial relations of mantle and plasma sheet. J. Geophys. Res. **104**(A7), 14639 (1999). doi:10.1029/1999JA900113

J.A. Slavin, E.J. Smith, P.W. Daly, T.R. Sanderson, K.-P. Wenzel, R.P. Lepping, Magnetic configuration of the distant plasma sheet: ISEE 3 observations, in *Magnetotail Physics* (1987), pp. 59–63. http://adsabs.harvard.edu/abs/1987magp.book...59S

J.A. Slavin, R.P. Lepping, J. Gjerloev, D.H. Fairfield, M. Hesse, C.J. Owen, M.B. Moldwin, T. Nagai, A. Ieda, T. Mukai, Geotail observations of magnetic flux ropes in the plasma sheet. J. Geophys. Res. Space Phys. **108**(A1), 1015 (2003). doi:10.1029/2002JA009557

J.A. Slavin, E.J. Smith, D.G. Sibeck, D.N. Baker, R.D. Zwickl, S.-I. Akasofu, An ISEE 3 study of average and substorm conditions in the distant magnetotail. J. Geophys. Res. **90**(A11), 10875 (1985). doi:10.1029/JA090iA11p10875

R. Smets, G. Belmont, D. Delcourt, L. Rezeau, Diffusion at the Earth magnetopause: enhancement by Kelvin–Helmholtz instability. Ann. Geophys. **25**(1), 271–282 (2007). doi:10.5194/angeo-25-271-2007. http://www.ann-geophys.net/25/271/2007/angeo-25-271-2007.html

P. Song, C.T. Russell, Model of the formation of the low-latitude boundary layer for strongly northward interplanetary magnetic field. J. Geophys. Res. **97**(A2), 1411 (1992). doi:10.1029/91JA02377

B.U.O. Sonnerup, Transport mechanisms at the magnetopause, in *Dynamics of the Magnetosphere*, ed. by S.-I. Akasofu (Springer, Berlin, 1980), pp. 77–100. 978-94-009-9519-2. doi:10.1007/978-94-009-9519-2_5

T.W. Speiser, Plasma density and acceleration in the tail from the reconnection model, in *Earth's Particles and Fields*, ed. by B.M. McCormac, New York, NY (1968)

H.E. Spence, M.G. Kivelson, Contributions of the low-latitude boundary layer to the finite width magnetotail convection model. J. Geophys. Res. **98**(A9), 15487 (1993). doi:10.1029/93JA01531

M. Stepanova, V. Pinto, J.A. Valdivia, E.E. Antonova, Spatial distribution of the eddy diffusion coefficients in the plasma sheet during quiet time and substorms from THEMIS satellite data. J. Geophys. Res. Space Phys. **116**(A5) (2011). doi:10.1029/2010JA015887

R. Strangeway, J.R.E. Ergun, Y.J. Su, C.W. Carlson, R.C. Elphic, Factors controlling ionospheric outflows as observed at intermediate altitudes. J. Geophys. Res. **110**(A3), 03221 (2005). doi:10.1029/2004JA010829

Y.-J. Su, J.L. Horwitz, T.E. Moore, B.L. Giles, M.O. Chandler, P.D. Craven, M. Hirahara, C.J. Pollock, Polar wind survey with the thermal ion dynamics Experiment/Plasma source instrument suite aboard POLAR. J. Geophys. Res. **103**(A12), 29305 (1998). doi:10.1029/98JA02662

Y.-J. Su, M.F. Thomsen, J.E. Borovsky, J.C. Foster, A linkage between polar patches and plasmaspheric drainage plumes. Geophys. Res. Lett. **28**(1), 111–113 (2001). doi:10.1029/2000GL012042

S. Taguchi, H. Kishida, T. Mukai, Y. Saito, Low-latitude plasma mantle in the near-Earth magnetosphere: Geotail observations. J. Geophys. Res. **106**(A2), 1949 (2001). doi:10.1029/2000JA900100

K. Takagi, C. Hashimoto, H. Hasegawa, M. Fujimoto, R. TanDokoro, Kelvin–Helmholtz instability in a magnetotail flank-like geometry: three-dimensional MHD simulations. J. Geophys. Res. **111**(A8), 08202 (2006). doi:10.1029/2006JA011631

K. Takahashi, B.J. Anderson, S.-i. Ohtani, G.D. Reeves, S. Takahashi, T.E. Sarris, K. Mursula, Drift-shell splitting of energetic ions injected at pseudo-substorm onsets. J. Geophys. Res. **102**(A10), 22117 (1997). doi:10.1029/97JA01870

S. Takahashi, T. Iyemori, M. Takeda, A simulation of the storm-time ring current. Planet. Space Sci. **38**(9), 1133–1141 (1990). doi:10.1016/0032-0633(90)90021-H. http://www.sciencedirect.com/science/article/pii/003206339090021H

S.W.Y. Tam, T. Chang, V. Pierrard, Kinetic modeling of the polar wind. J. Atmos. Sol.-Terr. Phys. **69**(16), 1984–2027 (2007). doi:10.1016/j.jastp.2007.08.006. http://www.sciencedirect.com/science/article/pii/S1364682607002428

T. Tanaka, Y. Saito, S. Yokota, K. Asamura, M.N. Nishino, H. Tsunakawa, H. Shibuya, M. Matsushima, H. Shimizu, F. Takahashi, M. Fujimoto, T. Mukai, T. Terasawa, First in situ observation of the Moon-originating ions in the Earth's magnetosphere by MAP-PACE on SELENE (KAGUYA). Geophys. Res. Lett. **36**(22), 22106 (2009). doi:10.1029/2009GL040682

M.G.G.T. Taylor, H. Hasegawa, B. Lavraud, T. Phan, C.P. Escoubet, M.W. Dunlop, Y.V. Bogdanova, A.L. Borg, M. Volwerk, J. Berchem, O.D. Constantinescu, J.P. Eastwood, A. Masson, H. Laakso, J. Soucek, A.N. Fazakerley, H.U. Frey, E.V. Panov, C. Shen, J.K. Shi, D.G. Sibeck, Z.Y. Pu, J. Wang, J.A. Wild, *Spatial Distribution of Rolled up Kelvin–Helmholtz Vortices at Earth's Dayside and Flank Magnetopause* (2012). http://eprints.lancs.ac.uk/60570/1/angeo_30_1025_2012.pdf

M.G.G.T. Taylor, B. Lavraud, C.P. Escoubet, S.E. Milan, K. Nykyri, M.W. Dunlop, J.A. Davies, R.H.W. Friedel, H. Frey, Y.V. Bogdanova, A. Å snes, H. Laakso, P. Trávnícek, A. Masson, H. Opgenoorth, C. Vallat, A.N. Fazakerley, A.D. Lahiff, C.J. Owen, F. Pitout, Z. Pu, C. Shen, Q.G. Zong, H. Rème, J. Scudder, T.L. Zhang, The plasma sheet and boundary layers under northward IMF: a multi-point and multi-instrument perspective. Adv. Space Res. **41**(10), 1619–1629 (2008). doi:10.1016/j.asr.2007.10.013. http://www.sciencedirect.com/science/article/pii/S0273117707010368

T. Terasawa, M. Fujimoto, T. Mukai, I. Shinohara, Y. Saito, T. Yamamoto, S. Machida, S. Kokubun, A.J. Lazarus, J.T. Steinberg, R.P. Lepping, Solar wind control of density and temperature in the near-Earth plasma sheet: WIND/GEOTAIL collaboration. Geophys. Res. Lett. **24**(8), 935–938 (1997). doi:10.1029/96GL04018

A. Teste, D. Fontaine, J.-A. Sauvaud, R. Maggiolo, P. Canu, A. Fazakerley, CLUSTER observations of electron outflowing beams carrying downward currents above the polar cap by northward IMF. Ann. Geophys. **25**(4), 953–969 (2007). doi:10.5194/angeo-25-953-2007. http://www.ann-geophys.net/25/953/2007/angeo-25-953-2007.html

A. Teste, D. Fontaine, P. Canu, G. Belmont, Cluster observations of outflowing electron distributions and broadband electrostatic emissions above the polar cap. Geophys. Res. Lett. **37**(3) (2010). doi:10.1029/2009GL041593

E.G. Thomas, J.B.H. Baker, J.M. Ruohoniemi, L.B.N. Clausen, A.J. Coster, J.C. Foster, P.J. Erickson, Direct observations of the role of convection electric field in the formation of a polar tongue of ionization from storm enhanced density. J. Geophys. Res. Space Phys. **118**(3), 1180–1189 (2013). doi:10.1002/jgra.50116

R.M. Thorne, Radiation belt dynamics: the importance of wave-particle interactions. Geophys. Res. Lett. **37**(22) (2010). doi:10.1029/2010GL044990

K. Torkar, A.I. Eriksson, P.-A. Lindqvist, W. Steiger, Long-term study of active spacecraft potential control. IEEE Trans. Plasma Sci. **36**(5), 2294–2300 (2008). doi:10.1109/TPS.2008.2003134. http://ieeexplore.ieee.org/lpdocs/epic03/wrapper.htm?arnumber=4663151

K.J. Trattner, J.S. Mulcock, S.M. Petrinec, S.A. Fuselier, Location of the reconnection line at the magnetopause during southward IMF conditions. Geophys. Res. Lett. **34**(3), 03108 (2007a). doi:10.1029/2006GL028397

K.J. Trattner, J.S. Mulcock, S.M. Petrinec, S.A. Fuselier, Probing the boundary between antiparallel and component reconnection during southward interplanetary magnetic field conditions. J. Geophys. Res. **112**(A8), 08210 (2007b). doi:10.1029/2007JA012270

K.J. Trattner, S.M. Petrinec, S.A. Fuselier, T.D. Phan, The location of reconnection at the magnetopause: testing the maximum magnetic shear model with THEMIS observations. J. Geophys. Res. Space Phys. **117**(A1) (2012). doi:10.1029/2011JA016959

B.T. Tsurutani, E.J. Smith, R.R. Anderson, K.W. Ogilvie, J.D. Scudder, D.N. Baker, S.J. Bame, Lion roars and nonoscillatory drift mirror waves in the magnetosheath. J. Geophys. Res. **87**(A8), 6060 (1982). doi:10.1029/JA087iA08p06060

N.A. Tsyganenko, Modeling the Earth's magnetospheric magnetic field confined within a realistic magnetopause. J. Geophys. Res. **100**(A4), 5599 (1995). doi:10.1029/94JA03193

J. Tu, P. Song, B.W. Reinisch, J.L. Green, Smooth electron density transition from plasmasphere to the subauroral region. J. Geophys. Res. **112**(A5), 05227 (2007). doi:10.1029/2007JA012298

C. Twitty, Cluster survey of cusp reconnection and its IMF dependence. Geophys. Res. Lett. **31**(19), 19808 (2004). doi:10.1029/2004GL020646

R.J. Walker, M. Ashour-Abdalla, T. Ogino, V. Peroomian, R.L. Richard, Modeling Magnetospheric Sources. Geophys. Monogr. **133** (2003). doi:10.1029/133GM03. http://192.102.233.13/books/gm/v133/133GM03/133GM03.pdf

B.M. Walsh, J.C. Foster, P.J. Erickson, D.G. Sibeck, Simultaneous ground- and space-based observations of the plasmaspheric plume and reconnection. Science **343**(6175), 1122–1125 (2014). doi:10.1126/science.1247212. http://www.sciencemag.org/content/343/6175/1122.short

C.-P. Wang, Modeling the transition of the inner plasma sheet from weak to enhanced convection. J. Geophys. Res. **109**(A12), 12202 (2004). doi:10.1029/2004JA010591

C.-P. Wang, L.R. Lyons, V. Angelopoulos, Properties of low-latitude mantle plasma in the Earth's magnetotail: ARTEMIS observations and global MHD predictions. J. Geophys. Res. Space Phys. **119**, 7264–7280 (2014). doi:10.1002/2014JA020060

C.-P. Wang, L.R. Lyons, T. Nagai, J.M. Weygand, R.W. McEntire, Sources, transport, and distributions of plasma sheet ions and electrons and dependences on interplanetary parameters under northward interplanetary magnetic field. J. Geophys. Res. **112**(A10), 10224 (2007). doi:10.1029/2007JA012522

C.P. Wang, L.R. Lyons, R.A. Wolf, T. Nagai, J.M. Weygand, A.T.Y. Lui, Plasma sheet Pv5/3and $n\nu$ and associated plasma and energy transport for different convection strengths and AE levels. J. Geophys. Res. Space Phys. **114**, 1–2 (2009). doi:10.1029/2008JA013849

C.-P. Wang, L.R. Lyons, T. Nagai, J.M. Weygand, A.T.Y. Lui, Evolution of plasma sheet particle content under different interplanetary magnetic field conditions. J. Geophys. Res. **115**(A6), 06210 (2010). doi:10.1029/2009JA015028

C.-P. Wang, M. Gkioulidou, L.R. Lyons, V. Angelopoulos, Spatial distributions of the ion to electron temperature ratio in the magnetosheath and plasma sheet. J. Geophys. Res. **117**(A8), 08215 (2012). doi:10.1029/2012JA017658

C.-P. Wang, M. Gkioulidou, L.R. Lyons, X. Xing, R.A. Wolf, Interchange motion as a transport mechanism for formation of cold-dense plasma sheet. J. Geophys. Res. Space Phys. **119**, 8318–8337 (2014). doi:10.1002/2014JA020251

D.T. Welling, M.W. Liemohn, Outflow in global magnetohydrodynamics as a function of a passive inner boundary source. J. Geophys. Res. Space Phys. **119**(4), 2691–2705 (2014). doi:10.1002/2013JA019374

D.T. Welling, A.J. Ridley, Exploring sources of magnetospheric plasma using multispecies MHD. J. Geophys. Res. **115**(A4), 04201 (2010). doi:10.1029/2009JA014596

D.T. Welling, S.G. Zaharia, Ionospheric outflow and cross polar cap potential: what is the role of magnetospheric inflation? Geophys. Res. Lett. **39**(23) (2012). doi:10.1029/2012GL054228

D.T. Welling, V.K. Jordanova, S.G. Zaharia, A. Glocer, G. Toth, The effects of dynamic ionospheric outflow on the ring current. J. Geophys. Res. **116**, 1–19 (2011). doi:10.1029/2010JA015642

J.M. Weygand, Plasma sheet turbulence observed by Cluster II. J. Geophys. Res. **110**(A1), 01205 (2005). doi:10.1029/2004JA010581

B.A. Whalen, S. Watanabe, A.W. Yau, Observations in the transverse ion energization region. Geophys. Res. Lett. **18**(4), 725–728 (1991). doi:10.1029/90GL02788

M. Wiltberger, W. Lotko, J.G. Lyon, P. Damiano, V. Merkin, Influence of cusp O+ outflow on magnetotail dynamics in a multifluid MHD model of the magnetosphere. J. Geophys. Res. **115**(June), 1–5 (2010). doi:10.1029/2010JA015579

M. Wiltberger, Review of global simulation studies of effect of ionospheric outflow on magnetosphere-ionosphere system dynamics, in *Magnetotails in the Solar System* (Wiley, Hoboken, 2015), pp. 373–392. Chap. 22. doi:10.1002/9781118842324

S. Wing, J.R. Johnson, P.T. Newell, C.I. Meng, Dawn-dusk asymmetries, ion spectra, and sources in the northward interplanetary magnetic field plasma sheet. J. Geophys. Res. Space Phys. **110**(A8), 08205 (2005). doi:10.1029/2005JA011086

S. Wing, J.R. Johnson, M. Fujimoto, Timescale for the formation of the cold-dense plasma sheet: a case study. Geophys. Res. Lett. **33**(23), 23106 (2006). doi:10.1029/2006GL027110

R.M. Winglee, Multi-fluid simulations of the magnetosphere: the identification of the geopause and its variation with IMF. Geophys. Res. Lett. **25**(24), 4441–4444 (1998). doi:10.1029/1998GL900217

R.M. Winglee, Global impact of ionospheric outflows on the dynamics of the magnetosphere and cross-polar cap potential. J. Geophys. Res. **107**(A9), 1237 (2002). doi:10.1029/2001JA000214

R.M. Winglee, E. Harnett, Influence of heavy ionospheric ions on substorm onset. J. Geophys. Res. **116**(A11), 11212 (2011). doi:10.1029/2011JA016447

G.Q. Yan, F.S. Mozer, C. Shen, T. Chen, G.K. Parks, C.L. Cai, J.P. McFadden, Kelvin–Helmholtz vortices observed by THEMIS at the duskside of the magnetopause under southward interplanetary magnetic field. Geophys. Res. Lett. (2014). doi:10.1002/2014GL060589

Y. Yao, C.C. Chaston, K.-H. Glassmeier, V. Angelopoulos, Electromagnetic waves on ion gyro-radii scales across the magnetopause. Geophys. Res. Lett. **38**(9) (2011). doi:10.1029/2011GL047328

A.W. Yau, M. André, M. Andre, Sources of ion outflow in the high latitude ionosphere. Space Sci. Rev. **80**(1-2), 1–25 (1997). doi:10.1023/A:1004947203046. http://link.springer.com/article/10.1023/A%3A1004947203046

A.W. Yau, W.K. Peterson, E.G. Shelley, Quantitative parametrization of energetic ionospheric ion outflow, in *Washington DC American Geophysical Union Geophysical Monograph Series* (1988), pp. 211–217

A.W. Yau, B.A. Whalen, A.G. McNamara, P.J. Kellogg, W. Bernstein, Particle and wave observations of low-altitude ionospheric ion acceleration events. J. Geophys. Res. **88**(A1), 341 (1983). doi:10.1029/JA088iA01p00341

A.W. Yau, B.A. Whalen, W.K. Peterson, E.G. Shelley, Distribution of upflowing ionospheric ions in the high-altitude polar cap and auroral ionosphere. J. Geophys. Res. **89**(A7), 5507 (1984). doi:10.1029/JA089iA07p05507

A.W. Yau, T. Abe, W.K. Peterson, The polar wind: recent observations. J. Atmos. Sol.-Terr. Phys. **69**(16), 1936–1983 (2007). doi:10.1016/j.jastp.2007.08.010. http://www.sciencedirect.com/science/article/pii/S1364682607002416

A.W. Yau, A. Howarth, W.K. Peterson, T. Abe, Transport of thermal-energy ionospheric oxygen (O+) ions between the ionosphere and the plasma sheet and ring current at quiet times preceding magnetic storms. J. Geophys. Res. **117**(A7), 07215 (2012). doi:10.1029/2012JA017803

I. Yoshikawa, Which is a significant contributor for outside of the plasmapause, an ionospheric filling or a leakage of plasmaspheric materials?: Comparison of He II (304 Å) images. J. Geophys. Res. **108**(A2), 1080 (2003). doi:10.1029/2002JA009578

Y. Yu, A.J. Ridley, Exploring the influence of ionospheric O+ outflow on magnetospheric dynamics: dependence on the source location. J. Geophys. Res. Space Phys. **118**(4), 1711–1722 (2013). doi:10.1029/2012JA018411

J. Zhang, M.W. Liemohn, D.L. De Zeeuw, J.E. Borovsky, A.J. Ridley, G. Toth, S. Sazykin, M.F. Thomsen, J.U. Kozyra, T.I. Gombosi, R.A. Wolf, Understanding storm-time ring current development through data-model comparisons of a moderate storm. J. Geophys. Res. **112**(A4), 04208 (2007). doi:10.1029/2006JA011846

Y. Zheng, T.E. Moore, F.S. Mozer, C.T. Russell, R.J. Strangeway, Polar study of ionospheric ion outflow versus energy input. J. Geophys. Res. **110**(A7), 07210 (2005). doi:10.1029/2004JA010995

X.-Z. Zhou, V. Angelopoulos, A.R. Poppe, J.S. Halekas, ARTEMIS observations of lunar pickup ions: mass constraints on ion species. J. Geophys. Res., Planets (2013). doi:10.1002/jgre.20125

L. Zhu, R.W. Schunk, J.J. Sojka, Polar cap arcs: a review. J. Atmos. Sol.-Terr. Phys. **59**(10), 1087–1126 (1997). doi:10.1016/S1364-6826(96)00113-7. http://www.sciencedirect.com/science/article/pii/S1364682696001137

Q.-G. Zong, B. Wilken, S.Y. Fu, T.A. Fritz, A. Korth, N. Hasebe, D.J. Williams, Z.-Y. Pu, Ring current oxygen ions escaping into the magnetosheath. J. Geophys. Res. **106**(A11), 25541 (2001). doi:10.1029/2000JA000127

S. Zou, M.B. Moldwin, A.J. Ridley, M.J. Nicolls, A.J. Coster, E.G. Thomas, J.M. Ruohoniemi, On the generation/decay of the storm-enhanced density plumes: role of the convection flow and field-aligned ion flow. J. Geophys. Res. Space Phys. **119**(10), 8543–8559 (2014). doi:10.1002/2014JA020408

R.D. Zwickl, D.N. Baker, S.J. Bame, W.C. Feldman, J.T. Gosling, E.W. Hones, D.J. McComas, B.T. Tsurutani, J.A. Slavin, Evolution of the Earth's distant magnetotail: ISEE 3 electron plasma results. J. Geophys. Res. **89**(A12), 11007 (1984). doi:10.1029/JA089iA12p11007

DOI 10.1007/978-1-4939-3544-4_6
Reprinted from *Space Science Reviews* Journal, DOI 10.1007/s11214-015-0184-5

Jupiter's Magnetosphere: Plasma Sources and Transport

Scott J. Bolton[1] · Fran Bagenal[2] · Michel Blanc[3] · Timothy Cassidy[4] · Emmanuel Chané[5] · Caitriona Jackman[6] · Xianzhe Jia[7] · Anna Kotova[8] · Norbert Krupp[8] · Anna Milillo[9] · Christina Plainaki[9] · H. Todd Smith[10] · Hunter Waite[1]

Received: 17 March 2015 / Accepted: 8 July 2015 / Published online: 7 October 2015

1 Introduction

Jupiter's plasma environment is one of the most interesting plasma laboratories in our solar system. Studying the plasma sources and sinks, as well as understanding the configuration and dynamics of the Jovian magnetosphere is key to the understanding of similar astrophysical systems in our galaxy. The study of Jupiter's plasma environment can be used as a template for exoplanets as well as examples of acceleration processes in protoplanetary discs.

The Jovian system is a world of superlatives: it is built around the largest planet in our solar system, more than 10 times bigger than the Earth (1 Jupiter radius (R_J) = 71492 km). Jupiter has the strongest magnetic field of all planets (its magnetic moment is 20000 times larger than Earth's, its surface magnetic field is 14 times larger compared to Earth), the largest magnetosphere (the radius of the terminator cross section is about $150R_J$) and the

✉ S.J. Bolton
sbolton@swri.edu

1 Southwest Research Institute, San Antonio, TX, USA

2 University of Colorado, Boulder, CO, USA

3 IRAP, CNRS-Université Paul Sabatier, Toulouse Cedex 4, France

4 LASP, University of Colorado, Boulder, CO, USA

5 Center for Mathematical Plasma Astrophysics, KU Leuven, Leuven, Belgium

6 Department of Physics and Astronomy, University of Southampton, Southampton, UK

7 Department of Atmospheric, Oceanic, and Space Sciences, University of Michigan, Ann Arbor, MI, USA

8 Max-Planck-Institut für Sonnensystemforschung, Göttingen, Germany

9 Institute of Space Astrophysics and Planetology, INAF, Rome, Italy

10 Applied Physics Laboratory, Johns Hopkins University, Laurel, MD, USA

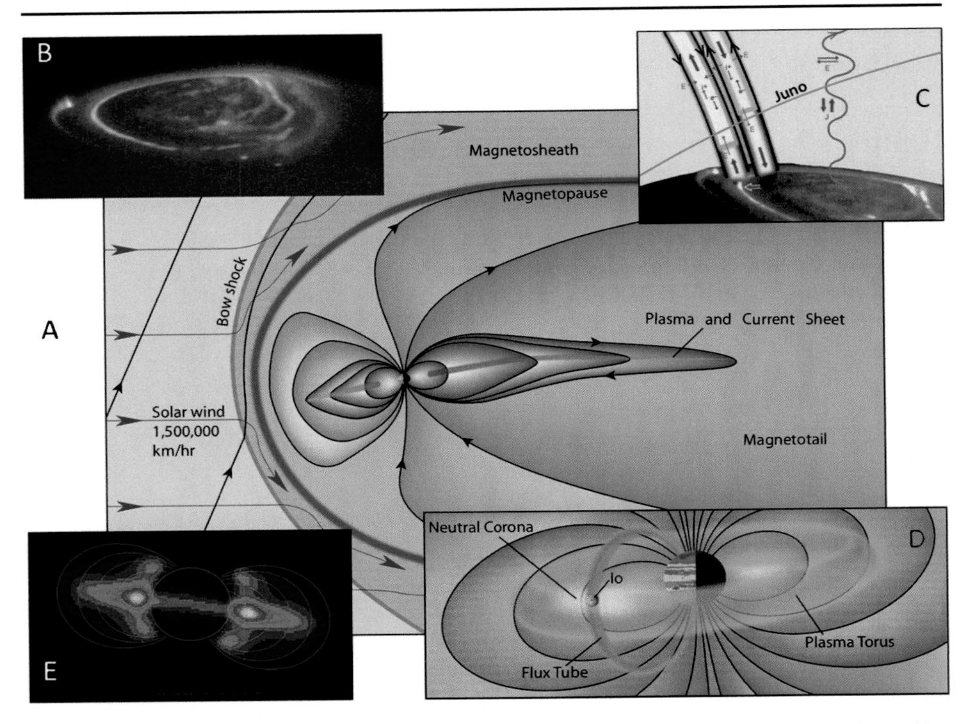

Fig. 1 (**A**) The magnetosphere of Jupiter extends 63–92 Jovian radii in the direction towards the Sun, with a tail that stretches beyond the orbit of Saturn > 4 AU, and occupies a volume over a thousand times that of the Sun. (**B**) Intense auroral emissions are signatures of the coupling between the planet and the magnetospheric plasmas. (**C**) Terrestrial experience suggests that there are regions within a few radii of the planet where the aurora-generating particles are excited. The *Juno* spacecraft will fly through these regions. (**D**) The magnetosphere is dominated by a ~ 1 ton/s source of plasma from Io's volcanic gases that forms a toroidal cloud around Jupiter. (**E**) Close to the planet are strong radiation belts comprising energetic (MeV) electrons that emit synchrotron emission. From Bagenal et al. (2014)

strongest radiation belts (see Fig. 1). Jupiter's auroral power is about 100 times stronger than at Earth. Jupiter is surrounded by 67 moons, the largest number of all planets. The moon Io is the body with the strongest volcanic activity in our solar system, Ganymede is the biggest moon of all in the heliosphere and the only moon with its own intrinsic magnetic field forming a unique mini-magnetosphere within the large Jovian magnetosphere.

Jupiter has been visited by a total of eight spacecraft in the last 40 years (see Table 1 and Fig. 2), but thus far the only dedicated orbiter has been the Galileo spacecraft, which orbited between 1995 and 2003 and is the source of most of our current knowledge about the Jovian system. Most recently (2006–2007), the New Horizons spacecraft traversed the Jovian tail to distances greater than $2500 R_J$ on its way to Pluto. The next chance to explore Jupiter will be with the arrival of the Juno mission in 2016 (Bolton 2010; Bolton et al. 2015) to be followed by the JUICE mission in 2030 (Grasset et al. 2013).

The measurements from these missions and telescopes have been used as input for global simulations of the entire magnetosphere as well as to derive new models of the magnetic field and the plasma environment. Based on these data and simulations our current view of the Jovian plasma environment is described below.

Table 1 Spacecraft exploration of the Jovian system

Spacecraft that encountered Jupiter	Year(s)	Type
Pioneer 10 (P10)	1973	Flyby
Pioneer 11 (P11)	1974	Flyby
Voyager 1 (VG1)	1979	Flyby
Voyager 2 (VG2)	1979	Flyby
Ulysses (ULY)	1992	Flyby
Galileo (GLL)	1995–2003	Orbiter
Cassini (CAS)	2000/2001	Flyby
New Horizons (NH)	2007	Flyby

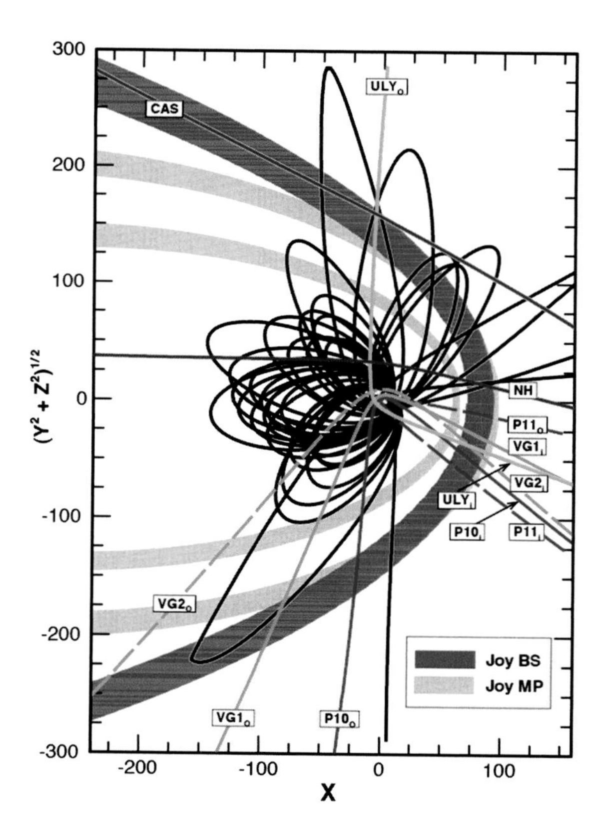

Fig. 2 Trajectories of the spacecraft that have visited Jupiter's magnetosphere (Vogt 2012). The *shaded areas* indicate the minimum and maximum distances of the magnetopause (MP, *light gray*) and bow shock (BS, *dark gray*) after Joy et al. (2002)

The pre-*Galileo* understanding of the Jovian magnetosphere is presented in Dessler's (1983) book *Physics of the Jovian Magnetosphere* and the advances made by the *Ulysses* and *Galileo* missions are reviewed in seven chapters of *Jupiter: The Planet, Satellites and Magnetosphere* (edited by Bagenal et al. 2004). Bagenal et al. (2014) reviewed the Jovian magnetosphere in anticipation of Juno's arrival in 2016, while the Jovian tail was reviewed by Krupp et al. (2015).

2 Global Configuration

The classical scale of a planet's magnetosphere, namely the Chapman-Ferraro radius R_{CF}, as derived by Chapman and Ferraro (1930), comes from a simple pressure balance between the ram pressure of the solar wind $(\rho V^2)_{\rm sw}$ and the magnetic pressure of a dipole field $(B^2 2\mu_0)$ assumed to represent the planetary magnetic field. This results in a weak variation in the dayside magnetopause distance $R_{\rm MP}$ such that $R_{\rm MP} \propto (\rho V^2)_{\rm sw}^{-1/6}$ (for a solar wind mass density $\rho_{\rm sw} = m_p n_{\rm sw}$ and speed $V_{\rm sw}$). While this Chapman-Ferraro magnetopause distance works well for Earth (except during periods of extremely unusual solar wind conditions, see Chané et al. 2012), it underestimates the sizes of the giant planet magnetospheres, particularly Jupiter. If the pressure P of the charged particle populations inside the magnetosphere dominates over the local magnetic field pressure $(B^2 2\mu_0)$, then $\beta = P/(B^2 2\mu_0) > 1$ and the particle pressure inflates and stretches out the magnetic field, generating strong currents in the equatorial plasma disk. In addition, the centrifugal force associated with the plasma rotating around the planet also stretches the magnetosphere. Figure 1 illustrates how the substantial internal plasma pressure as well as the centrifugal force at Jupiter expands the magnetosphere well beyond that of a dipole internal field. At Jupiter, values of β greater than unity are found beyond $\sim 15R_{\rm J}$, increasing to $\beta > 100$ by $45R_{\rm J}$ (Mauk et al. 2004). In addition to the plasma pressure dominating the magnetic pressure, the radial profile of plasma pressure is considerably flatter than the $R^{-1/6}$ variation in magnetic pressure for a dipole field. It is the high plasma pressure in the plasma disk as well as the centrifugal force that doubles the scale of Jupiter's magnetosphere from the dipolar stand-off distance of $\sim 42R_{\rm J}$ to over $90R_{\rm J}$.

Careful statistical analysis (combined with modeling) of how the magnetopause standoff distance at Jupiter varies with solar wind conditions by Joy et al. (2002) revealed a bimodal distribution with high probabilities at 63 and $92R_{\rm J}$. Furthermore, the observed magnetopause locations indicate a variation in $R_{\rm MP}$ with solar wind ram pressure $R_{\rm MP} \propto (\rho V^2)_{\rm sw}^{-\alpha}$ where α is found to be between 1/3.8 and 1/5.5, a stronger function than for a dipole (Slavin et al. 1985; Huddleston et al. 1997; Joy et al. 2002; Alexeev and Belenkaya 2005). A factor 10 increase in ram pressure at Earth reduces $R_{\rm MP}$ to 70 % of the nominal value, while at Jupiter a tenfold variation in solar wind pressure, often observed at 5 AU (Jackman and Arridge 2011; Ebert et al. 2014), causes the dayside magnetopause to move by a factor of ~ 2.

The overall configuration of the Jovian system has been very well described in the literature (see review articles from Khurana et al. 2004 and from Krupp et al. 2004b) and consists of an inner, middle and an outer magnetosphere, with transitions between those segments at approximately 10–$15R_{\rm J}$ and at 40–$60R_{\rm J}$. The major energy source of the system is derived from its fast rotation (with a rotation period of about 10 hours) and the major particle source is sputtered from Io's atmosphere and surface. Io is orbiting deep within the magnetosphere at $5.9R_{\rm J}$. The magnetic dipole axis is tilted about 10° from the rotation axis of the planet.

The inner magnetosphere ($< 15R_{\rm J}$) close to the planet is the region of trapped charged particles on dipolar-like field lines. This is the region of the harshest radiation belts in our solar system where electrons and ions reach energies of tens of MeV, with very high intensities (reviewed by Woodfield et al. 2014; Bolton et al. 2004). The sources of these populations include both galactic cosmic rays and radially inward drifting particles originating in the outer magnetosphere (see description below). The inner magnetosphere also includes the ring system of the planet (related to the moons Amalthea and Thebe) and the Galilean moon Io which is the major plasma source of the system. Gases escaping from Io's atmosphere form a neutral cloud extending along Io's orbit around Jupiter. Ionization of this neutral cloud produces a torus of plasma that emits over a terawatt of line emissions, mostly in the UV (reviewed by Thomas et al. 2004) powered by ion pickup in the rapidly rotating system.

The middle magnetosphere of Jupiter ($15 < r < 60 R_J$) is the region where the magnetic field stretches radially and significantly deviates from a dipole. Caused by the mass loading of the magnetic field lines with heavy ions from Io and due to the centrifugal forces in the rapidly rotating environment the entire magnetosphere is radially stretched forming a magnetodisc and associated current sheet close to the equatorial plane. Electrical currents flowing along the magnetic field couple the magnetodisc to the planet and transfer momentum from the neutral atmosphere to the magnetodisc. This momentum transfer is very efficient close to the planet and forces the plasma to rigidly corotate. However, farther from Jupiter, this coupling is not strong enough to accelerate the plasma to rigid corotation: the plasma sub-corotates (its angular velocity is lower than Jupiter's angular velocity). The region of the "corotation breakdown" is a function of local time and a function of time (see Bonfond et al. 2012). It is this current system which is responsible for the main auroral emission (see Hill 2001; Cowley and Bunce 2001) where mainly keV electrons are accelerated downward into the polar regions, hitting atmospheric particles and emitting radiation across the spectrum, from x-rays, UV and visible to IR and radio (reviewed by Clarke et al. 2004).

The outer magnetosphere beyond 40–60R_J is the region where the magnetic field lines are stretched further, until the magnetopause on the dayside, or several 1000R_J down the Jovian magnetotail on the nightside. While the Galileo trajectories covered only distances as far out as 150R_J near local midnight, the New Horizons spacecraft has sampled the coherent Jovian magnetotail in situ to distances from 1600R_J (McNutt et al. 2007) to 2500R_J. However, observations from the Voyager spacecraft suggest that the Jovian tail can stretch even as far as the orbit of Saturn (Kurth et al. 1982; Scarf et al. 1982), which would make the Jovian magnetosphere by far the largest coherent structure in our solar system (except for the heliosphere itself).

The whole system is fed by plasma sources that predominately come from inside the magnetosphere with external contributions. The volcanic moon Io is the strongest internal source, with minor contributions from the moon Europa and possibly other moons as well as Jupiter's ionosphere. Embedded in the inner Jovian magnetosphere, the icy moons experience a strong interaction with their surrounding plasma. Data from Galileo showed that these moons are continuously irradiated by energetic ions (H^+, C^{n+}, O^{n+} and S^{n+}) and electrons in the energy range from keV to MeV (Cooper et al. 2001; Paranicas et al. 2002). The effects of this intense irradiation on ice are the main drivers of the generation of tenuous atmospheres around these bodies and could be of crucial importance in generating the conditions of the ocean below the icy crust (Chyba 2000). However, the details of the surface processes and their impact on the environment are poorly known. External plasma sources for the Jovian magnetosphere are the solar wind and galactic cosmic rays. The details will be described in subsequent sections. We will first address Jupiter's atmosphere and ionosphere, followed by Io and the Io plasma torus. We then address Europa and Ganymede and finally the solar wind, and a general discussion on transport mechanisms.

3 Atmosphere and Ionosphere

Unlike the Earth and inner terrestrial planets, Jupiter does not have a solid surface. Altitude scales are generally referred to a reference pressure level, which is generally accepted to be the 1 bar level. This pressure level corresponds to a radial distance of about 71492 km from

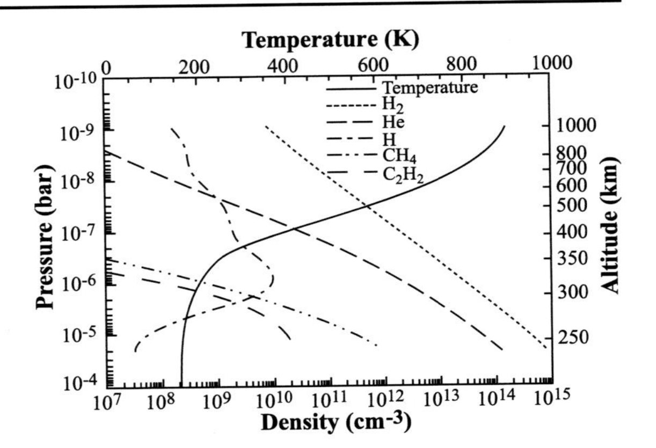

Fig. 3 A model of Jupiter's atmosphere, showing neutral gas densities and temperatures (from Schunk and Nagy 2009, courtesy of Tariq Majeed)

the center of Jupiter at the equator. Note that Jupiter, as all outer planets, is oblate due to the planet's rapid rotation rate. The atmosphere of Jupiter consists predominantly of molecular hydrogen and some lesser amounts of helium and atomic hydrogen. In the lower atmosphere CH_4 and other hydrocarbons are also present as minor constituents. The latest estimate of the thermospheric temperature at Jupiter is about 900 K. However, this value is rather uncertain. At present, the energy sources responsible for this relatively high temperature have not been established; candidate sources include Joule heating, gravity wave dissipation, and precipitating particle energy deposition. The latest estimates of the densities and the neutral gas temperature at Jupiter are shown in Fig. 3 based on a model of pressure and altitude profiles of the major neutrals in Jupiter's atmosphere.

The presently available direct information regarding the ionosphere of Jupiter is based on the Pioneer 10 and 11, Voyager 1 and 2 and Galileo radio occultation measurements. There is no direct information available concerning ion composition or plasma temperatures in Jupiter's ionosphere; our limited understanding is based on model calculations. Given that Jupiter's upper atmosphere consists mainly of molecular hydrogen, the major primary ion, formed by either photoionization or particle impact ionization, is H_2^+. In the equatorial and low to mid latitudes the electron-ion pair production is mainly due to photoionization, while at high latitudes impact ionization by precipitating particles is believed to become very important. The actual equilibrium concentration of the major primary ion, H_2^+, is likely to be very small because it undergoes rapid charge transfer reaction with neutral molecular hydrogen, resulting in H_3^+, which is believed to be a major ion and which is eventually lost by dissociative recombination with an electron, as indicated below:

$$H_2^+ + H_2 \rightarrow H_3^+ + H \tag{3.1}$$

$$H_3^+ + e \rightarrow H_2 + H \tag{3.2}$$

The presence of H_3^+ in Jupiter's ionosphere has been confirmed by ground-based measurements using the NASA Infrared Telescope Facility (IRTF) on Mauna Kea, Hawaii (Stallard et al. 2002). They estimated that the nighttime vibrational temperature is somewhere between 940 and 1065 K and the column density is of the order of $1 \times 10^{16}\ m^{-2}$.

Protons, H^+ are created at high altitudes by either the direct ionization of neutral atomic hydrogen or by the dissociative ionization of molecular hydrogen. H^+ can only recombine directly via radiative recombination, which is extremely slow. It was suggested years ago (McElroy 1973) and discussed in Chap. 2 (2.1.2b) (Seki et al. 2015) that it can be lost via

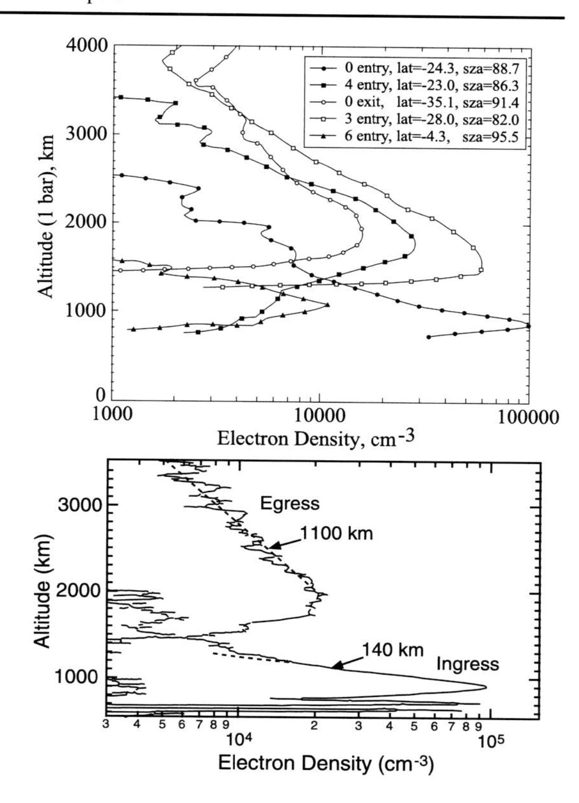

Fig. 4 Galileo radio occultation measurements of ionospheric electron densities (above) (from Schunk and Nagy 2009, courtesy of A.J. Kliore)

charge exchange with the fraction of molecular hydrogen which is in a vibrational state of 4 or higher. The uncertainty associated with the loss process of H^+ leaves open the question of the identity of the major ion near the ionospheric peak as a function of latitude. There are no measurements that constrain the vibrational distribution of molecular hydrogen, but some model calculations do indicate a significant population in the higher excited states (Cravens 1987; Hallett et al. 2005). Direct photoionization of hydrocarbon molecules at lower altitudes can lead to a relatively thin layer around 300 km (Kim and Fox 1994).

Figure 4 shows electron density profiles obtained by the radio occultation technique from the Galileo spacecraft. The top figure shows examples of egress and ingress for multiple latitudes. The observed peak electron densities are in the range of 10^4 to 10^5 cm^{-3}. By the nature of the encounter geometries all these results are very close to the terminator, thus representing similar solar zenith angles. There is great variability among the observed density profiles in the top panel and there seems to be no clear latitude dependence. The lower panel of Fig. 4 compares the two extreme cases of the altitude of the peak electron density and the topside scale height. These examples illustrate significantly different atmospheric profiles; the higher altitude peak is associated with a greater scale height. These differences may be the result of different major ionization source or loss mechanisms or different chemistries. A number of one and multi-dimensional models have been published to date (Majeed and McConnell 1991; Bougher et al. 2005; Millward et al. 2005), but none of these models have provided any clear explanation of these very significant variations.

The ionosphere may be a source of plasma for the magnetosphere. At Jupiter, the most convincing evidence comes from Hamilton et al. (1981) who report fluxes in the Jovian magnetosphere of He^+ and H_3^+ ions, which most likely come from Jupiter's ionosphere. The outflow of ionospheric plasma was proposed by Nagy et al. (1986) and estimated to be 2×10^{28} ions/s which is comparable in number density to the iogenic source (see next section) but, assuming the composition is mostly protons, the mass would be only 35 kg/s.

4 Io and the Plasma Torus

The magnetosphere of Jupiter is greatly influenced by strong internal sources of neutral particles and of plasma located deep inside the magnetosphere, i.e. the Galilean moons Io, and Europa, and to a lesser extent Ganymede and Callisto (see review by Thomas et al. 2004). While Io's atmosphere is dominated by sulfur dioxide (SO_2), Europa's atmosphere mostly contains molecular oxygen (O_2), but also molecular hydrogen (H_2) at higher altitudes. Particles from these atmospheres are constantly ejected into the Jovian magnetosphere, either directly as gas, or as plasma. Most of the neutral particles present in the magnetosphere stem from either charge-exchange processes or elastic collisions between the heavy ions in the magnetosphere and the atmosphere of Io or Europa. These neutral particles are then on a Keplerian orbit around the planet, forming the Io and Europa neutral clouds. These extended neutral gas clouds experience ionization processes and charge exchange collisions, making them the dominant source of plasma. The Io and Europa atmospheres are also a direct source of plasma, because their neutral particles are subject to electron impact (and to a lesser extent photo-) ionization, and to charge exchange collisions with the magnetospheric plasma. One these particles are charged they are accelerated by the Lorentz force, and start to corotate around the planet (i.e. they feel the effect of Jupiter's rotating magnetic field). When charge exchange occurs, the charged particles (which move at approximately the corotation speed) become neutralized and escape the torus.

Compared with the local plasma, which is nearly corotating with Jupiter at 74 km/s, the neutral atoms are moving slowly, close to Io's orbital speed of 17 km/s. When a neutral atom becomes ionized (largely via electron impact) it becomes subject to the ambient Jovian corotation electric field, resulting in a gyromotion of 57 km/s. Thus, new S^+ and O^+ ions gain 540 eV and 270 eV in gyro-energy, respectively. The new "pick-up" ion is also accelerated up to the bulk speed of the surrounding plasma by the magnetic Lorentz force. The necessary momentum comes from Jupiter's upper atmosphere and ionosphere via field aligned currents—the Jovian rotation being the ultimate source of momentum and energy for these (and most) processes in the magnetosphere. About one-third to one-half of the neutral atoms are ionized to produce additional fresh plasma, while the rest are lost via reactions in which a neutral atom exchanges an electron with a torus ion. When neutralized, the previously charged, corotating particle is no longer confined by the magnetic field and, since the corotation speed is well above the gravitational escape speed from Jupiter, flies off as an energetic neutral atom. This charge-exchange process adds gyro-energy to the ions and extracts momentum from the surrounding plasma, but it does not add more plasma to the system (even though it can add or remove mass to the system in case of asymmetric charge-exchange: when the neutral particle and the ion do not have the same mass).

Smyth and Marconi (2006) developed a model of the neutral clouds in the Jovian magnetosphere. The longitudinally-averaged column density of the neutral particles that they obtained is displayed in Fig. 5. This figure shows that close to Jupiter (less than $\sim 7.5 R_J$) the majority of the neutral particles are from Io, while farther away (beyond $\sim 7.5 R_J$) most

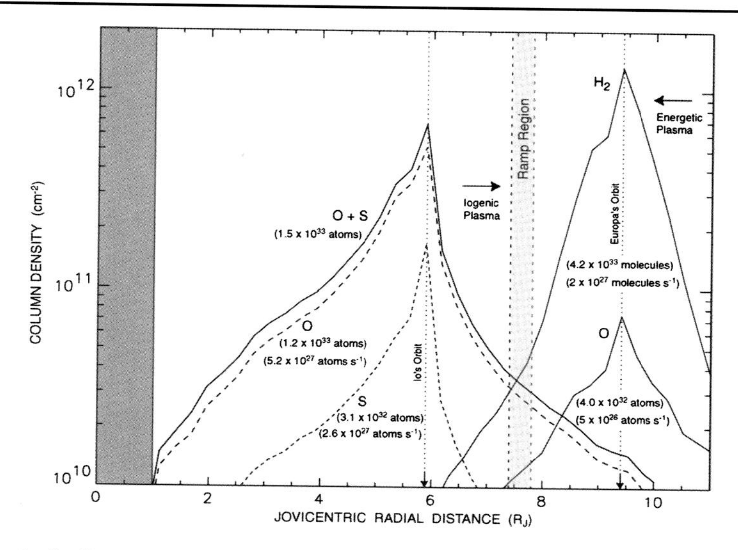

Fig. 5 Longitudinally-averaged radial column density of the neutral clouds of Io and Europa (from Smyth and Marconi 2006)

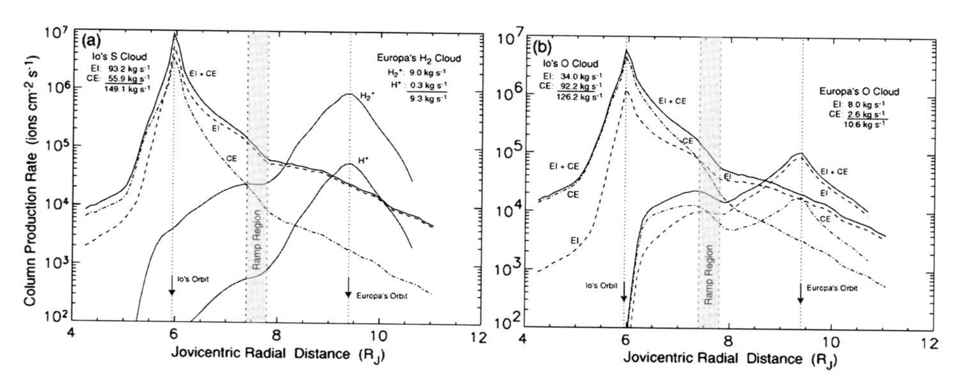

Fig. 6 Longitudinally-averaged column production of pickup ions generated by Io's and Europa's neutral clouds (from Smyth and Marconi 2006)

of them come from Europa. If one just counts the total number of surviving neutral particles in these clouds, the molecular hydrogen emanating from Europa (4.2×10^{33} molecules) is larger than the combined number of oxygen and sulfur atoms around the Io region (1.5×10^{33} atoms). The mass-loading associated with Io's and Europa's tori is shown in Fig. 6. According to Smyth and Marconi's (2006) model, the largest source of plasma is, by far, Io (particularly the extended neutral cloud), with a plasma production rate of $\sim$ 250 kg/s, while Europa's atmosphere and neutral cloud only generate $\sim$ 22–27 kg of plasma per second. Even though the number density of molecular hydrogen at Europa's orbit is higher than the density of sulfur atoms and oxygen atoms at the orbit of Io, the mass-loading rate at the orbit of Europa is more than an order of magnitude lower than at the orbit of Io. This is because: (1) the oxygen and sulfur atoms are heavier than molecular hydrogen, and (2) the electron temperature is much higher at smaller radial distances, meaning that electron impact ionization is more efficient at the orbit of Io.

Strong centrifugal forces confine the plasma near the equator. Thus, the densest plasma ($\sim 2000\ \text{cm}^{-3}$) forms a torus around Jupiter near the orbit of Io. A lighter population of H^+ ions (with a relative concentration of a few % and a temperature of a few 10s eV), less confined near the equator, has been inferred from radio (decametric, DAM) measurements (Zarka et al. 2001). The Io plasma torus has a total mass of ~ 2 megaton, which would be replenished by a source of ~ 1 ton/s in ~ 23 days. Multiplying by the typical energy of the ions ($T_\text{i} \sim 60$ eV) and electrons ($T_\text{e} \sim 5$ eV), we obtain $\sim 6 \times 10^{17}$ J for the total thermal energy of the torus. The observed UV power is about 1.5 TW, emitted via more than 50 ion spectral lines, most of which are in the EUV. This emission would drain all the energy of the torus electrons in ~ 7 h. Ion pickup replenishes energy, and Coulomb collisions feed the energy from ions to electrons but not at a sufficient rate to maintain the observed emissions. A source of additional energy, perhaps mediated via plasma waves, seems to be supplying hot electrons and a comparable amount of energy as ion pickup. The 20–80 day time scale (equivalent to 50–200 rotations) for the replacement of the torus indicates surprisingly slow radial transport that maintains a relatively strong radial density gradient.

It should be noted that these mass-loading rates vary on time scales of months to years. For the Io torus, Bagenal and Delamere (2011) estimated that at the time of Voyager 1 in March 1979 the neutral source and the plasma source were 800 kg/s and 260 kg/s, respectively. At the time of the Cassini flyby in September 2000, these sources were much higher: 3000 kg/s and 1400 kg/s, respectively. Bonfond et al. (2012) also argued that a major increase of the mass-loading rate happened in the spring of 2007, which had considerable repercussions for the configuration of the Jovian magnetosphere. These authors observed that the position of the main oval moved equatorward over a few months, which is consistent with an increased mass-loading rate.

As the Iogenic plasma moves outward, the conservation of angular momentum would suggest that the plasma should lose angular speed. In a magnetized plasma, however, electrical currents easily flow along magnetic fields and couple the magnetospheric plasma to Jupiter's flywheel. Hill (1979) argued that at some point the load on the ionosphere increases to the point where the coupling between the ionosphere and corotating atmosphere—manifested as the ionospheric conductivity—is not sufficient to carry the necessary current, causing the plasma to lag behind corotation. The main aurora is the signature of Jupiter's attempt to spin up its magnetosphere or, more accurately, Jupiter's failure to spin up its magnetosphere fully (see Cowley and Bunce 2001). The position of the corotation break-down, and thus the latitude of the main oval, depends on the mass-loading rate. Hill (1979), assuming that the magnetic field was a simple dipole, derived the following expression for the position of the corotation break-down:

$$R_0 = \sqrt[4]{\frac{2\pi\,\Sigma(\mu_0/4\pi M_p)^2}{\dot{M}}} \tag{4.1}$$

where Σ is the conductance of the ionosphere, μ_0 the permeability of free space, M_p the planetary magnetic moment, and M the total rate of production and outward transport of plasma mass. Analytical models (Hill 1979; Cowley and Bunce 2001; Nichols and Cowley 2003) are very useful for understanding the dynamics of the magnetosphere, but being axisymmetric, they cannot account for local time asymmetries. To study the three-dimensional structure of the magnetosphere, global simulations are more appropriate (Miyoshi and Kusano 1997; Ogino et al. 1998; Walker et al. 2001; Fukazawa et al. 2005;

Moriguchi et al. 2008; Chané et al. 2013). For instance, Chané et al. (2013) have shown that the discontinuity of the main oval in the pre-noon sector (discovered by Radioti et al. 2008) was caused by an asymmetry in the pressure distribution, due to the interaction between the rotating plasma and the magnetopause. It is known that the mass-loading rate affects the position of the main oval, but does it influence the intensity of the main oval? Nichols and Cowley (2003), using their axisymmetric analytical model, showed that, if one assumes that the magnetic field in the magnetosphere is dipolar, the peak value of the field-aligned currents in the ionosphere does not depend on the mass-loading rate:

$$(j_{\|}/B)_{max} \approx 0.1076\Sigma_P^* \,(\text{mho})\,\text{pA}\,\text{m}^{-2}\,\text{nT}^{-1} \tag{4.2}$$

On the other hand, using a more realistic magnetic field (the current sheet magnetic field model, see Connerney et al. 1981; Edwards et al. 2001) they found that the peak value of the field-aligned currents depends weakly on the mass-loading rate:

$$(j_{\|}/B)_{max} \approx 2.808\big(\Sigma_P^* \,(\text{mho})^{3.42}\dot{M}(10^3\ \text{kg/s})^{-0.71}\big)^{\frac{1}{2.71}}\ \text{pA}\,\text{m}^{-2}\,\text{nT}^{-1} \tag{4.3}$$

Using this formula, one finds that if the mass-loading rate increases by an order of magnitude, the field-aligned current peak value decreases by less than a factor of two. However, Nichols and Cowley (2003) did not take into account the fact that the magnetic field could be affected by the mass-loading rate. This effect was included in Nichols (2011), where a magnetic field model similar to the one from Caudal (1986) was used. Depending on the assumption made in this model (namely whether the cold plasma density depends on the mass-loading rate or not) they found that the peak value of the field-aligned currents is correlated or anti-correlated with the mass-loading rate; and this remains, as of today, an open question.

The above models of corotation breakdown assume the coupling is limited by the ionospheric conductivity. Studies by Ergun et al. (2009) and Ray et al. (2010, 2012, 2014) point out that the rarefaction of plasma between the plasma sheet and the ionosphere leads to small-scale regions of parallel electric fields ("double-layers") a few R_J above the ionosphere. They argue that the linear approximation to Knight's current-voltage relation (Knight 1973) (for more detail see Seki et al. 2015) commonly assumed for ionosphere-magnetosphere coupling, breaks down and that the currents flowing between the two regions become saturated, modifying the coupling between the magnetosphere and ionosphere. The Juno spacecraft will fly through the polar regions of Jupiter's magnetosphere with a suite of particles and fields instruments that will elucidate this key issue of magnetosphere-ionosphere coupling.

5 Europa

Europa is embedded in the radiation belt of Jupiter and it is not protected by an internal magnetic field; hence, it is subjected to energetic ion bombardment. The Jovian magnetospheric plasma, confined by Jupiter's magnetic field, slightly subcorotates anticlockwise at ~ 100 km/s at the orbit of Europa (Kivelson et al. 2009). Since the orbital velocity of Europa is 14 km/s anticlockwise, the bulk plasma flow is constantly overtaking the satellite. Mauk et al. (2004) showed that the energy deposited on the icy satellites by magnetospheric particles is carried principally by the particles at energies above 10 keV.

As a consequence of this deposited energy, Europa's surface releases particles that form a neutral gas envelope around the moon. Theoretical simulations (Johnson 1990;

Johnson et al. 2004; Shematovich et al. 2005; Smyth and Marconi 2006; Cassidy et al. 2007; 2010; Plainaki et al. 2010; 2012) predict that Europa's gas envelope consists mainly of three different populations (Fig. 5): (a) H_2O molecules, released through direct ion sputtering caused by the energetic ions of Jupiter's magnetosphere that impact the moon's surface; (b) O_2 and (c) H_2 molecules. The latter two species are produced through chemical reactions among different products of H_2O radiolytic decomposition. Sputtering also releases some minor surface species such as water group members (O, H, OH) and sodium or potassium (Brown and Hill 1996; Brown 2001; Leblanc et al. 2002, 2005; Cassidy et al. 2008 that populate the neutral gas envelope (for more details see Seki et al. 2015).

The presence of molecular oxygen in the exosphere of Europa has been proved only indirectly through either observations from the Earth or in situ measurements. The Goddard High-Resolution Spectrograph (GHRS) and the Advanced Camera for Surveys (ACS) on the Hubble Space Telescope (HST) observed the far-ultraviolet (UV) auroral emissions of atomic oxygen that were attributed to electron impact dissociative excitation of O_2 (Hall et al. 1995, 1998; Saur et al. 2011) with an estimated column density of $\sim 10^{14}$ to 10^{15} cm^{-2}. However, this column density estimate is quite uncertain since the Jovian magnetospheric electrons responsible for the observed emissions can be partially diverted and cooled through interactions with the atmosphere (Saur et al. 1998; Schilling et al. 2008). Kliore et al. (1997) estimated that the O_2 density (near the surface) required to produce the electron density observed by the Galileo spacecraft was $\sim 3 \times 10^{14}$ m^{-3}. Observations acquired in 2001 by the Ultraviolet Imaging Spectrograph (UVIS) on the Cassini spacecraft during its flyby of Jupiter (Hansen et al. 2005) confirmed, independently, the presence of an O_2 atmosphere at Europa with a comparable column density to the one obtained through the ground-based observations. McGrath et al. (2004), based on the HST/Space Telescope Imaging Spectrograph (STIS) observations of Europa's trailing hemisphere, evidenced an asymmetric auroral emission at Europa, with a surplus in the anti-Jupiter direction with column density in the range $2\text{–}5 \times 10^{15}$ cm^{-2}. Saur et al. (2011) analyzed HST/ACS observations of Europa's leading hemisphere and estimated an O_2 column density lower by a factor $2\text{–}3(1 \times 10^{15}\ cm^{-2})$ than the one calculated by McGrath et al. (2004). Moreover, Saur et al. (2011) observed a surplus of emission at the apex of Europa's leading hemisphere. Roth (2012) suggested that some of these oxygen emissions may result from electron impact of water vapor plumes. Roth et al. (2014) claimed that the simultaneous observation of emissions from both atomic oxygen and atomic hydrogen were further evidence of the existence of water plumes erupting from the moon's surface.

Although H_2O is the dominant sputter product from water ice, O_2 is the dominant exospheric constituent because, unlike the water molecules, it does not freeze to the surface after being sputtered and returned to the surface by the moon's gravity (Johnson et al. 1982b; Shematovich et al. 2005; Luna et al. 2005). The oxygen molecules, unlike the other major water-dissociation product, H_2, also lack sufficient energy to overcome Europa's gravity (Smyth and Marconi 2006). As a result a thin and almost homogenous exospheric envelope (with thickness of some hundreds of kms), consisting of thermal O_2 molecules, with relatively high density, accumulates around the moon (Plainaki et al. 2012; 2013). At higher altitudes non-thermal exospheric O_2 dominates. On the basis of the Kliore et al. (1997) density values, Plainaki et al. (2010) estimated that the O_2 mean-free-path in Europa's atmosphere ranges from 13 km to 78 km. The scale-height estimations vary from 17 km to ~ 26 km (Ip 1996; Plainaki et al. 2010). Therefore, Europa's O_2 environment can be considered as a transitional case between a (collisional) atmosphere and a (collisionless) exosphere. Nonetheless this neutral environment is so tenuous that it does not act as a significant obstacle to escaping particles released from the moon surface. Tenuous as it is, the neutral environment is still a barrier to magnetospheric bombardment: ionospheric conductivity results in the diversion of magnetospheric plasma flow around Europa (Saur et al. 1998). Europa's environment may work like a self-regulating system. The interaction may be self-limiting given that the ion bombardment with the surface creates the exosphere, but the exosphere (and ionosphere) limit ion bombardment by diverting plasma around Europa (Cassidy et al. 2013). The density of the overall oxygen exosphere is supplied until it reaches a steady state with exospheric loss processes (Johnson et al. 1982b, 1982a; Saur et al. 1998; Shematovich and Johnson 2001; Shematovich et al. 2005).

Saur et al. (1998) developed a 2D plasma model to study the interaction of the Jovian magnetosphere with the atmosphere/ionosphere of Europa and sources and sinks that maintain the neutral O_2 atmosphere. They concluded that the net mass balance between source and loss to/from the atmosphere is about $\sim 50\ \mathrm{kg\,s^{-1}}$. The equivalent O_2 escape rate of $8.5 \times 10^{26}\ \mathrm{s}^{-1}$ is dominated by the loss of fast neutrals, produced mainly via ion sputtering, rather than the loss of ionospheric O^{2+} pickup ions. The calculated ionospheric density, generated by electron impact ionization, was $\sim 10^4\ \mathrm{cm}^{-3}$, similar to measured values (Kliore et al. 1997). The Alfvénic current system closed by the ionospheric Hall and Pedersen conductivities carries a total current of 7×10^5 A in each Alfvén wing, which could contribute to the magnetic field disturbances observed by the Galileo spacecraft (Kivelson et al. 1997).

In contrast to O_2, the H_2 escape ratio is significantly higher and the hydrogen gas easily escapes from Europa's gravitational field (Plainaki et al. 2012). On the other hand, the H_2O escape rate is low because water molecules stick to the surface. The atmospheric density and residence time of H_2O in the exosphere are therefore considerably lower than those of O_2.

Different numerical, analytical and kinetic models have been developed to describe Europa's exosphere characteristics (Shematovich and Johnson 2001; Marconi 2003; Shematovich et al. 2005; Smyth and Marconi 2006; Plainaki et al. 2010, 2012, 2013). In particular, the Smyth and Marconi (2006) 2D axisymmetric kinetic model considered ion-neutral collisions in order to describe the physics in the lowest atmospheric layers above Europa's surface. Smyth and Marconi (2006) assumed that the source rates for the various species (H_2O, O_2, H_2 etc.) were determined by partitioning the O_2 source rate derived by the UV brightness of O emissions reported by Hall et al. (1995).

Recently, the Plainaki et al. (2012, 2013) 3D non-collisional Monte Carlo EGEON model described the main exospheric components that are directly generated by ion-sputtering and radiolysis. They found that the H_2O density due to ion sputtering is higher by a factor of ~ 6 on the trailing hemisphere, where the flux of Jupiter's energetic ions is higher on the leading hemisphere (Plainaki et al. 2012). Contrary to the H_2O case, the O_2 exospheric densities at high altitudes are higher on the sunlit hemisphere, thus having a periodic modulation during the moon's orbit around Jupiter (see Fig. 7 which illustrates the O_2 density spatial distribution due to magnetospheric ions impacting Europa). This happens because the temperature dependence (between 80 K at night and 130 K in the dayside) of yield values for O_2 release (Famà et al. 2008) is stronger than the effect of the enhanced trailing hemisphere bombardment. This model reproduces quite well the densities and illuminated/dark side asymmetries of the measured O_2 exosphere (Saur et al. 2011; McGrath et al. 2004). According to the EGEON model results, the observed surplus of OI emission at the 90° west longitude (leading hemisphere) (Saur et al. 2011) was due to the illumination of the leading hemisphere by the Sun that favors the radiolytic release of O_2 in the exosphere (Plainaki et al. 2013). Nevertheless, Cassidy et al. (2013) hypothesized that the yield, and therefore the actual release, has a delayed response to changes in temperature and therefore depends only on average ion precipitation. Although the Plainaki et al. (2013) model showed some global asymmetries in the O_2 density spatial distribution, it did

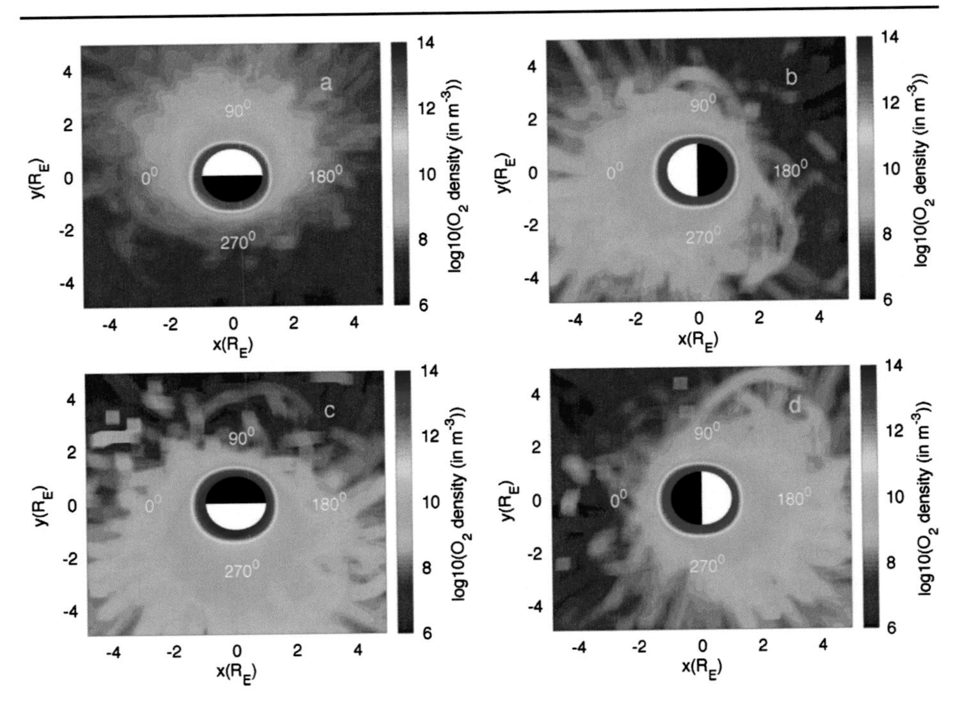

Fig. 7 Released O_2 density spatial distribution due to O^+ magnetospheric ions impacting the surface of Europa obtained by EGEON for 4 configurations along the Europa's orbit around Jupiter. In *all four panels* the xy-plane is Europa's orbital plane around Jupiter. Sunlit hemisphere is indicated with *white color* and dark hemisphere *with black*. Jupiter is *to the left* and trailing hemisphere is *down* in all four configurations (Plainaki et al. 2013)

not reproduce any local asymmetries consistent with the surplus of atomic oxygen UV emission, observed on Europa's trailing hemisphere towards Jupiter (McGrath et al. 2004). There are three possible explanations of the enhanced emission: non-uniform surface composition (resulting in anisotropic release of surface material to the exosphere); local surface activity (suggested by the recent water plume observation of Roth et al. 2014); and/or spatial variation of the impacting electron flux.

The material escaping from Europa's atmosphere is distributed along Europa's orbit forming an extensive neutral cloud. Charge exchange of inwardly-diffusing energetic ions with these neutrals generates energetic neutral atoms that were observed by the Cassini/INCA instrument (Mauk et al. 2003). The two most important escaping water group species are H_2 and O (Smyth and Marconi 2006), that are highly peaked about the satellite location and hence highly asymmetrically distributed around Jupiter, and have substantial forward clouds that extend radially inward to Io's orbit (Fig. 5). The H_2 and O neutral clouds provide a new source of molecular and atomic pickup ions for the thermal plasma; furthermore, the cooler iogenic plasma is transported radially outwards distributing from Io to Europa orbit. Smyth and Marconi (2006) estimated the spatially integrated instantaneous ion mass-loading rate for the H_2 cloud to be $\sim 9.3\ \text{kg}\,\text{s}^{-1}$ for the H_2 cloud and $\sim 10.6\ \text{kg}\,\text{s}^{-1}$ for the O cloud from electron impact and charge exchange processes. Estimates of ionization of the O cloud range from 4.4×10^25 O/s (Nagy et al. 1998) and 6.5×10^{25} O/s (2 kg/s) (Plainaki et al. 2013) to 2.6×10^{26} O/s ($\sim$ 5–10 $\text{kg}\,\text{s}^{-1}$) (Shematovich et al. 2005;

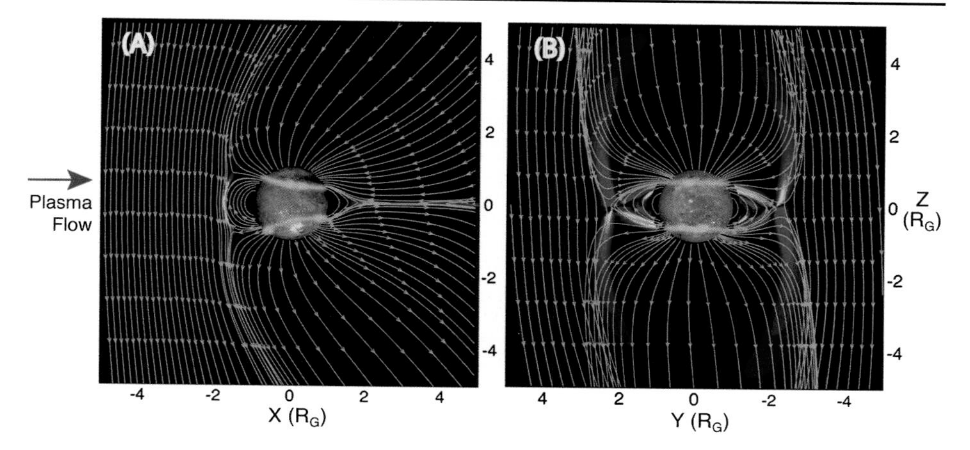

Fig. 8 Numerical model of the magnetosphere of Ganymede, with the satellite and the location of the auroral emissions superimposed (based on Jia et al. 2008). (**A**) The view looking at the anti-Jupiter side of Ganymede. (**B**) The view looking in the direction of the plasma flow at the upstream side (orbital trailing side) of Ganymede, with Jupiter to the left. *The shaded areas* show the regions of currents parallel to the magnetic field

Smyth and Marconi 2006). Plainaki et al. (2013) suggest that the O supply rate is modulated along Europa's orbit, being larger by a factor up to 4 when the trailing side is illuminated.

6 Ganymede

Jupiter's Galilean satellite Ganymede, with a radius of ~ 2634 km, is the largest moon in the solar system. It was discovered during the Galileo mission that Ganymede possesses an intrinsic magnetic field (Kivelson et al. 1996). The interaction between Jupiter's magnetospheric plasma and Ganymede's intrinsic magnetic field, whose equatorial surface field strength is about 7 times the background Jovian field, results in a mini-magnetosphere surrounding the moon (Fig. 8). Ganymede's magnetosphere is unique in that it is so far the only known satellite with an intrinsic field forming its own magnetosphere within a planetary magnetosphere (Jia et al. 2010a). The moon's magnetosphere has exhibited a variety of previously unknown phenomena revealed by the Galileo mission, including well-defined magnetospheric boundaries and magnetic perturbations associated with the intrinsic field (Kivelson et al. 1998), a rich subset of wave modes like those found within any planetary magnetosphere (Gurnett et al. 1996; Kurth et al. 1997), a significant population of charged particles associated with the moon (Frank et al. 1997a; Williams et al. 1997) and the existence of polar aurorae emitted from the atmosphere (Feldman et al. 2000; McGrath et al. 2013) shown in Fig. 9. Ganymede auroral emission has different morphologies dependent on the hemisphere of the moon and the interaction with the magnetospheric plasma.

At Ganymede's orbit, the corotating plasma of Jupiter's magnetosphere typically flows relative to the moon at speeds smaller than the ambient Alfvén speed. As a consequence, there is no bow shock formed in front of the magnetosphere. Instead, the incident Jovian plasma is slowed down by the interaction with magnetosonic waves that propagate upstream. The sub-Alfvénic interaction results in a magnetospheric configuration at

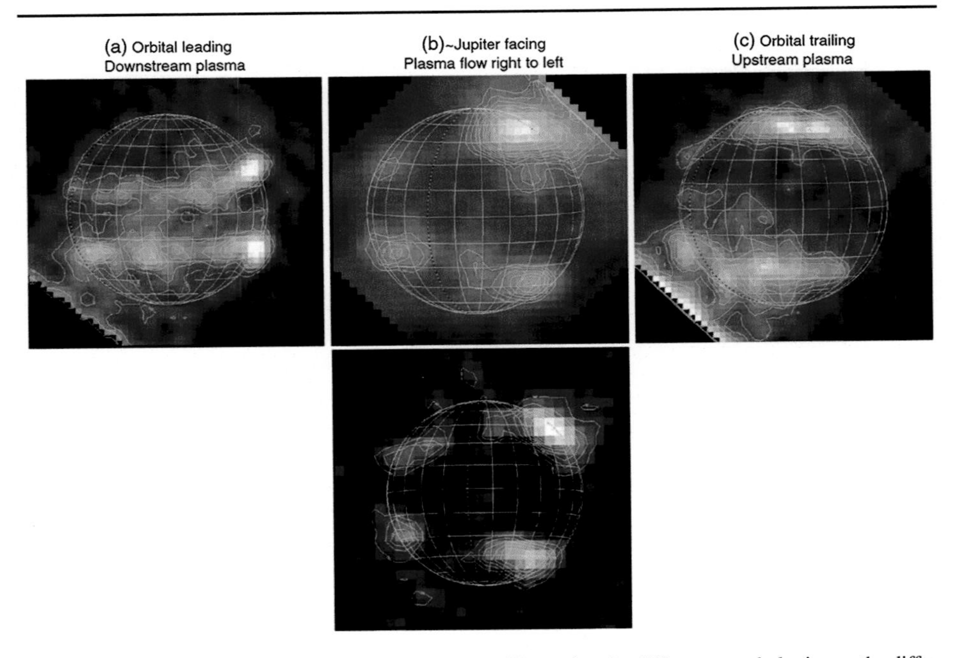

Fig. 9 Ganymede auroral emission from atomic oxygen illustrating the different morphologies on the different hemispheres of the satellite. The magnetospheric plasma flow is into the page for the trailing hemisphere, out of the page for the leading hemisphere, and approximately from right to left for the Jupiter-facing hemisphere (from McGrath et al. 2013)

Ganymede rather different from that of planetary magnetospheres arising from interactions with the super-Alfvénic and supersonic solar wind (except on extremely rare occasions when the solar wind is sub-Alfvénic, see Chané et al. 2012). A pair of the so-called Alfvén wings (Neubauer 1980, 1998; Southwood et al. 1980) form that extend almost vertically in the north-south direction, leading to a cylindrical shape of the magnetosphere in contrast to the bullet shape of planetary magnetospheres (see Fig. 8, Jia et al. 2008). While some of the incident flow diverts around the magnetosphere and is accelerated on the flanks (Frank et al. 1997a), the ambient plasma appears to gain significant access into the magnetosphere through magnetic reconnection, because Ganymede's intrinsic field is nearly anti-parallel to the external field near the equator at all times (Kivelson et al. 1998; Jia et al. 2010b). Plasma enters the Alfvén wings via magnetopause reconnection and is then convected across the polar caps towards the downstream region. Within the Alfvén wings, the plasma flow is significantly decelerated (Frank et al. 1997a; Williams et al. 1998) and the disturbances associated with the deceleration propagate away from the moon along the magnetic field lines via Alfvén waves that carry field-aligned currents. As with Io and Europa, the presence of field-aligned currents linking Ganymede to Jupiter's ionosphere has been confirmed by the discovery of ultraviolet emissions at the foot of Ganymede's flux tube in Jupiter's auroral images (Clarke et al. 2002; Grodent et al. 2009). Reconnection is expected to occur in Ganymede's magnetotail that eventually returns part of the flow back towards the moon and the upstream magnetosphere, and ejects the rest down the tail.

The plasma entering inside Ganymede's magnetosphere and impacting onto the surface, as in the Europa case, causes particle release generating a tenuous atmosphere/thick ex-

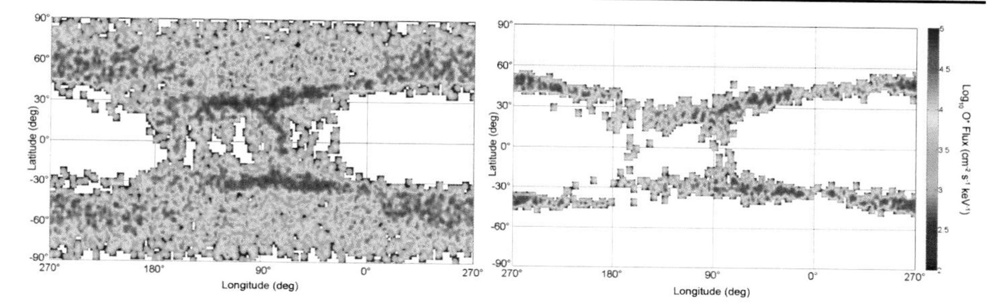

Fig. 10 Plainaki-1: Precipitation map of the O^+ differential flux ($cm^{-2}\,s^{-1}\,sr^{-1}\,keV^{-1}$) around Ganymede at initial energy equal to 10 keV in the hypothesis of full mirroring in the Jupiter's magnetosphere (*left*) and non-mirroring (*right*) assumption. Jupiter is at 0° longitude, leading at 90°. The colorbar scale is logarithmic

osphere. In fact, Jupiter's magnetospheric ions precipitating onto the surface cause sputtering, ionization and excitation of water-ice molecules, followed partially by dissociation; chemical reactions among the water-dissociation products result in the formation of new molecules (e.g. O_2, H_2, OH and minor species) that are finally ejected from the surface into Ganymede's exosphere. H_2 formed in ice diffuses and escapes much more efficiently than O_2 at the relevant temperatures in the outer solar system; moreover, H_2 escapes from the icy moons because of its low mass and the relatively weak gravitational fields. Therefore, the irradiation of Ganymede's surface can preferentially populate the magnetosphere with hydrogen, as is the case at Europa (Lagg et al. 2003; Mauk et al. 2003), leaving behind an oxygen-rich satellite surface (Johnson et al. 2004).

While the precipitation onto the surface is a loss process for Jupiter's magnetosphere, the ionization of the released exospheric particles provides a new source for Ganymede's ionosphere. These newly formed ions, after a chain of processes, could become again magnetospheric ions in Jupiter's magnetosphere.

Plainaki et al. (2015) showed that the plasma precipitation at Ganymede occurs in a region related to the Open/Closed magnetic Field line Boundary (OCFB) location, that is in good agreement with the Galileo magnetic field and plasma flow measurements (Gurnett et al. 1996: Kivelson et al. 1996, 1998). As shown in Fig. 10, the extent of the plasma precipitation regions depends on the assumption used to mimic the plasma mirroring in Jupiter's magnetosphere. In particular, in the hypothesis of efficient mirroring in Jupiter's magnetosphere, the O^+ precipitation takes place if the ions are assumed precipitating over the whole polar cap. If no mirroring is considered, the O^+ precipitation is confined to a latitudinal zone that is $\sim 10°$ wide and centered at the OCFB (i.e., at a latitude of $\sim 50°$ in the North trailing hemisphere). Moreover, in the latter case, the total rate of precipitating ions is lower (see Fig. 10). Nevertheless, the real ion-mirroring rate is expected to have an intermediate value between 0 and 100 %, since the ion population is confined inside the Jupiter Plasma Sheet (being partially reflected and partially lost). The sputtered H_2O density distribution mimics the morphology of the plasma impact to the surface as predicted by the global MHD model of Ganymede's magnetosphere (Jia et al. 2009) for the case that the moon is located close to the center of Jupiter plasma sheet. Indeed, both in the northern and southern hemispheres the sputtered H_2O exospheric density maximum is located at higher latitudes in the trailing hemisphere than in the leading one. Moreover, in the full mirroring assumption, the primary surface sputtering mechanism at the whole polar cap of Ganymede can alone explain the observed higher albedo of this region (Khurana et al. 2007); in the non-mirroring assumption the polar cap brightness above the OCFB ring can

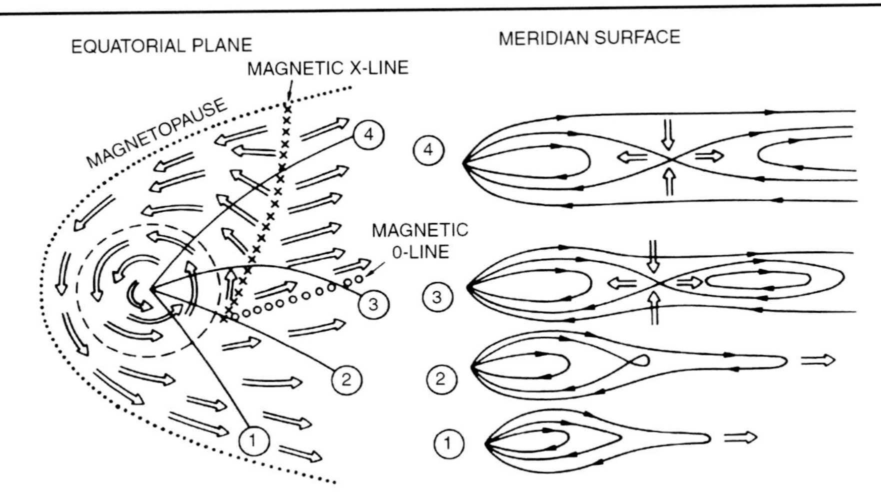

Fig. 11 Sketch of the global flow patterns of the Jovian magnetosphere known as the Vasyliunas-cycle (from Vasyliunas 1983)

be explained with the action of secondary sputtering due to ionized exospheric particles re-impacting the surface. A sublimated H_2O population adds to this sputtered population close to the subsolar point. Finally the estimated total surface release rate of sputtered H_2O molecules is $7 \times 10^{25}\ s^{-1}$ whereas the release rate of the sublimated H_2O is $7 \cdot 10^{29}\ s^{-1}$. The plasma effects on the exosphere generation are less evident in the O_2 density distribution, since this molecule does not stick onto the surface and thermalizes. Indeed, the energetic O_2 emission has a distribution that depends both on the morphology of the plasma precipitation to the surface and on the Sun illumination that determines the efficiency of the radiolysis mechanism, on the illuminated side (Plainaki et al. 2015).

The rates of the most important plasma-moon interactions leading to the loss of Ganymede's exosphere (and to a source for the magnetosphere) were calculated by Plainaki et al. (2015), who used previously published estimates of the plasma parameters (Kivelson et al. 2004; McNutt et al. 1981; Scudder et al. 1981; Gurnett et al. 1996; Eviatar et al. 2001) of the ambient magnetospheric environment at Ganymede, together with laboratory-based estimates of rate coefficients (for a review see Burger et al. 2010). They showed that the loss rate for H_2O in the polar caps is due to its charge exchange with ionospheric O_2^+ and is of the order of $10^{-5}\ s^{-1}$; in the closed field lines region, the H_2O loss rate is of the order of $10^{-6}\ s^{-1}$ and is mainly due to charge exchange between ionospheric O^+ and H_2O. The exospheric O_2 net loss rate in the polar caps is due to electron impact ionization and is in the range 9×10^{-8}–$9 \times 10^{-7}\ s^{-1}$ (the minimum value is where the electron density is lower, likely where the neutral density is higher); on the illuminated side the O_2 loss rate is of the order of $\sim 10^{-7}\ s^{-1}$ whereas on the night side of the closed field lines region it is of the order of $10^{-8}\ s^{-1}$.

Ions outflowing from Ganymede's ionosphere across the polar cap were detected during Galileo's polar flyby (Frank et al. 1997b). The ionospheric outflows were originally identified as hydrogen ions (Frank et al. 1997b; Paty et al. 2008) and later reinterpreted as atomic oxygen ions (Vasyliūnas and Eviatar 2000; Jia et al. 2009). In either case, it is suggested that there appears to be a polar wind similar to that observed in the terrestrial magnetosphere. Using Galileo's Plasma Spectrometer (PLS) measurements and assuming a circular area with radius of 1 Ganymede radius for the outflow region, Frank et al. (1997b) estimated the total ionospheric outflow rate to be $\sim 6 \times 10^{25}$ ions/s. While the fate of the ionospheric outflows is poorly known due to lack of observations, it is likely that some of the outflowing plasma will participate in the tail reconnection, through which a fraction of the ionospheric plasma will be recycled back into Ganymede's magnetosphere and the rest will be released down the tail to the ambient environment, providing a plasma source for Jupiter's magnetosphere albeit with a supply rate much smaller than from the moon Io.

In addition to the ionospheric outflows, the pickup of neutral particles originating from Ganymede's atmosphere (Hall et al. 1998) may also provide a plasma source to Jupiter's magnetosphere. Volwerk et al. (2013) recently analyzed the Galileo magnetometer measurements acquired during two upstream flybys, and found signatures of ion cyclotron waves near water-group ion gyro frequencies outside of the magnetosphere, which are indicative of pick-up of newly ionized particles from the moon's extended exosphere. Nonetheless, the estimated pickup rate of $\sim 5 \times 10^{23}$ ions s^{-1} is several orders of magnitude smaller than the ionospheric outflow rate, making the pickup ions from Ganymede's atmosphere a rather minor source of plasma for Jupiter's magnetosphere.

7 Solar Wind

There is evidence that the solar wind is a source of plasma to Jupiter's magnetosphere via magnetic reconnection, although the precise role of the solar wind in terms of driving a Dungey cycle at Jupiter is unclear. Other transport processes such as the Kelvin-Helmholtz instability can also be at work (e.g. Delamere and Bagenal 2010; Ma et al. 2014a, 2014b). The ion composition of the boundary layers, inside of the magnetopause, is consistent with mass transport at the magnetopause. At Jupiter Bame et al. (1992) reported ion composition in the boundary layer during the expansion of the magnetopause past the Ulysses spacecraft. The magnetopause was not a sharp spatial boundary, and rather magnetosheath and magnetospheric populations were observed to coexist within the boundary layer internal to the magnetopause. A boundary layer was clearly present for all but one of the Jovian magnetopause crossings. Similarly, Galvin et al. (1993) and Phillips et al. (1993) reported a mixed boundary layer composition and Galvin et al. (1993) suggested that transport across the magnetopause boundary can work both ways. A significant finding by Hamilton et al. (1981) from the Voyager 2 Low Energy Charged Particle instrument (LECP) data is that the plasma sheet composition beyond 60–80R_J in the tail is similar to that of solar wind energetic ions while the inner magnetosphere is dominated by iogenic material. Krupp et al. (2004a) discussed evidence of a boundary layer seen in the Cassini Magnetosphere Imaging Instrument/Low Energy Magnetospheric Measurement System (MIMI/LEMMS) energetic electron data when Cassini skimmed Jupiter's dusk magnetopause during the gravity assist flyby. They suggest that the leakage of energetic magnetospheric electrons to the magnetosheath is consistent with open field lines planetward of the magnetopause. Most recently, the particles measured by New Horizons as it traversed down the flanks of the magnetotail were increasingly dominated by light ions at farther distances down-tail (Haggerty et al. 2009; Hill et al. 2009; Ebert et al. 2010).

Hill et al. (1983) estimated the solar wind source by taking the fraction of solar wind leaking into the magnetosphere to be $\sim 10^{-3}$ and obtained a tiny source strength of 20 kg s^{-1} for a radius of cross-section of 100R_J. Bagenal and Delamere (2011) took a more realistic cross-section of the terminator of 150R_J, a local solar wind density of 1 cm^{-3} and speed of 400 km/s and estimated a solar wind flux of ~ 230 ton s^{-1}, which makes a source of

230 $\text{kg}\,\text{s}^{-1}$ for the Hill et al. (1983) 0.1 % leakage rate. Even with such low mass source rates, the enhanced density of protons will significantly alter the ion composition of the outer boundary layers.

8 Other Sources

At Saturn, the rings provide an important source of plasma for the magnetosphere. Although no such information is currently available regarding Jupiter, using the Saturn analogy the rings are also likely to be a source at Jupiter. The lack of relevant data so far probably implies that this source is small or negligible.

9 Transport Mechanisms

Jupiter is a rapidly rotating planet with the volcanic moon Io acting as a strong internal plasma source. In this case, the driving of magnetospheric dynamics by the (external) solar wind is thought to be secondary to the role of the (internally-driven) rotation (Hill et al. 1974; Michel and Sturrock 1974; Vasyliunas 1983; Kivelson and Southwood 2005). In what has become known as the Vasyliunas cycle shown in Fig. 11, the plasma created deep within the rapidly-rotating magnetosphere is accelerated by magnetic stresses from the ionosphere, gains energy, and moves outward from the planet. Centrifugal forces cause the field lines to stretch. These stretched field lines can form a thin current sheet, across which the closed field lines reconnect. This reconnection simultaneously shortens the field line and can release plasma down the tail in the form of a "plasmoid".

In order to observe the passage of plasmoids over the spacecraft, one should examine the north-south component of the magnetic field, to look for deflections from the radially stretched configuration. The Voyager flyby data gave a hint of reconnection processes in the Jovian tail (Nishida 1983), but it was only with the arrival of the Galileo orbiter in 1995 that the properties of reconnection at Jupiter could be probed in detail. One of the first studies to employ Galileo data to show evidence of plasmoid break-off was by Russell et al. (1998), and Fig. 12 shows an example of two characteristic magnetic field signatures. The sign of the change in the north-south component of the field provides information as to which side of the reconnection x-line the spacecraft is on and in this case the two events shown in Fig. 12 were on opposite sides of the x-line. In addition to magnetometer data, Galileo energetic particle detector data have been used to reveal evidence for both tailward and planetward plasma flows associated with magnetic reconnection, and thus to infer the position of the x-line in Jupiter's tail (Woch et al. 2002; Kronberg et al. 2008). The most comprehensive study to date was performed by Vogt et al. (2010), who surveyed all available Galileo data and identified 249 reconnection events, the locations of which are shown in Fig. 13. From this they extracted a statistical x-line extending from $\sim 90 R_J$ at dawn to $\sim 120 R_J$ downtail at dusk.

In order to estimate the effect of reconnection as a mechanism to remove mass from the magnetosphere, one must first obtain estimates of the size and composition of plasmoids. Kronberg et al. (2008) presented statistics on the length of Jovian plasmoids, based on measurements taken using the Galileo energetic particles detector. They found a typical length of $\sim 9 R_J$. In order to translate size estimates into mass calculations, Bagenal (2007) assumed that a typical plasmoid is a disk with a $25 R_J$ diameter and $10 R_J$ height, with density 0.01 cm^{-3}, and calculated that releasing one plasmoid per day (higher than the observed

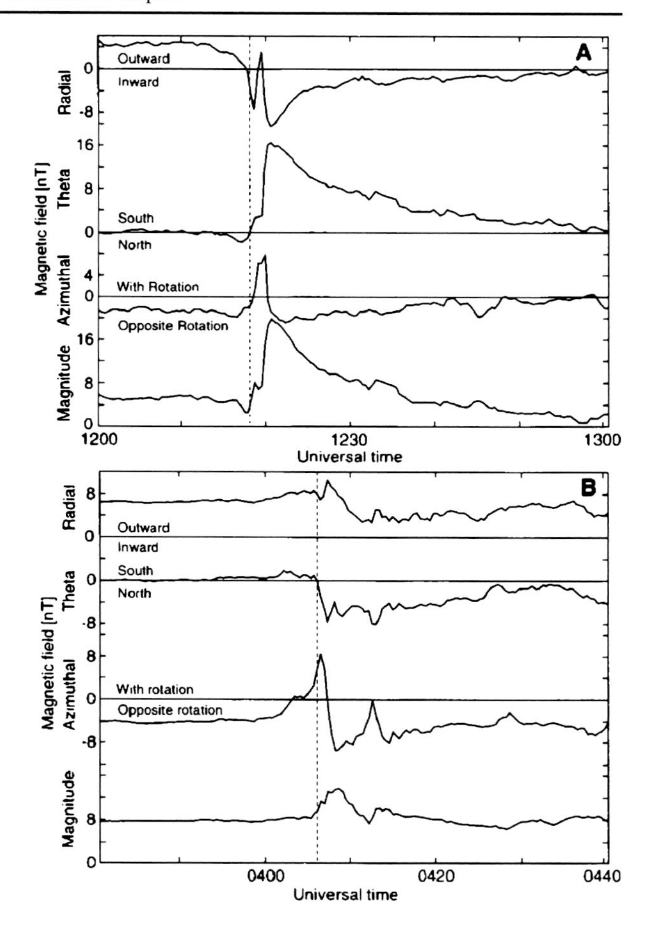

Fig. 12 Two events which display north-south and south-north turnings of the magnetic field in Jupiter's tail, indicative of the presence of reconnection event(s) planetward/tailward of the spacecraft (from Russell et al. 1998)

2–3 day recurrence period) is equivalent to a mass loss rate of only ∼ 30 kg/s. More recently, a survey of 43 plasmoids identified with the Galileo magnetometer found a mean length of $\sim 3R_J$ and a mass loss rate ranging from 0.7–120 kg/s (Vogt et al. 2014). It is clear that reconnection is active at Jupiter, and New Horizons data from deep in the Jovian tail confirm that iogenic material that has perhaps been broken off by reconnection is present many hundreds of R_J from the planet (Haggerty et al. 2009). However, regardless of the range of assumptions made, all studies indicate that the estimated rate of mass release supported by the observed plasmoids at Jupiter is far lower than the rate of plasma input from Io (260–1400 kg/s). Thus this has led authors such as Bagenal and Delamere (2011) to consider alternative mass loss pathways such as diffusive processes, or small-scale "drizzle" down the tail or loss across the magnetopause via small-scale intermittent reconnection. Although plasmoid ejection seems to play a relatively minor role in mass transport at Jupiter (hence opening up the possibility of important diffusive processes), it appears that tail reconnection is an important method of magnetic flux transport. For example, analysis of the observed plasmoids at Jupiter suggests an average flux closure rate of ∼ 7–70 GWb/day (Vogt et al. 2014), which closely matches the estimated rate of average flux opening through dayside reconnection, 18 GWb/day (Nichols et al. 2006).

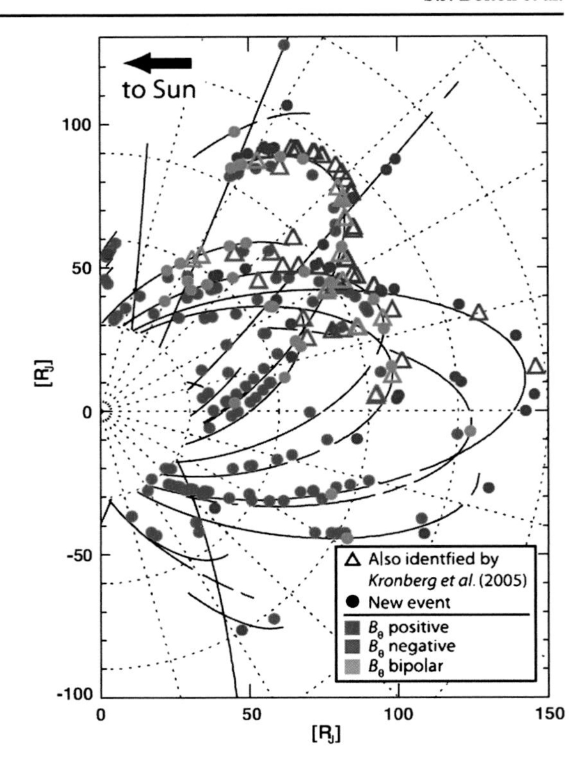

Fig. 13 Location of 249 reconnection events identified using the Galileo magnetometer (from Vogt et al. 2010)

10 Summary

The giant magnetosphere of Jupiter is fuelled primarily by the ionization of volcanic gases from Io, with additional minor sources from the other, icy, Galilean moons. There is likely a source of light ions from the atmosphere and ionosphere of Jupiter but it has neither been accurately measured nor modelled. The polar passes of the Juno mission will hopefully shed light on this possible contribution from the planet. While the radial transport mechanism in the plasma disc is described as flux tube interchange diffusion, the controlling factors, however, are not well quantified. Two major mysteries at Jupiter are the mechanism that heats the plasma as it moves outwards from the Io plasma torus, and the mechanism by which plasma is lost from the system. Some outstanding questions are as follows:

- What are the timescales for variability of the production of plasma at Io, as well as the other Galilean moons?
- What are the amounts of plasma that enter the Jovian plasma sheet from the atmosphere and ionosphere of Jupiter and from the solar wind? On what do these source rates depend?
- How do the plasma sheet properties (density, temperature, radial transport) respond to variability in the plasma sources?
- How is the main auroral emission affected by changes in the iogenic plasma production rate?

It is hoped that exploration of the Jovian system by NASA's Juno mission (2016–2018) and ESA's JUpiter ICy moons Explorer (JUICE) mission will answer these questions.

Acknowledgements The work of X. Jia was supported by the NASA Outer Planets Research Program through grant NNX12AM74G. The work of M. Blanc was done thanks to support by IRAP (CNRS and University of Toulouse). The work of F. Bagenal was funded by NASA JDAP program (grant number NNX09AE03G). C.M. Jackman was funded by a Science and Technology Facilities Council Ernest Rutherford Fellowship. The work by A. Kotova was financed by the International Max Planck Research School on Physical processes in the Solar System and Beyond (IMPRS) at the Max Planck Institute for Solar System Research (MPS). Work by N. Krupp at MPS was in part financed by the German BMWi through the German Space Agency DLR under contracts 50 OH 0301, 50 OH 0801, 50 OH 0802, 50 OH 1101, 50 OH 1502, 50 ON 0201, 50 QJ 1301, 50 OO 1206, 50 OO 1002 and by the Max Planck Society. This work was also supported by the European Research Infrastructure EUROPLANET RI in the framework program FP7, contract number: 001637. Work by T. Cassidy was NASA OPR Grant NNX13AE65G (through JPL subcontract 1440362). Work by E. Chané was funded by the Interuniversity Attraction Poles Programme of the Belgian Science Policy Office (IAP P7/08 CHARM) and by the Research Foundation-Flanders (FWO 12M0115N).

Conflict of interest The authors declare that they have no conflict of interest.

References

I.I. Alexeev, E.S. Belenkaya, Modeling of the Jovian magnetosphere. Ann. Geophys. **23**, 809–826 (2005)

F. Bagenal, The magnetosphere of Jupiter: Coupling the equator to the poles. J. Atmos. Sol.-Terr. Phys. **69** (2007)

F. Bagenal, P.A. Delamere, Flow of mass and energy in the magnetospheres of Jupiter and Saturn. J. Geophys. Res. **116**, A05209 (2011). doi:10.1029/2010JA016294

F. Bagenal, T. Dowling, W. McKinnon (eds.), *Jupiter: Planet, Satellites, Magnetosphere* (Cambridge University Press, Cambridge, 2004)

F. Bagenal, A. Adriani, F. Allegrini, S.J. Bolton, B. Bonfond, E.J. Bunce, J.E.P. Connerney, S.W.H. Cowley, R.W. Ebert, G.R. Gladstone, C.J. Hansen, W.S. Kurth, S.M. Levin, B.H. Mauk, D.J. McComas, C.P. Paranicas, D. Santos-Costa, R.M. Thorne, P. Valek, J.H. Waite, P. Zarka, Magnetospheric science objectives of the Juno mission. Space Sci. Rev. (2014). doi:10.1007/s11214-014-0036-8

S.J. Bame, B.L. Barraclough, W.C. Feldman, G.R. Gisler, J.T. Gosling, D.J. McComas, J.L. Phillips, M.F. Thomsen, B.E. Goldstein, M. Neugebauer, Jupiter's magnetosphere: Plasma description from the Ulysses flyby. Science **257**, 1539–1543 (1992)

S.J. Bolton, The Juno mission, in *Proceedings of the International Astronomical Union, IAU Symposium*, vol. 269 (2010), pp. 92–100

S. Bolton, R. Thorne, S. Bourdarie, I. DePater, B. Mauk, Jupiter's inner radiation belts, in *Jupiter: Planet, Satellites, Magnetosphere*, ed. by F. Bagenal, T.E. Dowling, W.B. McKinnon (Cambridge University Press, Cambridge, 2004)

B. Bonfond, D. Grodent, J.-C. G´erard, T. Stallard, J.T. Clarke, M. Yoneda, A. Radioti, J. Gustin, Auroral evidence of Io's control over the magnetosphere of Jupiter. Geophys. Res. Lett. **39**, L01105 (2012). doi:10.1029/2011GL050253

S.J. Bolton et al., The Juno mission, Space Sci. Rev. (2015, submitted)

S.W. Bougher et al., Jupiter's general circulation model (JTGCM): Global structure and dynamics driven by auroral and Joule heating. J. Geophys. Res. **110**, E04008 (2005). doi:10.1029/2003JE002230

M.E. Brown, Potassium in Europa's atmosphere. Icarus **151**, 190–195 (2001)

M.E. Brown, R.E. Hill, Discovery of an extended sodium atmosphere around Europa. Nature **380**, 229–231 (1996)

M.H. Burger, R. Wagner, R. Jaumann, T.A. Cassidy, Effects of the external environment on icy satellites. Space Sci. Rev. **153**(1–4), 349–374 (2010)

T.A. Cassidy, R.E. Johnson, M.A. McGrath, M.C. Wong, J.F. Cooper, The spatial morphology of Europa's near-surface O2 atmosphere. Icarus **191**, 755–764 (2007)

T.A. Cassidy, R.E. Johnson, P.E. Geissler, F. Leblanc, Simulation of Na D emission near Europa during eclipse. J. Geophys. Res. **113**, E02005 (2008)

T. Cassidy, P. Coll, F. Raulin, R.W. Carlson, R.E. Johnson, M.J. Loeffler, K.P. Hand, R.A. Baragiola, Radiolysis and photolysis of icy satellite surfaces: Experiments and theory. Space Sci. Rev. **153**(1–4), 299–315 (2010)

T.A. Cassidy, C.P. Paranicas, J.H. Shirley, J.B. Dalton III., B.D. Teolis, R.E. Johnson, L. Kamp, A.R. Hendrix, Magnetospheric ion sputtering and water ice grain size at Europa. Planet. Space Sci. **77**, 64–73 (2013)

G. Caudal, A self-consistent model of Jupiter's magnetodisc including the effects of centrifugal force and pressure. J. Geophys. Res. **91**, 4201–4221 (1986). doi:10.1029/JA091iA04p04201

E. Chané, J. Saur, F.M. Neubauer, J. Raeder, S. Poedts, Observational evidence of Alfvén wings at the Earth. J. Geophys. Res. Space Phys. **117**, A09217 (2012). doi:10.1029/2012JA017628
E. Chané, J. Saur, S. Poedts, Modeling Jupiter's magnetosphere: Influence of the internal sources. J. Geophys. Res. Space Phys. **118**, 2157–2172 (2013). doi:10.1002/jgra.50258
S. Chapman, V.C.A. Ferraro, A new theory of magnetic storms. Nature **126**, 129–130 (1930)
C.F. Chyba, Energy for microbial life on Europa. Nature **403**, 381–382 (2000)
J.T. Clarke, J. Ajello, G. Ballester, L.B. Jaffel, J. Connerneyk, J.C. Gerard, G.R. Gladstone, D. Grodent, W. PryorI, J. Trauger, J.H.W. Jr, Ultraviolet emissions from the magnetic footprints of Io, Ganymede and Europa on Jupiter. Nature **415**, 997–1000 (2002)
J. Clarke, D. Grodent, S. Cowley, E. Bunce, P. Zarka, J. Connerney, T. Satoh, Jupiter's aurora, in *Jupiter: Planet, Satellites, Magnetosphere*, ed. by F. Bagenal, T.E. Dowling, W.B. McKinnon (Cambridge University Press, Cambridge, 2004)
J.E.P. Connerney, M.H. Acuna, N.F. Ness, Modeling the Jovian current sheet and inner magnetosphere. J. Geophys. Res. **86**, 8370–8384 (1981). doi:10.1029/JA086iA10p08370
J.F. Cooper, R.E. Johnson, B.H. Mauk, H.B. Garrett, N. Gehrels, Energetic ion and electron irradiation of the icy Galilean Satellites. Icarus **149**, 133–159 (2001)
S.W.H. Cowley, E.J. Bunce, Origin of the main auroral oval in Jupiter's coupled magnetosphere-ionosphere system. Planet. Space Sci. **49**, 1067–1088 (2001)
T.E. Cravens, Vibrationally excited molecular hydrogen in the upper atmosphere of Jupiter. J. Geophys. Res. **92**, 11083 (1987)
P.A. Delamere, F. Bagenal, Solar wind interaction with Jupiter's magnetosphere. J. Geophys. Res. **115**(A), A10201 (2010). doi:10.1029/2010JA015347
A. Dessler (ed.), *Physics of the Jovian Magnetosphere* (Cambridge University Press, Cambridge, 1983)
R.W. Ebert, D.J. McComas, F. Bagenal, H.A. Elliott, Location, structure, and motion of Jupiter's dusk magnetospheric boundary from 1625 to 2550 RJ. J. Geophys. Res. **115**, 12,223 (2010)
R.W. Ebert, F. Bagenal, D.J. McComas, C.M. Fowler, A survey of solar wind conditions at 5 AU: A tool for interpreting solar wind-magnetosphere interaction at Jupiter. Front. Astron. Space Sci. **1**, 4 (2014). doi:10.3389/fspas.2014.00004
T.M. Edwards, E.J. Bunce, S.W.H. Cowley, A note on the vector potential of Connerney et al.'s model of the equatorial current sheet in Jupiter's magnetosphere. Planet. Space Sci. **49**, 1115–1123 (2001). doi:10.1016/S0032-0633(00)00164-1
R.E. Ergun, L. Ray, P.A. Delamere, F. Bagenal, V. Dols, Y.-J. Su, Generation of parallel electric fields in the Jupiter-Io torus wake region. J. Geophys. Res. **114**, 5201 (2009)
A. Eviatar, V.M. Vasyliunas, D.A. Gurnett, The ionosphere of Ganymede. Planet. Space Sci. **49**, 327–336 (2001). doi:10.1016/S0032-0633(00)00154-9
M. Famà, J. Shi, R.A. Baragiola, Sputtering of ice by low-energy ions. Surf. Sci. **602**, 156–161 (2008)
P.D. Feldman, M.A. McGrath, D.F. Strobel, H.W. Moos, K.D. Retherford, B.C. Wolven, HST/STIS ultraviolet imaging of polar aurora on Ganymede. Astrophys. J. **535**, 1085–1090 (2000)
L.A. Frank, W.R. Paterson, K.L. Ackerson, S.J. Bolton, Low-energy electrons measurements at Ganymede with the Galileo spacecraft: Probes of the magnetic topology. Geophys. Res. Lett. **24**(17), 2,159–2,162 (1997a)
L.A. Frank, W.R. Paterson, K.L. Ackerson, S.J. Bolton, Outflow of hydrogen ions from Ganymede. Geophys. Res. Lett. **24**(17), 2151–2154 (1997b)
K. Fukazawa, T. Ogino, R.J. Walker, Dynamics of the Jovian magnetosphere for northward interplanetary magnetic field (IMF). Geophys. Res. Lett. **32**, L03,202 (2005). doi:10.1029/2004GL021392
A.B. Galvin, C.M.S. Cohen, F.M. Ipavich, R. von Steiger, J. Woch, U. Mall, Boundary layer ion composition at Jupiter during the inbound pass of the Ulysses flyby. Planet. Space Sci. **41**, 869–876 (1993)
O. Grasset et al., JUpiter ICy moons Explorer (JUICE): An ESA mission to orbit Ganymede and to characterise the Jupiter system. Planet. Space Sci. **78**, 1–21 (2013)
D. Grodent, B. Bonfond, A. Radioti, J.-C. Gérard, X. Jia, J.D. Nichols, J.T. Clarke, Auroral footprint of Ganymede. J. Geophys. Res. **114**(A7) (2009). doi:10.1029/2009JA014289
D.A. Gurnett, W.S. Kurth, A. Roux, S.J. Bolton, C.F. Kennel, Evidence for a magnetosphere at Ganymede from plasma-wave observations by the Galileo spacecraft. Nature **384**, 535–537 (1996)
D.K. Haggerty, M.E. Hill, R.L. McNutt Jr., C. Paranicas, Composition of energetic particles in the Jovian magnetotail. J. Geophys. Res. **114**, A02208 (2009). doi:10.1029/2008JA013659
D.T. Hall, D.F. Strobel, P.D. Feldman, M.A. McGrath, H.A. Weaver, Detection of an oxygen atmosphere on Jupiter's moon Europa. Nature **373**, 677–679 (1995)
D.T. Hall, P.D. Feldman, M.A. McGrath, D.F. Strobel, The far-ultraviolet airglow of Europa and Ganymede. Astrophys. J. **499**, 475–481 (1998)
J.T. Hallett, D.E. Shemansky, X. Lin, A rotational level hydrogen physical chemistry model for general astrophysical application. Astrophys. J. **624**, 448 (2005)

D.C. Hamilton, G. Gloeckler, S.M. Krimigis, L.J. Lanzerotti, Composition of nonthermal ions in the Jovian magnetosphere. J. Geophys. Res. **86**, 8301–8318 (1981)

C.J. Hansen, D.E. Shemansky, A.R. Hendrix, Cassini UVIS observations of Europa's oxygen atmosphere and torus. Icarus **176**, 305–315 (2005)

T.W. Hill, Inertial limit on corotation. J. Geophys. Res. **84**, 6554–6558 (1979). doi:10.1029/JA084iA11p06554

T.W. Hill, The Jovian auroral oval. J. Geophys. Res. **106**, 8101 (2001)

T.W. Hill, A.J. Dessler, F.C. Michel, Configuration of the Jovian magnetosphere. Geophys. Res. Lett. **1**, 3–6 (1974). doi:10.1029/GL001i001p00003

T.W. Hill, A.J. Dessler, C.K. Goertz, Magnetospheric models, in *Physics of the Jovian Magnetosphere*, ed. by A.J. Dessler (Cambridge University Press, New York, 1983), pp. 353–394

M.E. Hill, D.K. Haggerty, R.L. McNutt Jr., C.P. Paranicas, Energetic particle for magnetic filaments. J. Geophys. Res. **114**, A11201 (2009). doi:10.1029/2009JA014374

D.E. Huddleston, C.T. Russell, G. Le, Magnetopause structure and the role of reconnection at the outer planets. J. Geophys. Res. **102**, 24,289–24,302 (1997). doi:10.1029/97JA02416

W.-H. Ip, Europa's oxygen exosphere and its magnetospheric interaction. Icarus **120**, 317–325 (1996)

C.M. Jackman, C.S. Arridge, Solar cycle effects on the dynamics of Jupiter's and Saturn's magnetospheres. Sol. Phys. **274**, 481–502 (2011). doi:10.1007/s11207-011-9748-z

X. Jia, R.J. Walker, M.G. Kivelson, K.K. Khurana, J.A. Linker, Three dimensional MHD simulations of Ganymede's magnetosphere. J. Geophys. Res. **113**, A06212 (2008). doi:10.1029/2007JA012748

X. Jia, R.J. Walker, M.G. Kivelson, K.K. Khurana, J.A. Linker, Properties of Ganymede's magnetosphere inferred from improved three-dimensional MHD simulations. J. Geophys. Res. **114**, A09209 (2009). doi:10.1029/2009JA014375

X. Jia, M.G. Kivelson, K.K. Khurana, R.J. Walker, Magnetic fields of the satellites of Jupiter and Saturn. Space Sci. Rev. **152**(1) (2010a). doi:10.1007/s11214-009-9507-8

X. Jia, R.J. Walker, M.G. Kivelson, K.K. Khurana, J.A. Linker, Dynamics of Ganymede's magnetopause: Intermittent reconnection under steady external conditions. J. Geophys. Res. **115**, A12202 (2010b). doi:10.1029/2010JA015771

R.E. Johnson, *Energetic Charged-Particle Interactions with Atmospheres and Surfaces, vol. X*. Phys. Chem. Space, vol. 19 (Springer, Berlin, 1990). 232 pp. (84 figures and 28 tables)

R.E. Johnson, J.W. Boring, L.J. Lanzerotti, W.L. Brown, Decomposition of ice by incident charged particles: The icy satellites and Rings, in *Lunar and Planetary Institute Science Conference Abstracts*, vol. 13 (1982a), p. 366

R.E. Johnson, L.J. Lanzerotti, W.L. Brown, Planetary applications of ion induced erosion of condensed-gas frosts. Nucl. Instrum. Methods **198**, 147–157 (1982b)

R.E. Johnson, R.W. Carlson, J.F. Cooper, C. Paranicas, M.H. Moore, M.C. Wong, Radiation effects on the surface of the Galilean satellites, in *Jupiter-the Planet, Satellites and Magnetosphere*, ed. by F. Bagenal, T. Dowling, W.B. McKinnon (Cambridge University Press, Cambridge, 2004), pp. 485–512 (Chap. 20)

S.P. Joy, M.G. Kivelson, R.J. Walker, K.K. Khurana, C.T. Russell, T. Ogino, Probabilistic models of the Jovian magnetopause and bow shock locations. J. Geophys. Res. **107**, 1309 (2002)

K. Khurana, M.G. Kivelson, V. Vasyliunas, N. Krupp, J. Woch, A. Lagg, B. Mauk, W. Kurth, The configuration of Jupiter's magnetosphere, in *Jupiter: Planet, Satellites, Magnetosphere*, ed. by F. Bagenal, T.E. Dowling, W.B. McKinnon (Cambridge University Press, Cambridge, 2004)

K.K. Khurana, R.T. Pappalardo, N. Murphy, T. Denk, The origin of Ganymede's polar caps. Icarus **191**, 193–202 (2007)

Y.H. Kim, J.L. Fox, The chemistry of hydrocarbon ions in the Jovian ionosphere. Icarus **112**, 310 (1994)

M.G. Kivelson, D.J. Southwood, Dynamical consequences of two modes of centrifugal instability in Jupiter's outer magnetosphere. J. Geophys. Res. **110**, A12209 (2005). doi:10.1029/2005JA011176

M.G. Kivelson, K.K. Khurana, C.T. Russell, R.J. Walker, J. Warnecke, F.V. Coroniti, C. Polanskey, D.J. Southwood, G. Schubert, Discovery of Ganymede's magnetic field by the Galileo spacecraft. Nature **384**, 537–541 (1996)

M.G. Kivelson, K.K. Khurana, S. Joy, C.T. Russell, D.J. Southwood, R.J. Walker, C. Polanskey, Europa's magnetic signature: Report from Galileo's pass on 19 December 1996. Science **276**, 1239–1241 (1997)

M.G. Kivelson, J. Warnecke, L. Bennett, S. Joy, K.K. Khurana, J.A. Linker, C.T. Russell, R.J. Walker, C. Polanskey, Ganymede's magnetosphere: Magnetometer overview. J. Geophys. Res. **103**(E9), 19963–19972 (1998)

M.G. Kivelson, F. Bagenal, W.S. Kurth, F.M. Neubauer, C. Paranicas, J. Saur, Magnetospheric interactions with satellites, in *Jupiter: The Planet, Satellites and Magnetosphere*, ed. by F. Bagenal, T.E. Dowling, W.B. McKinnon (Cambridge University Press, Cambridge, 2004), pp. 513–536

M.G. Kivelson, K.K. Khurana, M. Volwerk, Europa's interaction with the Jovian magnetosphere, in *Europa*, ed. by R.T. Pappalardo, W.B. McKinnon, K.K. Khurana (University of Arizona Press, Tucson, 2009)

A.J. Kliore, D.P. Hinson, F.M. Flasar, A.F. Nagy, T.E. Cravens, The ionosphere of Europa from Galileo radio occultations. Science **277**, 355–358 (1997)
S. Knight, Parallel electric fields. Planet. Space Sci. **21**, 741–750 (1973)
E.A. Kronberg, J. Woch, N. Krupp, A. Lagg, Mass release process in the Jovian magnetosphere: Statistics on particle burst parameters. J. Geophys. Res. **113**, A10202 (2008). doi:10.1029/2008JA013332
N. Krupp et al., Energetic particle observations in the vicinity of Jupiter: Cassini MIMI/LEMMS results. J. Geophys. Res. **109**, A09S10 (2004a)
N. Krupp, V. Vasyliunas, J. Woch, A. Lagg, K. Khurana, M. Kivelson, B. Mauk, E. Roelof, D. Williams, S. Krimigis, W. Kurth, L. Frank, W. Paterson, Dynamics of the Jovian magnetosphere, in *Jupiter: Planet, Satellites, Magnetosphere*, ed. by F. Bagenal, T.E. Dowling, W.B. McKinnon (Cambridge University Press, Cambridge, 2004b)
N. Krupp, E. Kronberg, A. Radioti, Jupiter's magnetotail, in *Magnetotails in the Solar System*, ed. by A. Keiling, C.M. Jackman, P.A. Delamere (Wiley, Hoboken, 2015). doi:10.1002/9781118842324.ch5
W.S. Kurth, J.D. Sullivan, D.A. Gurnett, F.L. Scarf, H.S. Bridge, E.C. Sittler Jr., Observations of Jupiter's distant magnetotail and wake. J. Geophys. Res. **87**(A12), 10,373–10,383 (1982). doi:10.1029/JA087iA12p10373
W.S. Kurth, D.A. Gurnett, A. Roux, S.J. Bolton, Ganymede: A new radio source. Geophys. Res. Lett. **24**(17), 2167–2170 (1997)
A. Lagg, N. Krupp, J. Woch, D.J. Williams, In situ observations of a neutral gas torus at Europa. Geophys. Res. Lett. **30**, 110000–110001 (2003)
F. Leblanc, R.E. Johnson, M.E. Brown, Europa's sodium atmosphere: An ocean source? Icarus **159**, 132–144 (2002)
F. Leblanc, A.E. Potter, R.M. Killen, R.E. Johnson, Origins of Europa Na cloud and torus. Icarus **178**, 367–385 (2005)
H. Luna, C. McGrath, M.B. Shah, R.E. Johnson, M. Liu, C.J. Latimer, E.C. Montenegro, Dissociative charge exchange and ionization of O_2 by fast H^+ and O^+ ions: Energetic ion interactions in Europa's oxygen atmosphere and neutral torus. Astrophys. J. **628**(2), 1086–1096 (2005)
X. Ma, A. Otto, P.A. Delamere, Interaction of magnetic reconnection and Kelvin-Helmholtz modes for large magnetic shear: 1. Kelvin-Helmholtz trigger. J. Geophys. Res. Space Phys. **119**, 781–797 (2014a). doi:10.1002/2013JA019224
X. Ma, A. Otto, P.A. Delamere, Interaction of magnetic reconnection and Kelvin-Helmholtz modes for large magnetic shear: 2. Reconnection trigger. J. Geophys. Res. Space Phys. **119**, 808–820 (2014b). doi:10.1002/2013JA019225
T. Majeed, J.C. McConnell, The upper ionospheres of Jupiter and Saturn. Planet. Space Sci. **39**, 1715 (1991)
M.L. Marconi, Structure and composition of Europa's atmosphere. Bull. Am. Astron. Soc. **35**, 943 (2003)
B.H. Mauk, D.G. Mitchell, R.W. McEntire, C.P. Paranicas, E.C. Roelof, D.J. Willims, S.M. Krimigis, Energetic ion characteristics and neutral gas interactions in Jupiter's magnetosphere. J. Geophys. Res. **109**, A09S12 (2003)
B.H. Mauk, D.G. Mitchell, R.W. McEntire, C.P. Paranicas, E.C. Roelof, D.J. Williams, S.M. Krimigis, A. Lagg, Energetic ion characteristics and neutral gas interactions in Jupiter's magnetosphere. J. Geophys. Res. **109**, 9 (2004)
M.B. McElroy, The ionospheres of the major planets. Space Sci. Rev. **40**, 460 (1973)
M.A. McGrath, E. Lellouch, D.F. Strobel, P.D. Feldman, R.E. Johnson, Satellite atmospheres, in *Jupiter. The planet, Satellites and Magnetosphere*, ed. by F. Bagenal, T.E. Dowling, W.B. McKinnon. Cambridge Planetary Science, vol. 1 (Cambridge University Press, Cambridge, 2004), pp. 457–483. ISBN 0-521-81808-7
M.A. McGrath, X. Jia, K. Retherford, P.D. Feldman, D.F. Strobel, J. Saur, Aurora on Ganymede. J. Geophys. Res. **118**(3) (2013). doi:10.1002/jgra.50122
R.L. McNutt Jr. et al., Energetic particles in the Jovian magnetotail. Science **318**, 220 (2007)
R.L. McNutt, J.W. Belcher, H.S. Bridge, Positive ion observations in the middle magnetosphere of Jupiter. J. Geophys. Res. **86**, 8319 (1981)
F.C. Michel, P.A. Sturrock, Centrifugal instability of the Jovian magnetosphere and its interaction with the solar wind. Planet. Space Sci. **22**, 1501 (1974)
G.H. Millward, S. Miller, T. Stallard, N. Achilleos, A.D. Aylward, On the dynamics of the Jovian ionosphere and thermosphere. IV. Ion-neutral coupling. Icarus **173**, 200–211 (2005)
T. Miyoshi, K. Kusano, MHD simulation of a rapidly rotating magnetosphere interacting with the external plasma flow. Geophys. Res. Lett. **24**, 2627–2630 (1997). doi:10.1029/97GL52739
T. Moriguchi, A. Nakamizo, T. Tanaka, T. Obara, H. Shimazu, Cur- rent systems in the Jovian magnetosphere. J. Geophys. Res. Space Phys. **113**, A05,204 (2008). doi:10.1029/2007JA012751
A.F. Nagy, A.R. Barakat, R.W. Schunk, Is Jupiter's ionosphere a significant plasma source for its magnetosphere? J. Geophys. Res. **91**, 351–354 (1986)

A.F. Nagy, J. Kim, T.E. Cravens, A.J. Kliore, Hot corona at Europa. Geophys. Res. Lett. **22**, 4153–4155 (1998)

F.M. Neubauer, Non-linear standing Alfvén wave current system at Io: Theory. J. Geophys. Res. **85**, 1171–1178 (1980)

F.M. Neubauer, The sub-Alfvénic interaction of the Galilean satellites with the Jovian magnetosphere. J. Geophys. Res. **103**(E9), 19843–19866 (1998)

J.D. Nichols, Magnetosphere-ionosphere coupling in Jupiter's middle magnetosphere: Computations including a self-consistent current sheet magnetic field model. J. Geophys. Res. Space Phys. **116**, A10232 (2011). doi:10.1029/2011JA016922

J.D. Nichols, S.W.H. Cowley, Magnetosphere-ionosphere coupling currents in Jupiter's middle magnetosphere: Dependence on the effective ionospheric Pedersen conductivity and iogenic plasma mass outflow rate. Ann. Geophys. **21**, 1419–1441 (2003). doi:10.5194/angeo-21-1419-2003

J.D. Nichols, S.W.H. Cowley, D.J. McComas, Magnetopause reconnection rate estimates for Jupiter's magnetosphere based on interplanetary measurements at ~ 5 AU. Ann. Geophys. **24**, 393–406 (2006)

A. Nishida, Reconnection in the Jovian magnetosphere. Geophys. Res. Lett. **10**(6), 451–454 (1983). doi:10.1029/GL010i006p00451

T. Ogino, R.J. Walker, M.G. Kivelson, A global magnetohydrodynamic simulation of the Jovian magnetosphere. J. Geophys. Res. **103**, 225 (1998). doi:10.1029/97JA02247

C. Paranicas, J.M. Ratliff, B.H. Mauk, C. Cohen, R.E. Johnson, The ion environment near Europa and its role in surface energetics. Geophys. Res. Lett. **29**, 050000–050001 (2002)

C. Paty, W. Paterson, R. Winglee, Ion energization in Ganymede's magnetosphere: Using multifluid simulations to interpret ion energy spectrograms. J. Geophys. Res. **113** (2008). doi:10.1029/2007JA012848

J.L. Phillips, S.J. Bame, B.L. Barraclough, D.J. McComas, R.J. Forsyth, P. Canu, P.J. Kellogg, Ulysses plasma electron observations in the Jovian magnetosphere. Planet. Space Sci. **41**, 877–892 (1993)

C. Plainaki, A. Milillo, A. Mura, S. Orsini, T. Cassidy, Neutral particle release from Europa's surface. Icarus **210**, 385–395 (2010)

C. Plainaki, A. Milillo, A. Mura, S. Orsini, S. Massetti, T. Cassidy, The role of sputtering and radiolysis in the generation of Europa exosphere. Icarus **218**(2), 956–966 (2012). doi:10.1016/j.icarus.2012.01.023

C. Plainaki, A. Milillo, A. Mura, S. Orsini, J. Saur, Exospheric O_2 densities at Europa during different orbital phases. Planet. Space Sci. **88**, 42–52 (2013). doi:10.1016/j.pss.2013.08.011

C. Plainaki, A. Milillo, S. Massetti, A. Mura, X. Jia, S. Orsini, V. Mangano, E. De Angelis, R. Rispoli, The H_2O and O_2 exospheres of Ganymede: The result of a complex interaction between the Jovian magnetospheric ions and the icy moon. Icarus **245**, 306–319 (2015)

A. Radioti, J. G´erard, D. Grodent, B. Bonfond, N. Krupp, J. Woch, Discontinuity in Jupiter's main auroral oval. J. Geophys. Res. Space Phys. **113**, A01,215 (2008). doi:10.1029/2007JA012610

L.C. Ray, R.E. Ergun, P.A. Delamere, F. Bagenal, Magnetosphere-ionosphere coupling at Jupiter: Effect of field-aligned potentials on angular momentum transport. J. Geophys. Res. **115**, 9211 (2010)

L.C. Ray, R.E. Ergun, P.A. Delamere, F. Bagenal, Magnetosphere-ionosphere coupling at Jupiter: A parameter space study. J. Geophys. Res. **117**, A01205 (2012)

L.C. Ray, N. Achilleos, Y.N. Yates, Including field-aligned potentials in the coupling between Jupiter's thermosphere, ionosphere, and magnetosphere. Planet. Space Sci. **62** (2014, submitted)

L. Roth, Aurorae of Io and Europa: Observations and modeling. Ph.D. thesis, University of Cologne (2012). Available at http://kups.ub.uni-koeln.de/4894

L. Roth, J. Saur, K.D. Retherford, D.F. Strobel, P.D. Feldman, M.A. McGrath, F. Nimmo, Transient water vapor at Europa's South pole. Science **343**(6167), 171–174 (2014). doi:10.1126/science.1247051

C.T. Russell, K.K. Khurana, D.E. Huddleston, M.G. Kivelson, Localized reconnection in the near Jovian magnetotail. Science **280**(5366), 1061–1064 (1998). doi:10.1126/science.280.5366.1061

J. Saur, D.F. Strobel, F.M. Neubauer, Interaction of the Jovian magnetosphere with Europa: Constraints on the neutral atmosphere. J. Geophys. Res. **103**, 19947–19962 (1998)

J. Saur, P.D. Feldman, L. Roth, F. Nimmo, D.F. Strobel, K.D. Retherford, M.A. McGrath, N. Schilling, J.C. Gérard, D. Grodent, Hubble space telescope/advanced camera for surveys observations of Europa's atmospheric ultraviolet emission at eastern elongation. Astrophysics **738**(2), 153–165 (2011)

F.L. Scarf, D.A. Gurnett, W.S. Kurth, R.L. Poynter, Voyager 2 plasma wave observations at Saturn. Science **215**, 587 (1982)

N. Schilling, F.M. Neubauer, J. Saur, Influence of the internally induced magnetic field on the plasma interaction of Europa. J. Geophys. Res. Space Phys. **113**(A3) (2008)

R.W. Schunk, A.F. Nagy, *Ionospheres*, 2nd edn. (Cambridge University Press, Cambridge, 2009)

J.D. Scudder, E.C. Sittler, H.S. Bridge, A survey of the plasma electron environment of Jupiter—A view from Voyager. J. Geophys. Res. **86**, 8157–8179 (1981)

K. Seki, A. Nagy, C.M. Jackman, F. Crary, D. Fontaine, P. Zarka, P. Wurz, A. Milillo, J.A. Slavin, D.C. Delcourt, M. Wiltberger, R. Ilie, X. Jia, S.A. Ledvina, M.W. Liemohn, B. Schunk, A review of general

physical and chemical processes related to plasma sources and losses for solar system magnetospheres. Space Sci. Rev. (2015). doi:10.1007/s11214-015-0170-y
V.I. Shematovich, R.E. Johnson, Near-surface oxygen atmosphere at Europa. Adv. Space Res. **27**, 1881–1888 (2001)
V.I. Shematovich, R.E. Johnson, J.F. Cooper, M.C. Wong, Surface-bounded atmosphere of Europa. Icarus **173**, 480–498 (2005)
J.A. Slavin, E.J. Smith, J.R. Spreiter, S.S. Stahara, Solar wind flow about the outer planets: Gas dynamic modeling of the Jupiter and Saturn bow shocks. J. Geophys. Res. **90**, 6275–6286 (1985)
W.H. Smyth, M.L. Marconi, Europa's atmosphere, gas tori, and magnetospheric implications. Icarus **181**, 510–526 (2006). doi:10.1016/j.icarus.2005.10.019
D.J. Southwood, M.G. Kivelson, R.J. Walker, J.A. Slavin, Io and its plasma environment. J. Geophys. Res. **85**, 5,959–5,968 (1980)
T.S. Stallard et al., On the dynamics of the Jovian ionosphere. II. The measurements of H3+ vibrational temperature, column density and total emission. Icarus **156**, 498 (2002)
N. Thomas, F. Bagenal, T. Hill, J. Wilson, The Io neutral clouds and plasma torus, in *Jupiter: Planet, Satellites, Magnetosphere*, ed. by F. Bagenal, T.E. Dowling, W.B. McKinnon (Cambridge University Press, Cambridge, 2004)
V.M. Vasyliunas, Plasma distribution and flow, in *Physics of the Jovian Magnetosphere*, ed. by A.J. Dessler (Cambridge University Press, Cambridge, 1983), pp. 395–453
V.M. Vasyliūnas, A. Eviatar, Outflow of ions from Ganymede: A reinterpretation. Geophys. Res. Lett. **27**(9), 1347–1349 (2000)
M.F. Vogt, The structure and dynamics of Jupiter's magnetosphere. Ph.D. thesis (2012)
M.F. Vogt, M.G. Kivelson, K.K. Khurana, S.P. Joy, R.J. Walker, Reconnection and flows in the Jovian magnetotail as inferred from magnetometer observations. J. Geophys. Res. **115**, A06219 (2010). doi:10.1029/2009JA015098
M.F. Vogt, C.M. Jackman, J.A. Slavin, E.J. Bunce, S.W.H. Cowley, M.G. Kivelson, K.K. Khurana, Structure and statistical properties of plasmoids in Jupiter's magnetotail. J. Geophys. Res. **119**, 821–843 (2014). doi:10.1002/2013JA019393
M. Volwerk, X. Jia, C. Paranicas, W.S. Kurth, M.G. Kivelson, K.K. Khurana, ULF waves in Ganymede's upstream magnetosphere. Ann. Geophys. **31**, 45–59 (2013). doi:10.5194/angeo-31-45-2013
R.J. Walker, T. Ogino, M.G. Kivelson, Magnetohydrodynamic simulations of the effects of the solar wind on the Jovian magnetosphere (2001)
D.J. Williams, B.H. Mauk, R.W. McEntire, E.C. Roelof, T.P. Armstrong, B. Wilken, S.M.K.J.G. Roederer, T.A. Fritz, L.J. Lanzerotti, N. Murphy, Energetic particle signatures at Ganymede: Implications for Ganymede's magnetic field. Geophys. Res. Lett. **24**(17), 2163–2166 (1997)
D.J. Williams, B. Mauk, R.W. McEntire, Properties of Ganymede's magnetosphere as revealed by energetic particle observations. J. Geophys. Res. **103**(A8), 17,523–17,534 (1998)
J. Woch, N. Krupp, A. Lagg, Particle bursts in the Jovian magnetosphere: Evidence for a near-Jupiter neutral line. Geophys. Res. Lett. **29**(7), 1138 (2002). doi:10.1029/2001GL014080
E.E. Woodfield, R.B. Horne, S.A. Glauert, J.D. Menietti, Y.Y. Shprits, The origin of Jupiter's outer radiation belt. J. Geophys. Res. Space Phys. **119**, 3490–3502 (2014). doi:10.1002/2014JA019891
P. Zarka, J. Queinnec, F.J. Crary, Low-frequency limit of Jovian radio emissions and implications on source locations and Io plasma wake. Planet. Space Sci. **49**, 1137–1149 (2001)

DOI 10.1007/978-1-4939-3544-4_7
Reprinted from *Space Science Reviews* Journal, DOI 10.1007/s11214-015-0172-9

Saturn Plasma Sources and Associated Transport Processes

M. Blanc[1] · D.J. Andrews[2] · A.J. Coates[3] · D.C. Hamilton[4] · C.M. Jackman[5] · X. Jia[6] · A. Kotova[7,8] · M. Morooka[9] · H.T. Smith[10] · J.H. Westlake[10]

Received: 3 March 2015 / Accepted: 3 June 2015 / Published online: 3 September 2015

Abstract This article reviews the different sources of plasma for Saturn's magnetosphere, as they are known essentially from the scientific results of the Cassini-Huygens mission to Saturn and Titan. At low and medium energies, the main plasma source is the H_2O cloud produced by the "geyser" activity of the small satellite Enceladus. Impact ionization of this cloud occurs to produce on the order of 100 kg/s of fresh plasma, a source which dominates all the other ones: Titan (which produces much less plasma than anticipated before the Cassini mission), the rings, the solar wind (a poorly known source due to the lack of quantitative knowledge of the degree of coupling between the solar wind and Saturn's magnetosphere), and the ionosphere. At higher energies, energetic particles are produced by energy diffusion and acceleration of lower energy plasma produced by the interchange instabilities induced by the rapid rotation of Saturn, and possibly, for the highest energy range, by contributions from the CRAND process acting inside Saturn's magnetosphere. Discussion of the transport and acceleration processes acting on these plasma sources shows the importance of rotation-induced radial transport and energization of the plasma, and also

✉ M. Blanc
michel.blanc@irap.omp.eu

[1] IRAP, CNRS/Université Toulouse III, Toulouse, France

[2] Swedish Institute for Space Physics, Uppsala, Sweden

[3] Mullard Space Science Laboratory, University College London, Holmbury St Mary, Dorking RH5 6NT, UK

[4] Department of Physics, University of Maryland, College Park, MD 20742, USA

[5] School of Physics and Astronomy, University of Southampton, Southampton, SO17 1BJ, UK

[6] Department of Atmospheric, Oceanic, and Space Sciences, University of Michigan, Ann Arbor, MI 48109-2143, USA

[7] Max Planck Institute for Solar System Research, Göttingen, Germany

[8] IRAP, CNRS/Université Paul Sabatier Toulouse III, Toulouse, France

[9] LASP, University of Colorado at Boulder, Boulder, CO, USA

[10] Johns Hopkins University Applied Physics Laboratory, Laurel, MD, USA

shows how much the unexpected planetary modulation of essentially all plasma parameters of Saturn's magnetosphere remains an unexplained mystery.

Keywords Saturn · Magnetosphere · Satellites · Plasma transport · Solar wind · Radiation belts

1 Introduction

The magnetosphere of Saturn is not observable from the surface of the Earth, because its main radio emission frequencies are below the terrestrial ionospheric cut-off. Therefore, most of what we know about the kronian magnetosphere was revealed to us by the investigations conducted by space probes, first Pioneer, Voyager and then Cassini. The first picture of Saturn's magnetosphere emerged from the plasma measurements performed by the particles-and-fields instruments on board Voyager 1 and 2 (Blanc et al. 2002).

One important feature of Saturn's magnetosphere is its near-perfect axisymmetry, at least if we look at the magnetic field only: the dipole axis of the magnetic field is nearly aligned to the rotation axis. Thus, as we will see, one of the biggest surprises of the studies of Saturn's magnetosphere is that nearly all magnetospheric parameters display a strong rotational modulation. The discovery and understanding of this rotational modulation is, still today, one of the greatest challenges of magnetospheric science at Saturn (Carbary and Mitchell 2013).

The size of the magnetosphere extends from 20 to 35 R_S on the dayside, depending on solar wind pressure upstream. The general shape of the magnetosphere is somewhat "stretched-dipole" like, similar to the Earth's. Inside the magnetosphere, Voyager was able to identify several plasma regimes. In particular:

- the inner plasma torus, a region of cold plasma located from the rings to about 8 R_S, which Cassini showed to be populated mainly by water-related ions; this region is in near-rigid corotation.
- the extended plasmasheet, a region of warm plasma extending out of the inner torus to about 15 R_S, populated by a more rarefied and hotter population of ions and electrons.

Beyond these populations, we find the classical boundary regions on the dayside (the region close to the magnetopause) and the different regions of the magnetotail on the nightside, with the extended plasmasheet and tail lobes, which have been visited extensively by Cassini (André et al. 2008; Arridge et al. 2011).

Overall, the magnetic field configuration of Saturn is rather Earth-like, but the plasma populations show the dominance of a source in orbit around Saturn, which happens to be Enceladus and to a lesser extent the ring system, as we shall see later.

The plasma flow regime seems to be close to the Jovian case. The inner torus is in near-rigid corotation, and this corotation is enforced by a system of field-aligned currents, which closes at the two ends of the field line, in the thermosphere/ionosphere and in the region of the equatorial plane. This current loop transports angular momentum outwards, from the thermosphere to the plasma torus, enforcing corotation in the inner part of the equatorial magnetosphere. However, beyond a certain distance (in the range of 10 R_S), the effect of corotation enforcement currents becomes insufficient, and a lag of the azimuthal plasma flow behind corotation progressively develops.

In this chapter, we are going to explore the different source regions of Saturn's magnetosphere, and the associated dynamical phenomena. Once all sources are visited, we will

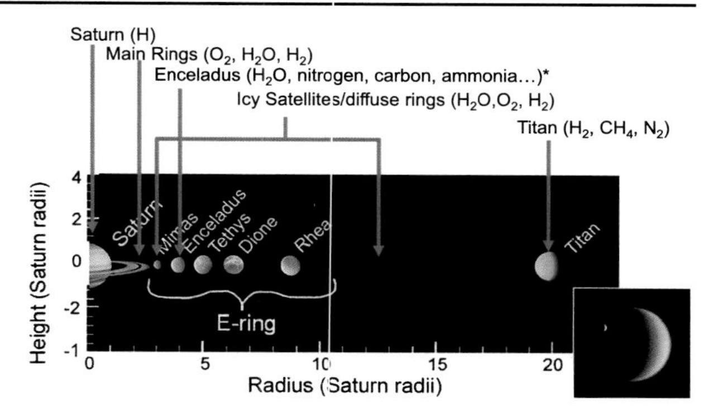

Fig. 1 Graphical representation of particle sources in Saturn's magnetosphere (not to scale) (from Smith et al. 2012)

wrap-up our exploration by summarizing the relative intensities of the different sources, and by placing the different sources in the context of a global description of Saturn's magnetosphere seen as an integrated system.

2 Enceladus: The Primary Source of Heavy Particles in Saturn's Magnetosphere

2.1 The Primary Enceladus Source

Unlike Jupiter where plasma dominates the magnetosphere, Saturn's magnetosphere is dominated by neutral particles by 1 to 2 orders of magnitude over charged particles. These magnetospheric particles originate from many sources including Saturn's atmosphere, rings and moons (Fig. 1). The dense atmosphere of Saturn's largest moon, Titan, was originally thought to be the primary source of these magnetospheric particles (Eviatar 1984), however more recent observations and data analysis indicate that the tiny icy moon, Enceladus, is actually the primary source (Dougherty et al. 2006; Porco et al. 2006). Thanks to Cassini observations, the source rates for all of these objects are much better defined. Figure 2 shows the current understanding of these source rates and illustrates that, with the exception of hydrogen, Enceladus clearly dominates over all of the other sources when producing magnetospheric particles. This surprising discovery is not yet completely understood, however it has large implications for the Saturnian system. While Enceladus has been known to be geologically active for some time, it was not expected that this tiny moon ($\sim$250 km radius) could be such a significant part of this planetary system. Thus understanding this phenomenon has been a subject of much research and debate.

With the dramatic increase in observations as a result of the Cassini mission, several studies were undertaken to better understand this primary source of magnetospheric particles. In particular, Smith et al. (2010), Tenishev et al. (2010), Dong et al. (2011) and Hansen et al. (2011) studied the Enceladus plumes using models and/or data analysis with relatively similar results. However, Smith et al. (2010) showed noticeable levels of plume source variability, which now appears to be consistent with the recent Cassini dust observations reported by Hedman et al. (2013). Therefore, we present the Smith et al. (2010) results in more detail here.

Smith et al. (2010) used a 3-D Monte Carlo particle-tracking, multi-species computational model (Smith et al. 2004; Smith 2006; Smith et al. 2007, 2008) to analyze the Ence-

Saturn:

H	300* x 10^{28} (~5000 kg)/s	*Shemansky et al. 2009 (rough upper limit) (Tseng et al. 2013 estimates ~20% of this value)*

Rings:

O_2	0.2 x 10^{28} (~100 kg)/s	*Tseng et al. 2012 (/100 at equinox)*
H_2	0.4 x 10^{28} (~13 kg)/s	*Tseng et al. 2012 (/100 at equinox)*

Enceladus:

H_2O	0.3-2.5 x 10^{28} (~100-760 kg)/s	*Hansen et al. 2006; Burger et al. 2007, Tian et al. 2007, Smith et al. 2010, Tenishev et al. 2010, Dong et al. 2010*
{N(C*)	0.01-0.1 x 10^{28} (~4-25 kg)/s	*Waite et al. 2006; Smith et al. 2007*

Titan

H_2	0.8 x 10^{28} (~27 kg)/s	*Cui et al 2008*
N_2	0.01-0.02 x 10^{28} (~5-10 kg)/s	*DeLaHaye et al. 2007*
CH_4	0.01 x 10^{28} (~3 kg)/s	*DeLaHaye et al. 2007*
{CH_4	0.2 x 10^{28} (~55kg)/s?	*Yelle et al. 2009; Strobel 2009}*

*Global detection (MIMI) of energetic Carbon ions 1.3% relative to water group (Mauk et al. 2009)

Fig. 2 Summary of current estimated Saturn magnetospheric source rate values with references (from Smith et al. 2012)

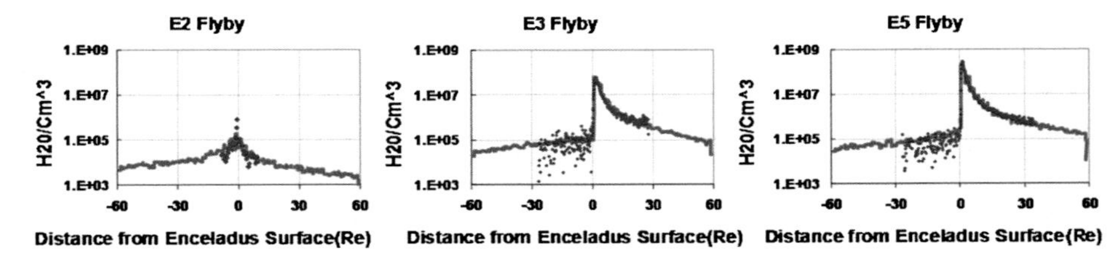

Fig. 3 Comparison of best fit model results (*red line*: 30° and 1.8 velocity ratio) with INMS measured neutral water densities (*blue circles*) for the E2 (*left panel*), E3 (*middle panel*) and E5 (*right panel*) Enceladus encounters. Results are displayed in water density (H_2O/cm^3) as a function of distance from Enceladus (in Enceladus radii or $\sim$ 252 km) with negative values for ingress and positive for egress. Source rates for each case are adjusted so model peak densities match peak INMS densities (from Smith et al. 2010)

ladus plumes. This validated model accounts for all gravitational effects of the planet and major satellites, as well as simulating particle interaction processes including electron impact ionization and dissociation, photo-ionization and photo-dissociation, recombination, charge exchange, neutral-neutral collisions, collision with the planet, satellites and the main rings as well as escape from the magnetosphere. They used Cassini Ion Neutral Mass Spectrometer (INMS) observations of neutral water particles during the E2, E3 and E5 Enceladus encounters to help constrain key plume parameters. By conducting a parametric set of simulation runs and extracting particle densities along each encounter trajectory, they determined that the data is best fit with plume velocities of $\sim$720 m/s and plume widths of $\sim$30 degrees. Figure 3 shows how these parameters allow the model results to coincide well with the observations.

Interestingly, Smith et al. (2010) were able to determine different source rates for each encounter. More specifically, they reported the following flowing source rates: E2 $\sim 2.4 \times 10^{27}$ H_2O/s; E3 $\sim 6.3 \times 10^{27}$ H_2O/s; E5 $\sim 25.0 \times 10^{27}$ H_2O/s; (E7 tentatively $\sim 9.5 \times 10^{27}$ H_2O/s). They report a factor of 3–4 variability between the E3 & E5 encounters (the E2 trajectory only passes through the outer edge of the plumes and thus they were not as

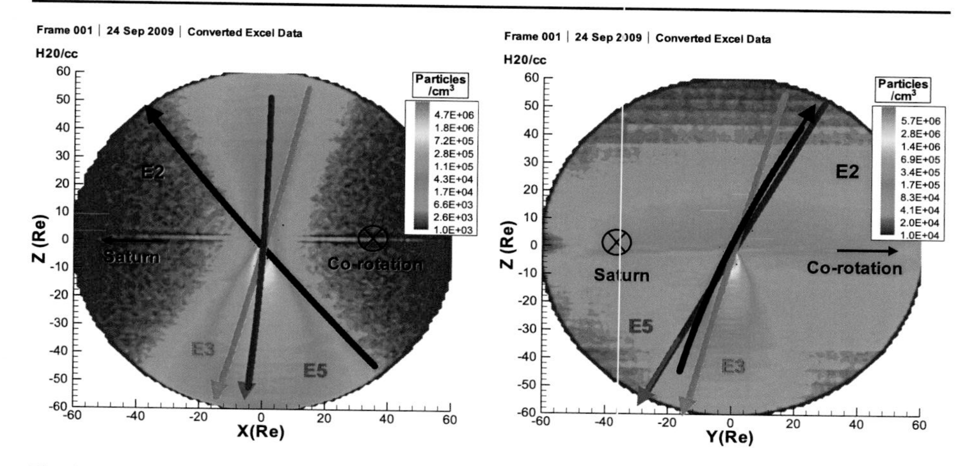

Fig. 4 *Contour plots* of H_2O density model results for the best fit Enceladus plume with 2× thermal velocity and ejection angle constrained to $+/-30°$ for the E3 source rate (divide densities by ∼15 for E2 and multiply by ∼4 for E5). Plots show Z axis (in Enceladus radii, or Re) based on the rotational axis, X axis (in Re: −X toward Saturn) in panel (**a**) and Y axis (in Re: +Y in co-rotational direction) in panel (**b**). E2 (*red*), E3 (*blue*) and E5 (*black*) trajectories shown in each panel (from Smith et al. 2010)

confident in that source rate). Figure 4 shows the resulting 3-D plume density distribution for the E3 trajectory.

As mentioned above, several other plume characterization studies were also conducted. In particular, Saur et al. (2008) constrained a neutral atmospheric model with Cassini magnetometer observations to report variable source rates of ∼1600 kg/s for the E0 encounter and ∼200 kg/s for the E1 and E2 encounters. Tenishev et al. (2010) applied a test particle model constrained by Cassini neutral particle (INMS) observations (with 8 variable plume sources to examine relative strength) to simultaneously fit the E3 and E5 encounters. They report source rates within a factor of 2 (E5 highest). Also, Dong et al. (2011) used an analytic model constrained by the same INMS observations of the E3, E5 & E7 encounters to determine source rates ranging from ~ 1.5 to $\sim 3.5 \times 10^{28}$ H_2O/s (factor of 2). However, Hansen et al. (2011) reported less than 20 % plume activity variability based on Cassini Ultraviolet Imaging Spectrograph (UVIS) observations of plume occultations in 2005, 2007 and 2010. Thus, the amount (if any) of plume variability was debated for several years and the limited number of observations made it difficult to resolve these inconsistent results.

More recent observations appear to support plume variability. Hedman et al. (2013) studied the relative brightness of dust particle observations in 252 Cassini Visual and Infrared Mapping Spectrometer (VIMS) images. Although these results involve dust particles, they provide a good proxy for relative vapor source rate. Interestingly, they find a factor of three variability in the plume source rate. Additionally, they find that this source rate is a function of the Enceladus mean anomaly with the highest source rates occurring when the satellite is furthest from Saturn, which supports the theory of enhanced source strength when tiger stripes are under tension. Figure 5 shows their results organized by orbital phase. This figure also shows that the Smith et al. (2010) plume variability results coincide well with the Hedman et al. (2013) results if one assumes orbit symmetry.

Therefore, our understanding of Enceladus as the primary source of heavy particles in Saturn's magnetosphere has noticeably increased. However, one must remember that these are neutral particles, which do not necessarily directly translate to a plasma source rate near

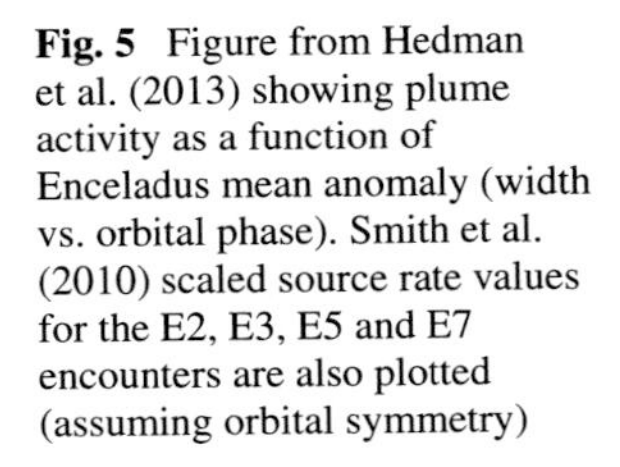

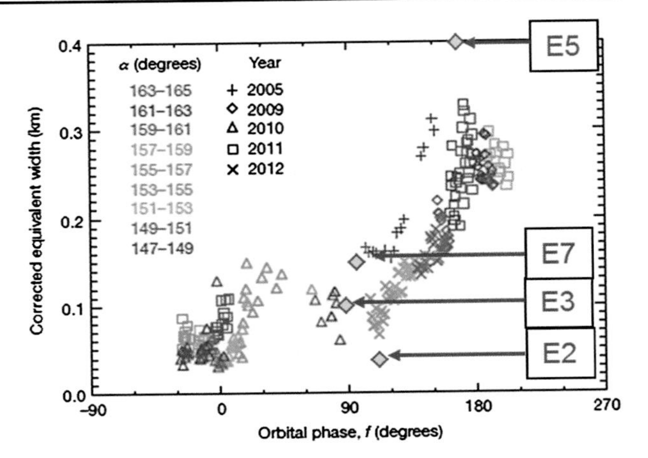

Fig. 5 Figure from Hedman et al. (2013) showing plume activity as a function of Enceladus mean anomaly (width vs. orbital phase). Smith et al. (2010) scaled source rate values for the E2, E3, E5 and E7 encounters are also plotted (assuming orbital symmetry)

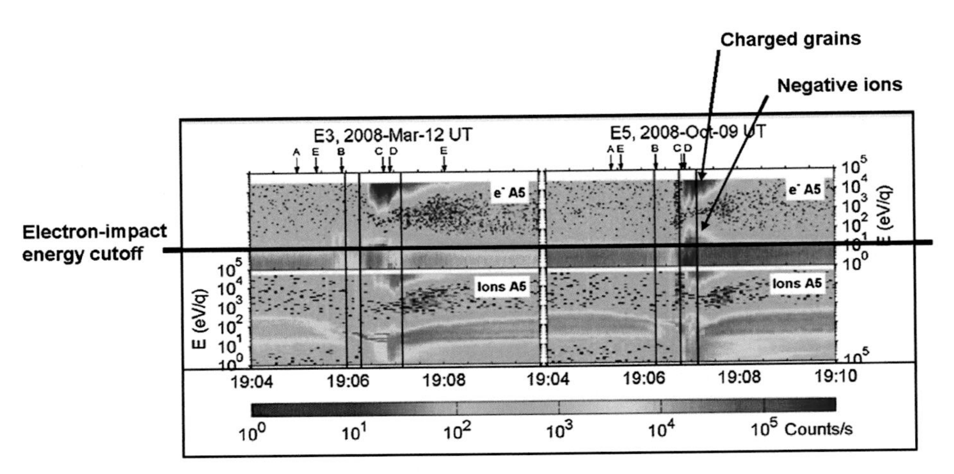

Fig. 6 Annotated plot from Tokar et al. (2009) showing Cassini Ion mass spectrometer and electron spectrometer spectra for the E3 and E5 flybys. Data is counts as a function of time (UCT) and energy (eV). The ionization cut-off energy is plotted on the electron spectra and the charged grain and negative ion signals are also annotated

this moon. The region near Enceladus is very complex and remains a topic of much debate. This is a region consisting of relatively large neutral particle densities, plasma (e.g. Tokar et al. 2009), negative ions (Coates et al. 2010), photoelectrons (Coates et al. 2013) and dust grains (Jones et al. 2009; Hill et al. 2012) where the plasma co-rotational ($\sim$36 km/s) speed is not much faster than the Keplerian orbital speeds ($\sim$ 12.5 km/s). In terms of plasma mass loading, photoionization lifetimes are very long (on the order of years) and the electron-impact ionization rates dramatically decrease in Saturn's inner magnetosphere because the core electron population is below ionization cut-off energies. Thus, the region near Enceladus is not well suited for ion production. Tokar et al. (2009) report a "plasma stagnation region" within a couple of satellite radii of Enceladus (Fig. 6) where the co-rotational plasma flow decreases significantly because of ion momentum loading (mostly through charge exchange). This should cause increased ionization, however Coates et al. (2013) report less than an order of magnitude ionization increase in this region. Additionally, this slower region should also dramatically increase the charge exchange rates, which would impede the process of creating new ions. Fleshman et al. (2010) use modeling to report that neutral particles should dominate over charged particles by a ratio of 40:1 near Enceladus and most of the electron impact ionization occurs from the higher energy electrons which should make up $<1\ \%$ of the total electron population. Thus, while Enceladus is a major source of neutral particles, the resulting charged particles are more likely to be produced further away from the moon.

2.2 Resulting Extended Sources

As mentioned above Enceladus is the main source of neutral water group particles throughout Saturn's magnetosphere. This extended source is modified by four main interactions:

- *Electron-impact ionization.* Electron-impact interactions can cause direct ionization as well as dissociation into daughter neutral and/or ionized species. In the case of dissociation, surviving neutral particles also have a small energy increase, which tends to spread out the neutral clouds.
- *Photo-ionization.* Neutral particles can also interact with solar UV photons, which interact similarly to the electron-impact processes. This process differs in that these rates vary with solar activity by up to an order of magnitude.
- *Neutral-ion collisions.* Neutral particles can interact with ions either through a simple collision or one where charge exchange occurs (effectively ionizing the neutral and neutralizing the ion). This process is much more efficient as plasma and neutral particle velocities approach each other. The net result of charge exchange is that while a fresh ion is created, an ion has become an energetic neutral particle so there is essentially no net increase in the ion population. Thus this process is not considered as plasma mass loading, but only as plasma momentum loading (i.e. the plasma flow tends to slow down as this fresh ion is picked up). This process dramatically spreads out the neutral clouds. Fleshman et al. (2012) indicate that gyrophase-dependant collisions have more of an impact on OH local densities than on the other water group species (i.e. collisions are more likely when the ion is moving in the same direction as the neutral particle).
- *Neutral-neutral collisions.* Some neutral particles can also interact with other neutral particles in the densest regions. This is a result of enhanced dipole interactions (Farmer 2009; Cassidy and Johnson 2010) and tends to spread out the neutral particles even more. This extension of the water and OH clouds then indirectly causes the daughter species (oxygen) to also extend.

Figure 7 shows relative contributions of each of these processes, which cause a spread in the neutral oxygen torus.

These processes interact in a complex manner with interaction lifetimes as the key driver in the dynamics of these distributed sources. Figure 8 shows the interaction lifetimes in Saturn's magnetosphere for water and hydrogen as a function of radial distance from Saturn (with Enceladus at $\sim 4\ R_S$). This figure helps to illustrate how photolysis is constant throughout all radial distances while electron impact is most significant (and dominant) at 7–9 R_S where the electron temperature and density is most conducive to interaction (consistent with Johnson et al. 2006). This figure also shows the charge exchange interaction rate dramatically increasing in the inner magnetosphere. These lifetimes combine to create and accelerate an extended source of particles with only $\sim$11–26 % of the original Enceladus particles actually being ionized, $\sim$23–43 % lost through collision (with Saturn, the moons

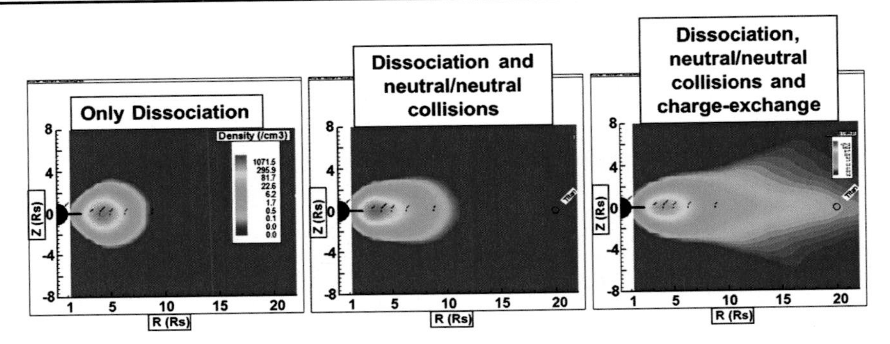

Fig. 7 Model results showing neutral oxygen distribution originating from the Enceladus plumes for differing theoretical cases: with only dissociation (*left*), with dissociation and neutral-neutral collisions present (*middle*) and with dissociation, neutral-neutral collisions and charge-exchange active (*right*). Data is oxygen density (cm^{-3}) in the R-Z plane (in Saturn radii) with Saturn *on the left of each panel* (from Smith et al. 2010)

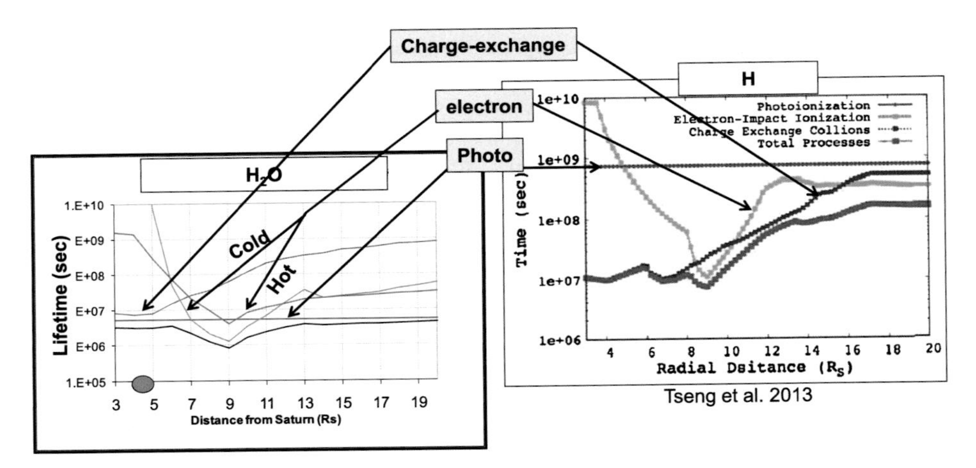

Fig. 8 Interaction lifetimes (seconds) for H_2O (*left panel*) and H (*right panel*) as a function of radial distance from Saturn (in Saturn radii). Charge-exchange, electron-impact and photolysis processes are annotated with the *right panel* offset so the values on the H lifetime y-axis are aligned with the H_2O lifetime values (from Smith et al. 2012)

and rings) and $\sim$ 31–66 % escaping the magnetosphere as a result of charge exchange (Cassidy and Johnson 2010; Fleshman et al. 2010).

The result is a series of co-orbiting neutral particles that form multiple Enceladus-generated neutral tori, which serve as a distributed source of plasma in the magnetosphere. Figure 9 shows the timescale and process that lead to tori formation. The Enceladus plumes provide the original water source species, which escape with relatively low velocity. The result is a relative confined torus of H_2O particles near Enceladus' orbit (this feature is expanded through neutral-neutral interactions). On the order of about 2.5 months, OH molecules are created through dissociation (primarily photo) creating a more extended torus. This torus is expanded via charge-exchange, neutral-neutral interactions as well as energy obtained from dissociation. Next, an even more extended O torus is created (on the order of $\sim$ 1.5 months) that is spread out through the same mechanisms. Finally, dissociated

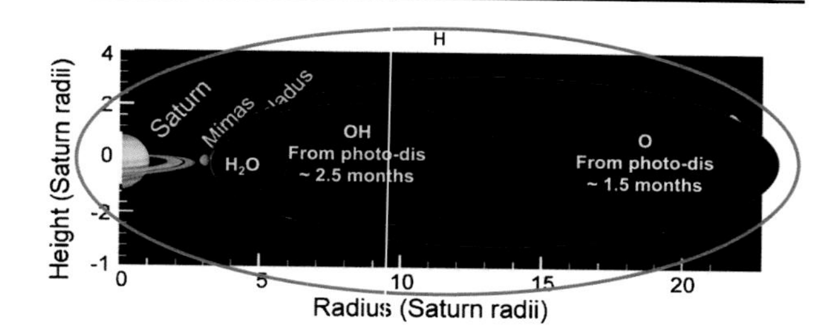

Fig. 9 Graphical representation of the rough distribution of resulting neutral H_2O, OH, O and H tori with the approximate times required to produce these species from the original water source (from Smith et al. 2012)

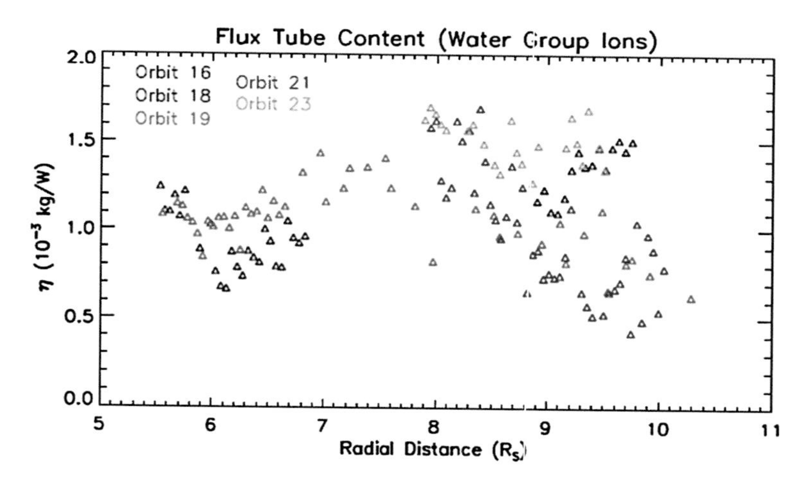

Fig. 10 Water group ion peak flux tube content (10^{-3} kg/W) as a function of radial distance from Saturn (from Chen et al. 2010)

hydrogen from these interactions (as well as from Saturn and Titan) forms a much more extended neutral torus. Thus Enceladus indirectly generates multiple distributed magnetospheric neutral sources. These sources in turn serve as plasma sources. As mentioned above, for typical magnetospheric conditions ions are most efficiently produced in the 7–9 R_S range which coincides with the peak in Saturn's electron and ion flux tube content (Sittler et al. 2008; Chen et al. 2010) as shown in Fig. 10. It is also important to consider that the heavy ions are fairly constrained to the equatorial plane while lighter ions are much more extended and peak densities likely actually occur off of the equatorial plane. Finally, plasma transport dominates over the other ion source and loss processes in the middle and outer magnetosphere, while ions in the inner magnetosphere tend to be lost through recombination (Fig. 11). Bagenal and Delamere (2011) also show that local sources/losses dominate over radial transport at distances closer than 8 R_S of Saturn. They also show that the ionization rate does not keep up with the neutral source rate variations.

3 Dusty Plasmas in Saturn's Plasma Torus

The water vapors and ice grains expelled from Enceladus' south pole are one of Cassini's most exciting discoveries (Dougherty et al. 2006; Porco et al. 2006; Waite et al. 2006). Furthermore the state of the plume and its surrounding plasma constitute a so called "dusty

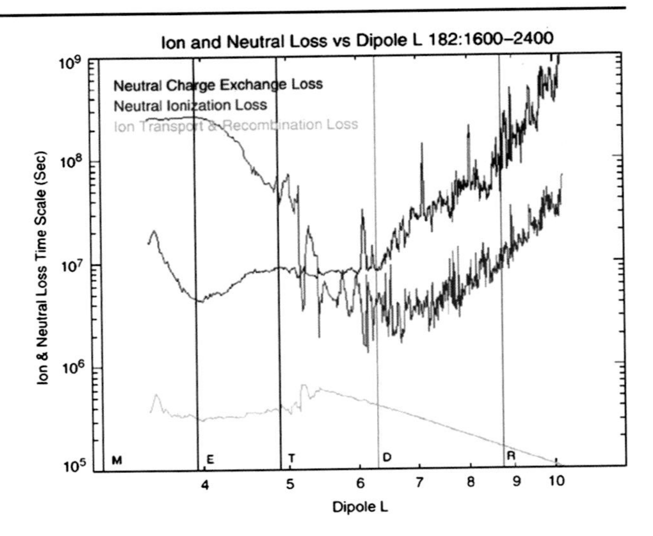

Fig. 11 Comparative lifetimes (seconds) for charge exchange (*brown*), neutral loss to ionization (*blue*) and combined ion loss to recombination and transport (*light green*) (from Sittler et al. 2008)

plasma" environment. In this section we describe briefly what is a dusty plasma, how the observation has been interpreted, and its possible consequences for the magnetosphere.

Dusty plasmas are plasmas containing charged particles (dust). The particle sizes are typically nanometers to micrometers and they are massive compared to the plasma elementary charged particles. The dust charge can be negative or positive and the grain charge number varies from single to several thousands. As a result the charged particles in dusty plasmas are under the control of both gravity as well as electro-magnetic forces.

In addition to this the charged dust particles are strongly coupled to the surrounding plasmas depending on the characteristic lengths. When the inter grain distance (a) is larger than the plasma Debye length (λ_D), the situation $r_d \ll \lambda_D < a$ holds, where r_d is the grain radius, the charged particles are considered as isolated among the plasma. This situation is referred to 'a dust-in-plasma'. A 'real' dusty plasma occurs when $r_d \ll a < \lambda_D$, wherein charged dust particles participate in the collective behavior (e.g., Shukla and Mamun 2002).

The dust and plasma conditions found near Enceladus are in the dusty plasma state. This has been suggested by the Langmuir probe that measures the electron and ion densities of cold plasmas (Fig. 12). The observed large ion densities (30000–100000 cm^{-3}) and low electron densities (2000–4000 cm^{-3}) in the plume region must be the result of the attachment of electrons onto the dust particles (e.g., Waite et al. 2006). Incidentally, the ion density obtained here exceeds the expected amount from just photo ionization (see Sect. 2 for ionization), indicating that an additional ionization mechanism is needed to generate this large amount of plasma. The electron densities are generally more than an order of magnitude smaller than the ion densities, as if more than 90 % of the electrons were missing from the region. This electron density dropout is due to the electron attachment to the dust grains (Farrell et al. 2009). It has been confirmed that the micrometer sized negatively charged dust observed by the Cosmic Dust Analyser [CDA] (Kempf et al. 2008) as well as the Radio and Plasma Wave Science (RPWS) signals (Kurth et al. 2006; Farrell et al. 2009) coincide within this region. Also both positive and negative nanometer sized small grains have been observed by the plasma particle detectors (Jones et al. 2009; Hill et al. 2012), which support the conclusion that the majority of the dust particles are negatively charged (Fig. 13). Since the electric charges of the grains are proportional to the

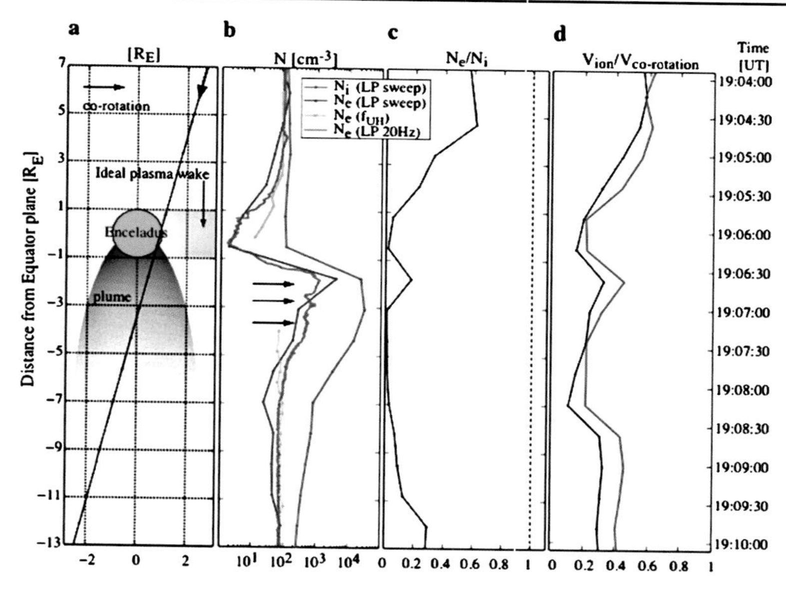

Fig. 12 Cassini RPWS/LP observation from the Enceladus E3 encounter showing a dusty plasma nature near the Enceladus plume. (**a**) Spacecraft trajectory in Enceladus frame. (**b**) Electron/ion densities. (**c**) LP electron to ion density ratio. (**d**) Ion speed relative to the rigid corotation speed in the spacecraft frame of reference. The striking features are: (1) a large electron/ion density difference ratio throughout the encounter (panels **b** and **c**), (2) plasma speed slowed down to near the Keplerian speed (panel **d**), and (3) there is no ion density depletion in the wake (no plasma wake signature, panel **b**). See Morooka et al. (2011) for detail

grain surface, large micrometer grains can hold several thousand charges while nanometer grains are often singly charged (e.g., Horanyi 1996; Yaroshenko et al. 2009).

Comparing the electron and ion density differences obtained from the Langmuir probe and the dust size distributions from CDA as well as CAPS (CAssini Plasma Spectrometer), a large amount of sub-micron grains are inferred to exist as the majority of the negative charge carriers (Fig. 14). Using the electron/ion density differences, the average grain potential, and the modeled dust size distributions, the dust density was estimated to be about 300 cm^{-3}, thus the inter grain distance is about 0.13 cm, which is smaller than the local Debye length ($\sim$ 6.04 cm) (Morooka et al. 2011). Furthermore, the estimated dust condition and plasma condition satisfy the Havnes condition for a dusty plasma (Havnes 1993).

The plasma characteristics in the geometric wake region also explain a strong electric coupling between the charged dust and plasma. In the wake of the moon Enceladus the speeds of the plasma are slowed down to the gravitational speeds (Tokar et al. 2009; Farrell et al. 2010; Morooka et al. 2011) and no wake signature of ion density has been observed (Morooka et al. 2011, see Fig. 12). If the plasma co-rotates with Saturn's magnetic field the moon should be an obstacle to the ions with large gyro radius and the magnetospheric ions could not enter behind the moon. However, the observations by the Langmuir probe showed that the wake region is filled with the cold ions. While the slow ion speeds can be interpreted as a result of charge exchange (Tokar et al. 2009), they can be also due to the fact that the plasma particles are electromagnetically coupled to the charged grains that have the gravitational speed. In the wake region the particle instruments observed that the ions were nearly at rest in the Enceladus frame and the electron densities were depleted (Tokar

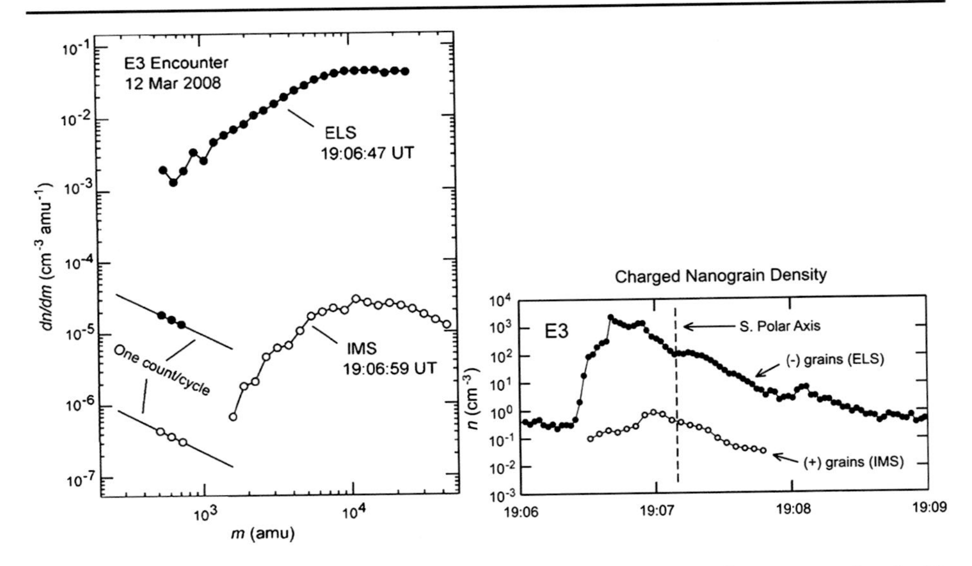

Fig. 13 CAPS observations of the charged nanometer grains during the Enceladus encounter described in Fig. 12. Both positive (*white circle*) and negative charged nanometer grains were observed by the electron/ion detectors in the high-energy ranges. *Left*: the energy is converted into mass. The densities of nano-grains were estimated in the plume region (*right*), showing the presence of a large amount of negative grains. They are expected to be a part of the dusty plasma. Mass distribution functions of the positive (*white circle*) and negative (*black dots*) charged particles. *Right*: densities of the positive and negative nano-grains (Hill et al. 2012)

et al. 2009), consistent with the fact that the electrons are attached to the charged dust and the surrounding plasma is in a state of coupling to the slow charged dust.

Dusty plasmas have been studied theoretically (e.g., Goertz and Ip 1984; Whipple et al. 1985; Havnes et al. 1987) and verified in the laboratory (e.g., Xu et al. 1993). Except for a few direct measurements in Earth's upper atmosphere (Reid 1990; Havnes et al. 1996), there have been no in situ observations of dust-plasma ensembles in space. Simon et al. (2011) suggested that the negatively charged dust grains in the plume act as a sink for "free" electrons and yield a reversal in the sign of the Hall conductivity, resulting in a slowdown of the ions (Kriegel et al. 2011). It is important to note that the negatively charged dust and its effect on the plasma dynamics appear not only in the plume but also in a large region around the moon Enceladus (Wahlund et al. 2009; Farrell et al. 2012). The magnetospheric plasma speeds are confirmed to be often slower than the co-rotating speed (Wilson et al. 2008; Thomsen et al. 2010; Holmberg et al. 2012), which could be associated with the negatively charged dust near the E ring (Holmberg et al. 2014).

4 Titan

Titan orbits Saturn at a distance of 20 R_S. With its extended atmosphere composed primarily of N_2 and CH_4 and apparent lack of internal magnetic field, it interacts with Saturn's magnetospheric plasma and the solar wind in a cometary fashion, producing an induced magnetosphere. Titan can act as a source of neutrals through atmospheric escape processes, which primarily eject H_2. It can also act as a source of ionized plasma through ion outflow

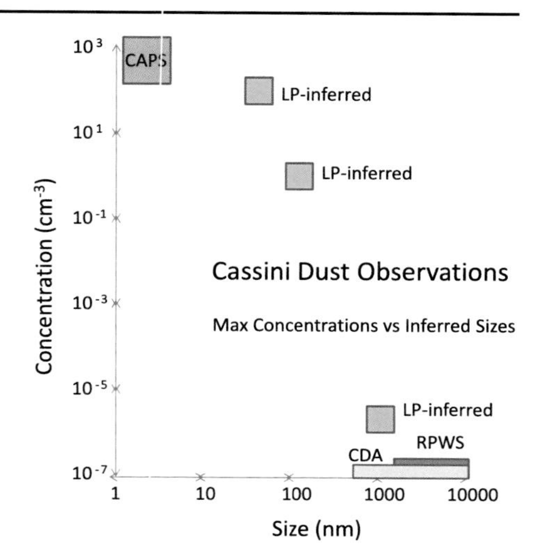

Fig. 14 Schematic view of the dust size distribution expected in the plume. Charged nanometer grains were detected by CAPS observations (Fig. 13) and large micrometer grains were detected by CDA and RPWS observations (Kempf et al. 2008; Farrell et al. 2009; Ye et al. 2014). By tracing the size distribution between nanometer and sub-micrometer grains from the different observations, one can infer that ten to hundred nanometer charged grains must exist to explain the large N_e/N_i differences obtained by the Langmuir probe. Courtesy of W. Farrell

and pickup-like processes. Titan is also a sink of magnetospheric plasma, albeit a small one compared to the major processes that act on a global scale like charge-exchange or tail reconnection.

Prior to the Cassini exploration of the Saturn system Titan was believed to be a significant plasma source with sufficient protection from the solar wind due to its magnetic shielding by Saturn's magnetosphere so as to develop an observable gas torus (Smith et al. 2004). However, Cassini observations have revealed that the region near Titan was more exposed to the solar wind and magnetosheath plasma than anticipated when Titan was on the dayside (Achilleos et al. 2008) and that the neutral source was more benign than previously expected (e.g. Tucker and Johnson 2009; Bell et al. 2011). From magnetospheric measurements Titan's orbit is expected to be in the solar wind for 5 % of the time, exposing it to enhanced solar electrons that cause greater losses of the neutrals (Achilleos et al. 2008).

Titan loses atoms and molecules from its atmosphere as ionized and neutral material. Several processes have been proposed as sources from Titan including thermal escape (Cui et al. 2008; Bell et al. 2011), polar wind and ambipolar electric field (Coates et al. 2007a, 2012, 2015), chemically-induced escape (De La Haye et al. 2007), slow hydrodynamic escape (Strobel 2008, 2009; Yelle et al. 2008), pick-up ion loss and ionospheric outflow (Ledvina et al. 2005; Wahlund et al. 2005; Sillanpää et al. 2006; Ma et al. 2006; Hartle et al. 2006; Coates et al. 2007a; Edberg et al. 2011; Westlake et al. 2012; Coates et al. 2012) and plasma-induced atmospheric sputtering (Shematovich et al. 2003; Michael et al. 2005; De La Haye et al. 2007).

The Saturn-Titan interaction is complex and highly variable. The magnetosphere presents different upstream conditions to Titan on timescales of hours (Simon et al. 2010). It is likely that the configuration of the moon-magnetosphere interaction affects the ion outflow rate (Sillanpää et al. 2006). However, it is interesting that Titan also has some self-shielding due to fossilized fields (Bertucci et al. 2008) and a robust ionosphere. Modeling work by Snowden et al. (2011a, 2011b) has also shown complex feedback processes that affect the configuration of the magnetosphere, especially when Titan is on the dayside (Fig. 15). When Titan is on the nightside the magnetosphere is easily compressed and the Titan neutral torus is eroded by magnetosheath and solar wind plasma (Snowden et al. 2011b).

Titan's neutral source comes primarily in the form of H_2, which readily escapes Titan's atmosphere (Cui et al. 2008). Energetic Neutral Atom (ENA) observations of Titan's ex-

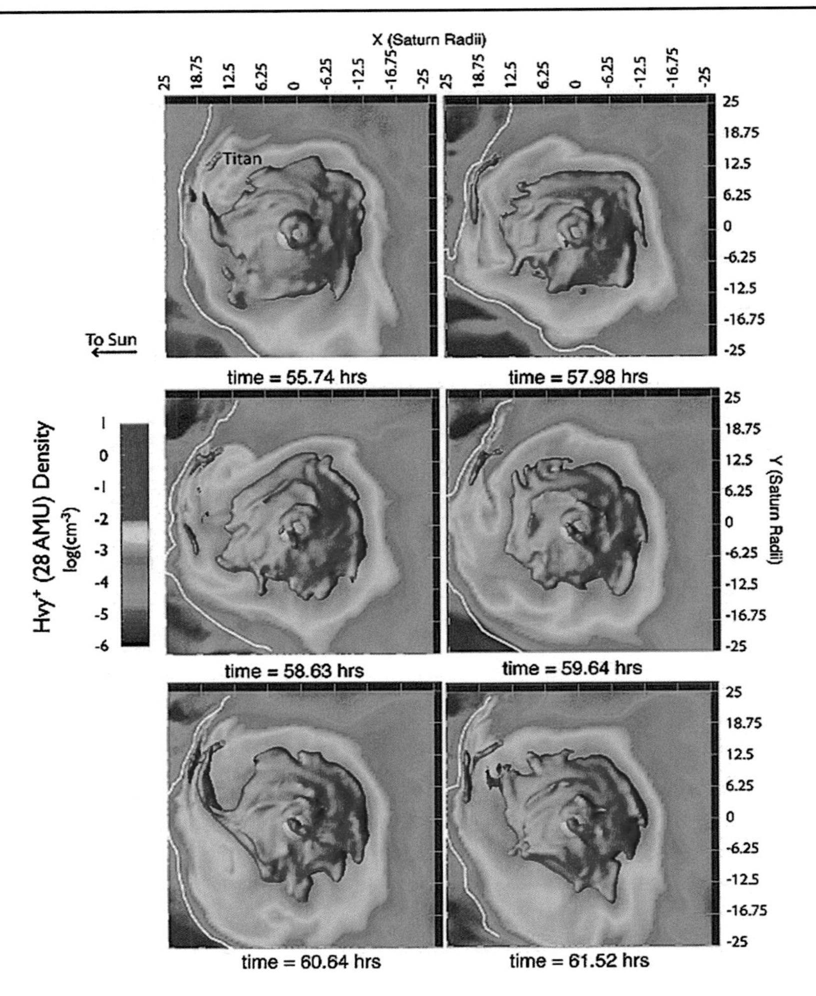

Fig. 15 Simulations by Snowden et al. (2011a, 2011b) showing the production and evolution of Titan's ion tail alongside Saturn's rapidly rotating magnetosphere. The *white line* is the magnetopause, which is affected by Titan's proximity

tended exosphere (Fig. 16) have found appreciable densities of H_2 radially out to the predicted Hill sphere radius of 60000 km (Brandt et al. 2012). Debates within the atmospheric modeling community have argued whether H_2 escape is limited by the Jeans escape rate (Bell et al. 2010, 2011) or is occurring at a significantly greater rate (Cui et al. 2008). Exospheric simulations using Direct Monte Carlo Simulation have found the H_2 escape rate to be consistent with Jeans escape (Tucker and Johnson 2009).

Titan is also a source of N_2 and CH_4, though their mass is closer to the mean mass of the atmosphere and they are thus much more benign in their escape rates. CH_4 and its escape rate have sparked heated debates in the modeling community as its escape rate implied from the atmospheric measurements depends on the assumptions regarding the eddy diffusion parameter in Titan's upper atmosphere and assumptions as to whether Titan is in hydrothermal equilibrium. Yelle et al. (2008) initially utilized the in-situ INMS measurements to show

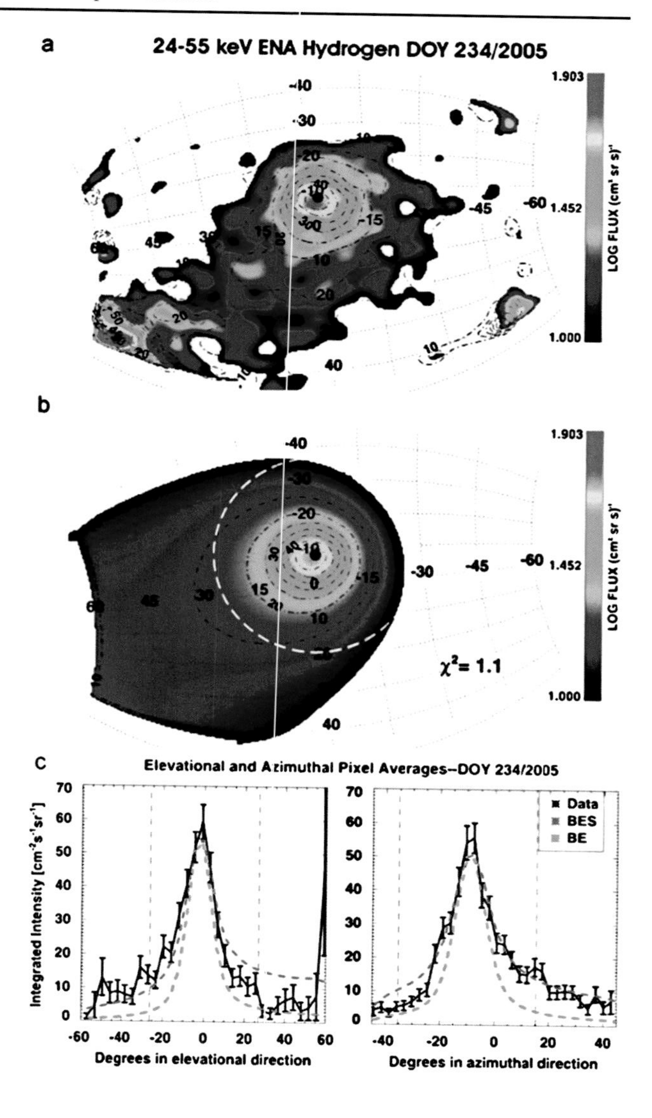

Fig. 16 Titan Energetic Neutral Atom (ENA) observations from Brandt et al. (2012) showing the extent of its exosphere. ENAs are produced through the interaction of Saturn's magnetospheric plasma with Titan's neutral atmosphere and are a sensitive probe of low-density environments. The *top figure* is the Cassini INCA observation, the middle one is the modeled flux using a Chamberlain exospheric neutral distribution with satellite, ballistic, and escaping distributions. The *bottom two plots* show the agreement between the model and the observations when using the ballistic, escaping, and satellite (BES) or just the ballistic and escaping (BE) distributions

that either the eddy diffusion rate at Titan is relatively high and the escape rate low or vice versa. Bell et al. (2014) have recently shown that the INMS constraints provided by simultaneously fitting the altitude profiles of N_2 (and its isotopes), CH_4 (and its isotopes), Ar, and H_2 are best met by a high eddy diffusion coefficient and a small CH_4 flux—a result that is reinforced by the relatively modest amount of carbon-bearing ions seen by CAPS in the near Titan environment (Smith et al. 2012).

Cassini and Voyager have passed through Titan's induced magnetospheric tail region allowing for measurements of Titan source plasma being accelerated into Saturn's magnetosphere. From these measurements the mass loss rate due to pick-up ion formation and sweeping combined with ionospheric outflow is compatible with a value of roughly 10^{25} amu/s (Wahlund et al. 2005; Hartle et al. 2006; Coates et al. 2007a, 2012). The Voyager plasma instrumentation did not have sufficient energy or mass resolution to determine the composition of these ions as they left Titan other than they consisted of a heavy and a light component. This led to the assumption that the ions were classical pick-up ions produced through the local ionization of CH_4 and N_2 that had escaped Titan's atmosphere. Cassini INMS observations in Titan's ionosphere found a plethora of hydrocarbon and nitrile ions produced through photoionization followed by ion-neutral reactions (Cravens et al. 2006; Vuitton et al. 2007). Cassini observations both with the CAPS and INMS instruments have found that the composition of the ions flowing from Titan is mainly ionospheric with significant amounts of CH_5^+, $C_2H_5^+$, and $HCNH^+$, that cannot be produced in Titan's sparse exosphere (Westlake et al. 2012; Coates et al. 2012). From this and the prevalence of "fossilized" magnetic fields in Titan's ionosphere it is clear that magnetospheric field lines penetrate deep into Titan's ionosphere where ion-neutral chemistry acts to mass load the field lines, which then carry Titan's ionospheric plasma away (Coates et al. 2012; Wellbrock et al. 2012). These ions have long lifetimes in Titan's exosphere due to the declining electron density and increasing electron temperature and can therefore remain as ions traveling downtail from Titan.

Combining these source rates one obtains a total rate that ranges between 0.03 and 0.5×10^{29} amu/s of primarily H_2 and lesser amounts of N_2 and CH_4 along with various compositions derived from these neutrals (Johnson et al. 2010). This source rate is significantly less than the Enceladus and rings source rates, and given Titan's proximity to the edge of the magnetosphere it is now clear that this material is readily picked up into the solar wind and is not a major source of particles to Saturn's magnetosphere.

5 The Solar Wind

In order to understand the influence of the solar wind, and its potential role as a plasma source for Saturn's magnetosphere, it is critical to study the large-scale structure upstream of Saturn to understand its impacts upon Saturn's magnetopause. The model of Parker (1958) predicted that as the solar wind evolves throughout the heliosphere, the magnetic field it carries winds into an Archimedean spiral. In addition, the radial magnetic field strength is expected to fall off with distance from the Sun, and compressions and rarefactions in the solar wind will develop into clear patterns. Several early studies using data from the Pioneer and Voyager spacecraft had an opportunity to test these claims. Thomas and Smith (1980) used data from Pioneer 10 and 11 to probe the solar wind between 1 and 8.5 AU. They found that the field directions conformed on average to those predicted by the Parker model to an overall accuracy of 1.1°. They also calculated the typical spiral angle at 8.5 AU to be $\sim 83°$, and thus suggest that substantial departures are unlikely to be found beyond these distances, as the field becomes almost azimuthal. As such, the nominal IMF at Saturn is in contrast to the case at the Earth, which has implications for the nature of solar wind-magnetosphere coupling. Similarly the analysis of Voyager and Pioneer 11 data by Burlaga and Ness (1993) shows that the radial variation of the magnetic field strength out to 19 AU is consistent with Parker's model when one considers the latitudinal and temporal variations of the source magnetic field strength and the solar wind speed.

Figure 17 shows the spacecraft that have visited Saturn's magnetosphere, compared to the phase of solar cycle. The three flybys missions (Pioneer 11, Voyager 1 and Voyager 2) visited at or near to solar maximum, while Cassini's approach and Saturn Orbit Insertion (SOI) took place during the declining phase of the solar cycle. Since then, Cassini's 10+ year mission

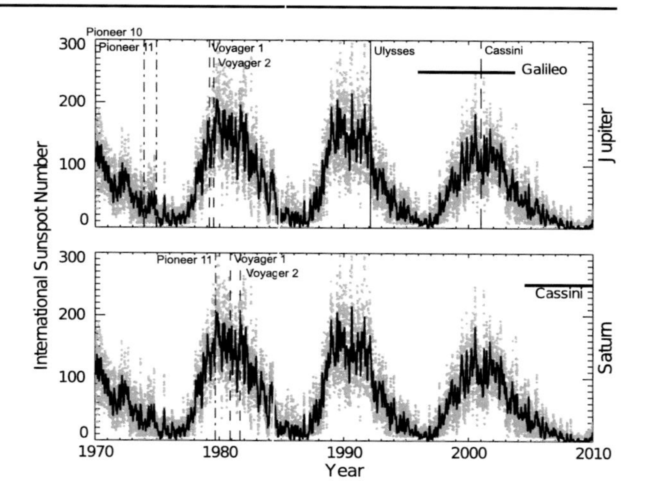

Fig. 17 Sunspot number from 1970–2010. The timings of the closest approach of spacecraft to Jupiter (*top panel*) and Saturn (*lower panel*) are marked (from Jackman and Arridge 2011)

has now encompassed a long and low solar minimum, and through the rising phase of the next solar cycle. The Cassini spacecraft approached Saturn's magnetosphere during early 2004, measuring the interplanetary magnetic field (IMF) continuously and sampling solar wind plasma properties when the pointing of the CAPS instrument was favorable. Jackman et al. (2004) then presented a comprehensive study of the structure of the solar wind during the Cassini approach, showing that the IMF was dominated by a clear pattern of corotating interaction region (CIR) compressions and rarefactions, caused in part by the tilt of the Sun's dipole during the declining phase of the solar cycle. This study was followed up by a survey of the IMF parameters (Jackman et al. 2008a) which indicated that the average spiral angle upstream of Saturn is $\sim 83^\circ$, agreeing very closely with the predictions of the Parker model. Data from Cassini (declining phase to solar minimum) were then combined with older data from Pioneer and Voyager to build a picture of the solar wind character across different stages of the solar cycle (Jackman and Arridge 2011).

Once the solar wind reaches Saturn's magnetopause it can shape the boundary and the IMF can merge with the planetary field via magnetic reconnection (illustrated schematically in Fig. 18), allowing the transfer of mass, energy and momentum to the system. The extent to which dayside reconnection operates is a topic of intense debate (e.g. Masters et al. 2014 and references therein). The reconnection rate is modulated in some manner by the orientation and magnitude of the IMF. Conflicting early studies have indicated that reconnection is (i) feasible and can be important (e.g. Huddleston et al. 1997; Grocott et al. 2009), or (ii) that reconnection can be suppressed by the high Mach number regime at Saturn (e.g. Scurry and Russell 1991). Some evidence for dayside reconnection has been presented (McAndrews et al. 2008), while other studies claim its role is rather limited (Lai et al. 2012). More recently, work has focused on the factors that may govern the reconnection rate, such as the plasma beta (Masters et al. 2012; Desroche et al. 2013), as well as cusp observations at Saturn (Jasinski et al. 2014). This is certainly an ongoing area of research. If/when dayside reconnection is active at Saturn's magnetopause, the theoretical work of Badman and Cowley (2007) has indicated that it can have a very important impact on the dynamics of the Saturnian system, particularly in the outer regions of the magnetosphere.

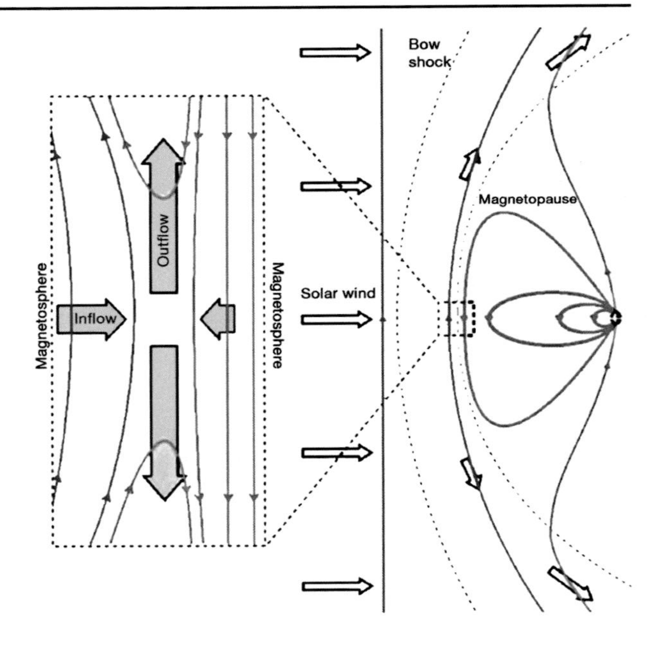

Fig. 18 An illustration of magnetic reconnection at Saturn's magnetopause for (rare) perfectly northward interplanetary magnetic field. Interplanetary (solar), planetary, and reconnected (open) magnetic field lines are shown in *blue*, *red*, and *green*, respectively (from Masters et al. 2014)

6 Sources for High Energy Particles

6.1 Introduction

The composition of suprathermal and energetic magnetospheric ions contains partial information on relative plasma source strengths, as does that for the thermal plasma. Although the terms thermal, suprathermal, and energetic are not associated with well-defined energy ranges, in this discussion we will apply them to <10 keV, 10–200 keV, and >200 keV particle populations, respectively. Energization frequently involves transport, so that the spatial distribution of the various energetic species is much less likely to pinpoint source locations than is that of the thermal plasma. Composition measurements at higher energies also have some advantages. In some cases, the mass per charge or mass resolution is better, and identification ambiguities that occur when two species have the same mass per charge can be resolved with techniques available at higher energies.

As is the case at other planets, the most energetic component of the high-energy charged particle population at Saturn is partly produced by the Cosmic Ray Albedo Neutron Decay (CRAND) mechanism, which is therefore not connected to the composition of these charged particles at lower energies. This CRAND source is described in Sect. 6.5.

Before Cassini entered into Saturn orbit in July 2004, composition measurements had only been made for the thermal plasma and at quite high energies ($\sim$ MeV) during the Pioneer 11 and Voyager 1 and 2 flybys. Pioneer 11 reported the presence of protons and helium nuclei at energies > 0.5 MeV/nucleon (e.g., Simpson et al. 1980). Frank et al. (1980) analyzed Pioneer 11 plasma data (100–8000 eV/e) and concluded from indirect evidence that, in addition to protons present throughout the magnetosphere, a torus of heavy ions existed inside the orbit of Rhea. They identified these ions as most likely being oxygen in the +2 and +3 charge states, but charge states > +1 were not confirmed by the Voyagers (Richardson and Sittler 1990). Voyager 1 and 2 reported the presence of both protons and an abundant heavy ion (either N^+ or O^+) at thermal energies (10–5950 eV/e) (Lazarus and

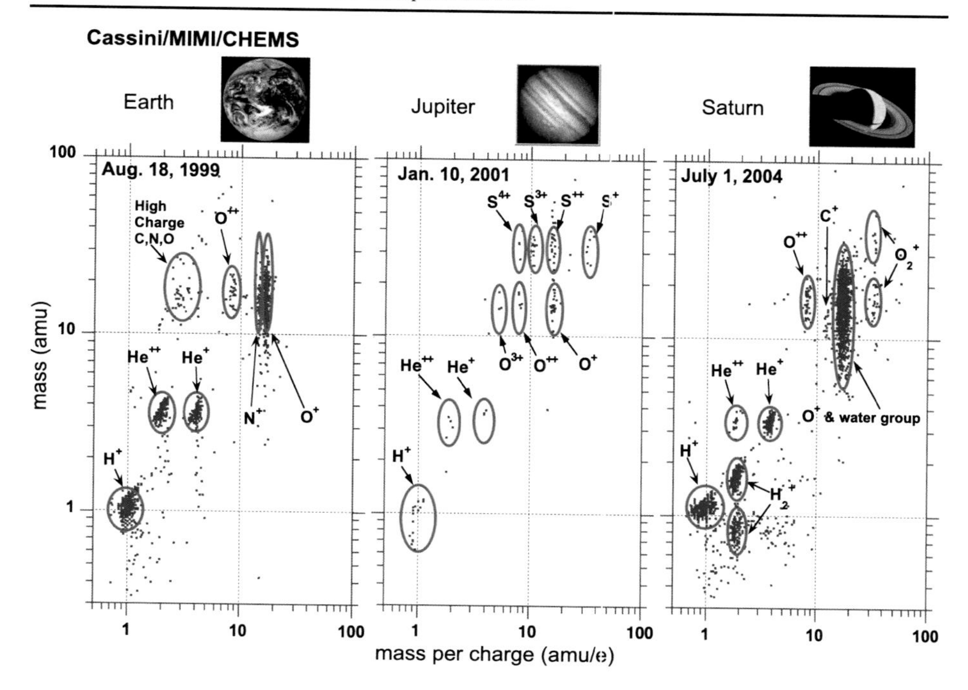

Fig. 19 Suprathermal ion composition data from the Cassini/CHEMS instrument for three planetary magnetospheres. The data were obtained during the Cassini flybys of Earth and Jupiter and during SOI at Saturn. The composition differs among the planets and is indicative of different plasma sources. The time, radial distances, and energy intervals for the three data sets are the following: Earth 0225–0250, Day 230, 1999, $R = 6.2$–$10.1\ R_E$, dayside, 55–167 keV/e; Jupiter 1916–2100, Day 10, 2001, $R = 204\ R_J$, dusk flank, 94–97 keV/e; Saturn 0800 Day 182–1610 Day 183, 2004, $R = 15.0\ R_S$ inbound to $10.8\ R_S$ outbound, 83–220 keV/e (from Hamilton et al. 2005)

McNutt 1983). Further analysis of the Voyager plasma data favored the O^+ interpretation (Richardson 1986). At higher energies above ~ 0.5 MeV/nucleon, Hamilton et al. (1983) identified H^+, H_2^+, H_3^+, and He along with a small amount of C, N, and O (without charge state identification for any of the species). Later, the HST detection of an OH torus (Shemansky et al. 1993) led to detailed modeling by Jurac and Richardson (2005) with the conclusion that Saturn's magnetosphere must have a strong source of water with a maximum around the orbit of Enceladus.

The Cassini measurements of suprathermal ion composition have been possible because of development of time-of-flight methods in space instrumentation (Gloeckler 1990) since the Pioneer and Voyager missions. The results reviewed in this section are largely measurements made in the suprathermal energy range by the Cassini Charge Energy Mass Spectrometer (CHEMS) instrument (instrument described by Krimigis et al. 2004). Suprathermal composition measurements were not made by either Pioneer 11 or the two Voyagers.

Data from Saturn Orbit Insertion, shown in the right panel of Fig. 19, immediately showed differences with observations made by the same instrument during Cassini's Earth and Jupiter flybys (left two panels). Although H^+, He^+, He^{++} and O^+ are present in all three magnetospheres, Saturn shows a broader O^+ distribution, indicative of the presence of additional water products (OH^+, H_2O^+, and H_3O^+) along with the molecular ions H_2^+ and O_2^+. The open question left from the Voyager flybys concerning the identity of the dom-

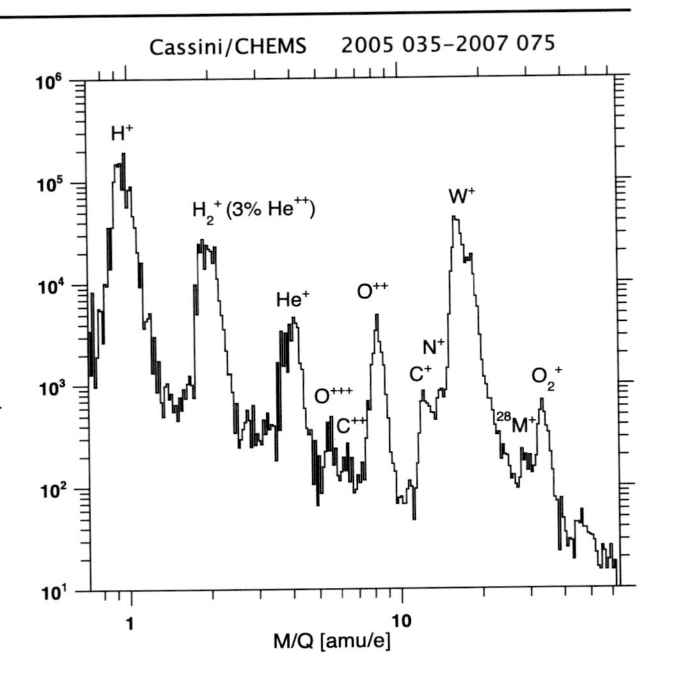

Fig. 20 Two-year sum of suprathermal ion composition in Saturn's magnetosphere from the Cassini/CHEMS instrument. Only events that include both a time-of-flight measurement and an energy signal from the solid state detectors (SSDs) are included (lowest background data). The minimum energy to trigger the SSDs increases with mass. Approximate energy ranges are 25–220 keV for H^+ and 55–220 keV for W^+. Apparent relative abundances are only qualitative (from Mauk et al. 2009)

inant heavy ion was immediately answered; it was O^+ and the other water products. Because of Titan's nitrogen atmosphere, the relative absence of N^+, found to be much less abundant in Saturn's magnetosphere than even in Earth's, was surprising to many. The lack of sulfur ions, present at Jupiter as a result of Io's volcanism, was not surprising.

6.2 Overview

Figure 20 presents a mass per charge (M/Q) histogram of Saturn data accumulated by the CHEMS instrument over a $\sim$ 2-year period showing multiple species. Since the detection efficiency and minimum energy both vary with mass (see figure caption), the apparent relative abundances in Fig. 20 are only qualitative. CHEMS uses electrostatic deflection to determine an ion's energy per charge (E/Q), along with a time-of-flight (TOF) measurement and a kinetic energy measurement (E) in solid state detectors (SSDs) to determine mass per charge M/Q and mass M. Although CHEMS' entire energy range is 3–220 keV/e, the SSDs do not trigger for incident ions in the lower portion of that range (see caption). When there is no energy signal, only M/Q is determined. All the events in Fig. 20 had an energy signal, so both M/Q and M are determined, and species such as H_2^+ and He^{++}, which have the same value of M/Q and coincide in Fig. 20, can be separated.

Even though quantitative comparisons cannot be made from Fig. 20, the three most abundant species in Saturn's magnetosphere, H^+, W^+ (the water group ions comprising O^+, OH^+, H_2O^+, and H_3O^+), and H_2^+, stand out. The dominant sources of these three species, according to current thought, are all local, but different. The water products originate predominately from the Enceladus plumes with a source rate of $1\text{–}4 \times 10^{28}\ \mathrm{s}^{-1}$ (Burger et al. 2007; Jia et al. 2010; Dong et al. 2011). The H^+ ion has several potential sources including the solar wind and dissociation of the Enceladus water, but the largest source appears to be ionization of the extensive neutral H cloud arising from Saturn's atmosphere (e.g., Melin et al. 2009). The strongest source of H_2 appears to be Titan's atmosphere ($\sim 1 \times 10^{28}\ \mathrm{s}^{-1}$,

Table 1 Average suprathermal ion abundances in Saturn's magnetosphere (dipole $L = 7$–16, near equatorial ($\pm 10°$), late 2004 to end of 2010). Abundances reflect relative partial number densities over the stated energy ranges. The energy range for W^+, H^+, H_2^+, O^{++}, He^+, He^{++} is 27–220 keV/e (DiFabio 2012). The energy ranges for some of the rarer or less well-resolved species are more restrictive: 83–167 keV/e for O_2^+, $^{28}M^+$ (N_2^+ and/or CO^+) (Christon et al. 2013); 127–220 keV/e for N^+, C^+ (DiFabio 2012); 36–167 keV/e for H_3^+ (Hamilton et al. 2013). The makeup of the water group ions W^+ was determined at 94–97 keV/e (DiFabio 2012)

Species	Relative to H^+	Relative to total
W^+ O^+ 53 % OH^+ 22 % H_2O^+ 22 % H_3O^+ 2.8 %	2.1	0.61
H^+	$\equiv 1$	0.30
H_2^+	0.17	0.050
O_2^+	0.037	0.011
O^{++}	0.032	0.0096
He^+	0.018	0.0054
N^+	0.016	0.0048
C^+	0.011	0.0034
$^{28}M^+$	0.0087	0.0026
H_3^+	0.0072	0.0021
He^{++}	0.0029	0.00086

Cui et al. 2008), although photo-dissociation of water can be important inside $\sim 6\ R_S$ (Tseng et al. 2011).

The relative abundances of the major and minor suprathermal species in the ring current ($L = 7$–16) are given in Table 1. The table lists ratios of each species to H^+ along with their fractional abundances in the total suprathermal population, listed from highest to lowest. The energy range is 27–220 keV/e for most of the species, including the most abundant, but differs for some of the rarer species. The makeup of the water group W^+ is given in the first column. We discuss several of the species in more detail in Sect. 6.4. The multi-MeV ions comprising Saturn's permanent radiation belts at $< 4\ R_S$ (Paranicas et al. 2008) are discussed in Sect. 6.5.

6.3 Spatial Variations

Figure 21 presents the radial profiles of the partial number densities (PNDs) of six suprathermal species (73–110 keV/e) (left panel) and their fractional abundances (FAs) (right panel) (DiFabio 2012). These ions comprise the more energetic portion of Saturn's ring current (Sergis et al. 2007) and have peak densities at $L \sim 10$ in this energy range. The similar profiles of the different species are an indication of common acceleration processes, probably involving a combination of outer magnetospheric and tail processes (flux tube interchange, reconnection, etc.) (e.g. Mauk et al. 2005; Mitchell et al. 2005; Rymer et al. 2009) and inward radial diffusion with rapid losses in the Enceladus neutral cloud causing the decreases inside $L \sim 9$ (Paranicas et al. 2008).

Differences among the species inside $L = 10$ are largely attributable to differences in charge exchange lifetimes and other loss processes (DiFabio 2012). The relative increase of

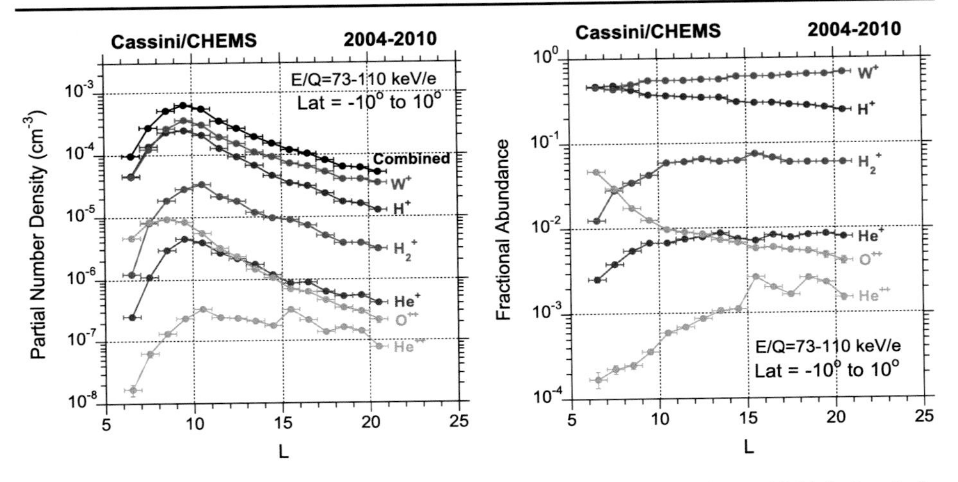

Fig. 21 Average (2004–2010) partial number densities (*left*) and fractional abundances (*right*) for ions in the 73–110 keV/e range are plotted versus dipole L (from DiFabio 2012)

the O^{++} FA at lower L-shells arises from its increased production from O^+ in the region of higher density neutrals and plasma electrons. The rapid decline in the He^{++} FA inside $L = 15$ is largely due to single electron capture from neutrals, producing more He^+.

6.4 Temporal Variations

The long-term temporal variations in suprathermal ion intensities from 2004 to the end of 2010 have been studied by DiFabio et al. (2011). Figure 22 presents PNDs (left panel) and FAs (right panel) of the various species. These measurements were made in the near-equatorial ring current (dipole $L = 7$–16). Overall variations are, in general, quite modest (∼factor 2) for the major species. In particular, the relative constancy of the W^+ PND led DiFabio et al. (2011) to conclude that the Enceladus plume source cannot have a variation much larger than that during this period when averaged over six months to a year. Shorter time variations are certainly possible (e.g., Smith et al. 2010; Jia et al. 2010) and would not show up in these long averages. The variations in He^{++} and He^+ are somewhat larger. They show decreases in 2009–2010 near solar minimum that are not seen in the other species.

Water Group Ions (W^+) The water group ions are the most abundant species throughout most of Saturn's magnetosphere at both thermal and suprathermal energies (Thomsen et al. 2010; DiFabio 2012). In the CHEMS data, the four species comprising the suprathermal water group have fairly broad distributions in measured M/Q. That fact combined with some instrumental issues has allowed an accurate determination of their relative abundances in only a limited energy range around 96 keV. Figure 23 taken from DiFabio (2012), indicates fits to the four W^+ components. As listed in Table 1 O^+ (53 %) dominates the suprathermal W^+ with OH^+ and H_2O^+ each present at somewhat less than half that amount (22 %). H_3O^+ is present in trace quantities (2.8 %), although its abundance is least well determined.

H^+, He^+, and He^{++} He^{++} has no known local source and comes from the solar wind. He^+ is thought to originate from interplanetary pickup ions along with a contribution from He^{++} charge exchanging with neutrals in the inner magnetosphere. Low magnetospheric

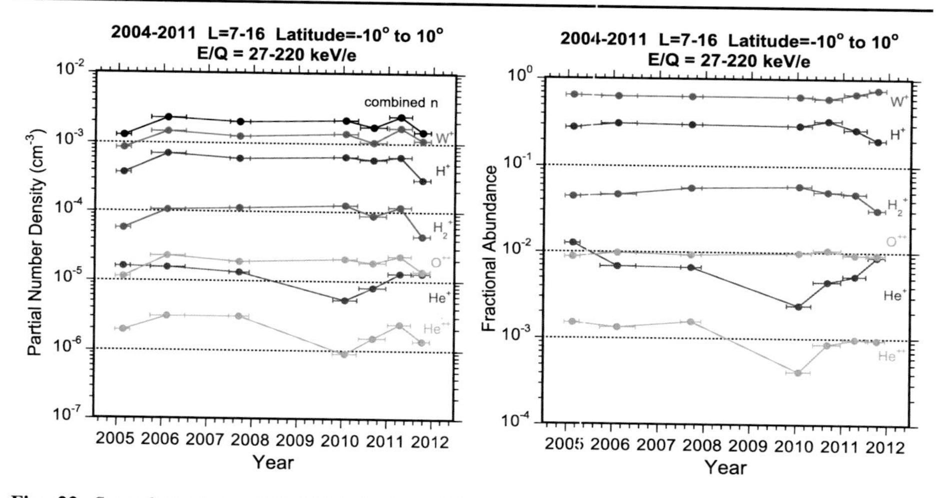

Fig. 22 Suprathermal ion (27–220 keV/e) partial number densities and fractional abundances from Cassini/CHEMS for five long averaging periods from late 2004 to the end of 2010. Statistical *error bars* are smaller than the *data point symbols* (from DiFabio et al. 2011)

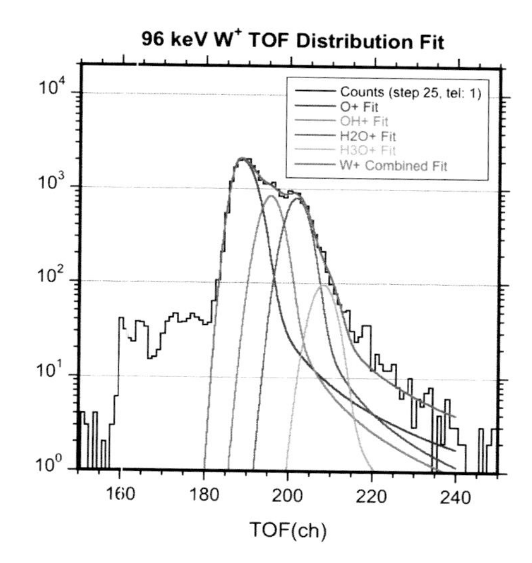

Fig. 23 Best fits to the water group ion distribution in the 94–97 keV energy range. The fits to each of the four individual species are shown along with the sum of the four fits *in red*. The data are from a long average from near equatorial (within 10°) passes through the ring current ($L = 7$–16) from late 2004 to the end of 2010. Cassini/CHEMS telescope 1 was used (from DiFabio 2012)

He^{++}/H^+ ratios, compared to 4–5 % in the solar wind, were noted at MeV energies during the Pioneer 11 and Voyager flybys (Simpson et al. 1980; Hamilton et al. 1983), with the conclusion that H^+ is mostly of local origin. Cassini's measurements at suprathermal energies lead to a similar conclusion. Table 1 lists that ratio as 0.0029 in the main ring current ($L = 7$–16), more than a factor of 10 less than the solar wind value. Figure 21 shows that this ratio has a higher value (0.0074) in the outer magnetosphere ($L = 15$–21), which is still a factor of 6 less than the solar wind value. That value is probably better for comparison since it avoids the inner region where He^{++} is preferentially lost. This would imply that about 84 % of the magnetospheric H^+ is of local origin and ~16 % originates in the solar wind.

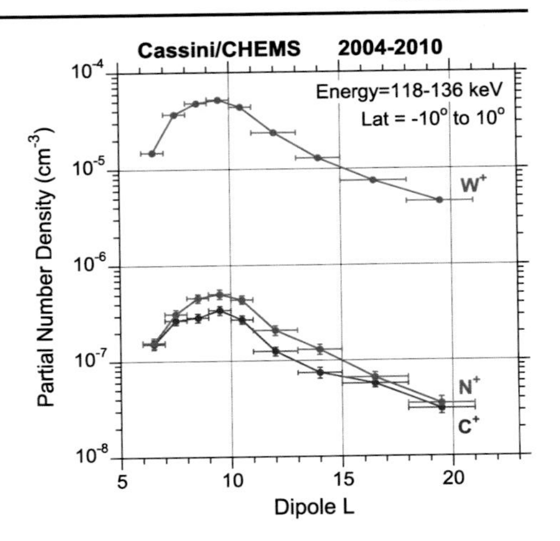

Fig. 24 Radial profiles of the 118–136 keV C^+, N^+, W^+ partial number densities. The data are averaged over the period late 2004 to the end of 2010 (from DiFabio 2012)

A crude estimate of the solar wind source rate can be made by assuming a small fraction of incident solar wind ions enter the magnetosphere (a range of 0.001 to 0.01 has been used at Earth). With a cross-sectional radius of 25 R_S, 400 km/s solar wind speed, and an average solar wind density of 0.05 $\mathrm{cm^{-3}}$, one obtains a solar wind H^+ source rate of $\sim 1.4 \times 10^{26}\ \mathrm{s^{-1}}$ to $1.4 \times 10^{27}\ \mathrm{s^{-1}}$. The calculated solar wind source rate for He^{++} is then $\sim 6 \times 10^{24}\ \mathrm{s^{-1}}$ to $6 \times 10^{25}\ \mathrm{s^{-1}}$. This implies a local source of H^+ in the range of $\sim 7 \times 10^{26}\ \mathrm{s^{-1}}$ to $7 \times 10^{27}\ \mathrm{s^{-1}}$.

To compare this estimate of the local H^+ source rate with Cassini measurements, we note that based on Cassini UVIS data, Shemansky et al. (2009) estimated the number of H atoms within Saturn's magnetosphere to be of $\sim 2 \times 10^{35}$. Photoionization, using a lifetime of $\sim 1.1 \times 10^9$ s (Melin et al. 2009), would produce only 2×10^{26} H^+ per second from that cloud. However, electron impact ionization is faster in parts of the magnetosphere. DiFabio (2012) used the fits of Schippers et al. (2008) to Cassini measurements of cold and hot electron populations to determine electron impact lifetimes from $L = 5$ to $L = 13$. He found a minimum of 1.5×10^7 s for H at $L = 9$, increasing to 10^8 s at $L = 6$ and $L = 12$. Using a 10^8 s lifetime, the H^+ source becomes $\sim 2 \times 10^{27}\ \mathrm{s^{-1}}$, which is within the range of our previous estimates. These very rough estimates await more detailed modeling.

C^+ and N^+ Singly charged carbon and nitrogen are interesting trace species (<1 % of W^+) in the suprathermal particle population. Before Cassini's arrival at the planet, it had been expected that N^+ might be a major species in Saturn's magnetosphere because N_2 constitutes 95 % of Titan's atmosphere. However, DiFabio (2012) found average ratios of C^+ and N^+ to W^+ of only 0.0055 and 0.0078, respectively, in the 127–220 keV range (see also preliminary CHEMS results in Table 11.3 of Mauk et al. 2009). Figure 24 shows the radial variations of the W^+, N^+, and C^+ PNDs from a slightly more restrictive energy range. The three species have similar radial profiles that offer little information about source locations. Although Titan would potentially be a source of both N^+ and C^+ (CH_4 is the second most abundant species in Titan's atmosphere), measurements of the thermal plasma have indicated that Enceladus is the more likely source of N^+ (Smith et al. 2007). C^+ has not yet been observed in the thermal plasma.

O_2^+ and $^{28}M^+$ Suprathermal molecular ions O_2^+ and $^{28}M^+$ (leading candidates for $^{28}M^+$ are N_2^+ and CO^+) have been investigated by Christon et al. (2013). Figure 25 (taken from that paper) shows the PNDs of 83–167 keV W^+, O_2^+, and $^{28}M^+$ in the top panel over the late

2004 to early 2012 time period. There are ~factor 2 variations in all three species. However, the ratios of O_2^+ and $^{28}M^+$ to W^+ (bottom panel) show smoother, better organized time variations. Christon et al. (2013) have interpreted the O_2^+/W^+ ratio variation as evidence for a varying ring source strength of O_2^+ that depends on the degree of solar illumination (insolation) of Saturn's rings. Tseng et al. (2010) have modeled such a seasonally varying O_2^+ source. The O^+/W^+ follows the dashed insolation curve from the beginning of mission until Saturn equinox, when sunlight strikes the rings edge on and insolation is minimum. Recovery of the ratio only begins after a year and a half at baseline values and falls below the 100 % (same as pre-equinox) insolation curve.

Although the $^{28}M^+/W^+$ ratio initially shows a seasonal decrease, by 2007 it hits a baseline minimum value that is maintained, with some variations, until 2012. The $^{28}M^+$, and even O_2^+ with its 1.5-year extended minimum, are not entirely seasonally varying and probably have multiple sources. Tseng et al. (2010) showed that O_2 in the ring atmosphere should scatter out into the magnetosphere to become a magnetospheric source of O_2^+ via photo- or electron impact-ionization. Their modeled magnetospheric source rate of O_2 decreases by about a factor of ~10 as the solar incident angle decreases from 24° to 4° with another factor 5 decrease as the angle decreases to 2°. In Fig. 25, the O_2^+ PND and $O_2^+/W+$ ratio both decline by a factor of ~ 5–6 from late 2004 to equinox, somewhat less than the model.

H_2^+ and H_3^+ Energetic (> 0.5 MeV/nuc) H_3^+, along with more abundant H_2^+, was discovered in the Jovian magnetosphere by Voyager 2 (Hamilton et al. 1980). Both Voyager 1 and 2 also detected H_2^+ in Saturn's magnetosphere and Voyager 2 detected a few counts of H_3^+ (Hamilton et al. 1983). At these high energies, at the very upper end of Saturn's ring current population, the H_2^+ abundance was similar to, or somewhat less than, that of helium nuclei, whose charge state was not measured by Pioneer 11 or the Voyagers, and was much less abundant than H^+ (<1 %). The situation is different in the suprathermal energy range in which the average H_2^+/H^+ ratio is ~ 0.17 (Table 1). The H_2^+/H^+ ratio is also high (tens of percent) at thermal energies in the outer magnetosphere (Thomsen et al. 2010).

The dominant source of H_2^+ is thought to be Titan (Cui et al. 2008) although other sources may play some role (Tseng et al. 2011). The only identified source for H_3^+ is Saturn's ionosphere, where it is produced by the reaction $H_2^+ + H_2 \rightarrow H_3^+ + H$. H^+ and H_3^+ dominate Saturn's ionosphere (see Nagy et al. 2009) and can be extracted from the auroral regions and accelerated to produce field-aligned beams and conics. Mitchell et al. (2009) reported the presence of H_3^+ in such a beam at ~10 % the level of H^+. Energy spectra of H^+ and H_3^+ from that event are shown in Fig. 26.

The abundance of H_3^+ in the more typical ring current population is much lower. Figure 27 shows PNDs of H^+ and H_3^+ (36–167 keV) from 37 Cassini passes through the ring current. The average H_3^+/H^+ ratio over this 2005–2012 period was 0.0072. Whether the sporadic auroral beams are sufficient to feed the ring current is not known. Another possible source is Saturn's polar wind. Glocer et al. (2007) have estimated its contribution to the magnetosphere to be in the range 2.1×10^{26} to 7.5×10^{27} ions/s. In their model the polar wind is comprised of H^+ and H_3^+ and, depending on the neutral temperature, either species can dominate. The fact that the H_3^+/H^+ ratio is about the same as the outer magnetospheric He^{++}/H^+ ratio discussed above would suggest an H_3^+ source rate of $\sim 7 \times 10^{24}\ s^{-1}$ to $7 \times 10^{25}\ s^{-1}$, assuming the loss rates of H_3^+ and He^{++} are not drastically different. This source rate range is a factor of 3 to 30 below the Glocer et al. (2007) lower limit.

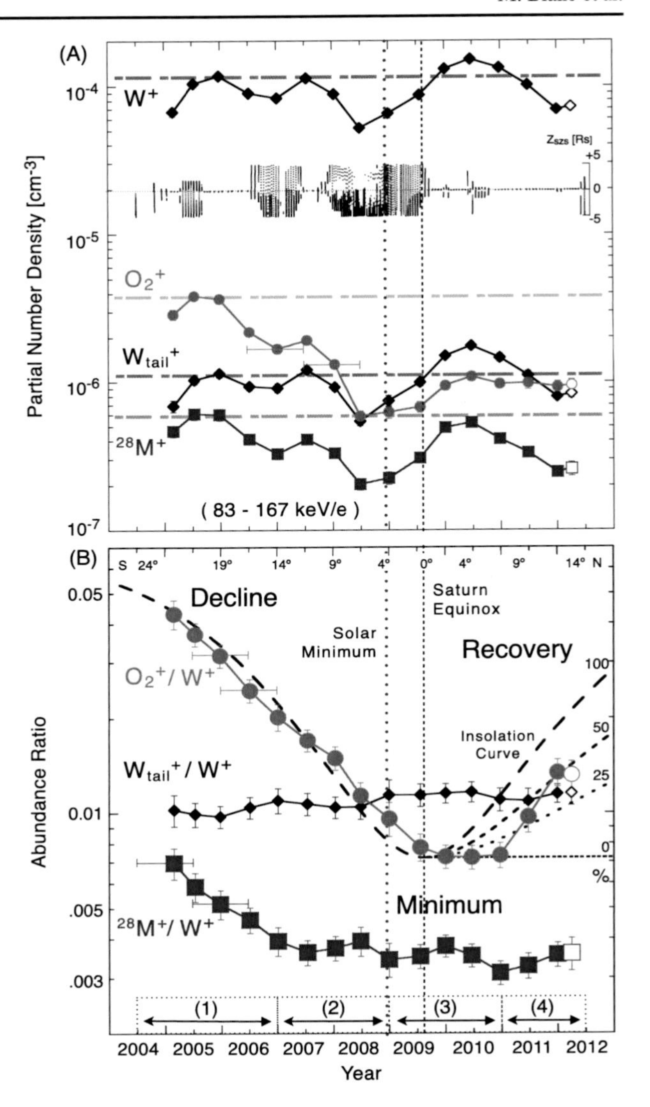

Fig. 25 (**A**) One-year moving averages of W^+, O_2^+, and $^{28}M^+$ partial number densities (along with W^+_{tail} to track possible time variations in the background contributions of the much more abundant W^+ ions to the rarer species) in the 83–167 keV/e energy range. (**B**) Abundance ratios relative to W^+ remove spatial/temporal variations common to all species. The *dashed lines* show the time variation of Saturn ring insolation. The curve is matched to the O_2^+/W^+ decline. During recovery curves representing 100 %, 50 %, and 25 % of pre-equinox insolation are shown (from Christon et al. 2013)

6.5 Contribution of the CRAND Source

Four missions visited the Kronian magnetosphere (Pioneer 11, the Voyagers and Cassini) and reported the existence of a significant population of energetic charged particles (>200 keV) in the radiation belts of Saturn (Cooper and Simpson 1980; Krimigis and Armstrong 1982; Krimigis et al. 2005). Analysis of the Cassini MIMI\LEMMS measurements of the differential energetic ion fluxes in the inner magnetosphere during the Saturn Orbit Insertion (SOI) and during the tens of posterior Cassini orbits around this planet shows the presence of stable radiation belts inside the Tethys orbit, which demonstrate sharp dropouts exactly at locations corresponding to the moons' L-shells (Janus, Epimetheus, Mimas, Enceladus and Tethys) and remain unchangeable during the large interplanetary events that influence significantly the middle and the outer magnetosphere.

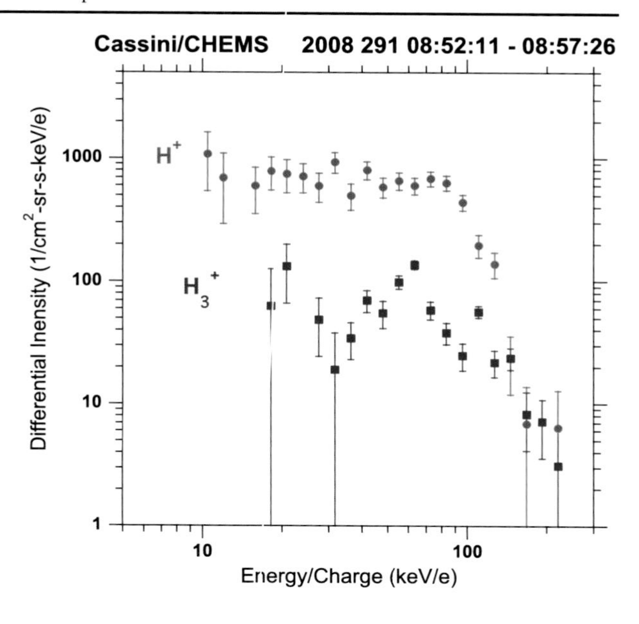

Fig. 26 Energy spectra of H^+ and H_3^+ observed during ~5 minutes of a Saturn auroral event during which CHEMS was favorably oriented to observe the nearly field-aligned beam (from Hamilton et al. 2013)

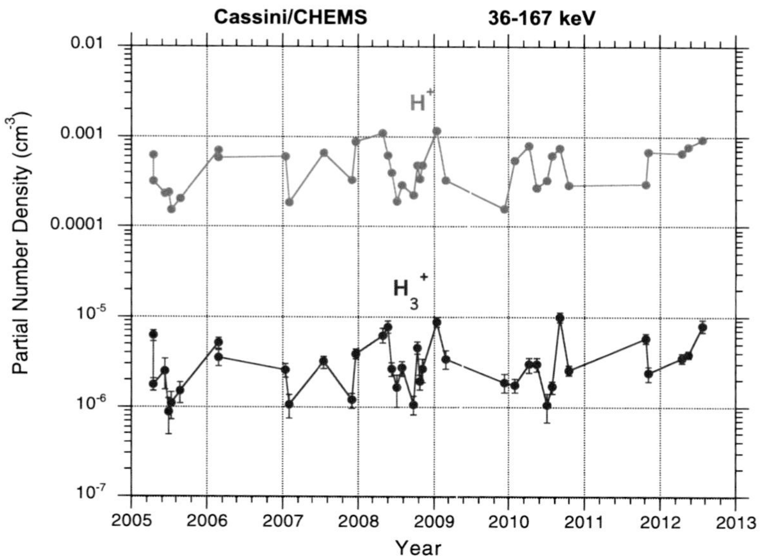

Fig. 27 Suprathermal (36–167 keV) partial number densities of H^+ and H_3^+ observed by Cassini/CHEMS during 37 passes through Saturn's ring current (from Hamilton et al. 2013)

In addition, the monitoring of Saturn's radiation belts during a half of solar cycle (2004–2010) shows weak intensification of the trapped proton component (>10 MeV) (Roussos et al. 2011) during solar minimum. All of this confirms that the moons Tethys and Dione in combination with the neutral gas cloud and dust prevent inward radial transport of energetic ions by their absorbing effect and isolate the inner ion radiation belts from the middle and outer magnetosphere (Roussos et al. 2008; Paranicas et al. 2008;

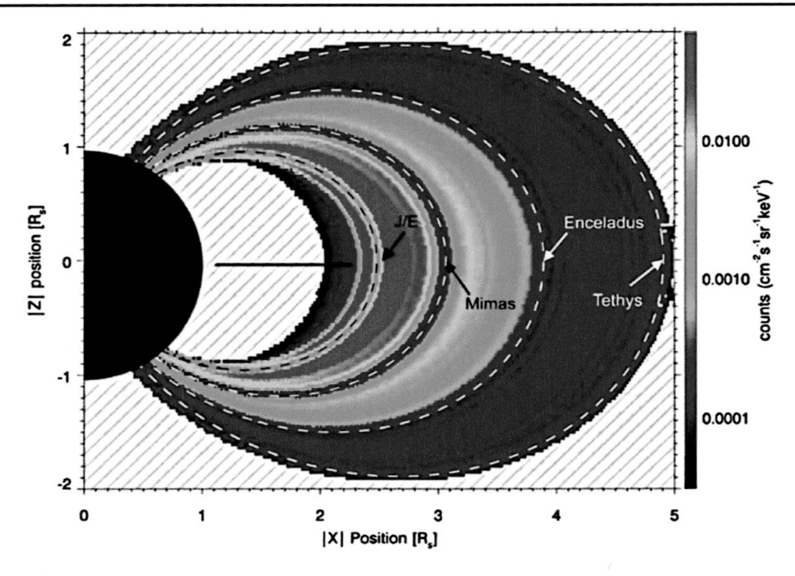

Fig. 28 Differential flux map of the stable belts inside Tethys' L-shell of the 25–60 MeV/nuc ions, based on LEMMS data from 36 orbits. The L-shells of the various moons are indicated. The partial flux dropout at the shell of the G-ring is also visible. Hatched regions above the main rings have particle flux lower than or equal to that of the *color bar* (from Roussos et al. 2008)

Roussos et al. 2011). Consequently the source for these energetic ions should be connected (directly or indirectly) to the access of galactic cosmic rays to the Saturnian system. Figure 28 represents the general structure of Saturn's radiation belts using the Cassini MIMI\LEMMS data.

Clear depletion in the energetic particle observations at the L-shells of the inner moons Janus, Epimetheus, Mimas and Enceladus (Fillius and Ip 1980; Krimigis and Armstrong 1982; Simpson et al. 1980; Vogt et al. 1982) due to the absorption effect is apparent in Fig. 29. As it was described by Roussos et al. (2008) and Paranicas et al. (2008), this indicates that the sources for the energetic component of the inner radiation belt cannot originate from the outer magnetosphere.

The likely mechanism responsible for populating the energetic ion belts is then the CRAND process. The possibility of the particular Galactic Cosmic Rays (GCR) to reach the atmosphere of the planet is usually described by the so-called "cut-off rigidity". Rigidity is a measure of the momentum of a charged particle in a magnetic field, used sometimes instead of energy because it is independent of the particle's charge. In this manner cut-off rigidity determines the minimum energy needed for a cosmic ray to reach the planet without being deflected by the planetary magnetic field. For protons, the equivalent energy for the cut-off rigidity values at Saturn depending on latitude varies from hundreds of MeV to ~65 GeV. GCRs with energies exceeding the cut-off rigidities required to reach the planetary atmosphere and/or the rings enter the magnetosphere, interact with the planetary atmosphere, rings, E ring and the extended neutral gas cloud, and create cascades of particles, partly at much lower energies, including neutrons, protons and also photons, electrons, pions, muons, and various antiparticles (Kollmann et al. 2013). Secondary charged particles will be almost immediately lost after bouncing back along the field lines to the location where they were produced. However, neutrons can travel away from their production region (since they are not bound by the magnetic field). The small fraction of those that will beta-decay within

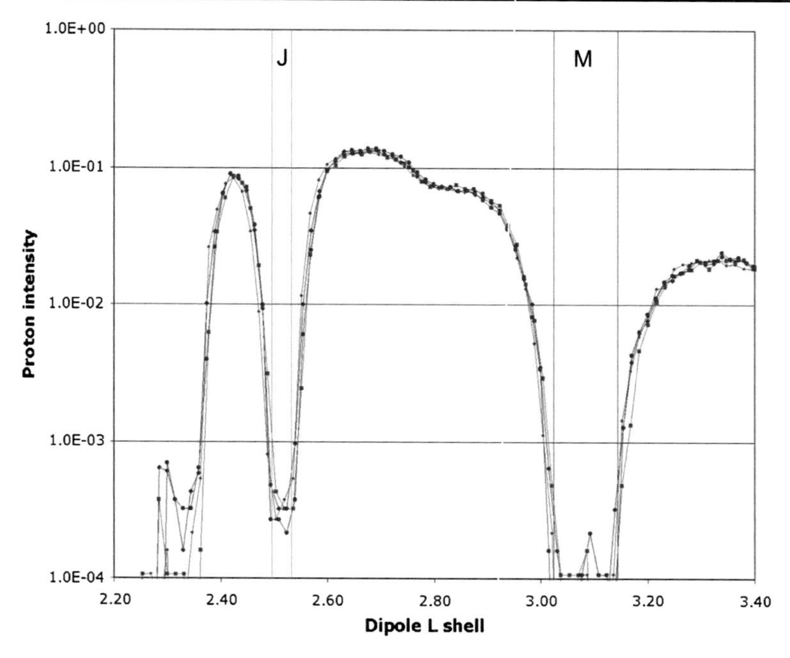

Fig. 29 Proton intensity (protons per cm^2 s sr keV) as a function of L shell during SOI. The *points* corresponding to equatorial pitch angles of 25–35° and 145–155° are shown as separate curves. *Blue points* were taken during the inbound and *red points*—during the outbound portion of the trajectory. Positions of the Satellites Janus ($L \sim 2.5$) and Mimas ($L \sim 3.1$) are shown as sweeping corridors (from Paranicas et al. 2008)

the strong dipole region will populate the radiation belts with energetic electrons (mostly below 1 MeV) and protons. This mechanism was discussed by several authors (Cooper and Simpson 1980; Fillius and McIlwain 1980; Van Allen 1983; Blake et al. 1983; Cooper 1983; Cooper et al. 1985; Randall 1994) and Fig. 30 illustrates it. Blake et al. (1983) proved theoretically that the high-energy component of the radiation belts originates from CRAND. The latest observations by Cassini confirmed this suggestion (Roussos et al. 2011). Kollmann et al. (2013) specify that atmospheric CRAND is the central process initially providing the protons (from 500 keV up to 40 MeV) and CRAND from the Main Rings contributes to some extent to the population, but only for >10 MeV, while other possibilities to supply the belts and exchange particles between them, such as diffusion and injections from outside the belts, or stripping of ENA's, can be excluded.

CRAND cannot be responsible for the presence of heavier species that have been detected in the belts (Armstrong et al. 2009). However, given that heavier species are distributed in the same way as MeV protons, their origin should also involve a stable, external cosmic ray source, without also excluding Anomalous Cosmic Rays.

6.6 Summary

Suprathermal ion abundances should broadly reflect the source rates of the various species, although some acceleration processes can change abundances compared to the low energy plasma as can differences in loss rates. The similar radial profiles of most suprathermal species in the 10–20 R_S range indicate they have probably undergone similar acceleration

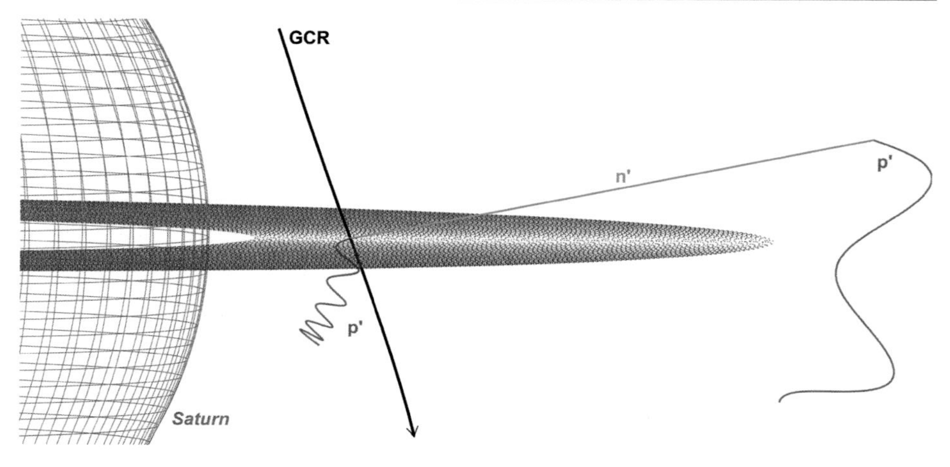

Fig. 30 Sketch of the CRAND process in the Saturn system. An incoming galactic cosmic ray (**GCR**) comes from above, passes through the planetary rings, and continues to infinity. A nuclear interaction creates cascades of secondary particles, including protons (**p**′), neutrons (**n**′), pions and muons. The created proton (**p**′) is trapped in the magnetic field of Saturn and is removed within a few bounces by repeated passages through the rings, consequently more neutrons are created. The first neutron (**n**′) successfully passes L-shells of the rings and decays in flight injecting an energetic proton (**p**′) into the radiation belt outside the rings (from Blake et al. 1983)

processes. Differences within 10 R_S are largely due to different charge exchange and other loss cross sections in the neutral cloud originating from Enceladus. Long term factor 2 constancy in the PNDs of the major suprathermal species from 2004 to 2012 indicates relatively stable plasma source rates. The strongest plasma sources are local as is evident from the low abundance of He^{++} and He^+, which originate from outside the magnetosphere. The very high H^+/He^{++} ratio indicates that only about 16 % of the magnetospheric protons come from the solar wind, with sizeable uncertainty. Trace species such as H_3^+ and O_2^+ indicate that Saturn's ionosphere and rings also contribute some plasma to the magnetosphere. The seasonal variation of the O_2^+/W^+ ratio confirms that the intensity of ring illumination largely controls the O_2^+ source rate. Trace species C^+ and N^+ probably arise from Enceladus, although Titan contributions have not been completely ruled out.

In the energetic ion range, at the top of the energy range, the CRAND mechanism is likely to produce a strong inner proton belt, as in other planetary magnetospheres. This process has been confirmed by Cassini studies. Recent studies and modeling results show that CRAND is a sufficient source process to generate the observed energetic protons flux in the inner magnetosphere of Saturn. In particular, for the energetic particles with energies of hundreds of keV to tens of MeV, atmospheric CRAND is most likely the central source process, while the CRAND from Main Rings plays an important role for producing protons with energies above 10 MeV and thereby amplifies the effect of atmospheric CRAND in this energy range.

7 Transport and Acceleration Processes and Related Losses

7.1 Interchange Signatures for Ions and Electrons

The inner regions of Saturn's magnetosphere are supplied with plasma produced by photoionization, charge exchange and electron impact ionization of a cloud of neutral water

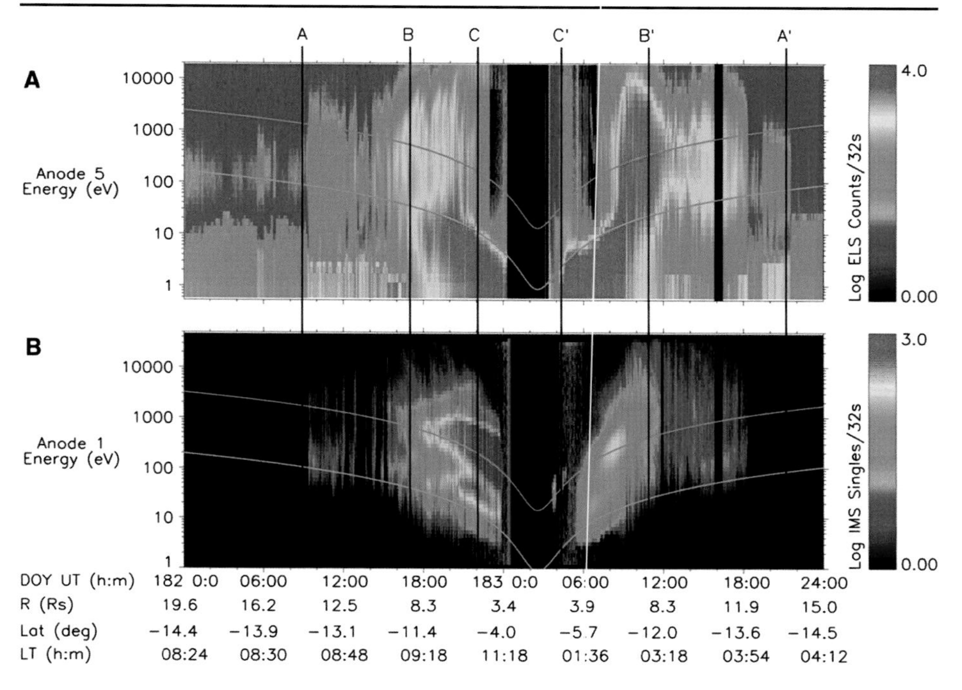

Fig. 31 Electron (*upper panel*) and ion (*lower panel*) spectrograms for two days of the Saturn insertion orbit, with proton (lower trace) and W+ (upper trace) energies overlaid (from Young et al. 2005)

molecules (Shemansky et al. 1993; Esposito et al. 2005; Young et al. 2005; Perry et al. 2010) which dominate the particle density in these regions between ~5 and 10 R_S. As developed in detail in Sect. 2, Cassini has shown that these neutrals emanate from Enceladus itself. This followed the discovery of a dynamic atmosphere (Dougherty et al. 2006) and vast plumes of neutrals (Waite et al. 2006), plasma (Tokar et al. 2006, 2009), water clusters (Coates et al. 2010) and neutral and charged dust particles (Spahn et al. 2006; Jones et al. 2009) emanating from tiger stripe features on the surface (Porco et al. 2006).

The first results from the CAPS instrument in Saturn's magnetosphere showed that the composition is dominated by the water group but also includes protons (Young et al. 2005). In addition, the density of plasma is much less than (~10 %) that of neutrals (Young et al. 2005; Jurac and Richardson 2007). The almost co-rotating inner magnetosphere, which includes hydrogen ions likely to be mainly from Saturn's ionosphere, is dominated by water-based neutrals (O, OH). Enceladus, supplemented by the rings and the associated neutrals, populates the outer magnetosphere as well (Smith et al. 2008; Thomsen et al. 2010; Arridge et al. 2011). Some of the remarkably complex chemistry at Titan appears to involve particles, oxygen in particular, originally from Enceladus (Coates et al. 2007b; Sittler et al. 2009).

The electron populations in the inner magnetosphere, as well as the water-rich composition, show remarkable structure and dynamics (e.g., Fig. 31, from Young et al. 2005). There is a cold component, the upper energy of which appears to be controlled by the proton corotation energy (Young et al. 2005) and a hot component, which appears separate and is transported from the outer to the inner magnetosphere by remarkable injections and interchange events. Further analysis of the hot and cold populations has been provided by Rymer

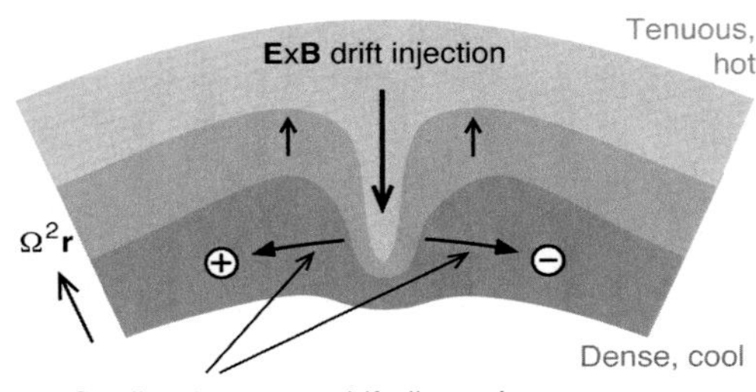

Fig. 32 Schematic of interchange event (from Hill et al. 2005)

et al. (2007) and Schippers et al. (2008). For the cold component, Rymer (2010) suggested that the cold electron tracking of the proton energy may be associated with Coulomb heating between pickup electrons and ions. This would require residence timescales of $\sim$100 s of hours, i.e. quiescent conditions for the cold electrons.

Interchange is a process known from Jupiter's rapidly spinning magnetosphere, where the mainly cold, dense, corotating plasma in the inner magnetosphere interchanges with hot, rare plasma in the outer magnetosphere, driven by radial transport. The resulting structure resembles 'fingers' interleaving the two populations according to models (e.g. Hill et al. 2005; Liu et al. 2010; see Figs. 32 and 33). Following the interchange event, the ions and electrons undergo gradient/curvature drift dispersion. Cassini observations revealed this type of structure in both ions and electrons (Hill et al. 2005) together with its magnetic counterpart (André et al. 2007; Leisner et al. 2005). Burch et al. (2005) associated small scale injections with interchange, while Hill et al. (2005) and Chen et al. (2010) showed that the distribution over longitude was uniform, indicating a rotationally driven process. In latitude, the interchange events appear close to the outer latitudinal boundary of the plasmadisc (André et al. 2005). However it is not yet clear if these structures are fingers or detached 'bubbles', i.e. flux tubes in 3 dimensions (cf. Pontius and Hill 1989; Pontius and Wolf 1990).

Electrons produced in Saturn's inner magnetosphere circulate with a combination of outward and inward motions driven by the centrifugal interchange instability, and azimuthal motion through gradient and curvature drifts (Rymer et al. 2008). Cool ($<$100 eV) electrons produced inside $L \sim 12$ move slowly outward. To balance the outflowing flux, inward transport occurs in small scale injection events. Electrons in these inwardly moving flux tubes are heated adiabatically to energies greater than 100 eV and their pitch angle distributions evolve from isotropic to "pancake" (peaked at 90°, see Fig. 34). The hot electron component in Saturn's magnetosphere is thus formed by the drift and dispersion of electrons from these small-scale inflow channels (Hill et al. 2005; Burch et al. 2005; Rymer et al. 2008).

7.2 Reconnection/Plasmoids

Magnetotail reconnection is a process which can allow large amounts of material to be broken off on the nightside and lost from the magnetosphere. Information about the products of reconnection can be gleaned either by direct encounters with plasmoids passing over the spacecraft, with travelling compression regions observed from the lobes, or with planetward-moving dipolarization structures. Much work has been done on these at the Earth and we refer the readers to reviews including those by Hesse and Kivelson (1998), Slavin (1998),

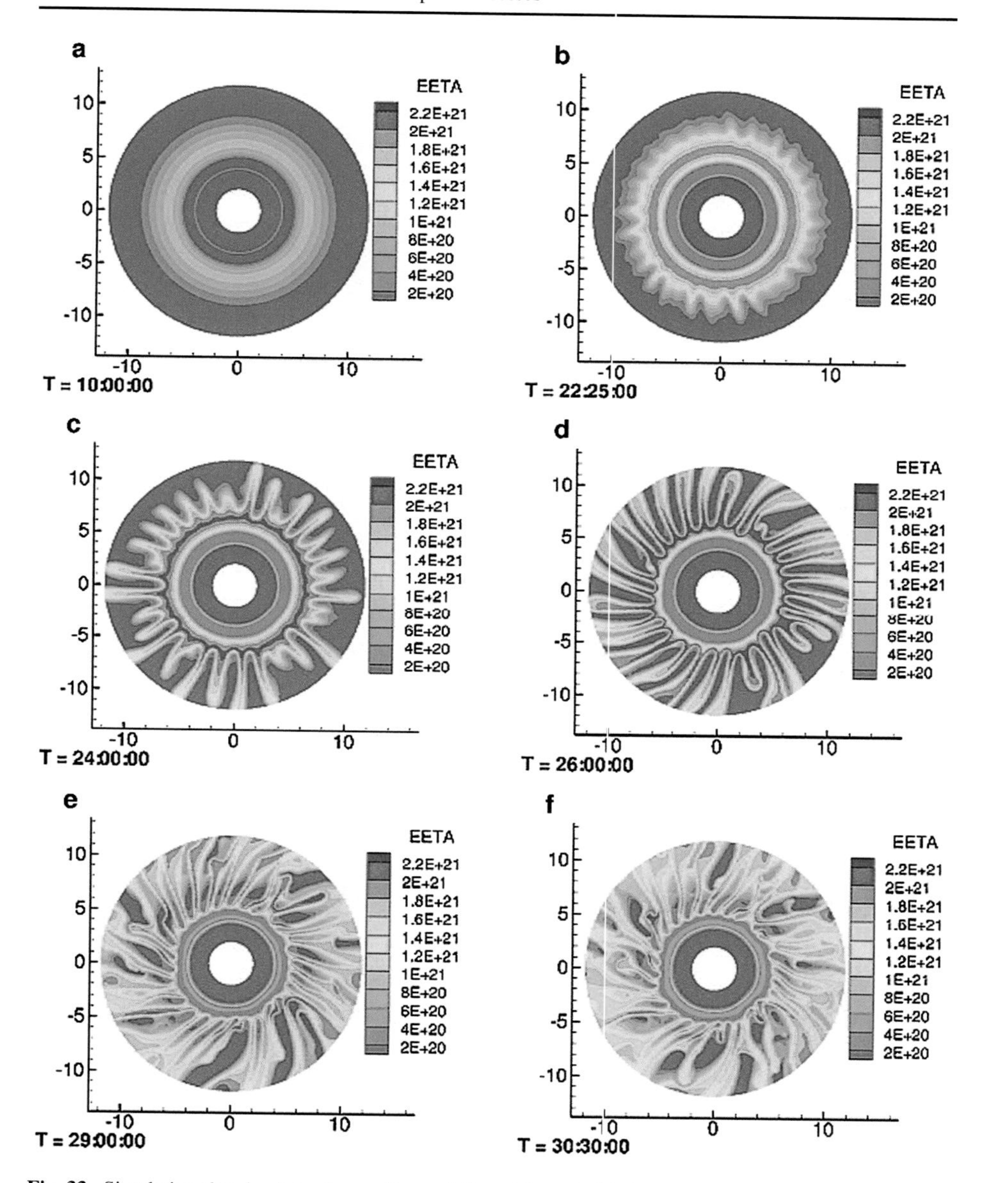

Fig. 33 Simulation showing interchange 'fingers' (from Liu et al. 2010)

Sharma et al. (2008) and Eastwood and Kiehas (2015) for more information. The first chance to search for evidence of the reconnection process at Saturn came with the arrival of Cassini in 2004, and particularly with the deepest tail orbits in 2006 when the spacecraft reached downtail distances of 68 R_S (1 $R_S = 60268$ km). Figure 35 shows the trajectory of Cassini during this time, with the timings of observed reconnection events marked. Events are identified initially by bipolar changes in the B_θ (north-south) component of the field, and, where plasma data are available, by concurrent changes in local plasma properties and plasma flow direction. The events displayed in Fig. 35 include a total of 69 plasmoids, 17 travelling

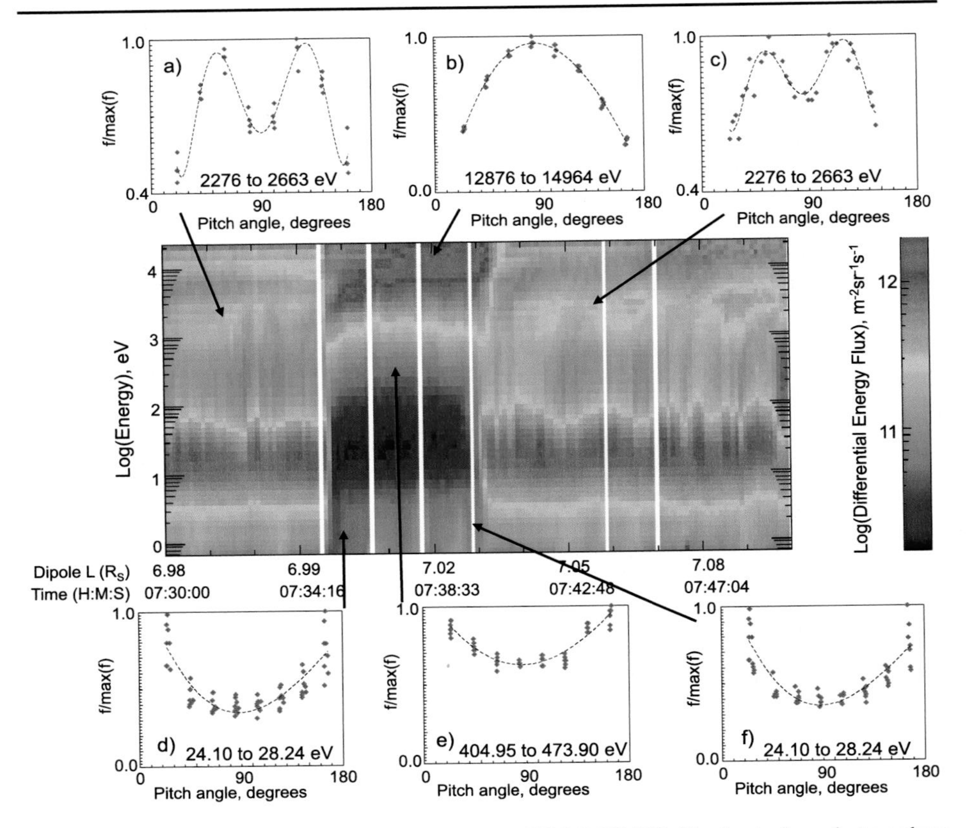

Fig. 34 Electron injection event observed on 30 October 2005 (DOY 303). The insets show electron phase space density versus pitch angle distributions derived from ELS data at the times and energies indicated by the arrows along with polynomial fits to the data (adapted from Rymer et al. 2008)

compression regions (TCRs) and 13 planetward-moving events (akin to terrestrial dipolarizations). The direction of motion is inferred in the first instance from the sign of the change in B_θ.

The properties of plasmoids and their effect on the local environment (in the form of TCRs) (e.g. Jackman et al. 2008b, 2009c, 2009d) and the global magnetosphere (in terms of changing plasma flows, flux closure, mass removal) (McAndrews et al. 2009; Jackman et al. 2011; Jia et al. 2012a; Thomsen et al. 2013) have been explored by a number of authors since the observation of the first three cases with magnetic field (Jackman et al. 2007) and plasma data (Hill et al. 2008). We refer the reader to these papers and to the reviews of Thomsen (2013) and Jackman (2015) for further comprehensive description of Saturn's magnetotail dynamics. For a comprehensive review of the dynamics of Saturn's magnetotail, compared and contrasted with Mercury, Earth and Jupiter, we refer to Jackman et al. (2014a). An example of two plasmoids and two TCRs is shown in Fig. 36. The interior structure of plasmoids is found to be primarily loop-like, as opposed to helical twisted flux ropes commonly seen in the Earth's magnetosphere. Some flux ropes are observed, but their relative scarcity may mean that the large-scale structure of Saturn's magnetotail field is less sheared than at other planets (Jackman et al. 2014b).

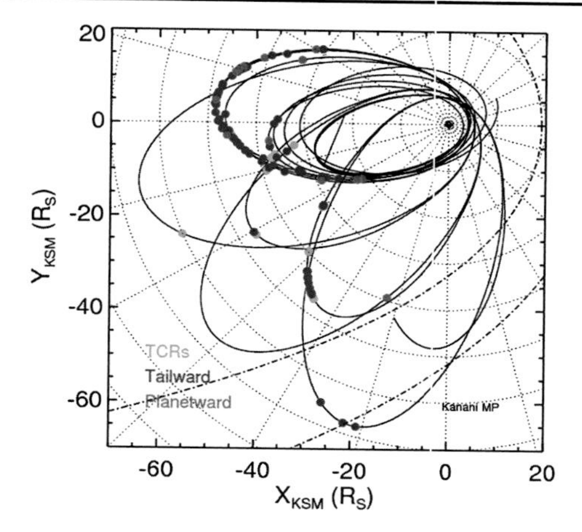

Fig. 35 X-Y projection of the Cassini trajectory for 2006 day 18–291 in the Kronocentric Solar Magnetospheric (KSM) co-ordinate system. KSM is the kronian analogue of GSM where the X axis coincides with the direction to the Sun, the XZ plane contains the planetary dipole axis, and the Y component is azimuthal, positive toward dusk. *Blue*, *red* and *green dots* show the locations of tailward and planetward-moving structures, and TCRs respectively. The Kanani et al. (2010) model magnetopause is overplotted for solar wind dynamic pressures of 0.1 and 0.01 nPa (from Jackman et al. 2014b)

7.3 Field-Aligned Acceleration and Current Generation

7.3.1 Field-Aligned Currents

Electric currents commonly arise in planetary magnetospheres as a result of plasma flow shears, pressure gradients, or inertial stresses (Baumjohann et al. 2010). These currents play a crucial role in magnetosphere-ionosphere coupling, and Saturn's magnetosphere is no exception. Like its giant planet sibling, Jupiter, Saturn is also a rapid rotator and contains significant internal plasma sources supplied by its satellites and rings, with Enceladus being the dominant contributor. Both the pick-up of newly produced plasma from the internal sources and the subsequent outward transport of the magnetospheric plasma tend to slow down the local flow, leading to the lag of plasma with respect to rigid corotation that consequently causes the magnetic field lines to bend backward. Corresponding to the bendback magnetic geometry is a radial current flowing through the equatorial plasma, which exerts a $J \times B$ force on the magnetospheric plasma that acts to accelerate it toward corotation. The radial, transverse current is then closed through field-aligned currents flowing between the ionosphere and the magnetosphere, forming an internally driven current system referred to as the corotation enforcement current (e.g., Hill 1979). The corotation enforcement current system is inherent in a rapidly rotating magnetosphere with a strong internal plasma source that is ultimately responsible for the momentum transfer between the planet and the mass-loaded magnetosphere. It is generally believed that the enforcement current system is responsible for the generation of the main auroral oval at Jupiter. While a similar situation might be expected for Saturn, it is unlikely to be the case because the region of corotation breakdown observed in Saturn's inner magnetosphere maps to too low latitudes compared to where the main oval is typically observed, and the upward field-aligned currents associated with the enforcement current system are too weak and do not appear to require the significant parallel acceleration which is needed for producing the observed auroral emissions (e.g., Cowley et al. 2004, 2008).

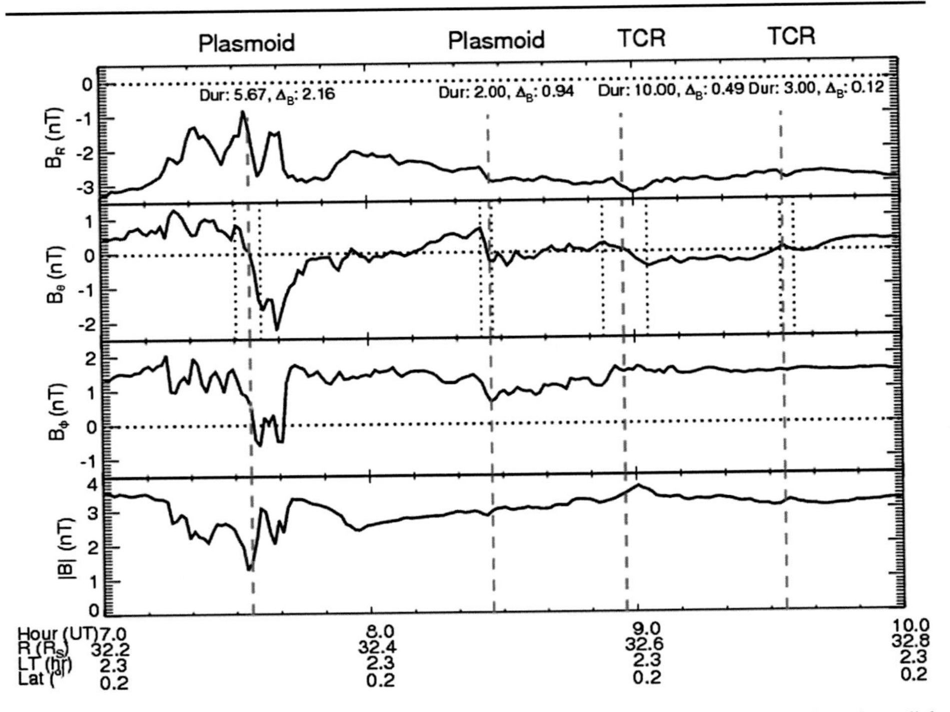

Fig. 36 Cassini magnetic field data in KRTP co-ordinates for 2006 day 60 07:00–10:00, where the radial component (B_r) is positive outward from Saturn, the theta component (B_θ) is positive southward, and the azimuthal component (B_ϕ) is positive in the direction of planetary corotation. The positions of two plasmoids and two TCRs are marked with *vertical lines*, and the amplitude and duration in minutes of the signatures are listed in the *top panel*. The plasmoids are identified by northward turnings of the field, while the TCRs are identified by localized compressions in the total magnetic field strength, and small northward turnings. It is inferred that the spacecraft penetrated most deeply into the first plasmoid, evidenced by the local decrease in $|B|$ (from Jackman et al. 2014b)

Instead, Saturn's main auroral oval has been suggested to be associated with processes occurring in the middle or outer magnetosphere. For instance, Sittler et al. (2006) proposed a model to explain Saturn's main auroral oval in which the source location of the oval maps to the outer edge of the plasma sheet located in the middle magnetosphere, where centrifugally driven interchange instability may produce significant particle acceleration and precipitation into the ionosphere leading to auroral emissions. Another model, proposed by Cowley and Bunce (2003) and Cowley et al. (2004), associated Saturn's main oval with the open-closed field line boundary in the polar ionosphere. Equatorward of the open-closed field line boundary are closed field lines mapped to the outer magnetosphere that contain plasma moving in the corotation direction at a fraction of the rigid corotation speed, whereas poleward of this boundary are open field lines whose motion is governed in combination by the solar wind from above and the ionosphere from below. The flow shear between the open and closed field lines implies the existence of a layer of upward field-aligned currents flowing near the boundary that are likely to require field-aligned electric fields to develop, in which case electrons are accelerated into the ionosphere producing aurora. In-situ observations of the magnetospheric conditions combined with remote observations of the aurora provide the best opportunity to test these scenarios. For example, Bunce et al. (2008) combined the HST observations of the aurora with simultaneous measurements of the particles and fields from one of Cassini's high-latitude passes through the dayside magnetosphere to show that the observed noon aurora lies close to the boundary separating the open and closed field lines. Talboys et al. (2011) later carried out a comprehensive survey of high-latitude field-aligned currents signatures using Cassini magnetic field data and compared their locations with the open-closed field line boundary inferred from particle data. The statistical results of Talboys et al. (2011) indicated that the upward field-aligned currents are typically seen not right at the open-closed field boundary, as predicted by the Cowley et al. (2004) theoretical model, but rather in a region equatorward of the boundary, which presumably maps to closed field lines in the outer magnetosphere. While the discrepancy between the observations and the theoretical model remains to be understood through future work, possible factors, such as the inhomogeneity of ionospheric conductivities and the effect of the magnetospheric periodicities, have been proposed that may account for the difference (Talboys et al. 2011).

7.3.2 Field-Aligned Acceleration

Regardless of their source, field-aligned currents require current carriers and electrons are usually the primary carrier due to their mobility. For a given amount of current demanded by any magnetospheric process, if there are not sufficient electrons available to carry the required current, then field-aligned electric fields normally develop to accelerate the current-carrying electrons (see the chapter by Seki et al. in this issue). In regions of downward flowing currents, a field-aligned potential drop may develop that would accelerate electrons out of the ionosphere forming field-aligned electron beams. Evidence for such electron beams has been found in Saturn's high-latitude region. Cassini frequently observed upward propagating whistler-mode hiss emissions in the auroral zone (Gurnett et al. 2009b), which are believed to be produced by upward moving electron beams associated with the downward flowing field-aligned currents (e.g., Kopf et al. 2010). In regions of upward flowing currents, parallel electric fields would accelerate electrons into the atmosphere/ionosphere where they can lead to significant magnetospheric consequences, such as the excitation of aurora, generation of radio emissions, and enhancement of ionospheric conductivity. For this reason, it is important to understand the acceleration process associated with field-aligned currents.

Parallel electric fields usually develop somewhere above the ionosphere; however, the exact location of the acceleration region depends on the electron distribution along the field lines. It has been suggested that the acceleration region at Saturn is likely to lie at ~ 0.5 Saturn radius above the ionosphere (Ray et al. 2013), which is below the lowest altitude that Cassini has thus far reached during its high latitude passes. In the absence of direct observations of the acceleration region, understanding of the acceleration process may rely on theoretical models developed for understanding similar processes occurring in other planetary magnetospheres, such as the current-voltage relation proposed by Knight (1973) for the terrestrial magnetosphere. However, the way in which the magnetospheric plasma is distributed in Saturn's magnetosphere is largely affected by the planetary rotation and the presence of strong internal plasma sources, a situation quite different from the terrestrial case. This aspect of the Saturnian system needs to be taken into account when considering the relationship between field-aligned current and field-aligned potential drop.

For a rapidly rotating magnetosphere like that of Saturn, centrifugal effects play an important role in determining the plasma distribution within the magnetosphere. The centrifugal acceleration tends to push plasma radially outward and to stretch magnetic field lines, which leads to equatorial confinement of the magnetospheric plasma. For a multi-species plasma, as is the case for Saturn's magnetosphere that consists primarily of heavy water-group ions, protons and electrons, the plasma distribution along magnetic field lines is determined by the centrifugal force, the gravitational force and the force associated with the ambipolar electric field (e.g., Sittler et al. 2008). It has been found that the heavier, water-group ions are more strongly confined to the equator while the lighter species (e.g., protons and electrons) are distributed more broadly along the magnetic field lines (e.g., Thomsen et al. 2010). However, it is possible that a low electron density region exists somewhere at mid latitude where the sum of the gravitational and centrifugal potentials exhibits a local minimum.

The latitudinal plasma distribution in Saturn's magnetosphere has important implications regarding the field-aligned current generation and associated particle acceleration. Ray et al. (2013) have used a one-dimensional Vlasov simulation to study Saturn's current-voltage relation taking into account the effect of centrifugal confinement of the magnetospheric plasma. They found that the relationship between the field-aligned potential drop and field-aligned current density derived from their simulations is essentially consistent with the prediction of the Knight (1973) kinetic theory. Their simulation results, however, emphasized the need of using plasma conditions at the top of the acceleration region, instead of those of the equatorial plasma sheet, in order to obtain an accurate estimate of the field-aligned potential drop.

7.4 Planetary Period Oscillations and Consequences

Oscillations with periods close to the estimated rotation period of Saturn ($\sim$ 10.6 h) have been detected in a multitude of magnetospheric parameters at Saturn, beginning with the initial detection in Voyager Planetary Radio Astronomy data of a strong modulation in the brightness of the Saturn Kilometric Radiation (SKR), a circularly polarized auroral radio emission with frequencies of tens to hundreds of kHz (Kaiser et al. 1980, 1981; Warwick et al. 1981; Desch and Kaiser 1981; Desch 1982). Subsequent further studies using data obtained during the Pioneer and Voyager flybys showed corresponding modulations to be present in the magnetospheric plasma populations and external magnetic field (e.g. Carbary and Krimigis 1982; Espinosa and Dougherty 2000). Detailed studies of modulations in the SKR and the magnetic field, from which quasi-continuous measurements of the oscillation parameters can be determined, have shown that the phase of the perturbations remains incredibly stable, with only slow drifts in period occurring on secular (seasonal) timescales (e.g. Galopeau and Lecacheux 2000; Kurth et al. 2007, 2008; Andrews et al. 2008; Provan et al. 2009; Andrews et al. 2012). Furthermore, apparent small differences between the periods of these phenomena have been reported, though it remains unclear whether such differences are physical, and to what extent they may be artifacts of different analysis techniques applied to the different data sets. For example, we note that there is evidence of weaker SKR emissions modulated by the period of the opposing hemisphere (e.g., "southern" period emission originating from the northern hemisphere, and vice-versa (Lamy 2011)). We refer the reader to the recent review of this topic by Carbary and Mitchell (2013) for a more complete introduction, and instead only discuss here those aspects of this phenomenon of direct relevance to this chapter.

Recent discovery of a weaker, apparently independent modulation in the SKR originating in the northern hemisphere (i.e., the opposite hemisphere to that preferentially illuminated during both the Pioneer-Voyager and early part of the Cassini mission epochs), corroborated by measurements made in the magnetic field and related plasma populations, has shown the system to be significantly more complex than first thought (Gurnett et al. 2009b, 2010; Andrews et al. 2010; Southwood 2011; Provan et al. 2011). Both atmospheric and magnetospheric sources have been proposed as possible origins of this system of large-scale, stable oscillations (Hill et al. 1981; Goldreich and Farmer 2007; Gurnett et al. 2007;

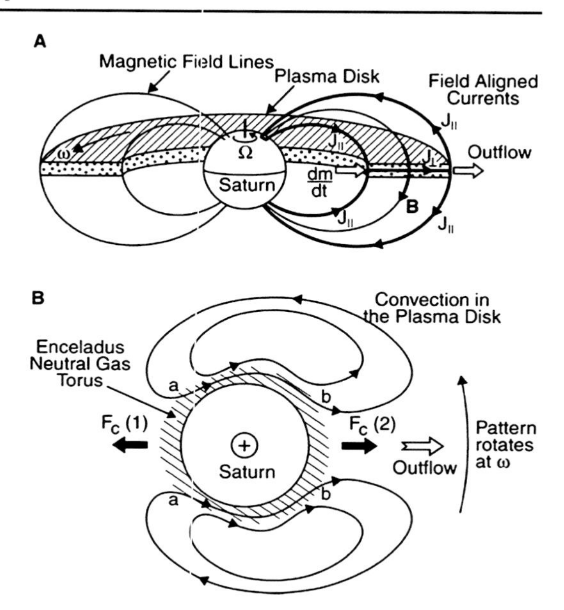

Fig. 37 Sketches of the rotating currents and associated plasma circulation streamlines within the rotating twin-cell convection pattern model proposed for Saturn. Ω is the angular rotation rate of Saturn, while ω is the corresponding rotation rate of the magnetospheric plasma (from Gurnett et al. 2007)

Southwood and Kivelson 2007; Smith 2011; Jia et al. 2012b). In Fig. 37, taken from Gurnett et al. (2007), a so-called rotating convection model is depicted, in which a stable outflow of plasma originating from Enceladus is established through a twin-cell convection pattern, having a single outflow and a single inflow sector. Gurnett et al. (2007) provided evidence for a rotating modulation in the equatorial plasma density within $\sim 5\ R_S$, while Burch et al. (2009) suggested that a corresponding systematic rotating modulation in the ion and electron count rates was present at larger radial distances, out to the magnetopause. However, the relative phasing between these apparent modulations in plasma density and the rotating magnetic field remains to be understood. Jia et al. (2012b) and Jia and Kivelson (2012) have developed a magnetohydrodynamic simulation of the coupled magnetosphere-ionosphere system that captures a host of observed magnetospheric periodicities with considerable fidelity. In their model, rotating vortical flows in the upper atmosphere, through coupling to the ionosphere and the magnetosphere, drive field-aligned currents that periodically modulate the entire magnetosphere.

The dual nature of these periodicities (comprised of independent modulations linked to the two hemispheres) presents some difficulty in envisaging a purely magnetospheric origin of the phenomena. Nevertheless, all theoretical models of these phenomena contain rotating systems of field-aligned currents with an $m = 1$ azimuthal wavenumber, closing to some extent through both the ionosphere and equatorial magnetosphere, so as to account for the observed modulations in the SKR and magnetic field. A schematic of one such system of currents, and the implied magnetic field perturbations, is illustrated in Fig. 38. The presence of equatorial closure currents is required as a consequence of the simultaneous presence of independent northern and southern modulations in both the SKR and magnetic field, and the high degree of apparent 'purity' in these modulations.

The extent to which this phenomenon drives, or is driven by, dynamical processes in the magnetospheric plasma populations remains to be demonstrated. Chen and Hill (2008) studied so-called 'injection events', in which plasma depleted flux tubes are interchanged with denser ones, a process by which transient radial plasma transport is achieved. Such

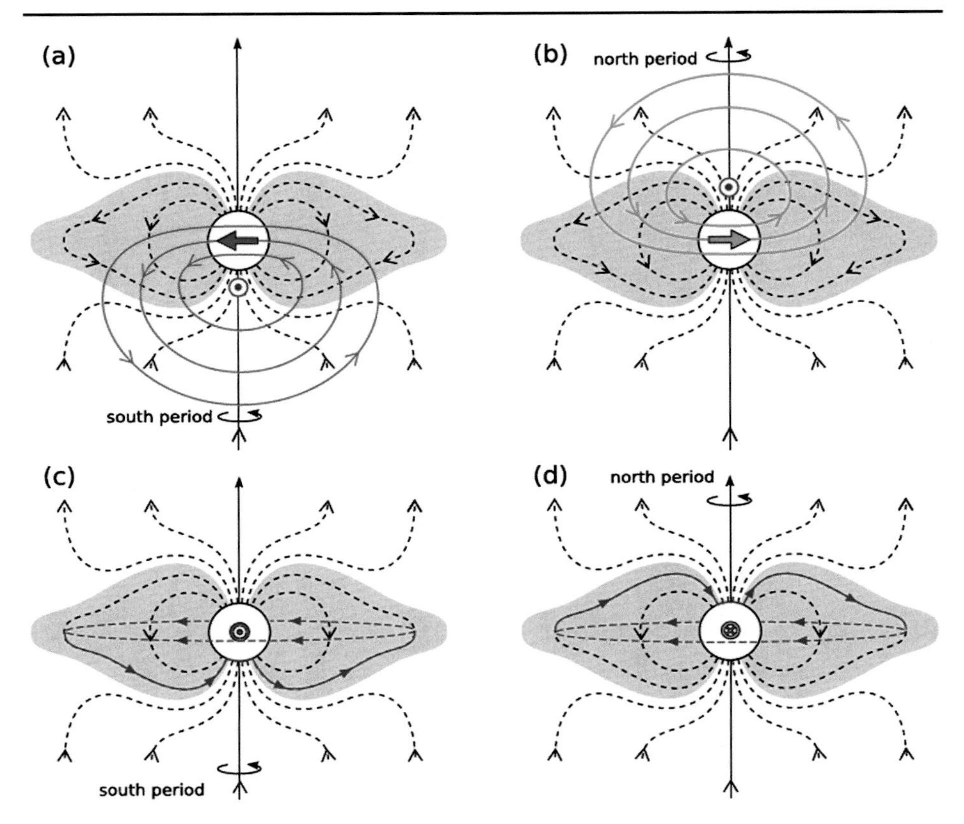

Fig. 38 Sketches of the form of the oscillating magnetic fields deduced from Cassini observations, and the implied rotating current systems. Panels (**a**) and (**b**) show the rotating magnetic field signatures of the southern and northern systems as *red* and *green solid lines*, respectively, superposed on an illustrative sketch of the static (symmetric) planetary background field. These patterns then rotate in the same sense but with subtly differing periodicities associated with the SKR modulation in the corresponding hemisphere. The rotating current systems implied by these field perturbations are sketched in panels (**c**) and (**d**) by the *blue solid lines*, as viewed in a plane orthogonal to panels (**a**) and (**b**) (from Andrews et al. 2010)

injection events are regularly detected at radial distances of 5–10 R_S. In their statistical survey of Cassini CAPS and MIMI data, Chen and Hill (2008) found no strong evidence for a periodicity to these injection events, or indeed any persistent organization by the rotating phase of the SKR modulation. Subsequent analysis of injection events detected in Cassini RPWS data has presented strong evidence that, within a restricted range of local times near midnight, the occurrence rate of injection events is indeed well organized by the phase of the SKR emission, specifically that originating from the hemisphere that is in polar night (Kennelly et al. 2013).

Meanwhile, the possible relationship between observations of tailward-moving plasmoids at larger radial distances and the magnetospheric period oscillations was studied by Jackman et al. (2009b, 2009c). In particular, Jackman et al. (2009d) found that while the repetition time between the losses of plasmoids into the magnetotail was likely much longer than the rotation period, the release of these plasmoids was nevertheless reasonably well ordered by the phase of the SKR, with plasmoids observed more frequently during a 'preferential' phase sector.

Many open questions remain regarding the relationships between the observed modulations in the SKR, magnetic field, and plasma populations. It is likely that the magnitude of the quasi-steady state convection velocity throughout the equatorial magnetosphere is sufficiently small (with respect to the bulk sub-corotational velocity) that direct detection of this systematic perturbation is essentially impossible with the available data.

8 Summary, Open Questions and Prospects for Future Studies

Saturn's magnetosphere appears, in the light of Cassini and previous missions, to display a variety of plasma sources, and these sources interplay with a host of dynamical phenomena to produce a large set of spatial structures and temporal behaviors. This chapter has illustrated this diversity of phenomena.

The first dominant feature of Saturn's magnetosphere examined from the viewpoint of its plasma sources is, just as for Jupiter, the dominance of one satellite source: to everybody's surprise, Cassini has revealed that the tiny icy satellite Enceladus and its southern hemisphere "tiger stripes" are the source of intense jets of water, called the Enceladus plumes, which provide the dominant source of water molecules for all the magnetosphere. This neutral water cloud spreads throughout the magnetosphere, in turn providing a source of plasma via a variety of ionization phenomena (e.g. UV photodissociation, UV and electron impact ionization, charge exchange). This source of plasma produces an ion torus which culminates somewhere outside the orbit of Enceladus and extends on either side of it. It is dragged into corotation via its coupling to Saturn's magnetic field and ionosphere/thermosphere.

One of the unique characteristics of this Enceladus ionized cloud system is that, near the location of Enceladus, the interaction of the charged particles with the water ice dust creates what one calls a "dusty plasma". So, Saturn is a unique place to study the behavior of this particular state of matter. Near the location of Enceladus, the dusty plasma indeed modifies the flow speeds of the plasma and the geometry of its interaction with the satellite.

Titan, which was suspected in the pre-Cassini years to be a major source of plasma for the magnetosphere, is in fact a minor source compared to Enceladus. It displays a cometary-type interaction with Saturn's corotating plasma, and is a limited source of both neutral and ionized particles. Neutral particle escape comes from the expanding exosphere of Titan, mainly H_2, and to a lesser degree N_2 and CH_4. Ion escape results from the plasma interaction of Titan's ionosphere with its induced magnetosphere and the kronian plasma. Ions of ionospheric origin, such as CH_5^+, $C_2H_5^+$ and $HCNH^+$ leave the Titan environment and feed the magnetosphere in the vicinity of Titan's orbit, however this process is strongly disturbed by the interaction with the solar wind on the dayside of the Titan torus. Overall, the Titan interaction and chemical complexity are unique and interesting for themselves, but provide only a minor source for the kronian magnetospheric plasma.

As for all other planetary magnetospheres, the solar wind is a likely source of plasma for Saturn, but the relative importance of this source is not quantified with great accuracy yet. The reason for this is that the efficiency of the dynamic coupling of the solar wind to the magnetospheric cavity, which depends on IMF orientation, plasma beta and Mach number of the interaction, is not known with certainty. Rather, opposing views on this subject are expressed, and more work is needed, if possible with direct measurements of the dynamic parameters in the vicinity of the magnetopause. In any case, and once more, the solar wind should remain a minor source compared to Enceladus, less than 5 % of the total supply according to most estimates.

In addition to the direct examination of the primary plasma reservoirs which feed the magnetosphere, another way of looking at plasma sources is to monitor the higher energy particles, which populate the magnetosphere after having been accelerated from the source regions. This chapter presents a comprehensive study of the distribution of these suprathermal and energetic particles, corroborating what we have learned from the examination of the main source regions.

The study of plasma sources cannot be separated from the one of the many dynamical phenomena acting on these sources, which tend to provide mechanisms for sources, transport and loss of plasma in each region, and which couple the different plasma reservoirs to the different dynamical modes of the magnetosphere as a whole. In this review chapter we gave an overview of some of these dynamical phenomena. In the tail, Cassini has unambiguously identified active magnetic reconnection producing plasmoids which, flowing downtail, evacuate plasma elements away from the magnetosphere and constitute an important plasma loss process. In the middle magnetosphere, flux tube interchange motions have been studied in considerable detail. These interchange motions contribute a lot to the radial redistribution of plasma. They are a key transport process for the magnetospheric plasma. On a more global scale, the kronian plasma is dynamically coupled throughout the magnetosphere to the magnetic field and high-latitude ionosphere. This coupling is the cause of the drag of the magnetospheric plasma into corotation. It operates via a current loop—the corotation-enforcement current—which connects the equatorial magnetosphere to the ionosphere. This process depends largely on the latitudinal distribution of the different plasma species, which it modifies in turn.

Finally, one of the strangest dynamical modes of Saturn's magnetosphere is the planetary period oscillation observed on most kronian magnetospheric parameters. The source of this rotational modulation, in a magnetosphere which should a priori be rotation-invariant, remains poorly known and a subject for future research. To solve this open question, there is no doubt that we need to elaborate a global comprehensive model of the dynamical behavior of the kronian magnetosphere, including its coupling to its plasma reservoirs. While Cassini is still flying around Saturn, this should be a major effort to accomplish in the coming years.

Acknowledgements M. Blanc wishes to thank Nicolas André for enlightening discussions, his careful reading of the manuscript and his suggestions for improvement; D.J. Andrews acknowledges support from the Swedish National Space Board (Rymdstylresan); A.J. Coates acknowledges support via the UCL-MSSL consolidated grant from STFC, UK; D.C. Hamilton is supported by the NASA Cassini mission through subcontract with JHU/APL; C.M. Jackman was funded by a Science and Technology Facilities Council Ernest Rutherford Fellowship; X. Jia is supported by the NASA Cassini Data Analysis Program through grant NNX12AK34G, and by the NASA Cassini mission under contract 1409449 with JPL; H.T. Smith is supported by the NASA Contract NAS5-97271 Task Order 003, the NASA Cassini Data Analysis Program and the NASA Outer Planets Research Program.

References

N. Achilleos et al., J. Geophys. Res. **113**, A11209 (2008). doi:10.1029/2008JA013265
N. André, C.T. Russell, J.S. Leisner, K.K. Khurana, Geophys. Res. Lett. **32**, L14S06 (2005). doi:10.1029/2005GL022643
N. André et al., Geophys. Res. Lett. **34**, L14108 (2007). doi:10.1029/2007GL030374
N. André et al., Rev. Geophys. **46**(4), RG4008 (2008)
D.J. Andrews et al., J. Geophys. Res. **113**, A09205 (2008). doi:10.1029/2007JA012937
D.J. Andrews et al., J. Geophys. Res. **115**, A12252 (2010). doi:10.1029/2010JA015666
D.J. Andrews et al., J. Geophys. Res. **117**, A04224 (2012). doi:10.1029/2011JA017444
T.P. Armstrong et al., Planet. Space Sci. **57**, 1723 (2009). doi:10.1016/j.pss.2009.03.008
C.S. Arridge et al., Space Sci. Rev. **164**, 1 (2011)

S.V. Badman, S.W.H. Cowley, Ann. Geophys. **25**, 941 (2007)
S.V. Badman et al., J. Geophys. Res. **110**, A11216 (2005). doi:10.1029/2005JA011240
S.V. Badman et al., Icarus **231**, 137 (2014). doi:10.1016/j.icarus.2013.12.004
F. Bagenal, P.A. Delamere, J. Geophys. Res. **116**, A05209 (2011). doi:10.1029/2010JA016294
W. Baumjohann et al., Space Sci. Rev. **152**(1–4), 99 (2010)
J.M. Bell et al., J. Geophys. Res. **115**, E12002 (2010). doi:10.1029/2010JE003636
J.M. Bell et al., J. Geophys. Res. **116**, E11002 (2011). doi:10.1029/2010JE003639
J.M. Bell et al., J. Geophys. Res. Space Phys. **119**, 4957–4972 (2014). doi:10.1002/2014JA019781
C. Bertucci, N. Achilleos, M.K. Dougherty, R. Modolo, A.J. Coates, K. Szego, A. Masters et al., Science **321**(5895), 1475–1478 (2008)
C. Bertucci et al., Geophys. Res. Lett. (2014, submitted)
J.B. Blake, H.H. Hilton, S.H. Margolis, J. Geophys. Res. **88**, 803 (1983). doi:10.1029/JA088iA02p00803
M. Blanc et al., Space Sci. Rev. **104**, 253 (2002)
P.C. Brandt et al., Planet. Space Sci. **60**, 107 (2012)
E.J. Bunce et al., Geophys. Res. Lett. **32**, L20S04 (2005). doi:10.1029/2005GL022888
E.J. Bunce et al., J. Geophys. Res. **113**, A09209 (2008). doi:10.1029/2008JA013257
J.L. Burch et al., Geophys. Res. Lett. **32**, L14S02 (2005). doi:10.1029/2005GL022611
J.L. Burch, A.D. DeJong, J. Goldstein, D.T. Young, Geophys. Res. Lett. **36**, L14203 (2009). doi:10.1029/2009GL039043
M. Burger et al., Geophys. Res. Lett. **112**, A06219 (2007). doi:10.1029/2006JA012086
L.F. Burlaga, N.F. Ness, J. Geophys. Res. **98**, 17451 (1993)
J.F. Carbary, S.M. Krimigis, Geophys. Res. Lett. **9**, 1073 (1982)
J.F. Carbary, D.G. Mitchell, Rev. Geophys. **51**, 1 (2013). doi:10.1002/rog.20006
T.A. Cassidy, R.E. Johnson, Icarus **209**, 696 (2010). doi:10.1016/j.icarus.2010.04.010
Y. Chen, T.W. Hill, J. Geophys. Res. **113**, A07215 (2008). doi:10.1029/2008JA013166
Y. Chen et al., Rate of radial transport of plasma in Saturn's inner magnetosphere. J. Geophys. Res. **115**, A10211 (2010). doi:10.1029/2010JA015412
S.P. Christon et al., J. Geophys. Res. **118**, 3446–3463 (2013). doi:10.1002/jgra.50383
A.J. Coates et al., Geophys. Res. Lett. **34**, L24S05 (2007a). doi:10.1029/2007GL030919
A.J. Coates et al., Geophys. Res. Lett. **34**, L22103 (2007b). doi:10.1029/2007GL030978
A.J. Coates et al., Icarus **206**, 618 (2010)
A.J. Coates et al., J. Geophys. Res. **117**, A05324 (2012). doi:10.1029/2012JA017595
A.J. Coates et al., J. Geophys. Res. Space Phys. **118**, 5099–5108 (2013). doi:10.1002/jgra.50495
A.J. Coates et al., Geophys. Res. Lett. **42**, 4676–4684 (2015). doi:10.1002/2015GL064474
J.F. Cooper, J. Geophys. Res. **88**, 3945 (1983)
J.F. Cooper, J.A. Simpson, J. Geophys. Res. **85**, 5793 (1980)
J.F. Cooper, J.H. Eraker, J.A. Simpson, J. Geophys. Res. **90**, 3415 (1985)
S.W.H. Cowley, E.J. Bunce, Ann. Geophys. **21**(8), 1691 (2003)
S.W.H. Cowley, E.J. Bunce, R. Prangé, Ann. Geophys. **22**(4), 1379–1394 (2004)
S.W.H. Cowley et al., Ann. Geophys. **26**, 2613–2630 (2008). doi:10.5194/angeo-26-2613
T.E. Cravens, I.P. Robertson, J.H. Waite, R.V. Yelle, W.T. Kasprzak, C.N. Keller, S.A. Ledvina et al., Geophys. Res. Lett. **33**(7) (2006)
J. Cui, R.V. Yelle, K. Volk, J. Geophys. Res. **113**, E10004 (2008). doi:10.1029/2007JE003032
V. De La Haye et al., J. Geophys. Res. **112**(A7) (2007)
M.D. Desch, J. Geophys. Res. **87**, 4549 (1982)
M.D. Desch, M.L. Kaiser, Geophys. Res. Lett. **8**, 253 (1981)
M. Desroche et al., J. Geophys. Res. **118**, 3087 (2013). doi:10.1002/jgra.50294
R.B. DiFabio Ph.D. thesis, Univ. of Maryland at College Park, College Park (2012), 214 pp.
R.B. DiFabio et al., Geophys. Res. Lett. **38**, L18103 (2011). doi:10.1029/2011GL048841
Y. Dong et al., J. Geophys. Res. **116**, A10204 (2011). doi:10.1029/2011JA016693
M.K. Dougherty et al., Science **311**, 1406 (2006)
J.P. Eastwood, S.A. Kiehas, in *Magnetotails in the Solar System*, ed. by e.A. Keiling, C.M. Jackman, P.A. Delamere (Wiley, Hoboken, 2015). doi:10.1002/9781118842324
N.J.T. Edberg et al., Planet. Space Sci. **59**, 788 (2011)
S.A. Espinosa, M.K. Dougherty, Geophys. Res. Lett. **27**, 2785 (2000)
L.W. Esposito et al., Science **307**, 1251 (2005)
A. Eviatar, J. Geophys. Res. **89**, 3821 (1984). doi:10.1029/JA089iA06p03821
A.J. Farmer, Icarus **202**, 280 (2009). doi:10.1016/j.icarus.2009.02.031
W.M. Farrell et al., Geophys. Res. Lett. **36**, L10203 (2009). doi:10.1029/2008GL037108
W.M. Farrell et al., Geophys. Res. Lett. **37** (2010). doi:10.1029/2010GL044768
W.M. Farrell et al., Icarus **219**, 498 (2012). doi:10.1016/j.icarus.2012.02.033

W. Fillius, W.H. Ip, Science **207**, 425 (1980). doi:10.1126/science.207.4429.425
W. Fillius, C.E. McIlwain, J. Geophys. Res. **85**, 5803 (1980)
B.L. Fleshman, P.A. Delamere, F. Bagenal, J. Geophys. Res. **115**, E04007 (2010). doi:10.1029/2009JE003372
B.L. Fleshman, P.A. Delamere, F. Bagenal, T. Cassidy, J. Geophys. Res. **117**, E05007 (2012). doi:10.1029/2011JE003996
L.A. Frank et al., J. Geophys. Res. **85**, 5695 (1980)
P.H.M. Galopeau, A. Lecacheux, J. Geophys. Res. **105**(A6), 13089 (2000)
A. Glocer et al., J. Geophys. Res. **112**, A01304 (2007). doi:10.1029/2006JA011755
G. Gloeckler, Rev. Sci. Instrum. **61**, 3613 (1990)
C.K. Goertz, W.H. Ip, Geophys. Res. Lett. **11**, 349 (1984). doi:10.1029/GL011i004p00349
P. Goldreich, A.J. Farmer, J. Geophys. Res. **112**, A05225 (2007). doi:10.1029/2006JA012163
A. Grocott et al., J. Geophys. Res. **114**, A07219 (2009). doi:10.1029/2009JA014330
D.A. Gurnett et al., Science **316**, 442 (2007). doi:10.1126/science.1138562
D.A. Gurnett et al., Geophys. Res. Lett. **36**, L16102 (2009a). doi:10.1029/2009GL039621
D.A. Gurnett et al., Geophys. Res. Lett. **36**, L21108 (2009b). doi:10.1029/2009GL040774
D.A. Gurnett et al., Geophys. Res. Lett. **37**, L24101 (2010). doi:10.1029/2010GL045796
D.C. Hamilton et al., Geophys. Res. Lett. **7**, 813 (1980)
D.C. Hamilton et al., J. Geophys. Res. **88**, 8905 (1983)
D.C. Hamilton et al. Paper SM11A-06, American Geophysical Union Meeting, New Orleans (May 2005)
D.C. Hamilton, S.L. Randall, R.D. DiFabio, in *7th Annual Cassini MAPS Workshop*, San Antonio, Texas (March 2013)
C.J. Hansen et al., Geophys. Res. Lett. **38**, L11202 (2011). doi:10.1029/2011GL047415
R.E. Hartle et al., Geophys. Res. Lett. **33**, L08201 (2006). doi:10.1029/2005GL024817
O. Havnes, Adv. Space Res. **13**(10), 153 (1993)
O. Havnes et al., J. Geophys. Res. **92**(A3), 2281 (1987). doi:10.1029/JA092iA03p02281
O. Havnes et al., J. Geophys. Res. **101**(A5), 10839–10847 (1996). doi:10.1029/96JA00003
M.M. Hedman et al., Nature **500**, 182–184 (2013). doi:10.1038/nature12371
M. Hesse, M.G. Kivelson, in *New Perspectives on the Earth's Magnetotail*, ed. by D.N.B.A. Nishida, S.W.H. Cowley. Geophys. Monogr. Ser., vol. 105 (Am. Geophys. Union, Washington, 1998)
T.W. Hill, J. Geophys. Res. **84**(A11), 6554 (1979). doi:10.1029/JA084iA11p06554
T.W. Hill, A.J. Dessler, L.J. Maher, J. Geophys. Res. **86**(A11), 9020 (1981)
T.W. Hill et al., Geophys. Res. Lett. **32**, L14S10 (2005). doi:10.1029/2005GL022620
T.W. Hill et al., J. Geophys. Res. **113**, A01214 (2008). doi:10.1029/2007JA012626
T.W. Hill et al., J. Geophys. Res. **117**, A05209 (2012). doi:10.1029/2011JA017218
M.K.G. Holmberg et al., Planet. Space Sci. **73**, 151–160 (2012). doi:10.1016/j.pss.2012.09.016
M.K.G. Holmberg, J.-E. Wahlund, M.W. Morooka, Geophys. Res. Lett. **41**, 3717 (2014). doi:10.1002/2014GL060229
M. Horanyi, Annu. Rev. Astron. Astrophys. **34**(1), 383 (1996)
D.E. Huddleston, C.T. Russell, G. Le, J. Geophys. Res. **102**, 24289 (1997). doi:10.1029/97JA02416
C.M. Jackman, in *Magnetotails in the Solar System*, ed. by A. Keiling, C.M. Jackman, P.A. Delamere (Am. Geophys. Union, Washington, 2015)
C.M. Jackman, C.S. Arridge, Sol. Phys. **274**, 481 (2011). doi:10.1007/s11207-011-9748-z
C.M. Jackman et al., J. Geophys. Res. **109**, A11203 (2004). doi:10.1029/2004JA010614
C.M. Jackman et al., Geophys. Res. Lett. **34**, L11203 (2007). doi:10.1029/2007GL029764
C.M. Jackman, R.J. Forsyth, M.K. Dougherty, J. Geophys. Res. **113**, A08114 (2008a). doi:10.1029/2008JA013083
C.M. Jackman et al., J. Geophys. Res. **113**, A11213 (2008b). doi:10.1029/2008JA013592
C.M. Jackman et al., Geophys. Res. Lett. **36**, L16101 (2009a). doi:10.1029/2009GL039149
C.M. Jackman et al., J. Geophys. Res. **114**, A08211 (2009b). doi:10.1029/2008JA013997
C.M. Jackman et al., J. Geophys. Res. **114**, A08211 (2009c). doi:10.1029/2008JA013997
C.M. Jackman et al., Geophys. Res. Lett. **36**, L16101 (2009d). doi:10.1029/2009GL039149
C.M. Jackman, J.A. Slavin, S.W.H. Cowley, J. Geophys. Res. **116**, A10212 (2011). doi:10.1029/2011JA016682
C.M. Jackman et al., Planet. Space Sci. **82–83**, 34 (2013). doi:10.1016/j.pss.2013.03.010
C.M. Jackman et al., Space Sci. Rev. **182**(1), 85 (2014a). doi:10.1007/s11214-014-0060-8
C.M. Jackman et al., J. Geophys. Res. **119**, 5465 (2014b). doi:10.1002/2013JA019388
J.M. Jasinski et al., Geophys. Res. Lett. **41** (2014). doi:10.1002/2014GL0593
X. Jia, M.G. Kivelson, J. Geophys. Res. **117**, A11219 (2012). doi:10.1029/2012JA018183
Y.-D. Jia et al., J. Geophys. Res. **115**, A12243 (2010). doi:10.1029/2010JA015534
X. Jia et al., J. Geophys. Res. **117**, A05225 (2012a). doi:10.1029/2012JA017575

X. Jia, M.G. Kivelson, T.I. Gombosi, J. Geophys. Res. **117**, A04215 (2012b). doi:10.1029/2011JA017367
R.E. Johnson et al., Astrophys. J. Lett. **644**, L137–L139 (2006)
R.E. Johnson et al., in *Titan from Cassini-Huygens* (Springer, Dordrecht, 2010), p. 373
G.H. Jones et al., Geophys. Res. Lett. **36**, L16204 (2009)
S. Jurac, J.D. Richardson, J. Geophys. Res. **110**, A09220 (2005). doi:10.1029/2004JA0106635
S. Jurac, J.D. Richardson, Geophys. Res. Lett. **34**, L08102 (2007). doi:10.1029/2007GL029567
M.L. Kaiser et al., Science **209**, 1238 (1980). doi:10.1126/science.209.4462.1238
M.L. Kaiser, M.D. Desch, A. Lecacheux, Nature **292**, 731 (1981)
S.J. Kanani et al., J. Geophys. Res. **115**, A06207 (2010). doi:10.1029/2009JA014262
S. Kempf et al., Icarus **193**(2), 420 (2008). doi:10.1016/j.icarus.2007.06.027
T.J. Kennelly et al., J. Geophys. Res. **118**, 832 (2013). doi:10.1002/jgra.50152
S. Knight, Planet. Space Sci. **21**, 741 (1973)
P. Kollmann et al., Icarus **222**, 323 (2013). doi:10.1016/j.icarus.2012.10.033
A.J. Kopf et al., Geophys. Res. Lett. **37**, L09102 (2010). doi:10.1029/2010GL042980
H. Kriegel et al., J. Geophys. Res. **116**, A10223 (2011). doi:10.1029/2011JA016842
S.M. Krimigis, T.P. Armstrong, Geophys. Res. Lett. **9**, 1143 (1982)
S.M. Krimigis et al., Space Sci. Rev. **114**, 233 (2004)
S.M. Krimigis et al., Science **307**, 1270 (2005). doi:10.1126/science.1105978
W.S. Kurth et al., Planet. Space Sci. **54**(9), 988 (2006). doi:10.1016/j.pss.2006.05.011
W.S. Kurth et al., Geophys. Res. Lett. **34**, L02201 (2007). doi:10.1029/2006GL028336
W.S. Kurth et al., J. Geophys. Res. **113**, A05222 (2008). doi:10.1029/2007JA012861
H.R. Lai et al., Reconnection at the magnetopause of Saturn: Perspective from FTE occurrence and magnetosphere size. J. Geophys. Res. **117**, A05222 (2012). doi:10.1029/2011JA017263
L. Lamy, in *Proc. VII Internat. Workshop on Planetary, Solar, and Heliospheric Radio Emissions*, ed. by H.O. Rucker, W.S. Kurt, P. Louarn, G. Fischer Graz, Austria (Austrian Acad. Sci., Vienna, 2011)
A.J. Lazarus, R.L. McNutt Jr., J. Geophys. Res. **88**, 8831 (1983)
S.A. Ledvina, T.E. Cravens, K. Kecskeméty, J. Geophys. Res. **110**, 6211 (2005)
J.S. Leisner et al., Geophys. Res. Lett. **32**, L14S08 (2005). doi:10.1029/2005GL022652
X. Liu et al., J. Geophys. Res. **115**, A12254 (2010). doi:10.1029/2010JA015859
Y. Ma et al., J. Geophys. Res. **111**, A05207 (2006). doi:10.1029/2005JA011481
A. Masters et al., Geophys. Res. Lett. **39**, L08103 (2012). doi:10.1029/2012GL051372
A. Masters et al., Geophys. Res. Lett. **41**, L059288 (2014). doi:10.1002/2014GL059288
B.H. Mauk et al., Geophys. Res. Lett. **32**, L14S05 (2005). doi:10.1029/2005GL022485
B.H. Mauk et al., in *Saturn from Cassini-Huygens*, ed. by M.K. Dougherty, L.W. Esposito, S.M. Krimigis (2009). doi:10.1007/978-1-4020-9217-6
H.J. McAndrews et al., J. Geophys. Res. **113**, A04210 (2008). doi:10.1029/2007JA012581
H.J. McAndrews et al., Plasma in Saturn's nightside magnetosphere and the implications for global circulation. Planet. Space Sci. **57**, 1714 (2009). doi:10.1016/j.pss.2009.03.003
R.L. McPherron, C.T. Russell, M.P. Aubry, J. Geophys. Res. **78**, 3131 (1973)
H. Melin, D.E. Shemansky, X. Liu, Planet. Space Sci. **57**, 1743–1753 (2009). doi:10.1016/j.pss.2009.04.014
M. Michael et al., Icarus **175**, 263 (2005)
D.G. Mitchell et al., Geophys. Res. Lett. **32**, L20S01 (2005). doi:10.1029/2005GL022647
D.G. Mitchell et al., J. Geophys. Res. **114**, A02212 (2009). doi:10.1029/2008JA013621
M.W. Morooka et al., J. Geophys. Res. **116**(A12), A12221 (2011). doi:10.1029/2011JA017038
A.F. Nagy et al., in *Saturn from Cassini Huygens* (Springer, Berlin, 2009)
C. Paranicas et al., Icarus **197**, 519 (2008). doi:10.1016/j.icarus.2008.05.011
E.N. Parker, Astrophys. J. **128**, 664 (1958)
M.E. Perry et al., J. Geophys. Res. **115**, A10206 (2010). doi:10.1029/2010JA015248
D.H. Pontius Jr., T.W. Hill, J. Geophys. Res. **94**, 15041–15053 (1989). doi:10.1029/JA094iA11p15041
D.H. Pontius Jr., R.A. Wolf, Geophys. Res. Lett. **17**, 49–52 (1990). doi:10.1029/GL017i001p00049
C.C. Porco et al., Science **311**, 1393–1401 (2006)
G.D. Provan et al., J. Geophys. Res. **114**, A02225 (2009). doi:10.1029/2008JA013782
G.D. Provan et al., J. Geophys. Res. **116**, A04225 (2011). doi:10.1029/2010JA016213
B.A. Randall, J. Geophys. Res. **99**, 8771 (1994)
L.C. Ray et al., J. Geophys. Res. **118**, 3214 (2013). doi:10.1002/jgra.50330
G.C. Reid, J. Geophys. Res. **95**, 13891 (1990). doi:10.1029/JD095iD09p13891
J.D. Richardson, J. Geophys. Res. **91**, 1381 (1986)
J.D. Richardson, E.C. Sittler, J. Geophys. Res. **95**, 12019 (1990)
E. Roussos et al., Geophys. Res. Lett. **35**, L22106 (2008). doi:10.1029/2008GL035767
E. Roussos et al., J. Geophys. Res. **116**, A02217 (2011). doi:10.1029/2010JA015954

A.M. Rymer, Pickup ions throughout the heliosphere and beyond, in *Proceedings of the 9th International Astrophysics Conference*, ed. by J. le Roux, G. Zank, A.J. Coates, V. Florinsky. AIP Conference Proceedings, vol. 1302 (2010), pp. 250–255
A.M. Rymer et al., Electron sources in Saturn's magnetosphere. J. Geophys. Res. **112**, A02201 (2007). doi:10.1029/2006JA012017
A.M. Rymer et al., J. Geophys. Res. **113**, A01201 (2008). doi:10.1029/2007JA012589
A.M. Rymer et al., Planet. Space Sci. **57**, 1779 (2009). doi:10.1016/j.pss.2009.04.010
J. Saur et al., Geophys. Res. Lett. **35**, L20105 (2008). doi:10.1029/2008GL035811
P. Schippers et al., Geophys. Res. Lett. **113**, A07208 (2008). doi:10.1029/2008JA013098
L. Scurry, C.T. Russell, J. Geophys. Res. **96**, 9541 (1991). doi:10.1029/91JA00569
N. Sergis et al., Geophys. Res. Lett. **34**, L09102 (2007). doi:10.1029/2006GL029223
A.S. Sharma et al., Ann. Geophys. **26**, 955–1006 (2008)
D.E. Shemansky, D.T. Hall, J. Geophys. Res. **97**, 4143 (1992)
D.E. Shemansky et al., Nature **363**, 329 (1993)
D.E. Shemansky, X. Liu, H. Melin, Planet. Space Sci. **57**, 1659 (2009). doi:10.1016/j.pss.2009.05.002
V.I. Shematovich et al., J. Geophys. Res. **108**, E8 (2003)
P.K. Shukla, A.A. Mamun, *Introduction to Dusty Plasma Physics* (Taylor & Francis, London, 2002)
I. Sillanpää et al., Adv. Space Res. **38**(4), 799 (2006)
S. Simon et al., Planet. Space Sci. **58**(10), 1230–1251 (2010). ISSN 0032-0633. doi:10.1016/j.pss.2010.04.021
S. Simon et al., J. Geophys. Res. **116**(A4), A04221 (2011). doi:10.1029/2010JA016338
J.A. Simpson et al., Science **207**, 411 (1980)
E.C. Sittler Jr., M.F. Blanc, J.D. Richardson, J. Geophys. Res. **111**, A06208 (2006). doi:10.1029/2005JA011191
E.C. Sittler Jr. et al., Planet. Space Sci. **56**, 3 (2008)
E.C. Sittler Jr. et al., Planet. Space Sci. **57**, 1547 (2009). doi:10.1016/j.pss.2009.07.017
J.A. Slavin, in *New Perspectives on the Earth's Magnetotail*, ed. by D.N.B.A. Nishida, S.W.H. Cowley. Geophys. Monogr. Ser., vol. 105 (Am. Geophys. Union, Washington, 1998)
C.G.A. Smith, Mon. Not. R. Astron. Soc. **410**(4), 2315 (2011). doi:10.1111/j.1365-2966.2010.17602.x
H.T. Smith, Ph.D. thesis, University of Virginia, Charlottesville, VA (2006)
H.T. Smith et al., Geophys. Res. Lett. **31**, L16804 (2004). doi:10.1029/2004GL020580
H.T. Smith et al., Icarus **188**(2), 356 (2007). doi:10.1016/j.icarus.2006.12.007
H.T. Smith et al., J. Geophys. Res. **113**, A11206 (2008). doi:10.1029/2008JA013352
H.T. Smith et al., J. Geophys. Res. **115**, A10252 (2010). doi:10.1029/2009JA015184
H.T. Smith et al., in *AAS/Div. Plan. Sci. Meeting Abstracts*, vol. 44 (2012)
D.R. Snowden, R. Winglee, A. Kidder, J. Geophys. Res. **116**, A08229 (2011a). doi:10.1029/2011JA016435
D.R. Snowden, R. Winglee, A. Kidder, J. Geophys. Res. Space Phys. **116**(A8), 1978–2012 (2011b)
D.J. Southwood, J. Geophys. Res. **116**, A01201 (2011). doi:10.1029/2010JA016070
D.J. Southwood, M.G. Kivelson, J. Geophys. Res. **112**(A12222), 2007 (2007). doi:10.1029/2007JA012254
F. Spahn et al., Science **311**, 1416 (2006)
D.F. Strobel, Icarus **193**, 2588 (2008)
D.F. Strobel, Icarus **202**(2), 632 (2009)
D.L. Talboys et al., J. Geophys. Res. **116**, A04213 (2011). doi:10.1029/2010JA016102
V. Tenishev, M.R. Combi, B.D. Teolis, J.H. Waite, J. Geophys. Res. **115**, A09302 (2010). doi:10.1029/2009JA015223
B.T. Thomas, E.J. Smith, J. Geophys. Res. **85**, 6861 (1980)
M.F. Thomsen, Geophys. Res. Lett. **40**, 5337 (2013). doi:10.1002/2013GL057967
M.F. Thomsen et al., J. Geophys. Res. **115**, A10220 (2010). doi:10.1029/2010JA015267
M.F. Thomsen et al., J. Geophys. Res. **118**, 5767 (2013). doi:10.1002/jgra.50552
R.L. Tokar et al., Science **311**, 1409 (2006)
R.L. Tokar et al., Geophys. Res. Lett. **36**, L13203 (2009). doi:10.1029/2009GL038923
W.-L. Tseng et al., Icarus **206**, 382–389 (2010). doi:10.1016/j.icarus.2009.05.019
W.-L. Tseng et al., J. Geophys. Res. **116**, A03209 (2011). doi:10.1029/2010JA016145
O.J. Tucker, R.E. Johnson, Planet. Space Sci. **57**(14), 1889 (2009)
J.A. Van Allen, J. Geophys. Res. **88**, 6911 (1983). doi:10.1029/JA088iA09p06911
R.E. Vogt et al., Science **215**, 577 (1982). doi:10.1126/science.215.4532.577
V. Vuitton, R.V. Yelle, M.J. McEwan, Icarus **191**(2), 722–742 (2007)
J.-E. Wahlund et al., Science **308**(5724), 986 (2005)
J.-E. Wahlund et al., Planet. Space Sci. **57**(1), 1795 (2009). doi:10.1016/j.pss.2009.03.011
J.H. Waite Jr., S.K. Atreya, A.F. Nagy, Geophys. Res. Lett. **6**, 723 (1979)
J.H. Waite et al., Science **311**, 1419 (2006)

J.W. Warwick et al., Science **212**, 239 (1981). doi:10.1126/science.212.4491.239
A. Wellbrock et al., J. Geophys. Res. **117**, A03216 (2012). doi:10.1029/2011JA017113
J.H. Westlake et al., Geophys. Res. Lett. **39**, L19104 (2012). doi:10.1029/2012GL053079
E.C. Whipple, T.G. Northrop, D.A. Mendis, J. Geophys. Res. **90**(A8), 7405 (1985). doi:10.1029/JA090iA08p07405
R.J. Wilson et al., J. Geophys. Res. **113**, A12218 (2008). doi:10.1029/2008JA013486
W. Xu, N. D'Angelo, R.L. Merlino, J. Geophys. Res. **98**(A5), 7843 (1993). doi:10.1029/93JA00309
V. Yaroshenko et al., Phys. Plasmas **12**, 093503 (2005). doi:10.1063/1.1947027
V.V. Yaroshenko et al., Planet. Space Sci. **57**(1), 1807–1812 (2009). doi:10.1016/j.pss.2009.03.002
S.-Y. Ye, D.A. Gurnett, W.S. Kurth, T.F. Averkamp, M. Morooka, S. Sakai, J.-E. Wahlund, J. Geophys. Res. Space Phys. **119**, 3373–3380 (2014). doi:10.1002/2014JA019861
R.V. Yelle, J. Cui, I.C.F. Müller-Wodarg, J. Geophys. Res. **113**, E10003 (2008). doi:10.1029/2007JE003031
D.T. Young et al., Science **307**, 1262 (2005). doi:10.1126/science.1106151

DOI 10.1007/978-1-4939-3544-4_8
Reprinted from *Space Science Reviews* Journal, DOI 10.1007/s11214-015-0176-5

Comparison of Plasma Sources in Solar System Magnetospheres

Norbert Krupp[1]

Received: 21 April 2015 / Accepted: 12 June 2015 / Published online: 10 July 2015

Abstract The plasma sources of Mercury, Earth, Jupiter, and Saturn have been described in this issue in great detail. Much less information exists about the plasma sources of Uranus and Neptune. Only one flyby of the Voyager 2 spacecraft through the highly complex and time variable magnetospheres of those ice giants gives us a limited snapshot of the main plasma sources in those systems. The basic knowledge derived from those flybys are described briefly in this paper for completeness. The main purpose of this paper is to summarize the plasma sources of all planetary magnetospheres and compare the similarities and differences of those huge plasma laboratories in our solar system.

Keywords Giant planet magnetospheres · Jupiter · Saturn · Uranus · Neptune

1 Introduction

The knowledge about the plasma sources in planetary magnetospheres in our solar system is mainly based on in-situ and remote sensing measurements onboard spacecraft during flybys or in orbit around the planet. The largest data set exists for Earth followed by Saturn, Jupiter, and Mercury. For Uranus and Neptune only one flyby for each planet exists. Table 1 summarizes the spacecraft exploration of the planetary magnetospheres in our solar system. For Earth only a few selected missions are listed. From those missions to the other planets than Earth performed so far only Messenger at Mercury, Galileo at Jupiter, and Cassini at Saturn went into orbit around the planet while all others were flyby missions, still major accomplishments of planetary exploration. The upcoming missions Bepi-Colombo (two spacecraft in orbit around Mercury), Juno at Jupiter, and Juice at Jupiter will orbit the central body. In the case of Juice, the spacecraft will even orbit Jupiter's satellite Ganymede in the later stage of its mission, and investigate this satellite's tiny intrinsic magnetosphere.

✉ N. Krupp
krupp@mps.mpg.de

[1] Max-Planck-Institut für Sonnensystemforschung, Justus-von-Liebig-Weg 3, Göttingen, Germany

Table 1 Spacecraft exploration of planetary magnetospheres

Mercury	Earth	Jupiter	Saturn	Uranus	Neptune
Mariner 10 (1974)	Geotail (1992–????)	Pioneer 10 (1973)	Pioneer 11 (1979)	Voyager 2 (1986)	Voyager 2 (1989)
Messenger (2008–2015)	Polar (1996–2008)	Pioneer 11 (1974)	Voyager 1 (1980)		
BepiColombo (2024–?)	Sampex (1992–2004)	Voyager 1 (1979)	Voyager 2 (1981)		
	CRRES (1990–1991)	Voyager 2 (1979)	Cassini (2004–2017)		
	ACE (1997–?)	Ulysses (1992)			
	FAST (1996–2009)	Galileo (1995–2003)			
	DoubleStar (2003–2007)	Cassini (2000/2001)			
	Themis (2007–?)	New Horizons (2007)			
	VanAllen probes (2012–?)	Juno (2016–2017)			
	Cluster (2000–?)	Juice (2030–2033)			
	Dynamic Explorer (1981–1991)				

Additional information about the plasma sources of the planetary magnetospheres come from Earth- or Earth-orbit based telescopes and from MHD- and Hybrid code simulations of the various systems.

A comparison of various parameters including the major plasma sources for all magnetospheres in our solar system can be found in Bagenal (2013). Basically internal plasma sources (atmosphere, ionosphere, moons, rings) can be distinguished from external sources (solar wind, cosmic rays). Usually internal and external sources differ in their energy distribution and their ion composition. However, the origin of protons in planetary magnetospheres is manyfold and more processes have to be taken into account. Mauk (2014) compared the intensity-energy spectra of ions measured in the ring current regions of Earth, Jupiter, Saturn, Uranus, and Neptune with each other and found that at Earth and Jupiter the spectra are most intense and limited by the Kennel-Petschek process (Kennel and Petschek 1966) while for Saturn, Uranus, and Neptune the most intense spectra are below that limit and limited by the interaction between the charged particles and the neutral gas and dust in the systems. This means that the ion composition and energy spectra are subject to interaction processes in the magnetospheres dependent on the strength of the plasma sources. In Mauk and Fox (2010) a similar approach has been used to describe and compare the electron energy spectra in the radiation belts of the magnetized planets (except Mercury). The limiting factors found are wave-particle interactions (e.g. scattering of whistler waves), missing acceleration processes (no injection-like processes in the Neptune case), and interaction with neutral gas clouds and dust.

In the previous papers of this issue the global configuration and the plasma sources of the magnetospheres of Mercury, Earth, Jupiter, and Saturn have been described in detail. The purpose of this paper is to add the very limited knowledge about the plasma sources of the Uranian and the Neptunian systems and compare them with the other systems.

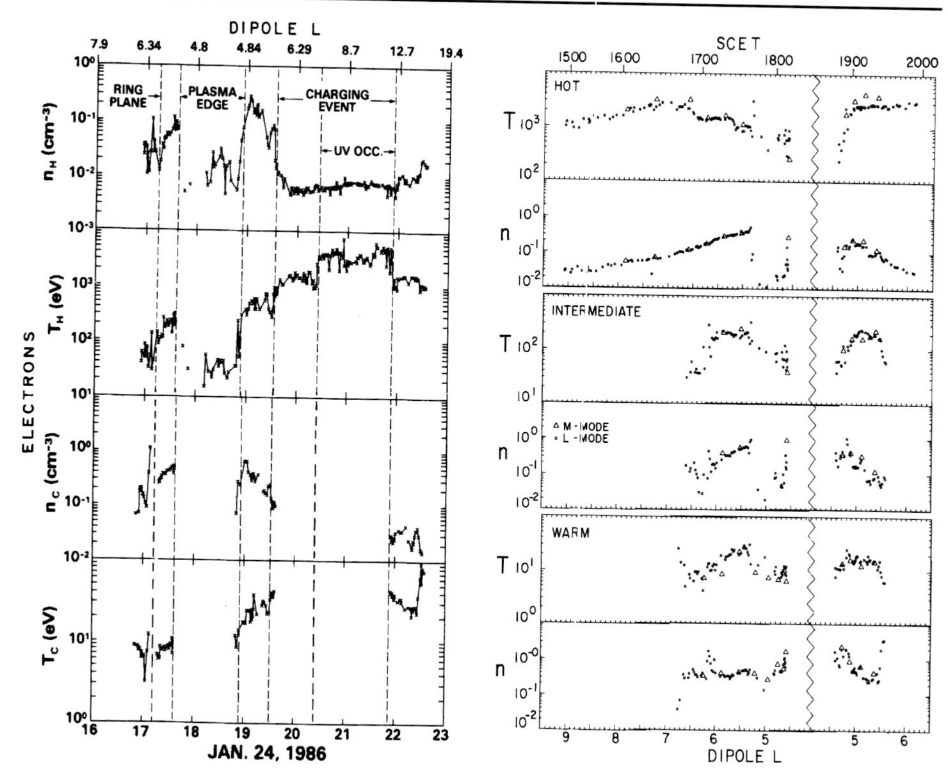

Fig. 1 Electrons and ions in the magnetosphere of Uranus. *Left*: density and temperature of cold (*upper panel*) and hot (*lower panel*) electrons as a function of Dipole L (from Belcher et al. 1991 after Sittler et al. 1987); *Right*: density and temperature of hot (*upper panel*), intermediate (*middle panel*), and cold (*lower panel*) ions vs. Dipole L (from Selesnick and McNutt 1987)

2 Plasma Sources of the Uranian and Neptunian Magnetospheres

For Uranus and Neptune the data sets are rather limited and based on the flybys of the Voyager 2 spacecraft in 1986 and 1989 only, respectively. Results of those flybys at these ice giants have been described in the books by Bergstralh et al. (1991) and Cruikshank et al. (1995).

The Voyager 2 Uranus encounter revealed that the plasma in the inner Uranian magnetosphere consists of electrons and subsonic protons (Belcher et al. 1991; Selesnick and McNutt 1987).

The distribution of electrons can be described with two Maxwellians, a cold part with temperatures $T_c = 7$–30 eV and a hot part ($T_h = 500$–2000 eV) (Belcher et al. 1991; Sittler et al. 1987). The left panel of Fig. 1 shows the measurements of density and temperatures in those two regimes from the Voyager 2 flyby. The transition energy between cold and hot electrons is around 100 eV but depends on the plasma density.

Ions found during the flyby were categorized by a warm, an intermediate, and a hot population as shown in the right panel of Fig. 1. For Uranus warm hydrogen ions originate in the neutral hydrogen corona or the ionosphere of the planet while the hotter, more energetic protons come from the magnetotail gaining energy during the inward convection motion as pointed out by Belcher et al. (1991) based on the observation that the variations of the density and temperature are quite small indicative of a localized source. Selesnick and McNutt (1987) described adiabatic compression in the sunward convection transport as an acceleration mechanism to explain the temperature and energy of the pickup ions from the neutral hydrogen cloud. On the other hand Cheng (1987) proposed ionospheric injection as a possible source of the warm ions in the inner Uranian magnetosphere.

Solar wind as a source of hot ions in the Uranian magnetosphere was ruled out because no helium ions above 600 keV/nuc have been detected inside the magnetosphere (Krimigis et al. 1986). Therefore it was concluded that the polar ionosphere may be the source of hot ions and electrons as well. Krimigis et al. (1986) also reported a minor fraction of H_2-molecules of about 10^{-3} besides the dominant proton component. Changes in the energy spectra of MeV ions as a function of dipole L suggested that cosmic rays through CRAND (Cosmic Ray Albedo Neutron Decay) processes may play a role as an additional particle source at least in the inner magnetosphere of Uranus. The CRAND source has been suggested and confirmed in the radiation belts of Jupiter (Bolton et al. 2004) (see also the Jupiter paper by Bolton et al. in this issue) and Saturn (Kollmann et al. 2011, 2013) (see also the Saturn paper by Blanc et al. in this issue).

Even though Uranus has 27 satellites no major plasma source on one of the moons could be identified during the Voyager 2 flyby. However, the possibility cannot be ruled out completely because of the highly variable inclination of the moons with respect to the magnetic equator (Mauk et al. 1987; Ness et al. 1991). This leads to highly variable interactions between the surfaces of the moons and the magnetospheric trapped population where also sputtering can occur. If water ice particles are released they could also contribute to the magnetospheric population, in a way similar to the moons in the Saturnian or Jovian magnetosphere. However, no evidence has been seen in the Voyager 2 data.

The picture of Neptune's magnetosphere and its plasma sources is based on the only flyby at the planet by Voyager 2 in 1989. The appearance of the Neptunian magnetic cavity is very similar to the Uranian magnetosphere. The dipole axis is also highly tilted by 47 degrees with respect to the planets' rotation axis. Therefore the system is also highly complex and time variable. A single flyby can only be a snapshot at a given time and a given geometry of the system.

The plasma instrument PLS onboard Voyager 2 found essentially three components of charged particles during the encounter: Protons, nitrogen ions and electrons (Zhang 1991). Figure 2 shows the density (left) and temperatures (right) of those three components during the Voyager 2 flyby as a function of distance from the planet.

From those measurements and the derivation of the quantity NL^2 where N is the flux tube content and L is the McIlwain parameter (equatorial distance of a fieldline from the planet assuming a dipole magnetic field) it has been concluded that the sources of protons and nitrogen are outside $L = 7$ (details can be found in Richardson et al. 1995). Protons are most abundant between $L = 8$ and $L = 10$ while the nitrogen ion maximum is beyond $L = 13$. Richardson et al. (1995) further pointed out that the observed densities and temperatures are consistent with a source in the outer magnetosphere of Neptune and gaining energy through radial inward motion. From those ideas and concepts it has been concluded that in Neptune's magnetosphere most probably the moon Triton is the major source of hot ions (Richardson et al. 1995; Mauk et al. 1995). With a diameter of 2700 km, Triton is the largest of the 13 moons of Neptune. It has been discovered that Triton has active geysers on its surface which most probably are the sources of the ions in the system, similar to the over one hundred geysers on Enceladus (Porco et al. 2014) in the Saturnian system or similar to the Jovian moon Io, the major plasma source of the Jovian system, the most active volcanic

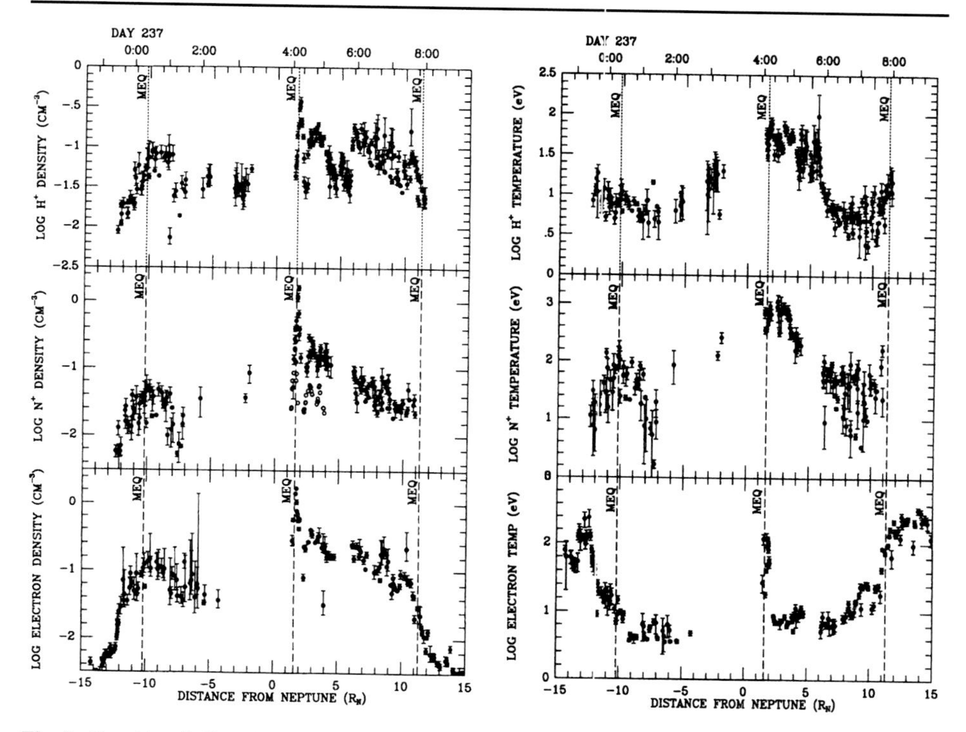

Fig. 2 Densities (*left*) and temperatures (*right*) of hydrogen (*upper panel*), nitrogen (*middle panel*), and electrons (*lower panel*) as a function of distance from the planet/time inside 15 R_N of the Neptunian magnetosphere as measured during the Voyager 2 encounter 1989 (from Richardson et al. 1995)

body in our solar system (see Bagenal 2013 and references therein). In addition there is also evidence in Hubble space telescope data that plumes may exist on the Jovian moon Europa (Roth et al. 2014).

3 Global Configuration and Scales Compared

The global configuration of planetary magnetospheres, their similarities and differences depend on a few basic characteristic parameters, e.g. distance from the Sun and variations of the solar wind parameters, intrinsic magnetic field strength, strengths and location of the plasma sources influencing the system's size, spatial and temporal scales on which processes and interactions occur, power- and energy source driving the systems.

Unique for the magnetospheres of Uranus and Neptune are their strange orientations of the spin axis with respect to their magnetic dipole axis (59 and 47 degrees, respectively) and with respect to the solar wind flow direction 8–172 and 60–120 degrees, respectively (Bagenal 2013), resulting in a temporarily changing plasma environment and a possibly changing plasma source strength. Also the variation of the obliquity of the planet's spin axis can produce strong seasonal variations in relationship of planetary and interplanetary magnetic field.

The spatial scales of the various systems vary by a factor of several hundreds between Mercury and Jupiter. The time scales in which processes occur inside the magnetospheres range from seconds/minutes in the case of Mercury to minutes/days for the Jupiter case.

Mercury has a small magnetosphere close to the Sun. The environment is highly complex and highly dynamic. As pointed out in the Mercury paper by Raines et al. (2015) in this issue, the Hermean plasma environment is driven by a weak internal magnetic field and the interaction between the solar wind, the planet's exosphere and its surface. In rather disturbed conditions the solar wind can directly impinge onto the surface of Mercury. The magnetosphere is small and has about the same size as the magnetosphere of the Jovian moon Ganymede. Magnetopause standoff distance typically is about 1500–1800 km or 0.5 $R_{Mercury}$ above the surface. The typical time scales of processes in the Hermean magnetosphere are on the order of seconds.

For the Earth the plasma sources are the solar wind and the ionosphere as described in the paper by Welling et al. (2015) in this issue.

Jupiter has by far the largest magnetosphere with a nose magnetopause distance of up to 90 R_J. Bolton et al. (2015) describes in the Jupiter paper of this issue that the Jovian system is driven by its fast rotation combined with the internal plasma source of the volcanic moon Io for oxygen and sulfur ions. Moon Europa contributes much less material but may play a role in the abundance of oxygen and hydrogen through sputtering off the surface or through active plumes as reported by Roth et al. (2014). The solar wind is a minor or negligible source of plasma and only relevant in the outer magnetosphere. Jupiter's tilt angle between dipole and rotation axis leads to periodic variations of the plasma density in the center of the plasma sheet.

Blanc et al. (2015) describes in the Saturn paper (this issue) that the entire system is dominated by the active moon Enceladus providing most of the material found in the Saturnian magnetosphere.

As described in this paper Uranus and Neptune are somehow special. The angle between the rotation axis and the sun-planet-line or the magnetic dipole axis is large introducing large asymmetries and geometry changes of the entire magnetosphere during one planetary rotation. For Uranus the extended hydrogen cloud around the planet was concluded to be the major plasma source while for Neptune the moon Triton plays the major role.

In Table 2 some of the relevant numbers for the major plasma sources of planetary magnetospheres from Bagenal (2013, 2009) are listed in combination with a few relevant parameters for each planet.

4 Similarities and Differences

Moons obviously are major plasma sources in outer planets' magnetospheres. They provide either material from their interior out of volcanoes, cryo-volcanoes or geysers or particles are released from their surfaces through sputtering and sublimation processes.

One major similarity is the fact that the Jovian, Saturnian, and Neptunian magnetospheres are mainly filled by the release of material of those active moons: Io and Europa for Jupiter, Enceladus for Saturn, and Triton for Neptune, respectively. Io is the most active volcanic body in the solar system; plumes of oxygen material have been observed at Europa and to a much larger extent at Enceladus. At Neptune's Triton activity in the form of released material has been observed and an active region is assumed. Those moons provide the magnetospheres with heavy ions oxygen, sulphur, nitrogen, watergroup ions. Due to the fast rotation of those planets most of these heavy ions are concentrated near the centrifugal or magnetic equator in a disk-like current sheet and associated plasma sheet. Away from the center of this plasma sheet the intensity of ions drops fast.

Table 2 Characteristic parameters of planetary magnetospheres compared (from Bagenal 2013, 2009). SW: solar wind; Ionos: Ionosphere; Atmos: Atmosphere; Eu: Europa; Enc: Enceladus; Tri: Triton; W^+: watergroup ions; MP: magnetopause

	Mercury	Earth	Jupiter	Saturn	Uranus	Neptune
Radius [km]	$1\ R_M = 2440$	$1\ R_E = 6373$	$1\ R_J = 71492$	$1\ R_S = 60268$	$1\ R_U = 25600$	$1\ R_N = 24.765$
MP nose distance	1.5 R_M	8–12 R_E	63–92 R_J	22–27 R_S	18 R_U	23–25 R_N
Dipole strength [nT]	195	30600	430000	21400	22800	14200
Dipole tilt [degrees]	3	9.92	9.4	0	59	47
Major plasma sources	SW	Ionos, SW	Io, Europa	Enceladus	Atmos	Tri
Major ion species	H^+	O^+, H^+	O^{n+}, S^{n+}, H^+	W^+, O^+, H^+	H^+	N^+, H^+
plasma source [kg/s]	5	5	260–1400	12–250	0.02	0.2
Life time	min	hours-days	20–80 days	30–50 days	1–30 days	days
Plasma $\beta = nkT/(B^2/2\mu_0)$		< 1	10–100	1–5	0.1	0.2

Another similarity is that the atmospheres and ionospheres of the planets, if existing, are also sources of plasma for the magnetospheres. In the case of Earth, Uranus, and probably Neptune the ionosphere plays a major role while for Jupiter and Saturn only minor contributions for the overall budget of plasma are assumed to originate in the atmosphere/ionosphere of those planets.

The solar wind is the major source for Mercury and for the Earth although some evidence of solar wind ions exists also for the other magnetospheres.

Sputtering or sublimation of material from the planet's surface through impinging solar wind (Mercury case) or due to impinging magnetospheric particles from the surfaces of the moons orbiting the planet play a role for the Galilean satellites at Jupiter as well as the bigger icy moons of the other outer gas giants.

Cosmic Ray Albedo Neutron Decay (CRAND) obviously plays a role as one of the sources of high-energy particles in the radiation belts of the planets. It is believed that high-energy protons in the inner radiation belts of Earth, Jupiter, Saturn, Uranus, and Neptune are the result of the CRAND process.

Overall the plasma source rates are different, highly variable and determine the global configuration and the dynamics of the magnetospheres. Therefore results from just one flyby through a highly complex magnetosphere can be quite tricky. As described in the Saturn paper in this issue by Blanc et al. (2015) Titan was thought to be a major source for nitrogen in the magnetosphere before Cassini went into orbit,stayed there for more than 10 years and ruled out the importance of Titan.

One of the major differences between the magnetospheres of Jupiter and Saturn compared with Uranus and Neptune is the value of the plasma β (ratio of plasma pressure to magnetic pressure). For Uranus and Neptune $\beta \approx 0.1$–0.2, similar values as in the Earth's case, while for Saturn β-values of 1–5 and for Jupiter 10–100 are reached. This means that the Jovian and the Saturnian magnetospheres are particle-driven where the magnetic field "follows" the particles and vice versa for Uranus and Neptune.

Another difference is the importance of solar wind convection in those magnetospheres and the importance of solar wind as a plasma source. Although all outer planets magnetospheres are rotationally dominated, Uranus is special in the sense that similar to Earth it is believed that particles entering the magnetosphere from the solar wind convect towards the planet gaining energy.

5 Future Missions to Mercury and to the Outer Planets and Their Capabilities to Study Plasma Sources

5.1 Mercury

Missions to Mercury close to the Sun are very demanding and challenging in terms of thermal properties. Thermal radiation from the Sun as well as from the surface of the planet itself have to be taken into account during the mission planning. The European Space Agency ESA and its Japanese counterpart JAXA will launch the two-spacecraft mission BepiColombo to Mercury in 2017 arriving in 2024. ESA will provide the Mercury Planetary Orbiter (MPO) to characterize the surface, exosphere, and interior of the planet. MPO will orbit Mercury in an elliptical polar orbit (400 × 1500 km). The JAXA spacecraft Mercury Magnetospheric Orbiter (MMO) will study the magnetosphere of the closest planet to the Sun in detail also in a polar orbit of 400 × 15000 km. The instrumentation onboard both spacecraft will enable to study the plasma sources in great detail. Compared to the NASA Messenger project BepiColombo has a more sophisticated package to measure neutral and charged particles. The biggest advantage compared to Messenger, however, will be the two-point measurements from the synchronized orbits of the two spacecraft inside the Hermean system.

5.2 Outer Planets

Missions to the outer planets are always a challenge because of their demanding radiation environments and the corresponding higher costs. In addition good launch opportunities are rather limited because of the need of special planet constellations for gravity assist maneuvers in order to reach the outer planets. Power onboard the spacecraft outside Jupiter's orbit requires extremely large or high-efficient solar panels or nuclear power generators (RTGs) increasing the complexity or the costs of these missions.

In the case of Jupiter two missions are on their way (JUNO) or in preparation (JUICE). While JUNO will mostly study Jupiter's atmosphere, ionosphere, and high latitude magnetosphere, JUICE will mainly characterize Ganymede, its magnetosphere, the Jovian plasma disk and will add new data from the environment of Europa and Callisto. Both missions have instrumentation onboard to study the plasma sources of the Jovian system.

Recently Jupiter's moon Europa got a lot of interest because of potential active plumes similar to the geysers of Enceladus (Roth et al. 2014). New mission concepts to study the moon Europa and its habitability in detail are currently on its way. Europa Clipper is the most advanced of those NASA mission concepts with up to 48 close flybys at Europa. There is a possibility that Europa Clipper and Juice could arrive at Jupiter at the same time.

At Saturn only Cassini will continue to study the ringed planet until its "Grand Finale" in 2017 when the spacecraft will finish its mission with a final dive into the planet. Cassini will continue to study Enceladus and Titan as plasma sources but will also investigate the region inside the D-ring where another trapped particle population has been inferred remotely. No other Saturn mission is currently in preparation but studies to go back to Enceladus or to Titan have been initiated.

Missions to Uranus and Neptune have been studied by ESA and NASA but so far no missions have been approved. As examples among the various other studies Uranus Pathfinder (Arridge et al. 2012) and Neptune-Triton-Kuiperbelt mission (Christophe et al. 2012) should be mentioned here. All studies included instruments specifically dedicated to measure and characterize the plasma sources of those systems.

Acknowledgements This work was in part financed by the German BMWi through the German Space Agency DLR under contracts 50 OH 0301, 50 OH 0801, 50 OH 0802, 50 OH 1101, 50 OH 1502, 50 ON 0201, 50 QW 0503, 50 QW 1303, 50 QJ 1301, 50 OO 1206, 50 OO 1002 and by the Max Planck Society. This work was also supported by the European Research Infrastructure EUROPLANET RI in the framework program FP7, contract number: 001637.

References

C.S. Arridge, C.B. Agnor, N. André, K.H. Baines, L.N. Fletcher, D. Gautier, M.D. Hofstadter, G.H. Jones, L. Lamy, Y. Langevin, O. Mousis, N. Nettelmann, C.T. Russell, T. Stallard, M.S. Tiscareno, G. Tobie, A. Bacon, C. Chaloner, M. Guest, S. Kemble, L. Peacocke, N. Achilleos, T.P. Andert, D. Banfield,

S. Barabash, M. Barthelemy, C. Bertucci, P. Brandt, B. Cecconi, S. Chakrabarti, A.F. Cheng, U. Christensen, A. Christou, A.J. Coates, G. Collinson, J.F. Cooper, R. Courtin, M.K. Dougherty, R.W. Ebert, M. Entradas, A.N. Fazakerley, J.J. Fortney, M. Galand, J. Gustin, M. Hedman, R. Helled, P. Henri, S. Hess, R. Holme, Ö. Karatekin, N. Krupp, J. Leisner, J. Martin-Torres, A. Masters, H. Melin, S. Miller, I. Müller-Wodarg, B. Noyelles, C. Paranicas, I. de Pater, M. Pätzold, R. Prangé, E. Quémerais, E. Roussos, A.M. Rymer, A. Sánchez-Lavega, J. Saur, K.M. Sayanagi, P. Schenk, G. Schubert, N. Sergis, F. Sohl, E.C. Sittler, N.A. Teanby, S. Tellmann, E.P. Turtle, S. Vinatier, J.-E. Wahlund, P. Zarka, Uranus Pathfinder: exploring the origins and evolution of Ice Giant planets. Exp. Astron. **33**, 753–791 (2012). doi:10.1007/s10686-011-9251-4

F. Bagenal, Comparative planetary environments, in *Heliophysics: Plasma Physics of the Local Cosmos* (2009), p. 360

F. Bagenal, Planetary magnetospheres, in *Planets, Stars and Stellar Systems*, ed. by T.D. Oswalt, L.M. French, P. Kalas (2013), p. 251. doi:10.1007/978-94-007-5606-9_6

J.W. Belcher, R.L. McNutt Jr., J.D. Richardson, R.S. Selesnick, E.C. Sittler Jr., F. Bagenal, in *The Plasma Environment of Uranus*, ed. by J.T. Bergstralh, E.D. Miner, M.S. Matthews (1991), pp. 780–830

J.T. Bergstralh, E.D. Miner, M.S. Matthews, *Uranus* (1991)

M. Blanc et al., Space Sci. Rev. (2015 this issue). doi:10.1007/s11214-015-0172-9

S.J. Bolton, C.J. Hansen, D.L. Matson, L.J. Spilker, J.-P. Lebreton, Cassini/Huygens flyby of the Jovian system. J. Geophys. Res. **109**(A18), A09S01 (2004). doi:10.1029/2004JA010742

Bolton et al., Space Sci. Rev. (2015 this issue)

A.F. Cheng, Proton and oxygen plasmas at Uranus. J. Geophys. Res. **92**, 15309–15314 (1987). doi:10.1029/JA092iA13p15309

B. Christophe, L.J. Spilker, J.D. Anderson, N. André, S.W. Asmar, J. Aurnou, D. Banfield, A. Barucci, O. Bertolami, R. Bingham, P. Brown, B. Cecconi, J.-M. Courty, H. Dittus, L.N. Fletcher, B. Foulon, F. Francisco, P.J.S. Gil, K.H. Glassmeier, W. Grundy, C. Hansen, J. Helbert, R. Helled, H. Hussmann, B. Lamine, C. Lämmerzahl, L. Lamy, R. Lehoucq, B. Lenoir, A. Levy, G. Orton, J. Páramos, J. Poncy, F. Postberg, S.V. Progrebenko, K.R. Reh, S. Reynaud, C. Robert, E. Samain, J. Saur, K.M. Sayanagi, N. Schmitz, H. Selig, F. Sohl, T.R. Spilker, R. Srama, K. Stephan, P. Touboul, P. Wolf, OSS (Outer Solar System): a fundamental and planetary physics mission to Neptune, Triton and the Kuiper Belt. Exp. Astron. **34**, 203–242 (2012). doi:10.1007/s10686-012-9309-y

D.P. Cruikshank, M.S. Matthews, A.M. Schumann, *Neptune and Triton* (1995)

C.F. Kennel, H.E. Petschek, Limit on stably trapped proton fluxes. J. Geophys. Res. **71**(1), 1–28 (1966)

P. Kollmann, E. Roussos, C. Paranicas, N. Krupp, C.M. Jackman, E. Kirsch, K.-H. Glassmeier, Energetic particle phase space densities at Saturn: Cassini observations and interpretations. J. Geophys. Res. **116**(A15), 05222 (2011). doi:10.1029/2010JA016221

P. Kollmann, E. Roussos, C. Paranicas, N. Krupp, D.K. Haggerty, Processes forming and sustaining Saturn's proton radiation belts. Icarus **222**, 323–341 (2013). doi:10.1016/j.icarus.2012.10.033

S.M. Krimigis, T.P. Armstrong, W.I. Axford, A.F. Cheng, G. Gloeckler, The magnetosphere of Uranus—Hot plasma and radiation environment. Science **233**, 97–102 (1986). doi:10.1126/science.233.4759.97

B.H. Mauk, Comparative investigation of the energetic ion spectra comprising the magnetospheric ring currents of the solar system. J. Geophys. Res. Space Phys. **119**, 9729–9746 (2014). doi:10.1002/2014JA020392

B.H. Mauk, N.J. Fox, Electron radiation belts of the solar system. J. Geophys. Res. **115**(A14), 12220 (2010). doi:10.1029/2010JA015660

B.H. Mauk, S.M. Krimigis, E.P. Keath, A.F. Cheng, T.P. Armstrong, The hot plasma and radiation environment of the Uranian magnetosphere. J. Geophys. Res. **92**, 15283–15308 (1987). doi:10.1029/JA092iA13p15283

B.H. Mauk, S.M. Krimigis, A.F. Cheng, R.S. Selesnick, Energetic particles and hot plasmas of Neptune, in *Neptune and Triton*, ed. by D.P. Cruikshank, M.S. Matthews, A.M. Schumann (1995), pp. 169–232

N.F. Ness, J.E.P. Connerney, R.P. Lepping, M. Schulz, G.-H. Voigt, in *The Magnetic Field and Magnetospheric Configuration of Uranus*, ed. by J.T. Bergstralh, E.D. Miner, M.S. Matthews (1991), pp. 739–779

C. Porco, D. DiNino, F. Nimmo, How the geysers, tidal stresses, and thermal emission across the South polar terrain of Enceladus are related. Astrophys. J. **148**, 45 (2014). doi:10.1088/0004-6256/148/3/45

Raines et al., Space Sci. Rev. (2015 this issue)

J.D. Richardson, J.W. Belcher, A. Szabo, R.L. McNutt Jr., The plasma environment of Neptune, in *Neptune and Triton*, ed. by D.P. Cruikshank, M.S. Matthews, A.M. Schumann (1995), pp. 279–340

L. Roth, J. Saur, K.D. Retherford, D.F. Strobel, P.D. Feldman, M.A. McGrath, F. Nimmo, Transient water vapor at Europa's South pole. Science **343**, 171–174 (2014). doi:10.1126/science.1247051

R.S. Selesnick, R.L. McNutt Jr., Voyager 2 plasma ion observations in the magnetosphere of Uranus. J. Geophys. Res. **92**, 15249–15262 (1987). doi:10.1029/JA092iA13p15249
E.C. Sittler Jr., K.W. Ogilvie, R. Selesnick, Survey of electrons in the Uranian magnetosphere—Voyager 2 observations. J. Geophys. Res. **92**, 15263–15281 (1987). doi:10.1029/JA092iA13p15263
Welling et al., Space Sci. Rev. (2015 this issue)
M. Zhang, Voyager II Plasma Observations at Uranus and Neptune. Ph.D. thesis, Massachusetts Institute of Technology (1991)